MONUMENTA SERICA MONOGRAPH SERIES
XXX
Sankt Augustin
Editor: ROMAN MALEK, S.V.D.

FERDINAND VERBIEST (1623-1688)

Jesuit Missionary, Scientist, Engineer and Diplomat

紀念南懷仁

逝世三百周年學術會議

一九八八年九月十二日

MONUMENTA SERICA MONOGRAPH SERIES
XXX

Ferdinand Verbiest (1623-1688) Jesuit Missionary, Scientist, Engineer and Diplomat

Edited by
JOHN W. WITEK, S.J.

Jointly published by
Institut Monumenta Serica, Sankt Augustin
and
Ferdinand Verbiest Foundation, Leuven

Steyler Verlag ▪ Nettetal
1994

Sumptibus Societatis Verbi Divini (S.V.D.)

Die Deutsche Bibliothek - CIP-Einheitsaufnahme

Ferdinand Verbiest : (1623-1688) Jesuit missionary, scientist, engineer and diplomat / jointly publ. by Institut Monumenta Serica, Sankt Augustin and Ferdinand Verbiest Foundation, Leuven. Ed. by John W. Witek. - Nettetal: Steyler Verl., 1994.
(Monumenta Serica Monograph Series; 30)
ISBN 3-8050-0328-5
NE: Witek, John W. [Hrsg.]; Institut Monumenta Serica <Sankt Augustin>; GT

Editorial assistance: CHRISTIANE VANDENBOGAERT (Leuven); BARBARA HOSTER and ROMAN MALEK, S.V.D. (Sankt Augustin)
Set by: Ferdinand Verbiest Foundation (Leuven)
and ROMAN MALEK (St. Augustin)
Distributed by: Steyler Verlag, Postfach 24 60, D-41311 Nettetal, Germany.

ISBN 13: 978-3-8050-0328-5 (hbk)
ISSN 0179-261X

Table of Contents

Verbiest: Missionary

Verbiest: Impact in China and in Europe

Abbreviations

AHSI	*Archivum Historicum Societatis Jesu* (Rome)
ARSI	Archivum Romanum Societatis Jesu (Rome)
BN	Bibliothèque Nationale (Paris)
C.I.C.M.	*Congregatio Immaculati Cordis Mariae*, Scheut
Jap. Sin.	*Japonica-Sinica*, Section ARSI
KH	K'ang-hsi (used with year, month and day)
O.F.M.	*Ordo Fratrum Minorum*, Franciscans
O.P.	*Ordo (Fratrum) Predicatorum*, Dominicans
S.J.	*Societas Jesu*, Jesuits

List of Illustrations

Gratitude is hereby expressed to each institution, which has kindly granted permission to reproduce the illustrations.

List of Tables

Acknowledgments

For a number of reasons the publication of this book has taken more time than was planned for initially. Yet we are proud to present it now as an important new contribution to research on the topic and as one more stepping-stone towards the publication of the first complete biography of Ferdinand Verbiest. We are grateful to all the experts who came from China and from the West to the conference, for the research and the time they spent in preparing such scholarly papers. This book is first of all the fruit of their efforts. Some participants did not deliver a paper but - as Prof. Gregory Blue of the Needham Institute (Cambridge) - contributed much by making important and much appreciated observations. Thanks to all this the Verbiest Conference brought altogether a lot of new findings on Verbiest, a key person in the history of the relations of Europe and China.

It was agreed upon during the conference that Fr. J. Witek, S.J., would do the editorial work for this book. With his well-known *Gründlichkeit* and perseverance he has done an excellent job. Thanks go to him first; and also to the sinologists and historians of the Verbiest Foundation who participated in the editorial work: An Migom, Guy Gommeren and later also Mieke Defoort and Sara Lievens. Their work was supervised by Staf Vloeberghs. The bulk of the painstaking work however for the typing and secretarial work towards the publication of this book has been done by Sr. Christiane Vandenbogaert, I.C.M. Without her competence, her patience and endurance the book would never have come out. It is mainly her merit. We are especially grateful to her.

Thanks go also to the members of the Organizing Committee - Staf Vloeberghs, Edmond Tang, Lieve Gullickers and Liu Qi - who assured a good organization of the conference as well as of the Verbiest exhibit that was set up parallel with the conference. They also organized the memorable visit to "China, Heaven and Earth", the successful exhibit in Brussels, initiated by the Ferdinand Verbiest Foundation and organized by the Flemish Community and the China-Europe Institute of Louvain.

Prof. Ulrich Libbrecht, director of the Verbiest Project, who had originally planned to organize an International Conference on Ferdinand Verbiest, called on the Verbiest Foundation to do this in his place. Funding agencies could no longer be approached in time. We therefore thank hereby the Chinese Province of the Congregation of the Immaculate Heart of Mary (Scheut) for providing all the financial means to organize the conference on "Ferdinand Verbiest, His Life and His Work."

We are grateful to the Belgian Foundation for Scientific Research (*Belgisch Fonds voor Wetenschappelijk Onderzoek*) which provided us with a grant towards the publication of this book. Thanks and acknowledgments go also to the Publishing Committee of the *Cihai zidian* (Shanghai) for allowing us to publish a copy of their item on Verbiest, to the much regretted Jorgen D. Scheel, who died recently, for allowing the publication of the illustrations on his Verbiest steam vehicle, and to Prof. Xi Zezong for allowing the publication of the diagrams. The Verbiest Foundation gratefully acknowledges the permissions granted by institutions and libraries to reproduce their illustrations.

After the 1988 Conference the Verbiest Foundation has continued its research program on Verbiest. Dr. Noel Govlers's scientific and much annotated English translation of the *Astronomia Europaea* is already published [Noel Golvers, *The "Astronomia Europaea" of Ferdinand Verbiest, S.J. (Dillingen, 1687). Text, Translation, Notes*. Monumenta Serica Monograph Series XXVIII (Sankt Augustin - Nettetal, 1993)]. Dr. Golvers has also undertaken the translation and annotation of the fifty-six letters of Ferdinand Verbiest, recently discovered by him. When he is ready with this, a complete and revised edition of the Josson and Willaert collection of Verbiest's letters should be prepared. Meanwhile, we also hope that Chinese colleagues continue their research on Verbiest in the rich collection of the Gu Gong archives. The vision of western historians on Verbiest must be completed with the vision of their Chinese colleagues. The study of the history of the relations of Europe and China is indeed a field where joint research has, from the beginning, been the goal of the Verbiest Foundation. Scholars from China and the West were invited to Louvain in 1986 for the Couplet Conference, in 1988 for the Verbiest Conference, and in 1990 for the Conference on the Historiography of the Chinese Catholic Church. It is our hope that even more cooperation in this research will follow.

JEROOM HEYNDRICKX, C.I.C.M.
Director, Ferdinand Verbiest Foundation K. U. Leuven

Preface

The late fifteenth and early sixteenth century European discovery of vast continents and civilizations and the subsequent literature about such areas that at first seemed too fantastic is a familiar story. Climate, people, food, clothing, and above all customs were at times so different to Europeans that the Jesuit Alessandro Valignano (1539-1606) referred to Japan as "a world the reverse of Europe."[1] Several years after these lines were written, Matteo Ricci (1552-1610) was in Macao preparing to enter China which he successfully did in 1583. The names of Ricci, along with those of Johann Adam Schall von Bell (1592-1666) and of Ferdinand Verbiest have become synonymous with the development of a nascent continuing Christian presence in China. Together they represent a century in which they, assisted by a number of confrères, were able to create lasting contacts with Chinese, and after the conquest in 1644, with Manchu officials in the imperial court and in the provinces.

A native of Pittem, Belgium, Verbiest was born on October 9, 1623. After attending the local Jesuit school, he entered the Society of Jesus on September 19, 1641 and soon showed his missionary interests by asking superiors to send him to South America. Twice his petition was not approved. In 1655, however, the Superior General of the Order urged Verbiest to consider the China mission as an alternative. Arriving in Macao on July 17, 1658, he was soon sent to Shaanxi province where he was an itinerant pastor for about a year. Called to Beijing to assist Schall in the Bureau of Astronomy, for the next five years he observed as an outsider such major events in the capital as the death of the Shunzhi Emperor (r. 1644-1661), the accession of the Kangxi Emperor (r. 1661-1722) and the control of the court by the regency council that ruled in the name of the new emperor. By 1664 Verbiest was defending Schall, whose health was declining, against the attacks of Yang Guangxian, an astronomer who challenged western methods of calendar-making and for a while headed the Bureau. After Yang's ouster, Verbiest's role in the Bureau was assured.

---

1 M. Cooper (ed.), *They Came to Japan* (Berkeley, 1965), p. vii.

According to the mind of Verbiest, to serve the emperor in such a capacity was to be a missionary for the Church. Frequently, fellow Jesuits and members of other religious orders would count on his presence at court to assist them in the provinces. He served as vice-provincial of China and in that position carefully considered the future of the mission. His correspondence shows that he definitely wanted not only more missionaries from Europe, but also at least one seminary in China to develop an indigenous clergy. Verbiest was very grateful to Louis XIV for sending five French confrères to China in order to carry out astronomical research. They arrived in Beijing ten days after Verbiest died on January 28, 1688.

This short sketch of Verbiest's life underscores the complexity of research in Sino-western cultural relations of his day. No complete, let alone definitive, biography based on the relevant European, Chinese and Manchu sources has yet appeared. Steps toward such an enterprise were first begun in 1982 when the Ferdinand Verbiest Foundation was established at the Catholic University of Louvain. Since its purpose is to make Chinese culture better known in the West and European culture better known in China, it has centered its efforts on historical research. The three hundredth anniversary of the death of Verbiest was a fitting occasion not just to honor the Jesuit after whom the Foundation is named, but above all to gather together scholars for an international conference that would examine several facets of the life and work of Verbiest. This was accomplished over a period of four days at the former Jesuit theologate in Louvain.

More than two dozen essays cover the multifaceted role Verbiest played in Europe and in China. His early years in Belgium in terms of family background and academic environment, Valère Arickx and Jan Roegiers have contributed. As the story moves to China, diversity arises. Roger A. Blondeau evaluates the kind of science Verbiest introduced, while U. Libbrecht offers a general assessment of his scientific efforts. Aspects of Verbiest's publications on European astronomy in China as these appeared in Latin are the focus of the study by Noël Golvers. This is carried further by Nicole Halsberghe in her comparison of Verbiest's astronomical instruments with those of Tycho Brahe. By focusing on Verbiest's astronomical instruments now located in the old Beijing Observatory, Isaia Iannaccone clarifies the syncretism between European and Chinese scholars. Another facet of Verbiest's research was geography and cartography. Chen Minsun places the work of Verbiest in the context of the Jesuit geographical writings before

1674, while Lin Tongyang discusses the impact of Verbiest on geography and cartography in China. In a similar vein Yi Shitong portrays the background of the celestial globe of the Kangxi Emperor whereas Xi Zezong evaluates Verbiest's contribution to Chinese science. Verbiest's role as a maker of cannons is related by Giovanni Stary and Shu Liguang from Manchu and Chinese sources. J.D. Scheel explores the engineering mind of Verbiest in his conception of an automotive machine.

Vitally interested in opening a possible overland route from Beijing to Europe via Russia, Verbiest was involved in Sino-Russian diplomatic contacts. V. Miasnikov and Hao Zhenhua consider the respective sides of that relationship, while Edward Kajdanski examines Verbiest's overtures to King John III of Poland, who was seen as an important link between the Russian and the other European courts. Jeroom Heyndrickx shows that the Chinese view of Verbiest as a spy for the Russians is an untenable position.

The Chinese court in Beijing and the Church were linked together through Verbiest as Ku Weiying indicates. As Lin Jinshui explains, Verbiest affected the development of the emperor's attitude towards Christianity. These efforts at influencing the imperial court are part of the larger picture that John E. Wills, Jr. portrays in the relationship of Christians and Chinese from the time of Ricci to Verbiest's day. The intellectual environment in which China faced Christianity Nicolas Standaert depicts in his study of the philosophical discussion between Chinese scholars and the Jesuits in that era. The catechetical writings Joseph Shih and the editor highlight as they explain how Verbiest used an approach to the Chinese that his predecessors found successful, but which he modified to reach larger audiences. Claudia von Collani describes how the emperor presented an inscription to the Jesuits at court which was later copied for placement in all the churches in China as a sign of imperial favor towards Christianity. G. Melis compares Martino Martini to Verbiest in their geographical works in which the former presented Chinese ideas of the five elements to his European readers while Verbiest described Aristotelian notions about the four elements to the Chinese. But all of his efforts to use astronomy as a means for the further establishment of Christianity were not always acceptable, even by his confrères. This led to his excommunication as the late Francis A. Rouleau wrote and Edward J. Malatesta readily explains. Exactly where Verbiest fits within the Chinese bureaucracy at the imperial court is the theme of Willy Vande Walle. In presenting several aspects of the mathematical and astronomical legacy of Verbiest, J. C. Martzloff and Catherine Jami explore some changes that occurred in the early eighteenth century. Knud Lundbaek shows the influence of Verbiest on T. S. Bayer.

These collected essays reflect the panorama of the life of Verbiest. Scientist, engineer, and diplomat, he was above all a missionary who contributed significantly to the growth of Christianity in China. His scientific and religious efforts at the imperial court were not a contradiction, but the culmination of his life. In his last memorial to the emperor he stated:

> Sire, I die content, since I have used almost all the moments of my life in the service of Your Majesty. But I very humbly beg Your Majesty to remember after my death that in all I have done, I have had no other view than to procure, in the person of the grandest king of the Orient, a protector for the holiest religion in the universe.[2]

His dying wish became a short-lived reality through the Edict of Toleration in 1692, but internal controversies on the mission changed the status of Christianity, not scientific inquiry, during the later Kangxi period.

JOHN W. WITEK, S.J.
Georgetown University

---

2 H. Josson and L. Willaert, *Correspondance de Ferdinand Verbiest* (Brussels, 1938), p. 551.

Verbiest
in
Europe

NEW FINDINGS ON THE LIFE AND FAMILY OF FERDINAND VERBIEST[1]

VALERE ARICKX

Valère Arickx has been librarian at the National Library of Belgian Congo and the Republic of Congo, from 1957 to 1961. From 1965 till 1978 he was national inspector of Public Libraries of Belgium. Until 1986 he was active as producer and director of the Belgian Radio and Television in West Flanders. He is the author of many articles on history. He also wrote a history of Pittem (his and Verbiest's place of birth). At present he is president of the Flemish Association for Family History.

The history of Ferdinand Verbiest begins with that of the Verbiest family, namely that of his parents, his grandparents and, if possible, his ancestors. The family history of Verbiest can help us to understand much better the personality and the intellectual background of Verbiest. As the Kangxi emperor wrote to Verbiest in 1676, everyone is grateful that an important part of his personality depends on what his parents and ancestors had given him in mental and physical talents.

---

1 This essay is based upon the results of an exhaustive study of the archives in various depositories. These archives are, among others: 1) Parish registry (baptism, marriage, death) of the period 1590-1797 from the following villages and towns: Bruges, Kortrijk, Tielt, Roeselare, Oudenaarde, Aalst, Pittem, Wingene, Wakken, Kortemark, Vinkt, etc. 2) County clerk's office registry (births, marriages, deaths) of the period 1797-1871 from the following places: Bruges, Pittem, Wingene, Evergem, etc. 3) The old and new archives of Bruges, for instance the Bogaardenschool, Tielt, Pittem, Wingene, Roeselare, the orphanage of Kortrijk, etc. Some material has been found in the *Archives départementales du Département du Nord* in Lille and in the *Chambre des Comptes*. Valère Arickx published a comprehensive history of the Verbiest family: "De familie van Ferdinand Verbiest S.J., missionaris, mandarijn en astronoom (1623-1688)", *Vlaamse Stam. Tijdschrift voor Familiegeschiedenis*, 26 (1990), 185-221 (first part); 28 (1992), 337-374 (second part).

That subject, part of Verbiest's biography, has not been studied until now, although the archives - dispersed though they are - are still preserved. When, at the end of June 1988, I wrote the catalogue of the Verbiest-exhibition held in Pittem - the town where Verbiest was born - I also wrote a large chapter about the ancestry of Verbiest and the descendants of his brothers and sisters as far as possible. That chapter was the result of about fifty years - half a century - of historical research in the records preserved in Pittem, Tielt, Bruges, Ghent, Brussels, Kortrijk, Roeselare, Lille (northern France), and other towns and villages.

The first biographer of Verbiest, the Reverend Canon Charles Carton of Bruges - a priest and historian also born in Pittem - published in that biography in the year 1839 (150 years ago !) a short genealogical outline of the Verbiest family. Carton, however, did not himself make that genealogical outline; he only corrected certain names, dates and filiations - but even these corrections were errors. Later, in 1885, a booklet concerning the Verbiest family was published. The author, a young local historian of the town of Tielt in West Flanders, writing these pages, did make some efforts to situate Verbiest's ancestors, as well as the house where he was born, and the descendants of that family.

Concerning Verbiest's parents, namely Judocus (or Just) Verbiest, and his wife Ann Van Hecke, there was no doubt. The records of baptism and other papers mention them many times. But concerning the grandparents and their residence, there was no certainty. Ferdinand himself caused the error about his grandparents. In his book *Astronomia Europaea*, published in 1687 by Philippe Couplet, Verbiest included a copy of the titles of honor, issued in 1678 by the Kangxi emperor, exalting the merits of Verbiest, his parents and also his grandparents. The names of these grandparents, written on order of Verbiest, no doubt, were really those of his great-grandparents. Verbiest surely knew that, for he personally knew his grandmother in Bruges. That was the beginning of all genealogical errors concerning Verbiest.

The great-grandparents of Verbiest were Peter Verbiest and Paschasia (or Paschale) Dewolf, known only by the titles of honor from 1678. After long historical research, we found that the real grandparents of Verbiest were from the town of Bruges, namely Paschasius (or Paschal) Verbiest, who in 1592 married Judoca (or Justa) Van der Straete. Paschal Verbiest died, rather young, in or before 1603. He had three children: Judocus (or Just) Verbiest - the father of Ferdinand, born in 1593/1594, Peter, born in 1595, and a girl, Paschasia (or Paschale), born in 1598. Three times I found references to "Judocus filius Paschasius" (or Just, son of Paschal), and once Just Verbiest wrote "I, Just Verbiest, son of Paschal". So there was no more doubt about the filiation of Paschal and Just Verbiest.

In the beginning of 1603 Paschal Verbiest's widow, Justa Van der Straete, was appointed governess or first house-mistress of the famous school for orphan-boys in Bruges, the municipal Bogaardenschool. There she was responsible for the lodging, the nourishment, the laundry, and so on, for more than one hundred boys between nine and nineteen years of age, the teachers, and the director or the prefect of studies. It was a great exception that the board of governors of the school allowed the three little children to be lodged with their mother in the school. The little girl, Paschale, was placed in an orphanage for girls from 1606 to 1615. In 1617 she married the tailor Cornelius Van Hecke, also a former pupil of that school. They had eight children. She died in 1654 and he in 1657, both in Bruges.

Just and Peter Verbiest were admitted as pupils in the school in 1603. In accordance with the rules of that school, the boys were poor, nine to ten years old, citizens of Bruges, orphans, and having a normal or good intelligence. In that famous orphanage, they were instructed and educated from 1603 to 1611, where they got their elementary and secondary instruction. At the end of their education in the school, the board of governors, acting on the proposal of the school-director, decided on the future of these young men. They were either to begin their practical training (five to six years) as skilled laborers (carpenter, tailor, butler, printer, secretary, clerk, and so on), or were allowed to continue their studies (in a monastery - e.g. Franciscan - for philosophy and theology, and so to be priests). The very best pupils among them were sent to the University of Louvain or to Douai (northern France), all of them at the expense of the school.

Important for the knowledge of the background of Verbiest is the family of Judoca (or Justa) Van der Straete, the mother of these three children. She had two close relatives - her two brothers, I believe - both former pupils of the Bogaardenschool. The first, John, became a Franciscan priest, and eventually the prior of the monastery in Bruges; the second died as governor of the house of the archbishop of Reggio (southern Italy). John had visited Jerusalem and Mount Sinai. The young Ferdinand often heard his father, Just, telling about the life and travels of these two uncles. Another fact in that background is the close relation between the Franciscan monastery in Bruges and the Bogaardenschool: many of the pupils of this school became Franciscans.

Peter Verbiest, the second son of Paschal and Justa Van der Straete, pupil in the Bogaardenschool from 1603 to 1611, started in 1611 his practical training as judicial and administrative secretary (or town-clerk) in the town of Oudenburg in Flanders. Later he lived in Bruges as clerk and notary, and died there in 1641. His last descendant died in 1760.

The oldest son was Judocus (or Just) Verbiest, born in Bruges in 1593/1594 and admitted to the Bogaardenschool in 1603. In October 1611 he was sent, as a very talented student, to the University of Douai (actually in northern France. That town belonged, up to 1713, to the seventeen provinces of the Netherlands). There he studied philosophy and became, after two years (in October 1613) Master of Arts, *magister in artibus*. The man who made the decision to send Just to Douai was his school-director in Bruges, Christian Janssens, himself a former student in Louvain. Along with Verbiest, other students from Bruges were also in Douai. They all became priests. Why, we can ask ourselves, did Just return to Bruges at the end of 1613? Indeed, he was appointed teacher, probably in Latin, in his former Bogaardenschool for two years, under the direction of his protector and spiritual father, Christian Janssens. He lived in that school, where his mother was also governess.

At the end of 1615 or at the beginning of 1616, Just Verbiest left his school and home in Bruges and, at his request, was appointed teacher - *scholasticus* - in Tielt, in the well-known Latin school of the church. At the same time he was appointed sexton, or sacristan, of that church. Together with a second sexton, he helped with the many religious services. Just Verbiest had much to do in Tielt in the church and in the school. So in 1620 he accepted a young assistant, also a pupil from Bruges, at his own expense, for the services in the church. Meanwhile, Verbiest got a good name and reputation in the town as an intelligent and learned man with a good pen and very nice writing! Master Verbiest found, in this manner, his way into the administration, and the municipal government asked him many times to draw up and write official papers and letters for the municipality, the church, and the public assistance committee.

Meanwhile, Master Just Verbiest had married Ann Van Hecke in 1617 (probably in Bruges). This young woman was closely related to several interesting families in Bruges, namely the Veggelman's and Van Wel's, both known in juridical circles in Bruges, and also in connection with the Bogaardenschool. Perhaps that relationship with the municipality of Tielt, his time at the university, and the family background of his wife encouraged Master Verbiest in his ambition to mount the social ladder. In 1620 he was admitted by the Council of Flanders in Ghent as notary, and at the end of 1622 he was appointed bailiff, or revée, of the rather large village of Pittem, near Tielt. There he was the agent or representative on a local scale, as well as the tax collector for the lord of Pittem, the Spanish nobleman Don Ferdinando de Zuñiga from Madrid. It was at the birth of Dona Inés de Zuñiga in Madrid in 1639 (she was the granddaughter of this Don Ferdinando) that the young student Ferdinand Verbiest wrote a long poem in her honor.

As bailiff, or revée, in Pittem, Master Verbiest was the number one man of the village, as well as of other places. He was president of the local aldermen, judge, notary, and tax collector. His direct chief in Flanders was the notary and lawyer, Master Ferdinand Van der Schuere[2], alderman in Kortrijk, licentiate in law from Douai, and jurist. It was, I believe, at the proposal of Van der Schuere, that Verbiest was appointed bailiff in Pittem and other places in West Flanders. It is thus understandable that this man became the godfather of Ferdinand Verbiest in 1623. We can also mention that Van der Schuere's son became a Jesuit in 1641 - at the same time as Verbiest.

Master Just Verbiest settled in Pittem, at the end of 1622, in the large house - the Verbiest house - by the market place near the church. In that house, Ferdinand was born on October 9, 1623. Between the church and the house of his birth, a statue of Ferdinand was erected in 1913.

The appointment of Master Verbiest in Pittem was also the start of a new career. In May 1623 he was appointed free-alderman of the castellany of Kortrijk, which means that he was one of the eleven administrators of that large territory surrounding the town of Kortrijk, about sixty towns and villages. He occupied that position until his death in 1651. His wife died from the pestilence in 1665. Both were buried in the church in Pittem.

Space limitations do not allow a presentation of detailed information about the years of Ferdinand Verbiest's youth in Pittem from 1623 to 1635, when he went to school in Bruges. In those years, he saw the rebuilding of the church and the houses in Pittem, but he also saw the

---

2 See Valère Arickx, "De rechtsgeleerde, notaris en schepen Ferdinand Van der Schuere L.U.J. (1597-1654) uit Kortrijk en zijn familie", *Vlaamse Stam. Tijdschrift voor Familiegeschiedenis*, 25 (1989), 127-144.

pestilence, the war, and the endless stream of foreign soldiers in the village: Spanish, French, Dutch and so on. His father, Master Verbiest, often traveled to Kortrijk, Ghent, Bruges, and Brussels for the affairs of the castellany and the village. The little Ferdinand went to the small school of the village, where either the parish priest or the sexton gave him his first individual lessons.

A last word about Justa Van der Straete. She died in the Bogaardenschool in Bruges in 1642, where she had been governess for about forty years. Without a doubt, her grandson Ferdinand, who studied in Bruges in 1635-1636, knew her and visited her in that school.

In the family of Master Just Verbiest and Ann Van Hecke were born four girls and three boys - three of them in Tielt and four in Pittem:

1. Judoca (or Justa), born in 1618, died in Pittem in 1634 from the pestilence.
2. Katherin, born in 1620, married the merchant R. Van de Vinck in Kortrijk in 1658.
3. John Baptist, born in Tielt, in March 1622, died in Kortrijk in September 1659. Thanks to his father, he got the opportunity to occupy several functions as bailiff, tax collector, free alderman, and judge; but John Baptist did not have the qualities to become a good administrator. He moved many times, and finally died poor in Kortrijk as tax collector of the Hospital of Our Lady, where his sister Ann was a nun. He was simply not a financial manager.
4. Ferdinand, born in Pittem on October 9, 1623, was baptized by Father Adrian van Brackle, assisted by the sexton Louis Van de Sande (or Ludovico d'Arenas, a Spaniard born in Brussels). This sexton was also Ferdinand's teacher in Pittem.
5. Jemina, born in Pittem in 1626, married John Lust, secretary and later mayor of the village, in 1650. She died in the large house of her parents in 1678. With her husband, she was buried in the church in Pittem.
6. Albert was born in Pittem in 1628 and died in Roeselare from the pestilence in 1694. He got his degree of licentiate in law, probably in Douai. After the death of his father, Albert was for a short time bailiff in Pittem in 1651. In 1658 he married in Bruges Katherine Isenbaert, the daughter of a leather-dresser. Albert Verbiest was a lawyer who lived in Bruges and in Ghent before returning to Kortrijk about 1662, to care for the orphans of his brother John Baptist. In Kortrijk he became alderman of the town, but in 1667 he was in Tielt. There he became a citizen, and was appointed alderman and

mayor. But what happened to him? A few years later, Albert moved to Roeselare, another little town in West Flanders, where he became a citizen. There he was also lawyer, notary and alderman. He lost his wife in 1688. In that family, at least seven children were born in Kortrijk, Tielt, Roeselare, and other places. The question I ask is this: Why did Albert Verbiest move so many times? In a normal situation, an alderman, notary, and lawyer does not do that. Had he financial troubles? We do not know. When he died as a virtual pauper, however, only two children were still alive. The son of Albert named Georges-Bernard was born in Kortrijk in 1663 and died as poor as his father in Wingene, an important village near Tielt in 1740. He owned several portraits of the Verbiest family. We do not know what happened to them. Georges-Bernard Verbiest married three times and had several children. In different places he was secretary or mayor. He ended his days as inn-keeper and brewer in Wingene. There are several descendants - from a daughter - still living to this day in Flanders. A grandson in the female lineage was mayor in Pittem until 1840.

7. The last child of Master Just Verbiest was the daughter Ann, born in 1630, who died as a nun in the Hospital of Our Lady in Kortrijk in 1681.

Conclusions

Today is the first time that the scientific world takes interest in the family background of Ferdinand Verbiest.

1. He belongs to a family whose origins we found in Bruges. That Verbiest family was poor and lived in a municipal orphanage school.
2. The oldest known ancestors are Ferdinand's great-grandparents, namely Peter Verbiest and Paschale Dewolf. They are incorrectly called his grandparents in the Chinese titles of honor, given by the Kangxi emperor.
3. The intellectual background of Verbiest is the famous Bogaarden-school in Bruges, closely related with the Franciscan monastery.
4. Just Verbiest, Ferdinand's father, was Master in Arts from the University of Douai, and was successively teacher in Latin, sexton, bailiff, judge, notary, free-alderman and tax-collector. For his time and country, he was a learned and important man.
5. Ferdinand Verbiest was born into an intellectual family, having as background the world of Bruges, the university, the church, and the

administrative and juridical experience of Master Just Verbiest. His mother also had connections in these circles in Bruges.

When Ferdinand Verbiest went to the secondary school in Bruges in 1635, he was not a mere peasant. He knew what a book and a map were, he had seen Bruges, and he had heard about far-away countries. Letters to his family are not preserved.

THE ACADEMIC ENVIRONMENT OF THE UNIVERSITY OF LOUVAIN AT THE TIME OF FERDINAND VERBIEST

JAN ROEGIERS

Jan Roegiers, Ph. D. in Modern History and M.A. in Moral and Religious Sciences, became first archivist and, since 1981, the head librarian of the K.U. Leuven. He published articles on the history of the University of Louvain and on the relation between Church and State in the eighteenth century.

One week before his seventeenth birthday, October 1, 1640, Ferdinand Verbiest began his studies at the Arts Faculty of the University of Louvain. Precisely eleven months later, September 2, 1641, he left the university and joined the Society of Jesus. Before examining some facts and situations in the university that might have influenced his decision, this contribution will give a general overview of the organization and situation of the University of Louvain at the time and try to answer the question whether this short stay at the university could have influenced his later scientific career.

For the university this year 1640 ended a period of half a century of restoration, development and reorganization that gave the two centuries' old institution a shape that would last till the end of the eighteenth century. Between 1572 and 1585 the town of Louvain was flooded several times, sustained three sieges, was struck by a terrible epidemic of pestilence that decimated the population and drove away students and professors, except those of the Faculty of Medicine who all, except one, died of the plague. A Spanish garrison pillaged the deserted houses and burned them down. The severe winter of 1584-1585 was accompanied by a general famine. The city, the university and the colleges were ruined and in 1583 the remaining professors seriously considered closing the university definitively. According to them the University of Louvain had become useless since the foundation of new universities at

Douai in 1562 and Leyden in 1575[1].

Despite all problems the university survived this heaviest crisis in its history. After 1585 lecturing and college life resumed little by little their former course. Pope Gregory XIII and King Philip II worked for a financial restoration. Already before the archdukes Albert and Isabella ascended the throne of the Netherlands in 1598, the restoration of the University of Louvain had become one of the major objectives of the civil and religious restoration policy of the government. This restoration was crowned by the so-called Visitation, a general inspection of the whole university, ordered by the archdukes in 1606. It resulted in a new organic law for the university issued in 1617[2]. This visitation put an end to the complaints about a certain number of abuses within the university, e.g. about the discipline among the students, the organization of teaching, the functioning of the university courts of justice, the application of the university privileges, the problems of election and nomination, the rights and duties of the clerical staff, etc. But at the same time the government had imposed its supreme authority in all fields of university life, at the expense of papal authority and university autonomy. For many reasons this visitation of 1617 is often called the first law on higher education issued in this country.

Most of the prescriptions of the visitation confirmed older rules, often fallen into disuse, or more recent changes and additions to the university laws. It cannot be called a real reorganization[3]. The head of the university, the rector, was elected for six months from the members of the congregation, the academic senate, rotating among the five faculties. The rector presided over all assemblies of the congregation and the deputation and was the representative of the university at all occasions. His most important task was to preside over the *curia rectoralis*, the first of the five university courts of justice. As such he

---

1 On the fate of the University of Louvain in this period: F. Camerlynckx, "L'Université de Louvain depuis le dernier quart du XVIe siècle jusqu'à la 'visite' des Archiducs Albert et Isabelle (1575-1617)", in L. Van der Essen (ed.), *L'Université de Louvain à travers cinq siècles* (Brussels, 1927), pp. 31-64.

2 On this *Visitatio*: H.J. Elias, *Kerk en Staat in de Zuidelijke Nederlanden onder de Regeering der Aartshertogen Albrecht en Isabella (1598-1621)* (Antwerp, 1931), pp. 268-287. The text of the new rules was published several times, e.g. in V. Brants (ed.), *Recueil des Ordonnances des Pays-Bas. Règne d'Albert et Isabelle* (Brussels, 1912), 2:326-339.

3 An overview of the institutions of the University, with a further bibliography, is found in *The University of Louvain, 1425-1985* (Louvain, 1989).

could even give sentence of excommunication. Because of this jurisdiction over clergy as well as over lay people he had always to belong to the clergy himself. For the daily government the rector was assisted by the deputation, the deputies and the deans of the faculties. Although the authority of the rector was virtually unlimited, the real supremacy belonged to the congregation, the assembly of all professors, doctors of the higher faculties and some officials. This congregation elected not only the rector but also, from its members, the other officials such as the receptor, for one year in charge of the university finances, the dictator, who drafted the correspondence and other official documents, and those who were in charge of the defence of the university rights and privileges. The second dignity in the university was the office of chancellor. By papal authority he granted the university degrees to those presented by the faculties. The foundation bull of the university had attached this office of chancellor to the already existing dignity of provost of St. Peter's chapter, the first clerical community of the city. The dean of the chapter, most often a professor, acted as vice-chancellor.

From the first years of its existence, the University of Louvain enjoyed a series of privileges that defined the relationship of the university to the outside world. To share in the advantages of these privileges one had to become a regular member of the academic community, *suppositus* as it was called. This quality was obtained by the matriculation, inscription on the roll, and required effective residence at the university. Not only students and professors, but also some officials and savants were matriculated and considered as *suppositi*. The first of the four great privileges was complete juridical autonomy. In virtue of this privilege the university governed itself, disposed of its own judicial organization with five university courts of justice, had its own police and even a university prison. The second great privilege was fiscal immunity. The members of the university were free of some taxes, as e.g. the city beer and wine excises. In virtue of this immunity the university had its own wine and beer cellar and a tax free shop for its members. Another privilege was the right of nomination, granted to the university by Pope Sixtus IV and to the Arts Faculty by Leo X. In virtue of this right, the university and the faculty could nominate professors and graduates to clerical benefices in the whole Netherlands and some neighboring dioceses. The fourth great privilege was a monopoly of higher education. Within the Spanish Netherlands only the University of Louvain was qualified for the public teaching of philosophy, theology, law and medicine and could grant academic degrees. The only exception to this rule that had been accepted by the university was the foundation of the University of Douai in 1562. This privilege, several times recognized by the sovereign, gave occasion to a long series of conflicts with the Jesuit college, the only clerical institution in Louvain that was never incorporated into the university[4]. The Jesuit Order enjoyed its own papal privileges which authorized the Order to teach philosophy and theology and to grant degrees, even to those who were not members of the Society. When the Louvain Jesuits tried to organize public lectures in philosophy and to grant academic degrees, but refused to submit themselves to the authority of the university, an open conflict arose. After a first attempt of the Jesuits in 1594, Pope Clement VIII (1592-1605) decided against them. After his death the conflict flared several times: in 1607, 1612, 1618, 1622, 1624 and finally in 1626 when a second embassy of the theology professor Cornelius Jansenius on behalf of the university to the Spanish court put an end to the disputes. By this time this conflict, more than some doctrinal disagreements, had poisoned the relationship between the Jesuits and the university.

Within the university the five faculties enjoyed a high degree of autonomy. Each faculty was presided over by a dean, elected the same day as the rector. The highest authority was the faculty council. In the Arts Faculty this council consisted of the professors, the regents of the four pedagogies and some senior graduates. In the four so-called higher faculties, theology, canon law, civil law and medicine, a distinction was made between the *stricta* and the *lata facultas*. The *stricta facultas* or *strictum collegium* consisted of a fixed and rather limited number of members, in the faculties of law and medicine the holders of the most important chairs, in the Theology Faculty eight doctors elected by cooptation who were not necessarily professors. They governed the faculty, took the exams, decided upon the degrees and divided among themselves the income of the faculty. The *lata facultas* included all doctors and professors but did not have any real authority.

All students, except the members of the religious orders started their academic career as a student at the Arts Faculty. For two years they studied philosophy, natural sciences and some mathematics. Mainly in the sixteenth century, but also in the first half of the seventeenth

---

4 On the conflicts between the University and the Jesuit college: A. Poncelet, *Histoire de la Compagnie de Jésus dans les anciens Pays-Bas* (Brussels, 1926), 2:188-228, 252-297; F. Claeys-Bouuaert, "Jansénius en Espagne", in *Hommage à Dom U. Berlière* (Brussels, 1931), pp. 73-98.

century, Latin language and literature formed a part of the curriculum. During these two years the student stayed in one of the four pedagogies of the faculty; there he lived and attended the lectures given by the professors of pedagogy. A pedagogy was called after the name of the house in which it had been founded in the fifteenth century: the Falcon, the Pig, the Castle and the Lily.

The program and the method of teaching was laid down by the visitation. The new faculty constitution, adopted in 1639, confirmed these regulations[5]. Nine months were spent on the logic of Aristotle and the rules of dialectics, eight months on natural philosophy (the *Libri physicorum* and *De anima*), four months on metaphysics and three final months on a general repetition of the whole course. Ethics was taught on Sunday and on holidays. The teaching was based upon the *corpus Aristotelicum*, the writings of Aristotle and the other treatises ascribed to him. The professor had to follow the doctrine of Aristotle in all points that did not contradict the Christian faith. In fact they did not comment in their lectures on the original text of Aristotle but used systematic compendia of his doctrine, elaborated by the medieval Scholastics and the faculty professors themselves.

The teaching was provided by a *professor primarius*, who lectured in the morning on fundamental problems, and a *secundarius*, who lectured in the afternoon on more particular questions. These two professors followed the students during their two years in the pedagogy; two other professors lectured for those of the other year. In the early seventeenth century a pedagogy could have more than four professors; these *extraordinarii* were specialized in other fields such as ethics or literature.

After two years the students participated in a series of exams that formed a general *concursus* between the students of the four pedagogies. The winner of this competition was proclaimed the *primus* of that year. He was celebrated in his pedagogy and by the whole faculty, royally received in his birthplace and snowed under with presents. Most *primi* followed a brilliant career at the university, in the Church or the

---

5 Edition of the faculty constitution by P.F.X. de Ram in *Codex veterum statutorum Academiae Lovaniensis*, as appendix of *Les quatorze livres sur l'histoire de la ville de Louvain du Docteur et Professeur en Théologie Jean Molanus* ... (Brussels, 1861), pp. 1089-1128. On the seventeenth-century study program: G. Vanpaemel, *Echo's van een Wetenschappelijke Revolutie. De Mechanistische Natuurwetenschap aan de Leuvense Artesfaculteit (1650-1747)* (Brussels, 1986), pp. 39-43.

judiciary[6].

Once graduated from the Arts Faculty, the student had to make a choice between the higher faculties. He could continue his studies in theology, law or medicine; canon and civil law were mostly combined. The lectures of the higher faculties were held in the main building of the university, the old city cloth hall in which each faculty had its auditorium. The studies in medicine lasted for three years, in law for four and in theology for seven more years before a student could graduate and obtain the *licentia*. The *Magisterium* of the higher faculties or doctoral degree was only taken by those who, as a university professor, wanted to obtain a seat in the *strictum collegium* of their faculty. To be appointed as a professor, to practise in medicine or as a lawyer, only the licence degree was required. During his studies at the higher faculty the student could live in a private house or in a college either as *bursalis*, with a scholarship, or as a *commensalis*, on his own expenses.

Most of the some forty colleges of the university had been founded by rich benefactors for the housing of students belonging to their family or to natives of the same region as the founder, or they had been erected for poor students. Some colleges were intended for the members of a religious order and others served as a seminary for the clergy of a region or a diocese. These colleges were only intended for housing, and no teaching was provided, except for some repetitions, disputations or pastoral training. The only exceptions were the famous *Collegium Trilingue*, the college of the three languages, in which students of all faculties could attend lectures on ancient history, Latin, Greek and Hebrew languages and literature, and the *Collegium Vaulxianum*, in fact a secondary school attached to the university. As a part of the restoration of the university, some eighteen new colleges had been founded between 1592 and 1633[7].

The celebrations of the second centenary of the university foundation in 1626 affirmed the prosperity of the restored university. The number of students had again reached 1,800-1,900. Foreign students came from all parts of the German Empire and from countries under protestant rule: Dutch, English and Irish students lived in colleges and convents specially founded for their needs. The finances of the university, the faculties and the colleges were healthy and many buildings had recently

---

6 On the celebration of the *primus*: Cl. Bruneel, "Le 'Primus' de Louvain au XVIIIe siècle", *Revue du Nord*, 69 (1987), 575-588.

7 A history of all the colleges: E. Reusens, *Documents relatifs à l'histoire de l'Université de Louvain 1425-1797*, 5 vols. (Louvain, 1881-1903).

been renewed or enlarged. Nine years later, in 1635, a siege by the united French and Dutch troops threatened the academic prosperity. The city was defended by citizens, students and professors and the rector Jansenius himself fired them with fighting songs of his own poetry. When the news spread that the imperial army was on the way to defend the city, the enemy left the place, leaving behind a new epidemic of the plague[8].

These reverses did not really endanger the new prosperity, which can be seen from the many new initiatives, such as the foundation of a central library in 1636 and the many books published by the professors of this period. The university of the 1630s-1640s included a number of professors who became famous for their learning and publications; to recall only the best known: the latinist Erycius Puteanus, historian, philologist, poet and also interested in natural sciences; the lawyer, philosopher and playwright Nicolaus Vernulaeus; the professor of law and Hebrew, historian and bibliographer Valerius Andreas, in 1636 appointed as the first librarian of the university; the physicist astronomer and theoretician Libertus Fromondus; the medical doctor Vopiscus Fortunatus Plempius, who became one of the best known adversaries of René Descartes; several professors of civil and canon law such as Theodorus Tuldenus, Antonius Perezius, Andreas Vallensis and Petrus Stockmans, the theologians Jansenius, in 1636 appointed bishop of Ypres, and Johannes Sinnich, an Irishman who became one of the most firm defenders of Jansenius[9].

It was a renowned, rich and self-conscious university in which Verbiest was enrolled in the crucial year 1640. He was inscribed in the pedagogy of the Lily on October 2, of that year, the day after the new academic year opened and the lectures began[10]. He was matriculated on January 7, 1641 with the other students of the pedagogy[11]. As usual all the students of the pedagogy matriculated on the same day. This had to

---

8 A short account of this siege with a full bibliography: J. Dauwe, "Allegorie op het Beleg van Leuven in 1635. Een Kopergravure van P. Franchoys", *Arca Lovaniensis*, 1 (1972), 95-109.

9 Short biographies and further bibliography on these professors in the exhibition catalogue *550 Jaar Universiteit Leuven 1425-1975* (Leuven, 1976).

10 *Liber Commensalium* of the pedagogy of the Lily, 1636-1644, *ff.* 140v, Brussels, General Archives of the Realm, *Old University of Louvain*, 1164.

11 A. Schillings (ed.), *Matricule de l'Université de Louvain* (Brussels, 1962), p. 377, n° 324.

be done before the end of the term of the present rector, in other words before February 28. The date of matriculation of an arts student tells nothing about when he precisely arrived.

To the students of the Lily who matriculated the same day, who attended the lectures together with Verbiest and lived in the same house, belonged two other people who afterwards followed an academic career: Henricus Scaille, later theology professor, president of the College of the Holy Spirit and dean of St. Peter's chapter, a convinced Jansenist, and the better known Arnold Geulincx (1624-1669), in 1646 appointed professor in the Lily, a famous Cartesianist, who left the university in 1658, became a Calvinist and philosophy professor at the University of Leyden; he is generally considered as the greatest Dutch philosopher after Baruch Spinoza[12].

I could not establish who exactly were the *primarius* and the *secundarius* whose lectures Verbiest attended. The names of all professors of the Lily in this period are known, but it is not clear when they precisely finished their career at the pedagogy and left the university[13]. Nothing special is known about the regent of the pedagogy, Baldric Buterne, rector of the university until August 1640, nor about the professors Joannes Carolus a Quercu, alias Van Eyck, Franciscus Du Chasteau and Joannes Baptista de Bettencourt. They all left the university for a prebend in some chapter, nominated in virtue of the university privileges. A more interesting man was Joannes Antonius de Locquet, a nobleman who taught only a few years, and left the university to become a member of the royal councils and ended as president of the Great Council of Mechelen (Mechlin), the supreme court of the country. A brilliant career was foreseen for Sebastianus Stockmans, the younger brother of the famous lawyer, who became a philosophy professor in 1637, only one year after his graduation and while still a student at the Theology Faculty. In 1644 he became a professor of theology. He died in 1650, only thirty-three years old, at the moment he was acting as rector[14].

The most important professor of the Lily at the time of Verbiest was

---

12 On Scaille (1624-1689): J.L. Bax, *Historia Universitatis Lovaniensis*, Brussels, Royal Library, *Ms. 22172/III*, pp. 366-367; Reusens, *Documents*, 3:360. On Geulincx as professor at Louvain: G. Monchamp, *Histoire du Cartésianisme en Belgique* (Brussels, 1889), *passim*.

13 Details on the pedagogy and its professors: Reusens, *op. cit.*, 3:168-198.

14 *Ibid.*, 3:207-208.

undoubtedly William Philippi[15]. In 1618, some twenty years old, Philippi had been appointed professor at the Lily. He remained there for thirty-two years. As a professor of philosophy he studied law and medicine. In 1635 he was appointed professor of fundamental medicine, but remained professor of philosophy at the same time. He had some troubles because he married, a very unusual situation in the Arts Faculty and in fact, he was the last professor of this faculty to whom this was allowed, against the will of a reluctant faculty. Like many of the faculty members, he had received the tonsure and was a cleric, but he had never been ordained. In 1650 he left the Arts Faculty but remained professor of medicine until his death in 1665. As a professor of philosophy Philippi was one of the first to introduce Cartesianism in the university. In 1658-1664 he published a series of four handbooks of logic, metaphysics and physics in which the Aristotelian program was interpreted in a Cartesian way. It is clear that Verbiest's classmate Geulincx was a pupil of Philippi and was introduced by him into Cartesianism. It is not impossible that Verbiest heard from Philippi some details of Descartes and his publications. In 1637, immediately after it had appeared, Descartes had sent copies of the *Discours de la méthode* to Plempius, the famous professor of medicine with whom he had become acquainted in Amsterdam, and to Fromondus. The two professors of Louvain criticized the book and a controversy followed for many years[16].

In 1639 Gerard van Gutschoven, Cartesianist from the first moment, was appointed supply professor of mathematics[17]. With his appointment the mathematical discoveries of Descartes, his mechanistic and astronomical ideas, were introduced in the faculty; but it is also clear that Van Gutschoven in his successful but rather elementary teaching could not demonstrate his real qualities as a mathematician. In 1659 he became a professor at the Faculty of Medicine and in 1664 he published the posthumous *Traité de l'homme* by Descartes. Verbiest might have attended the lectures of Van Gutschoven, but this is not sure at all.

If we are not certain who were Verbiest's professors, it is clear that he stayed at the university at the very moment when the new philoso-

---

15 On this man: *Ibid.*, 3:262-264; Monchamp, *op. cit.*, pp. 317-336 and *passim.*

16 Monchamp, *op. cit.*, pp. 43-94.

17 *Ibid.*, pp. 113-114, 295-317; Reusens in *Biographie Nationale*, 8 (1884-1885), 556-558.

phical ideas first appeared in Louvain. We can also be sure that, even if he heard something about the new ideas, the teaching Verbiest received was traditionally Aristotelian. We even know what must have been the handbook he used during his first eight months in the Lily. These months were devoted to the study of logic, based upon Porphyrius's introduction to Aristotle, the *Praedicamenta*, *De interpretatione* and the *Analytica priora et posteriora*. In 1627 Laurentius Ghiffene, professor at the Lily, had published a volume of *Prodidagmata sive logicae partem primam introductoriam ad organum Aristotelis*. Since it was reprinted in 1641, we can be sure that it was in use in the Lily when Verbiest spent his first year there. Unfortunately no copy of the book could be retraced up to now[18].

After eight months of logic, the *primarius* tackled Aristotle's *Libri physicorum*. The first books deal with the fundamental ideas of Aristotelian physics, the definitions of *materia*, *forma*, *natura* and *causa*. In the meantime the *professor secundarius* spent several months on the *Tractatus de sphaera* by Johannes de Sacrabosco. This treatise formed a short introduction to cosmography and presented the occasion to discuss some elements of mathematics, geography and calendar computation. The problems posed by Copernicanism were discussed at the occasion of this part of the curriculum.

The basic ideas of Copernicanism were known very early at Louvain and accepted in the 1540's by Gemma Frisius, professor of medicine and mathematics, constructor of globes and astronomical instruments[19]. The traditional Ptolemaic system was attacked several times in the first half of the seventeenth century, e.g. by Fromondus in his *quotlibetica* of 1615. After the Roman condemnation of Nicolaus Copernicus in 1616 and even more after the condemnation of Galileo Galilei himself in 1633, the Louvain professors became more cautious. Already in 1619 Fromondus and the medical Fienus rejected the Copernican system. Puteanus made a clear choice in favor of the system of Tycho Brahe. Most seventeenth-century Louvain professors presented the systems of Ptolemy, Copernicus and Brahe as three hypotheses, without making a real choice[20]. It is an open question whether Verbiest ever heard the

---

18 Vanpaemel, *op. cit.*, p. 40, n° 7.

19 E.H. Waterbolk, "The 'reception' of Copernicus' teachings by Gemma Frisius (1508-1555)", *Lias*, 1 (1974), 225-242.

20 G. Monchamp, *Galilée et la Belgique. Essai historique sur les vicissitudes du système de Copernic en Belgique* (Saint-Trond, 1892).

comments of the professors of the Lily on these problems: about this time he left the university.

Why? Since Verbiest himself never clearly indicated a clear reason for his decision to join the Jesuits, we can only guess. But some events of the year 1640-1641 can help us to understand what happened.

In 1640 the Jesuits celebrated the first century of their foundation. In all Jesuit colleges of the Netherlands public solemnities were held. As a pupil of the Jesuits at Courtray, Verbiest must have assisted in these celebrations and have heard the praise of the order, its members and its achievements. Once arrived at the university he must have heard of the splendid ceremonies performed at Louvain with the collaboration and the assistance of the university[21]. It is also very likely that Verbiest, pupil of the Jesuits, became a member of the student sodality founded by the Jesuits in the university town in 1585. It is even most likely that he was already a member of the sodality at the Courtray college. This religious confraternity became such a success among the students of Louvain that very soon it was split into four sections: one for philosophy students, a second for theology students, a third for students of law and medicine. In 1640 the Emperor Ferdinand III was enrolled as a member of this third section, called after the Purification. A fourth section, of All Saints, was a secret one: it consisted of the elite of the three others; its members made special vows. A similar secret sodality had been founded in Courtray in 1630[22]. It is possible that Verbiest was a member of both, in Courtray and in Louvain. In any case, there were many ways in which Verbiest could have been associated with the Jesuits before he joined the order.

The festive joy of the Louvain Jesuits was broken off the same year. In June 1640 they learned that the Louvain printer Jacob Zegers was finishing a book in which the theological ideas of the Jesuits were openly attacked, the *Augustinus* of Jansenius. They immediately took action to prevent the publication of the book. When they did not succeed, they worked with more success for a Roman prohibition and even a condemnation of the book. In a few months an open conflict saw the majority of the university opposed to the Jesuits. This was the start of a continual struggle that lasted for nearly a century[23].

---

21 L. Willaert, *Les origines du jansénisme dans les Pays-Bas catholiques* (Gembloux, 1948), 1:227-229.

22 On these sodalities: Poncelet, 2:326-335.

23 On the beginnings of the Jansenist controversy at Louvain: L. Ceussens,

This is not the place to give a detailed account of this controversy, its background, its origin and its early developments. Others have done this before me and I would only like to stress how both parties considered this dispute a life and death struggle. Its origin lay already in the sixteenth century, when the Louvain theologians had in 1587 officially condemned the teaching of Leonardus Lessius, the most famous of the Jesuit theology professors, as semipelagian. An examination of the Louvain problems and similar questions that had arisen in Spain by a special Roman Commission in 1598-1607, nearly led to the official condemnation of the Jesuit doctrine on grace and free will. Finally nothing was decided, except that the solution had to be found in the right comprehension of the works of St. Augustine, the *doctor gratiae*. After the death of Lessius (1623), the Louvain Jesuits tried to justify his doctrine by his canonization as a saint. In 1640 the history of the Louvain Jesuit college mentions a series of miracles and favors obtained through the intercession of Lessius.

The already mentioned dispute with the university about the monopoly of teaching worsened the situation. The theology professor Jansenius, who not only defended the Augustinian doctrine of the faculty, but who also traveled twice to the Spanish court on behalf of the university to defend its teaching monopoly, was considered by the Louvain Jesuits as their main enemy. When this man undertook the justification of the Augustinian doctrine by a voluminous interpretation of the works of the Church Fathers, the Jesuits considered this publication as directed immediately against them. Jansenius, who was appointed bishop of Ypres in 1636 and died within two years, could not publish his magisterial study. This was done by the executors of his last will; one of these two was his successor at the Theology Faculty, Fromondus, well-known also as a physicist and astronomer, who, as dean of St. Peter's chapter, had celebrated the official mass for the Jesuit jubilee[24].

Among the many attacks of the Jesuits on the *Augustinus* during the years 1640-1641, the most sensational involved the defence of voluminous theses against the *Augustinus* on March 22, 1641. Many

---

"De Allereerste Ontwikkeling van de Strijd tegen Jansenius", in *Jansenistica*, I (Mechlin, 1950), 7-165; *idem*, "Les débuts du jansénisme et de l'antijansénisme à Louvain", in *Facultas S. Theologiae Lovaniensis 1432-1797. Bijdragen tot haar Geschiedenis. Contributions to its History. Contributions à son Histoire* (Louvain, 1977), pp. 381-431.

24 On Fromondus: *idem*, "Le janséniste Libert Froidmont (1587-1653)", *Bulletin de la Société d'Art et d'Histoire du Diocèse de Liège*, 43 (1963), 1-46.

students attended this defence, which lasted for a whole day and was rather tumultuous. The Jesuits obtained a Roman prohibition of the book, the first in 1641, a second in 1643. The main argument for the second prohibition, in which the book was already condemned, was the commotion the book had provoked in the country. To be honest one must admit that this commotion was more the result of the action of the Jesuits than of the publication of the book.

It is impossible that Verbiest would not have followed the developments in this open struggle, especially if he remained in contact with the Jesuits who mobilized all their friends against Jansenius. In his pedagogy, however, he heard very different opinions. The regent of the Lily, Buterne, acted as rector when the affair broke out. He did not do very much to prevent the publication, although he was asked to by the papal nuncio. His successor in the rectorate, the man who enrolled Verbiest in his own handwriting, was the theologian Gerard van Werm, the most firm defender of Jansenius. Among the professors at the Lily, at least Philippi and Sebastianus Stockmans are known to have been convinced Jansenists. The latter was the brother of Jansenius's great friend Peter Stockmans, a lawyer, who with the theologian had composed a pamphlet against the politics of Richelieu; as a member of the Privy Council, Stockmans successfully defended the memory of his friend Jansenius. Van Gutschoven was a Jansenist too[25].

We could continue to enumerate names of people Verbiest met daily at the university, who were by now open enemies of the Jesuits. It is my opinion that Verbiest had the feeling that he had to make a choice and that he made this choice as soon as possible, not awaiting the end of his course at the Arts Faculty. From the very beginning he must have been a very convinced Jesuit.

In September 1641 Verbiest left Louvain for Mechlin where he remained for two years in the novitiate of the Jesuit province. Two years later he came back to Louvain to finish his philosophical studies, not at the university but at the Jesuit college. There he attended the lectures of the famous mathematician André Tacquet. It is clear that his encounter with this great scientist and the two years he spent at the Louvain college in 1643-1645 were much more important for the

---

25 Beside inclusion in the publications quoted in note 23, all these people are mentioned in *idem*, *Sources relatives aux débuts du jansénisme et de l'antijansénisme* (Louvain, 1957); *idem*, *La première bulle contre Jansénius. Sources relatives à son histoire (1644-1653)*, 2 vols. (Brussels and Rome, 1961-1962).

scientific training of Verbiest than his eleven months in the Lily. It is even most probable that he hardly kept up any contact with the academic milieu in which the majority lived in open conflict or even at enmity with the Jesuit college. Another world, much more challenging than an academic career in the closed world of the university, was waiting for him.

Verbiest

Scientist

DID THE JESUITS AND FERDINAND VERBIEST IMPORT OUTDATED SCIENCE INTO CHINA?

ROGER A. BLONDEAU

Roger A. Blondeau, historian of sciences, is a member of the Belgian Commission for the history of sciences. He is the author of a biography of Ferdinand Verbiest. In 1974 he received the J. Gillis award of the Belgian Royal Academy of Sciences. During the academic year 1991-1992 he was awarded the Sartonmedal from the University of Ghent (Belgium).

Some present-day historians have advanced the idea that the seventeenth-century Jesuits, and, in particular, Father Ferdinand Verbiest, imported outdated scientific knowledge into China, especially in the field of astronomy. The accusation originates in the West[1]. It has been pointed out that Verbiest never mentioned the innovative work of Nicolaus Copernicus, Galileo Galilei, Johann Kepler and Isaac Newton, but kept leaning on Aristotle and Ptolemy, only because the Jesuits were "enemies of modern science" and "led the battle against science and its practitioners"[2]. It has even been said that "the import of such outdated scientific knowledge has contributed to the arrears of Chinese culture"[3].

---

1 The well-known Dutch sinologist Prof. Dr. J.J.L. Duyvendak (d. 1954) published a review of *Galileo in China*, by the Italian Jesuit Prof. Pasquale M. d'Elia (*T'oung Pao* 38 [1947-1948], 321-329). At the end of his review, Duyvendak wrote: "The impossibility for the Jesuits, the mediators of western science in China, to accept Galileo's heliocentric theory, is a matter of immense cultural significance. It meant that China, when it received western science, received it in a form that, in principle, was already antiquated." In later years, this proposition was adopted by a number of other historians, cf. R.A. Blondeau, *Ferdinand Verbiest S.J. als wetenschapsmens, 1688-1988* (Roesbrugge, 1987), p. 99.

2 Bai Shouyi, *An Outline History of China* (Beijing, 1982), p.427: "Naturally, the Society of Jesus would stand on the front line against science and scientists ..."

3 J.J.L. Duyvendak, "De Ming-dynastie", p. 159 in *Wereldgeschiedenis*, De

Some have also alleged that the instruments Verbiest built for the Imperial Astronomical Observatory were mere copies of outdated types which had been designed and described by Tycho Brahe three quarters of a century earlier. It is my intention in the present paper to provide a decisive answer to these accusations.

During the last few centuries, Copernicus has been presented as a visionary thinker, as the very pioneer of modern scientific thinking, and the creator of our present world-view. These are, however, mere sympathetic overstatements about a man who developed a certain hypothesis, which only much later proved to be fundamentally correct, and whose own uncertainty can be gathered from his age-long hesitation about the hypothesis.

Is it not true that Copernicus had already developed his heliocentric hypothesis before 1514 in a discourse that was never issued in print[4]? His *De revolutionibus orbium caelestium* appeared no earlier than 1543, the year of his death, and only at the constant insistence by a few of his friends. And it is not at all hard to understand his uncertainty. Copernicus had not got scientific evidence whatsoever for his hypothesis. He was able to point out that it would be much easier to explain the heavenly movements on the contention that the earth and the planets were revolving around the sun, and that the earth was spinning in its own axis every twenty-four hours. His further arguments were of a purely aesthetic nature, and he expressed them as follows: "Whoever would in this beautiful temple want to position this lamp (the sun!) otherwise than at a place where it is able to illuminate the whole." And also: "By virtue of the present system, we are able to discover an admirable balance in the world, and a harmonious relationship, which would otherwise be unattainable"[5].

These and similar viewpoints would not suffice to convince everybody. Scholars were his first critics. Copernicus's hypothesis was contrary to daily experience. The thesis of a spinning earth could never agree, in their view, with the stability of buildings. Moreover, a bird in the sky, detached from the spinning earth and flying in the sense of the earth's rotation, would not be able to advance - it would, in fact, move backwards - because the turning speed would exceed the bird's flying

---

Haan Zeist (Antwerp, 1963).

4 H. Kesten, *Copernic et son temps* (Paris, 1951), pp. 7, 226.

5 A. Pannekoek, *De Groei van ons Wereldbeeld* (Amsterdam and Antwerp, 1951), pp. 152-162; A. Koyré, "La Révolution copernicienne", in *Histoire générale des Sciences* (Paris, 1958), 2:52-74.

speed ... and more such arguments. Brahe (1546-1601), alleged to be the best observer of his days, rejected Copernicus's system. He felt that the heavy, material earth could not possibly be a fast-moving heavenly body. The sun and the planets, however, being regarded at the time as practically weightless luminous phenomena, were much more likely to perform such a role[6]. Indeed, if the earth were to draw a gigantic circle annually, as Copernicus claimed it did, the latter would necessarily have to be reflected in the stars - in the so-called parallax. As this was not the case, Brahe devised, forty years after the publication of Copernicus's book, a compromise solution, with his own system of the world, which eliminated the complications of Ptolemy, but saved the earth as the motionless center, viz. by having the sun revolve around the earth and the planets around the spinning sun.

Contemporary scholars were not willing to adopt the heliocentric position for lack of scientific evidence. Christopher Clavius would not only rage against the physical absurdities of Copernicus's system, but also against the fact that the system conflicted with several passages of the Bible[7]. This placed Copernicus under quarantine for the time being. But, as the belief in authority had largely been disposed of, Ptolemy's system was also left. This meant triumph for Brahe's vision of the world.

At present, now that we have all been turned into confirmed heliocentrists, Brahe's system is only mentioned in passing, as if it had never been of any importance, and was to be regarded as a curiosity that served only very temporarily to bridge the mental gap between Ptolemy and Copernicus. In a broad historical context this is undoubtedly true; but during the seventeenth century, Brahe's presentation of the world was a dominant one, which was being taught at the universities, as well as by the Jesuits! This made them second to none. Giovanni Domenico Cassini (1625-1712), professor of astronomy at Bologna, who in 1669 was invited to Paris as a member of the *Académie des Sciences* by minister Jean-Baptiste Colbert to build an official observatory - the first professional observatory in Europe - where he made some quite remarkable discoveries, remained faithful to Ptolemy's system until his

---

6 Pannekoek, *op. cit.*, p. 181; N. Max Wildiers, *Wereldbeeld & Theologie van de Middeleeuwen tot Vandaag* (Antwerp and Amsterdam, 1973), p. 164.

7 Koyré, *op. cit.*, p. 70; P. Thuillier, *D'Archimède à Einstein* (Paris, 1988), p. 157.

death[8]. As late as 1689, Jan Luyts, professor at the Utrecht University, issued a treatise, *Astronomica institutio*, in which he rejected Copernicus's theory. Not earlier than 1726, James Bradley would demonstrate the aberration of the starlight: the first piece of evidence for the revolution of the earth around the sun. And only in 1791, Giovanni Battista Guglielmini (1763-1817) would be able to deliver convincing evidence for the spinning of the earth through an experiment with a metal sphere which was thrown from the 78 meter Asinelli Tower at Bologna and clearly deviated from a perpendicular course[9]. In this connection, the question may be duly asked whether Verbiest, who died in 1688, or his successors, who were banned in 1732, can be reproached for not having propagated Copernicus's hypothesis in China.

Galileo had been the first, in 1609, to aim some primitive kind of telescope at the firmament, and to see a few quite remarkable things. He had been able to notice the relief structure of the moon surface and to discover that some nebulae and the Milky Way were, in fact, groupings of stars. And, near to Jupiter, he had found four satellites turning around the planet. He had made all this public in his *Sidereus nuncius* of March 1610. Shortly after that, he discovered as well that Saturnus had some weird appendages, and that Venus showed different phases. In this connection, we have to mention that neither the telescope, nor Galileo's discoveries have remained unknown to China, exactly thanks to the Jesuits! As early as 1615, the Portuguese Jesuit Manuel Dias, Jr. mentioned the discoveries that Galileo had made with his telescope in his booklet *Tianwen lüe* (On the Firmament), and in 1621 the German Father Johann Terrenz Schreck brought the first telescope from Europe to China, while his compatriot Father Johann Adam Schall von Bell wrote, in 1626, a Chinese treatise, *Yuanjing shuo* (On Telescope-Optics), which included the first Chinese picture of a telescope. Father Schall also projected, on September 1, 1644, a solar eclipse on a sheet of paper, using a telescope in behalf of two Chinese mandarins, and the solar eclipse of April 29, 1669 was observed by the emperor through a telescope the fathers had put at his disposal[10].

---

8 It is clear from a booklet he offered Pope Alexander VII in 1659, that G.D. Cassini held the earth to be in the center, *Systema revolutionum superiorum planetarum circa terram, etc.*; also see F. Bouquet, *Histoire de l'astronomie* (Paris, 1925), p. 367.

9 J. van der Bilt, *Sterrenkunde* (Amsterdam, 1927), p. 261.

10 Blondeau, *op. cit.*, pp. 67-69; J. Needham, "Astronomy" in *Science and Civilisation in China* (Cambridge, 1972), 3:444.

Although Kepler (1571-1630) had published his theorems on the movements of planets as early as 1609 and 1619, his ideas were but very slowly assimilated. When Galileo died in 1642, he had not even taken notice of Kepler's law, although both scholars had maintained a regular correspondence. They had, for example, communicated their respective findings on the 1613 comet. Moreover, Kepler had sent Galileo a copy of his *Astronomia nova* (on its appearance in 1609) containing his first two laws concerning the movements of the planets. But even then, Galileo wrote in his 1633 *Discorsi* (Conversations), that the actual shape of the planetal orbits was as yet unknown[11]! Galileo was not unique in being ignorant of Kepler's findings. Cassini, the above-mentioned professional astronomer and first director of the Paris Observatory, had never known Kepler's laws at his death in 1712. That Kepler's work took so long to be assimilated was certainly due to the fact that his writings were imbued with mythical contemplations. His "laws" did not have, at first, the straight formulation which, at present, is to be found in every booklet dealing with cosmography: they were buried under an abundance of calculations and argumentations aimed at the construction of a harmonious universe on a geometrical basis. Only his practical work, his tables of astronomy issued in 1627, enjoyed an immediate success.

Kepler had demonstrated how the planets moved. More than half a century later, Newton started wondering why they were moving that way. The story goes that already around 1666, Newton would first have suspected that the force that keeps the moon in its orbit is the same that makes a stone fall, and that a central power, originating in the sun, keeps all planets within their respective orbits. But these ideas were yet to grow to maturity, and only years later were they shaped in their mathematical form.

Although they were ignorant of the above insights of Newton, who had published nothing about his presumptions yet, the British astronomer Edmund Halley (1656-1742) and the London professor of mathematics Robert Hooke (1635-1703), discussed the assumed power that was to cause the movements within the solar system in the Royal Society in London in 1683. Christopher Wren, who had been president of the Society since 1680, offered a small prize to those who would be able to demonstrate the existence of the power in question. Wren's proposal did not get much of a response. In August 1684, Halley

11 Pannekoek, *op. cit.*, p. 198.

traveled to Cambridge in order to see Newton on the subject, because it appeared impossible to state properly the problems relating to the orbit of heavenly bodies without the mathematical methods that Newton had developed in his fluxion calculus. Newton confessed to Halley that, in fact, he had already solved the problem of the mechanics of heavenly bodies, but that his findings had not yet been prepared for publication. Halley was able to convince Newton to do this, and to put the treatise at the disposal of the Royal Society in 1686. Because he did not want this remarkable piece of work to be buried in the Society's archives for too long, Halley spared himself no effort to raise the funds needed to get it printed. He even made part of his private means available to that end. The treatise was issued in 1687, a few months before Verbiest died, at 250 copies, entitled *Philosophiae naturalis principia mathematica*[12].

This mathematical deduction from Kepler's laws was, at first, only moderately successful, no doubt because the treatise was rather inaccessible: only a few scholars had the necessary mathematical background to follow Newton's reasoning. Moreover, in his 1644 *Principia philosophiae*, the French philosopher René Descartes (1594-1650) had promoted the idea that universal space was filled with some thin substance whirling around the sun and dragging the planets along, and only the earth remained motionless relative to its immediate surroundings, although carried by the whirling substance as well. This system was accepted fairly readily at the time. Only the great Dutch mathematician and physicist Christiaan Huygens (1629-1695) did not agree, but he did not accept Newton's mutual attraction principle either, because he was unable to see which kind of mechanism might underlie universal gravitation. Even Gottfried Wilhelm Leibniz (1646-1716), one of the founders of differential and integral calculus, remained negative towards Newton's theory of attraction throughout his life[13].

Among the first continental advocates of Newton's theory were professors Willem Jacob 's Gravesande (1688-1742) and Pieter van Musschenbroeck (1692-1761), who respectively in 1723 and about 1730, focused attention on the theory of gravitation. At about the same time, Pierre Louis Moreau de Maupertuis (1698-1759) introduced Newton's theory into the Paris *Académie des Sciences*. The new insights were at last widely spread by François-Marie Arouet Voltaire, who not

---

[12] J. Lévy, "Le succès différé de la loi de la gravitation universelle", in *Histoire générale des sciences* (Paris, 1958), 2:478-479.

[13] *Ibid.*

only declared himself a Newton supporter in his *Lettres philosophiques* (1734), but, four years later, issued a popularized booklet *Éléments de la philosophie de Newton*. In 1766, a French translation of the *Principia* was issued by the Marquise Emilie du Châtelet, with a preface by Voltaire.

Meanwhile, a first proof for Newton's theory had been given. Cassini, the director of the Paris Observatory, had alleged - departing from the French degree-measurements by Jean Picard and Philippe de La Hire (in 1669 and 1683) - that the earth was flattened around the equator, whereas Newton's theory of gravitation predicted a flattening at the poles. In order to decide on this matter, expeditions were sent to Peru (1735) and to Lapland (1736) (with, a.o., Maupertuis and Alexis Clairaut), to measure the length of one degree on the meridian circle. The results of both expeditions corroborated the Newtonian gravitation theory. But a more spectacular proof for Newton's insights came in 1758. In 1705 Halley had predicted, using Newtonian mechanics, the return of the 1681 comet in 1759. This was a pretty risky prediction at the time, because comets had been assumed, until then, to be purely accidental phenomena, showing no regularity whatsoever. The actual return of the comet in 1758 provided additional evidence in favor of Newton's theory. But only at the end of the eighteenth century, when Pierre Simon de Laplace (1749-1827) managed to prove the stability of the solar system (*Traité de méchanique céleste*, 1799), would Newton's influence on physics be generally acknowledged.

Again, we can ask the same question. Can Verbiest, who died in 1688, or his successors, who were banned in 1732, be reproached for not having spread the Newtonian and Keplerian laws in China? It has been pretended that the instruments that Verbiest had built for the Beijing Imperial Observatory between 1670 and 1673 were of an archaic type described by Brahe in his *Astronomia instauratae mechanica* (1598), and that they were not equipped with sights or micrometers, although these aids had already been developed in Europe at that time. We have to observe, however, that the fact that something has been invented and described, does not usually mean that it is immediately taken into general use. When the *Observatoire de Paris* became operational in 1672 - i.e., about the time that Verbiest's instruments were ready - there were no instruments with sights either. The five foot radius sighted quadrant was put into use only in 1683[14]. It is true that in

14 Pannekoek, *op. cit.*, pp. 230-231.

1676, in Greenwich, there was a sextant with two sights, each with a hair cross, but the real instruments were still of the Tycho-type and made - just like in China - by craftsmen, albeit that they attested to a finer mechanical engineering and a more precise graduation. We should not forget that even in Europe, professional astronomy originated only in the last quarter of the seventeenth century. It is, of course, not Verbiest's fault that, when Europeans later managed to improve their instruments, the Chinese did not.

When Verbiest is dealing in his writings with the center of gravity (Simon Stevin), the inclined plane (Galileo), the refraction of light (Willebrordus Snellius), the Scheiner pantograph, the Bürgi proportional dividers, the alcohol thermometer (introduced in Europe about 1650), and the string hygrometer (Hooke), they are all discoveries or creations of his own seventeenth century[15]. When Verbiest drew a map of the world, taking into account the newest cartographical work published in the Netherlands, and when his successor Antoine Thomas mapped China following Snellius's triangulation method, can this then be called outdated science - or its practitioners, enemies of science? That the Jesuits in Beijing followed the European scientific developments is evident from the list of books available in the fathers' library at Beijing, which includes the newest scientific acquisitions of that time[16]. Verbiest and the seventeenth-century Jesuits offered China everything which Europe had in store, and which they were able to realize. Their activities in China were marked by their love for the Chinese people, without any other purpose than to ensure their salvation. Why should they have deceived China?

---

[15] L. Van Hee, S.J., *Ferdinand Verbiest, écrivain chinois* (Brugge, 1913), pp. 6-7; R.A. Blondeau, *Mandarijn en Astronoom, Ferdinand Verbiest S.J. (1623-1688) aan het Hof van de Chinese Keizer* (Brugge, 1970), p. 301.

[16] A partial list of the works that were available is to be found in Blondeau, *Mandarijn en Astronoom*, pp. 494-501.

GENERAL EVALUATION OF THE SCIENTIFIC WORK OF FERDINAND VERBIEST

ULRICH LIBBRECHT

Professor Ulrich Libbrecht holds a Ph.D. degree in Oriental Philology and Sinology, from the State University of Leiden (Netherlands). At present he is professor of Classical Chinese, History of Chinese Science, Chinese Philosophy and Comparative Philosophy at the K.U. Leuven. Since 1984 he is head of the Department "Modern Asian Studies" and director of the Ferdinand Verbiest Project, at the K.U. Leuven. In 1989 he founded the "School for Comparative Philosophy" in Antwerp (Belgium). Professor Libbrecht is the author of several books and articles.

In some popular writings Ferdinand Verbiest is depicted as a great man of science, a kind of emigrated Isaac Newton, who not only brought European science to China, but contributed much to the development of technology in that country. Of course some more spectacular inventions such as the automobile appeal to our imagination, but we have the impression that Verbiest was not so much an inventor as a transmitter of western inventions such as the thermometer and the hygrometer; that he made good use of western science as a vehicle for his missionary ideals, but that not a single important and original astronomical observation was made by the observatory in Beijing; that he was a most clever engineer, not only constructing new astronomical instruments, but also making interesting mechanical tools, the information for which he found in the many technological books later accessible in the Beitang Library.

On the other hand, when the romanticized image of Verbiest was created in the West, almost nothing was known about indigenous Chinese science. In recent times many studies on this topic have been published, culminating in Joseph Needham's *Science and Civilisation in China*. When it became clear that the Chinese sciences were not all that underdeveloped, the few scholars interested in the matter could hardly change their mind. So Louis Van Hee, e.g., a Flemish Jesuit writing between the two World Wars on Chinese mathematics, was convinced of the fact that the Chinese borrowed everything from India.

Verbiest was surely endowed with some engineering talent, so that he was able to make such complicated constructions as the astronomical instruments and cannons, simply relying on descriptions in western books. It is clear, however, that for the casting itself, he had to rely on the traditional and age-old skill of Chinese bronze and iron founders. The evaluation we will try to give in this essay restricts itself to the scientific work, and does not deal either with religious works, or with the highly informative and historically interesting letters or petitions to the emperor. As to Verbiest's missionary methods, this is a subject belonging to the field of missiology.

I. Verbiest's Education

The seventeenth century was the time of the scientific revolution in Europe. One could imagine that Verbiest was a product of this scientific movement and, consequently, that he was the ambassador for the new cosmological insights. I believe that the only real man of science - before Verbiest's time - who left Europe for China, leaving behind a scientific career, was Johann Terrenz Schreck, member of the *Academia di Lincei* and personal friend of many famous scientists, among them Galileo Galilei. The others were mostly occasional scientists, mainly self-taught men who made use of science and technology within the framework of their missionary strategy.

In 1640 Verbiest enrolled at the University of Louvain for one year, but interrupted his studies by entering the Jesuit Order in 1641; in 1643-1645 he again came to the *Collegium Societatis Jesu* in Louvain, attending lectures at one of the Faculties of Arts of the university. Of the twenty-four months he had to spend on these studies, only eight were devoted to physics. As the lecture notes of those times are preserved in the archives of the university, we know exactly what we have to understand by that "physics." It was almost nothing other than the *Physica* of Aristotle, dealing with such concepts as *materia*, *forma*, *natura*, *causa*, etc. Even the *De Coelo* contains almost no astronomy: it is devoted to such topics as the perfection of the universe, the influence of the celestial bodies on terrestrial events, etc. In other words, "physics" was nothing but *philosophia naturalis*. An analysis of Verbiest's geographical works and of his treatise on the thermometer and the hygrometer shows that he never rose above his Aristotelian education. Moreover, when, from 1650 on, the controversy between Cartesianism and Aristotelianism began at the universities, Verbiest was already engaged in totally other activities. Let us not forget that, till 1655, Verbiest tried to get permission to go to the missions in America, so that there was no direct reason for a special interest in scientific studies.

As for astronomical studies at the university, the basic work was still the *Sphaera Mundi* of Johannes de Sacrabosco, written in 1233, and actually based on the books of al-Farghani (c. 800-861) and al-Battani (before 858-929). Some professors made modern comments, and most of them also gave descriptions of the new systems of Nicolaus Copernicus and Tycho Brahe. As a whole, though, the lecture notes on astronomy give a real medieval impression.

The universities were saddled with a heavy scholastic inheritance. As late as 1675 the official dictum of the university was: "Magistri et scholares teneantur defendere doctrinam Aristotelis, nisi ubi ea Fidei nostrae repugnaverit." (Professors and students are obliged to defend the doctrine of Aristotle, except where it contradicts our faith.) However, this did not prevent the greater part of the professors from being Cartesians.

From these data we may conclude that Verbiest did not become a man of science at the University of Louvain, as the scientific evolution took place in academies outside the universities, and was guided not by the faculties, but by scientific periodicals. This left the universities in a rather marginal position, so that they became more and more strongholds of conservatism - a situation continuing up to the nineteenth century.

The Jesuits, who tried to keep pace with every modern evolution, were well aware of this low level of scientific life at the universities; consequently they tried to organize colleges themselves. Their plans, however, were thwarted by the universities, which claimed the monopoly of higher education. As the level of scientific research was rising very fast in the academies, the Jesuits understood the importance of coping with the evolution, and so founded a special school for mathematics in Antwerp in 1615. The first professor was Gregorius of St. Vincent (1584-1667), who studied mathematics in Rome at the *Collegium Romanum*, where the famous Christopher Clavius was his teacher. The courses of mathematics were accessible for Jesuits who had finished the course of physics at the university. In 1621 the school was transferred to Louvain, where Jan Ciermans and Willem Boelmans, students of St. Vincent, were successively professors of mathematics. André Tacquet, himself a student of Boelmans, held the chair for one year in 1644 and, as chance would have it, Verbiest was one of his students.

One of the problems I was unable to solve, concerns the subject matter of Tacquet's course of 1644-45. No lecture notes are preserved, and Tacquet's *Opera Mathematica* was of a much later date, and thus not useful for reconstruction. Ciermans, however, published the program of his course in 1640, and this can give us at least a general idea about the contents and level of Verbiest's studies at the *Collegium Lovaniense S.J.* Until now, we had no opportunity to make an analysis of Ciermans's work, but we hope to do so in the near future.

In 1656 Verbiest was appointed for the mission in China. On the way from Genoa to Lisbon the ship was hijacked, so that the missionaries arrived in Lisbon too late, and the fleet for the Far East - which left only once a year - had already set off. So they had to stay for a whole year in Portugal. Verbiest was sent to Coimbra, and in a letter to Athanasius Kircher, dated December 18, 1656 he writes: "I alone was sent to Coimbra, in order to teach mathematics to our people; it is clear that I learned more mathematics than I taught." This year in Coimbra could have been very important for Verbiest's mathematical education, because at that moment he knew very well the importance of mathematics for the missionary work in China.

The fact that Verbiest was appointed mathematics teacher for other Jesuits - and he alone among the ten young missionaries stranded in Portugal - proves that he was considered as a mathematician of some ability. On the other hand, as he says that he learned a lot, we are curious about the matter he studied there. But this is also a precarious reconstruction. Skimming the well-known standard works on the history of the Jesuits in Portugal by Sebastian Gonçalves and Francesco Rodrigues, we did not become much wiser. The situation of mathematical education in Portugal is in some way comparable to that of Louvain. It was surely a period of decline. The sixteenth century was undoubtedly the century of Pedro Nuñes, but after his death in 1578 mathematics came into decadence for several reasons: the expulsion of the Jews, the institution of the tribunal of Inquisition, with its condemnation of the systems of Galileo and Copernicus, the astrology which hindered scientific astronomy, and the decline of Portuguese navigation. Indeed, when Antoine Thomas arrived in Coimbra in 1678 as mathematics teacher, he complains in a letter of 1680 about the lack of printed books on mathematics there.

We actually have no other information about Verbiest's stay in Coimbra than the two letters he wrote to Kircher. More research has to be done on the spot.

Verbiest was ordered by Adam Schall von Bell to come to Beijing in 1660, and this on account of his mathematical skill. He writes in a Dutch letter dated July 5, 1660 to Philippe Couplet: "... me, as being (as he writes) well experienced in the science of mathematics ..." One wonders, from where Verbiest had got this reputation. But, once he was in Beijing, he clearly knew his task, and with Schall as teacher, a library at hand, and the tables of Argoli at his disposal, he could master the calender computation in a short time. Let us not forget that Verbiest was, during the controversy with Yang Guangxian and the subsequent house arrest, immobilized for many years, till he became official astronomer in 1669. During these nine years he could, as a self-taught scholar, devote his time to the study of astronomy, and from 1666 to the construction of new instruments.

A reconstruction of the contents of his home study can only be done by an analysis of his letters, and by special research on the Beitang Library. Until now, our Verbiest Project confirmed our idea that almost all Verbiest's work is in one or another way tributary to this library. But, although we have H. Verhaeren's catalogue at our disposal, a first-hand comparison of Verbiest's work with Beitang sources was never done.

Some of Verbiest's works were simple copies of older works written by his predecessors. So, for instance, all his geographical works (with exception of his world map) are copies of older works by Giulio Aleni; his philosophical work - offered to the emperor, but never printed - is a compilation of older texts. But some of his works are undoubtedly original works: his maps, his *Yixiang zhi*, his *Grammatica Tartarea* and others. However, the fact that he had to devote so much time to his astronomer duties is often forgotten. We mostly know his (never analyzed) calendar works, but in the Qing Archives in Beijing, 185 documents are preserved, containing the astronomical reports Verbiest had to offer once or twice a month to the throne in Chinese and in Manchu (we have copies of a few of them in the Project). I believe that, as imperial official, he had little time for much original scientific work. And it would be a total misunderstanding of his goal if we forgot that for him science was only a vehicle for religion.

Another relation never analyzed was that between Schall and

Verbiest. A superficial skimming of Schall's bibliography shows that this latter wrote many original works in Chinese, as well as translations of western works, much more than Verbiest did. It is a well-known fact that Verbiest made use of Schall's work for teaching the emperor. This indicates what a highly important source of information it was for Verbiest.

No matter how hard we try to bring together the pieces of the puzzle, we shall never be successful as long as we do not dispose of the very important mission records of Tomasz Ignacy Dunin-Szpot, finished circa 1700, and only existing in manuscript form in ARSI. I hope that the Jesuits will take the initiative themselves for the publication. The same can be said about the *Ajuda* Archives in Lisbon and Madrid.

Making a restricted analysis of the Beitang catalogue, I found that, while Verbiest had indeed not much scientific contact with Europe, he was not entirely isolated, as suggested by Henri Bernard-Maitre. But in general, we may conclude that he could not keep up with the fast evolution of science in Europe. This brings us to the controversial question: Did the Jesuits bring obsolete science to China? And if so, was it done deliberately?

II. The Question of Obsolete Science

J.J.L. Duyvendak reproached the Jesuits for having brought an antiquated science to China, because they did not introduce the Copernican system. This idea is taken over by Needham and by many Chinese historians of science. Objectively speaking, it is true that the Jesuits brought obsolete science to China - not only concerning the cosmological system, but also such Aristotelian interpretations as the four elements system. We state, indeed, that Verbiest relied on the Tychonian system, and that his explanations of the thermo- and hygrometer were purely Aristotelian.

For the historian of science, history is more than a mere registration of facts; it is also the investigation of the causal web that resulted in these unscientific deeds performed by men of science. In this essay we have to restrict ourselves to the problem of the non-introduction of the system of Copernicus.

We have to start our discussion from the sufficiently known Galileo controversy. When Galileo got a telescope in 1609, and made observations of the moons of Jupiter, he became convinced that the Copernican system was right, but he was condemned by the Church of

Rome in 1616, and sentenced again in 1633.

It is obvious that this fact had a strong repercussion on the universities and on the Order of the Jesuits. Scholars only had the choice between the antiquated system of Ptolemy and the intermediate system of Brahe; progressive scholars all became Tychonians. In all the countries subject to the Inquisition - and these included the homelands of Schall, Giacomo Rho and Verbiest - this was the choice made; in Protestant countries, on the other hand, the Copernican system was publicly propounded.

Verbiest left the university in 1645, not long before the struggle over the world system flared up in 1650. When we read the *argumenta contra systema Copernicanum* in the lecture notes, we see that the *argumenta ex ratione* were not of much value, so that they were unable to convince good scholars, who were, after all, impressed by the great simplicity of the system. The *argumenta ex fide* were devastating for Copernicanism, because in the Holy Scripture there are at least ten passages "proving" that the earth is immobile and the sky turning around it. As a consequence, some professors in Louvain, who were at first Copernicanists, "converted" to Ptolemaism, or at least to Tychonianism. We can hardly understand this dissidence, but there was a general principle stating that science had no right to interfere in theological questions. The idea that non-falsifiable judgments are not scientific, and consequently that religious belief belongs to another category than science, was not known in those days. Hence, a conclusion of natural science could not cancel a verdict of the pope, because this latter had an absolute character, whereas science is built on hypotheses. And indeed, we see that from this time on the Copernican system is called an hypothesis. However, those who were so bold as to say that Copernicanism was the only truth - as Professor M. Van Velden did - were suspended. And this in 1691! (Verbiest died in 1688). Copernicanism was to remain censured till 1822!

And what about the Jesuits? We have to restrict ourselves to a few Belgian Jesuits to illustrate the case. In 1650 the General Congregation of the Jesuits condemned sixty-five propositions, but did not reject the possibility of these being true. The thirty-fifth was: "Terra movetur motu diurnu ... firmamentum stat." Consequently, all the Jesuits had to reject the Copernican system. But were they all scientifically convinced that the Copernican system was false?

St. Vincent, disciple of Clavius, attended the triumphal reception of Galileo in Rome. He was clearly an adherent of the new astronomical ideas, and he did not hide this, even in the period between the two condemnations of Galileo. But the most interesting case is surely that of Tacquet, Verbiest's teacher, who declared that he rejected the Copernican system, but only because of his vow of obedience, and not because of scientific arguments. In Tacquet's *Opera Mathematica* we find a full description of "De Hypothesi Terrae Motae" (p. 223), but he clearly states that all arguments against the Copernican system are without any value. Moreover, he says that the Holy Scripture must be interpreted in a figurative sense, if there seems to be any contradiction with natural reason.

This was the schizophrenic situation in which all Catholic scholars were living. It is true that a heroic attitude such as that of Galileo - who, after all, renounced his scientific conviction before the tribunal of the Church - appeals more to one's imagination; but it is very unhistorical to forget that the Jesuits were in the first place priests, and only secondarily men of science.

The situation of the Jesuits in China can only be understood against the European background. Before the condemnation of Galileo, the Polish missionaries Michal Boym and Nicolas Smogulecki were - out of a kind of natural pride - adherents of the system of their compatriot, Copernicus. And Wenceslas Pantaleon Kirwitzer (1586-1626), who arrived in China in 1620, was a real Copernican.

Matteo Ricci died in 1610 - before the condemnation of Galileo; but the discoveries of Galileo were already described in Chinese by Manuel Dias, Jr., in 1614-1615, which proves that scientific inventions could circulate relatively fast, even in such remote times.

The case of Galileo brought the Jesuits into a nasty situation, not only because he had been received as a hero by the *Collegium Romanum*, but also because, as a member of the *Academia di Lincei*, he was on good terms with Schreck. As Schreck was not an astronomer *stricto sensu*, he turned to Galileo for assistance. After his condemnation, this latter was unwilling to help any more, and so the Jesuits got in touch with Johann Kepler who, although himself a Protestant, was willing to help. The simplest solution for the Jesuits in China and elsewhere was to accept the system of Brahe, even when they were Copernicans in their hearts. So in China, the Copernican system was not made known until the French Jesuit Michel Benoît introduced it in 1761.

Needham (3:445f.) says that "after news of the condemnation (of Galileo) had reached China, a curtain descended and a return to the

Ptolemaic view took place." This is not a correct representation of facts: the missionaries in China chose the Tychonian system, as clearly proved by the writings of Verbiest.

Later on, the transition to the Copernican system in China became difficult, for missionary reasons. The Jesuits were afraid to raise doubts about the teachings of Ricci. The purpose of the Jesuits was purely religious; science was only a means to an end, and so it had to stand aside for religion. This was clearly proved by the fact that as late as 1716 the official visitor of the Chinese mission forbade Father Jean-François Foucquet to switch to the new system. It is a fact that the introduction of Copernicanism into China was fully dependent on the religious situation in Europe: Dutchmen brought the system to China in the nineteenth century.

Another fact to keep in mind is that the Jesuit astronomers in Beijing had no other task than drawing up the calendar. For that purpose there is no difference between the system of Brahe and that of Copernicus. World systems are geometrical constructions, whereas the calendar is an algebraic construction, which harmonizes periodicities and predicts recurrences. This requires only an algorithm.

That the Jesuits tried to obstruct the introduction of modern science is a very black-and-white representation of historical facts. For instance, the telescope was invented circa 1600 in Holland, Galileo constructed one in 1609, Dias described it in Chinese in 1615, Schreck brought the instrument to China in 1618, Schall devoted a whole booklet to its description in 1626, and one was offered to the emperor in 1634.

Schall, writing before 1650 (condemnation of Copernicanism by the General Congregation of the Jesuits), gives a description of the three cosmological systems in his *Lifa xizhuan*, and does not try to hide anything from the Chinese. Verbiest, writing after 1650, describes only the Tychonian system.

III. Conclusion

1. That Verbiest was Tychonian is perfectly normal. As a missionary, it was not his place to take part in scientific controversies. Moreover, he was bound by his vow of obedience to Rome.
2. On the other hand, it is clear that he did not try to keep track of the scientific evolution in Europe. He surely did not try to introduce modern science.
3. Although he got a good education in mathematics and astronomy, he

was not a scholar as Schreck was. He never had relations with men of science in Europe, as Schreck had.

4. As astronomer, he was surely a competent official. But it is clear that the observatory in Beijing never contributed anything to the development of astronomy as a science.
5. On the other hand, in the case of medicine, he complains in several letters to his superiors that the books in the library were so highly antiquated as to be almost useless.
6. Although he was not completely isolated from the West, I have the impression that he had not much scientific contact, as is proved by his letters, which are mostly concerned with missionary problems. In his times, only a few new scientific books came to the library.
7. Verbiest was surely a talented engineer, which is proved in the first place by the new instruments he built, and by the description of them he gives in the *Yixiang zhi*. Its technical drawings are of high value, even for the history of science in Europe. The casting of cannons also gives evidence of remarkable technological skill, although we know that many of the howitzers he built at the end of his life were not a great success. The invention of the automobile appeals to the imagination of modern men.
8. Verbiest was in the first place a missionary, and if we do not keep that in mind, we not only give a wrong image of his scientific-technological realizations, we simply give a biased view of this great missionary by reducing him to the predecessor of Flanders technology. That he was indeed; but he was also much more than that.

FERDINAND VERBIEST ON EUROPEAN ASTRONOMY IN CHINA: FROM THE *COMPENDIA* TO THE *ASTRONOMIA EUROPAEA* (1687)

A Historical Philological Analysis

NOEL GOLVERS

Noël Golvers studied classical philology in Louvain, where he obtained his Ph. D. in 1984 with a dissertation concerning ancient Italic onomastics. Since 1985, he turned to the study of Ferdinand Verbiest and the contemporary Jesuit mission in China within the frame of the Verbiest Project. Since 1988 he is a part-time scientific researcher granted by the Verbiest Foundation. He is mainly involved in the edition of Verbiest's Latin written works on astronomical matters, and in the new, critical edition of the correspondence of Verbiest and his contemporaries.

Since I am a classicist, it was not until the summer of 1985 that I came into close contact with Ferdinand Verbiest who, by his honest personality and through his intriguing work, has since added a stimulating and rich experience to my personal education, thus far determined mainly by western classics. As my own competence is situated in the field of Latin and Romance philology and linguistics, my approach to Verbiest was necessarily restricted to that part of his work written in western languages. From the whole of this oeuvre, I first chose as my domain the Latin treatises on western astronomy in China, i.e. the *Astronomia Europaea* (AE), with its preliminary stages, the so-called *Compendium libri observationum* (CLObs.), the *Compendium libri organici* (CLOrg.) and the *Compendium Latinum* (CL), which were afterwards incorporated - with only minor adaptations - in the AE.

A thorough understanding of this text and its subsequent translation involves many problems. Besides the general fact that Verbiest's phraseology is often ambiguous and his intentions obscure, there are the more specific difficulties connected to the subjects treated. As the AE gives an elaborate account of the Jesuit achievements in fifteen different mathematical sciences such as astronomy, gnomonics, ballistics, mechanics, optics, catoptrics, etc., a correct understanding of the many descriptions of often bizarre-looking instruments and their working requires a technological imagination rather uncommon for a classicist. Verbiest's technical Latin vocabulary, of course, is substantially tributary to contemporary research literature in all those different fields; but this, unfortunately, has not yet been inventoried[1]. On the other hand, there is - especially in the first part of the AE - a sinological reality hidden behind an inadequate Latin terminology[2]. Apart from this, there is the problematic collation of the many widely dispersed copies of this treatise (except the AE), and the complicated development of the text itself.

In May 1988, after three years of study, I was able to give a first account of my work in a Dutch publication: *Astronoom van de Keizer. Ferdinand Verbiest en zijn Europese Sterrenkunde* (Leuven, 1988)[3]. It is, in a certain sense, a propaedeutic to an English *editio major* (see Noel Golvers, *The Astronomia Europaea of Ferdinand Verbiest, S.J. (Dillingen, 1687). Text, Translation, Notes.* Monumenta Serica Monograph Series XXVIII [Sankt Augustin - Nettetal, 1993]). In anticipation, I take this opportunity to explain some of my ideas concerning the complex growth-process which led from Verbiest's first (Chinese) descrip-

---

1 A very helpful reference book was J.-H. Zedler, *Grosses vollständiges Universallexicon aller Wissenschaften und Künste*, 68 vols. (Halle-Leipzig, 1732-1754; reprint, Graz, 1961), which offers a synopsis of seventeenth-century scientific research in Europe, and thus incorporates a good part of Verbiest's direct or indirect sources, containing, in particular, numerous Latin terminological entries concerning technology also found in Verbiest. In addition, there are several specific studies which give good information both by text and illustrations, as, e.g., E. Zinner, *Deutsche und niederländische astronomische Instrumente des 11.-18. Jahrhunderts* (München, 1956), or AA.VV., *De Landt-meeters in onze Provinciën van de 16de tot de 18de Eeuw* (Brussels, 1971).

2 The most adequate approach to these questions was, as far as possible, a systematic confrontation with other contemporary sources for the same topics, both western and Chinese: Verbiest's own letters to Europe (published or newly-discovered), the letters and works of fellow missionaries in China and, last but not least, the Chinese works of Verbiest himself, accessible to me only in rudimentary translations at the Verbiest Project.

3 This book contains: a) a general introduction by Prof. Ulrich Libbrecht (pp. 11-84), b) a specific introduction to the *Astronomia* (pp. 85-108), and c) an annotated Dutch translation (pp. 109-252), both by the present author.

tion of his astronomical observations of 1668-1669 in the so-called *Ceyan ji lüe*, to his final synoptical treatment in the *Astronomia Europaea*, which was finally published at Dillingen in Swabia in 1687.

My first remarks are of a purely terminological nature, and they are intended to clear up some uncertainty, even confusion, in "modern" literature when referring to the preliminary Latin stages of the AE, due, in part, to the confusing multiformity of the preserved copies and their title pages[4] and, in part, to a certain "negligence" on the part of some scholars. In accordance with Verbiest's own practice in the AE and in his letters, and comparing these with the copies at my disposal[5], I think it is possible to put forward some specifications.

The first report on his spectacular astronomical observations of the winter 1668-1669 for a European readership is the so-called *Compendium libri observationum* (CLObs.) containing six folios of Latin text explaining a set of twelve drawings which are exactly the same as those in the previously written Chinese *Ceyan ji lüe*. When referring to these drawings, Verbiest always speaks of his *Liber observationum* (Book of Observations)[6]. Since the preserved copies in Europe, to my knowledge, bear on their first page a vignette with the inscription *Observationes astronomicae* (Astronomical Observations), I think both expressions are an abbreviation of a fuller title *Liber observationum astronomicarum* which, indeed, is attested once in a hitherto unobserved letter of Verbiest to Giovanni Filippo de Marini dated August 24, 1678:

> I send on this occasion my two volumes on astronomical instruments (= *Yixiang tu*), together with a *Liber observationum astronomicarum*, which is the fundament of our mathematics, or astronomy, introduced again ...[7].

---

4 See the remarks of H. Walravens, in *Börsenblatt für den deutschen Buchhandel - Frankfurter Ausgabe*, n° 87 (1 November 1983), A 348-A 350.

5 The copy of the *Compendium Latinum* (including the CLObs. and the CLOrg., with twelve and eight illustrations respectively) is preserved at the Royal Library in Brussels (V.H. 31075), and that of the *Astronomia Europaea* is at the University Library of Ghent (A 1339/8).

6 Cf. "figuras in *Libro observationum* ... expressas; ... in *Libro observationum* designata ...," AE, p. 41; "figura ejusdem *Libri observationum* ...," p. 43; "figura 8va *Libri observationum*," "figura 12a *Libri observationum*," p. 45; "*Libellum* meum *observationum* figura 1a," p. 49; "... ex hac rudi instrumentorum adumbratione et ex libello observationum praemisso ...," p. 53.

7 The original letter is now in Madrid, Archivo Histórico Nacional, *Jesuitas, leg. 270* n° 87 (see J. F. Schütte, *Documentos del "Archivo del Japón" en el*

On the other hand, when occasionally speaking of the text, Verbiest uses the word *compendium*[8]; but the title *Compendium libri observationum* is not found, as far as I can see, in his own writings, nor in other contemporary sources. At any rate, *Compendium* here cannot mean "summary" or "abstract," as is generally believed. The so-called *Compendium libri observationum* is not a summary of the *Ceyan ji lüe*, as is made clear by comparison[9], but actually is a brief Latin explanation (of the drawings) in the *Liber observationum astronomicarum*.

In 1673-1674, for the benefit of his Chinese assistants and, indeed, of the emperor himself, Verbiest published a series of 117 drawings together with a commentary in Chinese, mainly on the construction and use of a comprehensive "mathematical" instrument: the well-known *Yixiang tu* in two "volumes," and the *Yixiang zhi* in fourteen "volumes." Many copies found their way to Europe, with or without a Latin title page. In addition, there is a Latin text of eleven folios, accompanied by the same drawings as in the *Yixiang tu* (either the first eight [seven + one] showing the newly erected Beijing Observatory with its instruments, or by the whole first *juan* [53 ff.; 58 pict.], or even by the complete set in two *juan* [106 ff.; 117 pict.]). This Latin text contains a description of the first eight figures (6 ff.), but is preceded by a brief chronological synopsis of Verbiest's previous accomplishments (2 ff.), and followed by a long digression (with separate subtitle in the margin of f. 8) describing substantially the emperor's enthusiastic interest in European mathematics (4 ff.).

When referring to these drawings, Verbiest speaks consistently of his *Liber organicus / Libri organici* (Instrument Book[s]), a term extended only exceptionally to the text[10]. The identity of *Liber organicus* and *Yi-*

Archivo Histórico Nacional de Madrid, [Madrid, 1978-1979], p. 32), but an eighteenth-century copy is in Lisbon, *Biblioteca da Ajuda*, *Jesuitas na Asia*, 49-V-17, N. 110. The paragraph in question reads as follows: "Mando com esta occasiao meus dous libros dos instrumentos astronomicos juntamente com hum Libro Observationum Astronomicarum, que he o fundamento de nossa mathematica ou astronomia de novo introducida. Mando tambem hum compendio latino das figuras primeiras etc." (f. 1).

8 Cf. "... et hujus compendii aliud brevissimum (compendium) Latino idiomate postea scriptum sit, quod in fine hujus tractatus subijcio ...," AE, p. 16. The phrase *hujus compendii* refers to the *Ceyan ji lüe*.

9 For this purpose I used the draft of a Dutch translation by P. De Riemaecker, now in the library of the Verbiest Project. Verbiest's own words (cited in note 8) are a bit misleading here.

10 Cf. "figuras in *Libro observationum et organico* expressas," AE, p. 41; "eorum autem omnium (sc. instrumentorum) figuras duobus libris quos

xiang tu is proved not only by Verbiest's description in AE (p. 47), but also by the copies of the *Yixiang tu* with a Latin title page *Liber organicus astronomiae Europaea restitutae* ... Finally, we have two precisely located references of Verbiest himself to some figures in his *Liber organicus*: one in the CLOrg. (*sub figura* 8a) to figures 43-55 "showing a set of all kinds of mathematical instruments," and a second one in his AE (p. 70) to figure 97, showing a geodetic operation concerning the diversion of a river outside Beijing. Both these references are found again exactly in the indicated place, i.e. on folios 43-55 (= fig. 46-61) and on folio 97 (= fig. 106) of the *Yixiang tu* - *Liber organicus*. When referring to his Latin text, on the other hand, Verbiest speaks of his *Compendium libri organici* (CLOrg.), i.e. the "Short (Latin) commentary of his *Liber organicus*;" indeed, a translation as "Summary of the Liber organicus" makes no sense, because the *Liber organicus* - *Yixiang tu* is, in essence, a sketchbook; actually, it is the short explanation of its first eight figures. This title is apparently the abbreviation of a fuller expression, quoted in AE (p. 59) as *Compendium octo priorum figurarum libri organici* and, in his letter to Marini - in a somewhat careless (and abridged) Portuguese translation - as *hum compendio latino das figuras primeiras* etc.. Quite unexpectedly, there is one instance of *Compendium* applied to a volume of drawings: indeed, my copy bears under the first picture of the *Liber organicus* - representing the Beijing Observatory - a vignette with the Latin inscription *Compendium astronomiae organicae*, which we have to understand as the abbreviation of an hypothetic *Compendium libri organici astronomiae Europaeae* ... Finally, there is the *Compendium Latinum* (CL) or, with its full title, *Compendium latinum proponens XII posteriores figuras libri observationum necnon priores VII figuras libri organici*. This title is attested - in Verbiest's own (calligraphic) handwriting - in the preserved copies, as well as in his AE, with the minor correction of VII to VIII[11]. This *Compendium*, or at least the copies known to me, has the complete text of the two aforementioned *Compendia*, including their drawings.

Of course other combinations were possible and, indeed, occasion-

'*Organicos*' appello," p. 47; "in *Libro* meo *Organico* ... nudis lineamentis adumbravi," p. 58; "figuram binorum horum instrumentorum in *Libro meo Organico* ..." opposed to "rationem et usum in *Libris Mathematicis*," p. 96. Only on p. 72, the term *Libri organici* refers to the entire *Yixiang tu* and the *Yixiang zhi*.

11 "Compendium proponens XII posteriores figuras ..." with omission of the qualifier "Latinum!"

ally produced. One example is the *Astronomia organica*, quoted in Verbiest's letter to Innocent XI of August 15, 1678 as the title of one of the books he is presenting to the pope: "... a second present is the *Instrumental Astronomy of the Chinese-Manchu Emperor*, printed in three volumes"[12]. This title - found nowhere else - is, in my opinion, an *ad hoc* indication for the incidental joining of distinct items: this shipment consisted essentially of drawings, as is well-illustrated by the pope's answer[13], though not exclusively! After comparison with Verbiest's more elaborate description in his letter to Marini[14], we see that his presents to the pope consisted of (a) the *Yixiang tu* - *Liber organicus* (two volumes) and (b) the *Liber observationum astronomicarum* (one volume) - but also of (c) the *Compendium libri organici*! Referring to the *three* volumes ("... *tribus libris expressis*"), Verbiest must have considered the eleven folios of this *Compendium* not as a separate volume, but as an integral part of the *Liber organicus*, of which it is, in a certain sense, the Latin introduction.

The important point in all this is that, although the indications at first glance seem complicated and confusing, there is a clear system - with only a few exceptions - behind it in Verbiest's own practice. Speaking of a volume of drawings, he uses the term *Liber*, so any unclear reference to a *Liber* must be understood in that sense. *Compendium*, by its very nature, refers to a volume of text, though occasionally accompanied by a set of drawings. In the case of the three astronomical treatises under investigation, it is not a question of a "summary" or "abstract" in the usual sense, but rather of an independent "short explanation," for the benefit of European users of the sets of drawings.

In the next part of this contribution, I will discuss the date of these preliminary stages of the AE, for (particularly in the case of the *Compendium libri organici*) there is some useful internal evidence on this matter, so far not - or not sufficiently - explored. As to the Latin notes

---

12 Cf. Henri Josson and Léopold Willaert, *Correspondance de Ferdinand Verbiest de la Compagnie de Jésus (1623-1688), Directeur de l'Observatoire de Pékin* (Brussels, 1938), pp. 227-228.

13 *Ibid.*, p. 369 (December 3, 1681): "... imagines astronomicas Sinensi item more a te affabre delineatas"; Verbiest (*ibid.*, p. 255, 265) speaks of "tabellae (astronomicae)", i.e. small pictures/paintings.

14 "Estos mesmos libros (= dous libros dos instrumentos astronomicos; hum libro Observationum Astronomicarum) juntamente com o ditto Compendio (= *Compendio libri organici*) do mesmo modo encadernados" (Madrid, Arch. Hist. Nac., *Jes.*, *leg. 270*, n. 87, f. 1r-v).

on the drawings of the *Ceyan ji lüe* - the so-called *Compendium libri observationum* - there are insufficient indications to date its origin except in a very vague way. It seems reasonable - and it can also be inferred from Verbiest's own words in AE (p. 16)[15] - that the Chinese text was the original one, and that the Latin version was drawn up after it. However, we have no exact idea how much later it came into being; one hint might lurk in Verbiest's own words on the first folio of this *Compendium*, referring to the reopening of the Christian churches in China[16], which happened in January 1671, the only *terminus post quem* for the writing of this Compendium.

For the *Compendium libri organici*, on the contrary, we have some useful internal evidence. Indeed, in his introductory note to this *Compendium* (ff.1-2), Verbiest refers to the *Astronomia perpetua imperatoris Cam Hy*, speaking explicitly of "thirty-two printed volumes," under one general title[17]; therefore, there can be little doubt that it is the complete, printed edition that is meant here. The date of its completion will be a reliable *terminus post quem* for the writing of the CLOrg.! We know that the enormous printing work of the *Astronomia perpetua* could not have started before early 1676, since the memorial of Verbiest, asking for the necessary paper, was only presented to the emperor on January 23 of that year[18]; the finishing of the first sixteen volumes was expected around the end of 1677[19], and the solemn presentation to the emperor took place on August 27, 1678, as is demonstrated by another memorial. It seems reasonable to fix the actual completion of the whole work not too far before this last date, and it could, after all, roughly be

---

15 See note 8.

16 "Tandem Religio Christiana quinto persecutionis suae anno in publicum iterum prodiit, templis in omnibus fere vastissimi hujus Imperii provinciis, uti prius, apertis," CL, f.1 = AE, p. 41.

17 "Libris 32 impressis, quorum titulus *Astronomia perpetua*," CL, f. 8 = AE, p. 47.

18 For these and the following memorials, I am indebted to a kind communication by Prof. Willy Vande Walle, to whom I extend my thanks.

19 A hitherto unexploited letter of Verbiest to Father General Oliva, only preserved in a transcript of Dunin-Szpot, *Collectanea pro Historia Sinica*, ARSI, *Jap.-Sin.* 109, pars 2, p. 128, refers to the first sixteen volumes in print: "... me suscepi opus sedecim librorum qui iam sub proelo sudant et hoc ipso anno, ut spero, absolvendum, omnibus scilicet discipulis qui multi sunt calculantibus et publicos sumptus praebente Rege" (July 16, 1677).

situated in the first half or in the middle of 1678[20]. This, in turn, is a *terminus post quem* for the *Compendium libri organici* which, in its introduction, refers to this *Astronomia* in its completed form.

Another indication in the text of the CLOrg. concerning Verbiest's *Manchu Grammar*[21], which is mentioned under figure 8 of the same *Compendium*, leads us to the same conclusion. Reading the preface (*ad lectorem*)[22] of that work, we learn that its composition took place after the Beijing mission of the Russian ambassador Nikolai Govrilovich Spathary Milescu, which is recorded here, i.e., after September 10, 1676[23]. This is, *a fortiori*, a second *terminus post quem* for the redaction of the CLOrg., but a date not too close after September 1676 seems preferable, because the note in the CLOrg. presupposes - we do not know to what extent - the finishing of the *Manchu Grammar*, which was itself only started after September 1676. In this approximate sense, it reinforces the reliability of the aforementioned evidence. The absence, on the other hand, of a clear reference to the solemn presentation of the *Astronomia perpetua* to the emperor in the same introduction of the CLOrg. could be significant as well, especially if we take into account

20 This seems to be confirmed by Verbiest's own statement in Josson and Willaert, *op. cit.*, p. 228 (15.8.1678): "Nuper 32 alios libros sinico item idiomate edidi quorum titulus est *Astronomia perpetua* ..." (edition "recently" in relation to August 15). Other western sources give additional evidence, but only in vague terms: Verbiest in AE, p. 34; LA, 1678 (ARSI, *Jap.-Sin.* 117, f. 184v): "prodiit hoc anno ... et quidem extensis ad bis mille annos venturos, triginta tribus (!) omnino justis tomis illas complexus"; Dunin-Szpot, *Collectanea pro Historia Sinica*, ARSI, *Jap.-Sin.* 104, f. 299v.

21 "Cum Magnates passim idiomate Tartarico utantur, illudque in omnibus fere tribunalibus magnum usum habeat, mihi quoque singulari benevolentia ex domesticis suis magistrum dedit, qui me linguam Tartaricam doceret, *cujus nunc grammaticam composui*." AE, p. 55.

22 See M. Thevenot, *Relations de divers voyages curieux*, nouvelle édition (Paris, 1696), vol. 2, 4ième partie, p. 4: "Ego certe, cui singularis Imperatoris Sinico-Tartari benevolentia de industria magistrum dedit, quo idioma et litteras Tartaricas addiscerem, horis quibusdam per diem, quibus per otium licuit, animum ita applicui, ut idiomatis hujus praecipuos et magis necessarios loquendi modos ad certas quasdam leges Grammaticae Latinae similes conatus sim reducere, ex quo hunc saltem fructum brevi tempore percepi, ut legati Magni Ducis Moscoviae Latino-Tartarum interpretem per tres et amplius menses non incommode agere potuerim, tam apud supremum Imperii hujus Consilium quam apud ipsum Imperatorem Tartarum. Nunc itaque ad exemplar Latinae grammaticae leges grammaticae Tartaricae, seu Grammaticae Tartaricae speciem aliquam adumbratus, hunc tractatum ..."

23 Cf. Paul Pelliot, *T'oung Pao* 21 (1922), 385.

Verbiest's constant and deliberate attention to such "offerings" in his writings to Europe (see AE and his letters), for obvious apologetic reasons. The solemn presentation on August 27, 1678 would be, in that case, a *terminus ante quem* for the CLOrg. This is confirmed by newly discovered external evidence. Recent investigations in the archives of the Ajuda and Madrid have brought to light the already quoted letter of Verbiest to Marini dated August 24, 1678, and containing the first positively datable mention of the *Compendium* in question, cited as: "... *hum compendio latino das figuras primeiras* etc." The text of this compendium must therefore have been finished and available, at the latest, on August 24, which explains why the "offering" of the *Astronomia perpetua* on August 27 is not mentioned! By combining all available evidence (*terminus ante/post quem*), we can safely conclude that the *Compendium libri organici*, in this version, was written somewhere between the first half of 1678 and August 24 of that same year. This fits very well with the final note of its introduction (f. 2), according to which Verbiest, at the same moment this *Compendium* (or its introduction) was completed, was engaged in "... *the introduction of the European dialectics and philosophy in China* ...," and this is quoted thrice as a "work in (advanced) process" in some letters of ... August 1678[24]!

This dating to 1678 for the CLOrg., however, does not agree with conventional opinion on the matter, which accepts for this *Compendium* a date as early as 1676 or even 1668! Therefore, I will present my conclusions against the arguments of earlier Verbiest scholars. Dating the work in 1668 seems to rest exclusively on the fact that this year is mentioned at the bottom of the title page of the *Liber organicus*[25]. A date, however, in this very spot, must not be interpreted in the same way as one in a modern publication! It has to be integrated syntactically in the wording of the entire title which, according to European conventions of that time, is elaborated in one long periodic sentence. In

---

24 "*Nunc* vero habeo prae manibus Dialecticam et Philosophiam nostram, sub pallio sinico introducendam, praetextu inquam Astronomiae, revera autem ad religionis nostrae evidentiam clarius demonstrandam." CLOrg. f. 2 (in fine). Cf. "Ipsa dialectica et philosophia ... magna ex parte impressa sunt." Josson and Willaert, *op. cit.*, p. 217; see also Verbiest's letter of August 15, 1678, *ibid.*, pp. 220, 228.

25 *Liber organicus / astronomiae apud Sinas / restitutae / sub Imperatore Sino-Tartarico / Cam Hy appellato* / Auctore P. Ferdinando Verbiest / Flandro-Belga Brugensi / e Societate Jesu / Academiae Astronomicae / in Regia Pekinensi / Praefecto / Anno Salutis M.DC.LXVIII.

this view, the date "1668" must be connected with the syntagm: "... *ex Umbra in Lucem Revocata* ..." ("from the darkness called back into full daylight"), preceding the author's name and qualifications, which refers to the restoration of European astronomy in China (the starting point of which was, as is well-known, Christmas Day 1668) - not to the publication of this compendium.

A dating to 1676, on the other hand, is proposed by Henri Bosmans[26] on the basis of one piece of external evidence, i.e., a section in Verbiest's letter of the same year to the Russian tsar, in which the famous scholar has recognized a reference to the *Compendium Latinum*[27], thus postulating the completion of the *Compendium libri organici* in September 1676, at the latest - or even before, because the CLOrg. is an integral part of the CL (cf. 1.3.). A close examination of this text - with the kind assistance of a few slavicists of Leuven University[28] - has yielded the following translation and interpretation:

> ... The volumes - which I recently wrote in the Chinese and Manchu languages, and offered to the (Chinese) emperor, and in which (I represented) European astronomy restored by myself among the Chinese people as a sort of short presentation with pictures - those humble volumes, I repeat, I offer to Your Highness the Tsar, explaining the Chinese language and tongue in Latin.

When we compare this description of these Chinese volumes with Verbiest's own works, it is hard not to recognize here the *Yixiang tu* and the *Yixiang zhi*, finished and presented to the emperor in 1674 (i.e., relatively "recent," in comparison with 1676), dealing with European astronomy in Beijing, and containing pictures (a sort of short presentation) and text. In that case, the Latin explanation mentioned must refer to the *Compendium libri organici*, the European counterpart of the *Yixiang tu - Yixiang zhi* collection; the Spathary embassy - which

---

26 See Henri Bosmans, *Revue des questions scientifiques*, 74 (1913), 275.

27 See Henri Bosmans, "Le Problème des relations de Verbiest avec la Cour de Russie", *Annales de la Société d'émulation pour l'étude de l'histoire et des antiquités de la Flandre*, 63 (1913), 193-223. He gives his translation of the pericope concerning this work on p.201: "Et les livres écrits en chinois et en tartare que j'ai offerts récemment au tsar (chinois), dans lesquels j'ai reconstruit l'astronomie des Chinois en la représentant brièvement en images, ces livres dis-je, je les offre maintenant humblement à la Majesté Tsarienne, expliquant en latin la langue et le discours chinois". For the "original" Russian text, see Josson and Willaert, *op. cit.*, p. 193.

28 I want to express my gratitude to Prof. Dr. E. Waegemans and to Prof. Dr. E. Böcker for their kind help.

brought Verbiest's offering to its destination in 1676 - thus becomes a *terminus ante quem* for the CLOrg. This, finally, is in apparent contradiction to our previous conclusions!

A hint for a solution to this puzzling contradiction between internal and external indications, however, is to be found in the structure (and behind it, in the elaboration) of the *Compendium libri organici* itself. As I have described above, it is composed of a nucleus with systematic descriptions of the first seven or eight pictures in the *Yixiang tu*, preceded by a short introductory note and concluded by an *addendum* on the emperor's studies in European mathematics. One of the two references to 1678 - incompatible with a dating to 1676 or even earlier - occurs in the introduction, the second in the concluding digression. This configuration suggests the possibility of two distinct editorial stages: a first and original one during or before 1676, which was restricted to the nucleus, giving Latin commentary to the first eight instrument drawings[29], and a subsequent one, which added to the original text an introduction and a final *addendum*, containing also the references to the year 1678. This explanation by a two-stage draft is purely conjectural, and looks like a mere construction to reconcile two irreconcilable indications; on the other hand, because both are well-founded, it seems the only imaginable way to overcome a hermeneutic impasse. The fact that a process of growth through distinct editorial stages is not uncommon in the works of the Jesuits in China, reinforces the plausibility of this assumption. Besides, in this case, such a conjecture would find more positive support in the historical circumstances which characterize the relations between China and Europe in the 1670s. There are, indeed, several explicit references, including some in Verbiest's own writings of that time, which prove an interruption in the mutual contacts between Beijing and Macao-Europe in a period from 1673/1674 to 1677, due to the internal uprising known as the "Rebellion of the Three Feudatories"[30]; during the same period the way to Russia was still

---

29 On this assumption, the original *Compendium libri organici* would have had the same structure as the *Compendium libri observationum*.

30 See Verbiest in his pro forma letter of September 1, 1678 (Josson and Willaert, *op. cit.*, p. 255): "Nam iam a multis annis, im(m)o ab ipso illo tempore quo Procurator nostrae Missionis (= Intorcetta) ex Europa redux (1674) penetravit in Sinam, totum hoc imperium intestino bello semper exarsit, quod vel hoc nomine nobis est valde ingratum quia nullum interim gratitudinis signum ... monstrare permisit, clausis scilicet omnibus portis et intercepta litterarum mittendarum via." See further Josson and Willaert, *op. cit.*, p. 257; the LA for 1677-1678 (ARSI, *Jap.-Sin.* 116, f. 214); a letter of

open, as is sufficiently shown by the Spathary mission in 1676 itself. A supposed first stage of the *Compendium libri organici*, written for a European readership (or for the tsar himself?) and already available in September 1676, could have been presented to the Russian emperor but, because of the Rebellion mentioned, not to the other European countries until after the restoration of contacts in 1677. At that moment, however, an update had become necessary, containing references to Verbiest's most recent publications, not yet available in 1676, such as the *Manchu Grammar* (which, on internal grounds, must be situated after Spathary had left Beijing) and the printed edition of the *Astronomia perpetua* (presented on August 27, 1678).

In any event, the restoration of these mutual contacts released a torrent of letters from Verbiest to his European superiors and benefactors dated in August/September 1678[31], most of them accompanying packages with copies of one or more of his works. It is in one of these letters, the already quoted item addressed to Marini (which, by the way, illustrates nicely how voluminous such shipments could occasionally be), that we find the first direct and unequivocal reference to the *Compendium libri organici*; this *Compendium*, or its supposed first draft, was drawn up not long before that moment, as I have tried to demonstrate.

Consequently, the *Compendium Latinum* remains to be discussed. This, quite evidently, cannot be dated earlier than its most recent component, which apparently was the *Compendium libri organici*, composed about the summer of 1678. The first (and only) external quotation of the title *Compendium Latinum* is to be found in the *Astronomia Europaea* (p. 40), written between the end of 1678 and mid-1680. Because other bindings of the same type are part of shipments dated in August/September 1678, it seems not unreasonable to assume

de Haynin from Macao on February 24, 1675 (cf. Henri Bosmans, *Analectes pour servir à l'histoire ecclésiastique de la Belgique*, 35 (1908), 26). The first letters after this period of silence are: Verbiest's letter cited in note 19, written in 1677; an (unknown) letter of the same to Father General Gianpaolo Oliva of 1677, quoted in Josson and Willaert, *op. cit.*, p. 358; the letter of Simon de Almeida from Macao dated July 8, 1677 and arrived in Beijing on December 30, 1677 (Josson and Willaert, *op. cit.*, p. 197).

31 Letters of 15.8.1678 to S. de Almeida; to the General; to Pope Innocent XI; to the fathers in Europe; the letter of 24.8.1678 to Giovanni Filippo de Marini (in Macao; see note 7); the formula-letter of 1.9.1678 to his European benefactors and the copy for Ferdinand Maria, duke of Bavaria (Munich, Bayerische Staatsbibliothek, CLM, 27.323 ff. 14r-15r); the letter of September 7, 1678 to Afonso VI, king of Portugal.

that the CL was also prepared on the same occasion, i.e., directly after the CLOrg. was finished. All these facts (the second draft of the *Compendium libri organici*; the binding together of several copies of text and different sets of drawings) can be situated in or about August 1678. Together with the above-mentioned torrent of letters in the same period - some of them accompanying copies of these bindings - they are part of one and the same offensive on the part of Verbiest towards Europe, restoring, in this way, many contacts which had been broken off for several years by internal Chinese conflicts. The summer of 1678 seems to have been the first in years with a prospect of safe and regular mail service between Beijing and Macao, at least safe enough to trust it with such important and massive shipments. Looking at the dates of the other letters for Europe, August/September seems to be the usual period for Verbiest to prepare his annual correspondence for Europe, depending on the normal mail traffic conditions between the two cities, and on the approximate sailing of the fleet towards Europe from Macao in December/January. I believe, however, that this is only the external explanation for this surprising concentration of parcels and letters to Europe, the internal one being the dramatic situation of the mission in the Chinese interior for some years, partly due to a shortage of priests and partly to grave financial problems. This is sufficiently illustrated by the letter of Verbiest dated on August 15, 1678 to the Jesuits in Europe, which is colored by an unusually pathetic language with bitter undertones.

The final part of this contribution will deal with the *Astronomia Europaea* (Dillingen, 1687), and discuss some evidence concerning the chronology of its redaction. But first, let us recall that the *Compendium Latinum* (i.e. CLObs. and CLOrg. without drawings) was incorporated into this book with only minor adaptations - which I cannot illustrate here - , making up its twelfth chapter. The *Astronomia Europaea* from now on refers to the newly written part that I have in mind, i.e., its preface, chapters 1-11 and 13-18. From the following evidence we can deduce the initial and terminal dates of its drafting. The initial moment can be inferred from Verbiest's explicitly announced intention to describe, in Chapters 14-27, the European technological achievements at the Beijing court in the decade 1668/1669-1678/1679[32], and the actual

32 "Nunc autem quaedam ex iis, quae unaquaeque illarum jam intra decem hosce annos, postquam scilicet cum Astronomia restituta in Aulam ingressae sunt, Imperatori obtulerunt, in medium afferam." (AE, p. 58). This chronological demarcation has to be taken seriously, as is proved a) by the

realization of this intention only seems conceivable from the end of 1678. On the other hand, it is generally accepted (though not proved) that the manuscript of this work, with many others, reached Europe (where it was published in 1687) with Father Philippe Couplet, who arrived in Europe on October 8, 1683. A hitherto virtually unobserved passage in Verbiest's *Responsum apologeticum* of 1681[33] makes this assumption even more probable[34]. In that case, Couplet's departure from Macao on December 4, 1681 is the last possible *terminus ante quem* for the finishing of the manuscript.

Actually, we do not know exactly how the manuscript got from Verbiest's study to Macao. There are two possible ways. First, it could have been taken by Couplet himself, who had already left his mission station in the Chinese interior at the beginning of August 1680 (as we learn from a quite interesting letter of the Visitator Giovanni Domenico Gabiani[35]); in that case, the AE must have been finished in the course of the first half of 1680. On the other hand, it is also possible that the manuscript traveled independently of Couplet, directly from Beijing to

---

fact that all the achievements mentioned in Chapters 14-27 are indeed datable to 1669 or later, and b) by Verbiest's explicit refusal, on the same chronological ground, to go into detail about a most remarkable achievement of Schall "... because it was in relation to the 'pulleys' in use in the European mechanic before those ten years" (AE, p. 80: "... Haec modo non refero, quia ad Mechanicae trochleas ante 10 annos adhibitas spectant").

33 Except in Josson and Willaert, *op. cit.*, p. 341, note 32.

34 Josson and Willaert, *op. cit.*, p. 341: "Verum de his fusius Pater noster procurator Romam nunc iturus in tractatu illo in quo ejusmodi PP. (= Patrum) Pekinensium in rebus mathematicis inventa ex professo refert"; the Procurator, of course, is Father Couplet, the "tractatus" on the discoveries of the Beijing fathers on "mathematical topics" is the *Astronomia Europaea* manuscript, which would be brought by Father Couplet. Elsewhere, we find only general references to the many books and manuscripts Couplet took with him as, for instance, in a letter of Gabiani (December 11, 1681) in ARSI, *Jap.-Sin.* 163, f. 161: "Alle 4 di decembre parti di questo Macao ... il N. Patre Procuratore Philippo Couplet con molti scritti sopra le negotij spettanti alla N(ostr)a V(ice) Prov(inci)a della China ...", and in another of Maldonado (November 9, 1680), published by H. Bosmans, *Analectes pour servir à l'histoire ecclésiastique de la Belgique*, 35 (1910), 83-84: "... Spero brevi appulsurum in Flandriam Philippum Couplet Mechliniensem qui multa secum deferet Plantinianis typis imprimenda."

35 See Gabiani in a letter of December 20, 1681 (ARSI, *Jap.-Sin.* 163, f. 165): "Procuratorem ipsum nostrum P(atr)em Ph. Couplet ... ab antecessore meo V(ice) Provinciali Ferdinando Verbiest legitime electum anno superiore 1680, ineunte mense Augusto, ex interioribus Sinis itineri sese commisisse, et post dimidiatum Octobrem Macaum pervenisse ..."

Macao, where it arrived at the beginning of December 1681 at the latest. A purely internal indication, however, leads to the acceptance of the first assumption as most probable. Indeed, treating of his ballistic achievements in one of his most circumstantial chapters[36], Verbiest does not speak of the emperor's new order of 320 cannons around the end of 1680[37]; this silence would be rather strange, especially if we think of the full description of his previous exploits in this area, unless the manuscript itself had already been finished before that moment! All this means that the redaction of the AE can be situated, on fairly good grounds, between late 1678 and mid-1680.

These are, of course, only the *terminus ante* and the *terminus post quem* respectively, but there are some indications in the text itself which support the idea that Verbiest actually worked on it during this whole period. The two long passages from Chapters 7 and 28, that recur literally in a letter of August 1678[38], have probably not much evidential value to prove Verbiest's occupation with the AE manuscript by the end of 1678, but there are other, very special facts and achievements which, in the *Astronomia*, are directly related to the moment of writing and, on the other hand, can be dated positively on the basis of external evidence available in the *Litterae annuae* (LA) for 1676-1679[39].

- When Chapter 25 (*Musica*) is situated "about six years after the emperor sent for Tomas Pereira in Macao" (arrival in Beijing in January 1673), it refers to a period shortly before or after January 1679 and, at the latest, to the first half of that year[40].
- Claudio Filippo Grimaldi's cylindrical clock, presented to the emperor in the same year the chapter on gnomonics was drafted (*hoc*

---

36 AE, ch. 15, pp. 61-68.

37 Josson and Willaert, *op. cit.*, p. 357: "Sub finem autem anni superioris, id est 1680, rebellibus iam devictis et provinciis omnibus recuperatis ... Imperator ... alias 320 levioris formae bombardas ... conficiendi curam et praeceptum mihi imposuit."

38 AE, ch. 7, pp. 19-20 = Josson and Willaert, *op. cit.*, pp. 244-245; AE, ch. 28, p. 98 = Josson and Willaert, *op. cit.*, p. 248. Both passages occur in the letter to the fathers in Europe of August 15, 1678 which existed in a xylograph edition (see C.R. Boxer, *JRAS*, 1947, p. 203), being at Verbiest's disposal not only on the moment of its redaction but also after it.

39 Cf. ARSI, *Jap.-Sin.* 116, f. 214 etc. (1677-1680) and *ibid.*, 117, ff. 183r-188r (1678 and 1679).

40 AE, pp. 88-89: "P. Thomas Pereyra Lusitanus, qui iam a 6 circiter annis ab Imperatore Pekinum Macao evocatus est ..."

anno: in this year), is also mentioned in the LA of 1678-1679 as a part of the achievements of 1679[41]; Chapter 14, thus, can be situated in that same year.

- After seven or eight months of experiments, Grimaldi's *Hydrargyrologium* (a hydraulic clock using quicksilver instead of water) was - according to AE (p. 91) - almost finished and ready for presentation to the emperor at the moment this passage was written down; this reflects the situation after the LA of 1678-1679 was completed[42], i.e., late 1679 or even early 1680.
- The same can be said about the spectacular organ which Pereira constructed on behalf of the Nantang in Beijing. According to the end of Chapters 25 and 26[43], the completion and the actual introduction of this highly promising instrument was ardently expected in the near future at the moment these chapters were finished. Because the same question is also quoted in the LA of 1678-1679, we know that Pereira had only started his work on it in 1679, at the moment the LA was completed; the organ's completion is only promised here for the *Litterae* of the next year, "because it is uncertain how much time the duties towards the emperor would leave for its construction"[44]. Therefore, Verbiest's repeated reference to the (expected) imminent introduction refers to a moment (early) in 1680, rather than 1679.
- Finally, there is one explicit reference to 1680, more precisely in Chapter 9, mentioning this very year explicitly in connection with an episode of the "Rebellion of the Three Feudatories" which, in my

---

41 Cf. AE, p. 61 and *Jap.-Sin.* 117, f. 187r, "Qua occasione P. Philippus Grimaldi eidem obtulit novum quoddam genus horologij cylindracei, inventum sane perquam rarum quod et Imperatori cumprimis placuit, quippe usui Tartaro perquam appositum et accomodatum."

42 Cf. AE, p. 87 and *Jap.-Sin.* 117, f. 186r, "Pater item Philippus Grimaldi cum alio novo invento, eoque perquam curioso prodiit, quod hydrargyrologium appellat ... Alia nunc (1679) praeparat Imperatori ejusque fratri majori, de quorum successu et gratia dicetur in annuis sequentibus."

43 "Brevi, ut spero, collocandum," AE, p. 91; "organo ... brevi confecto," p. 93.

44 "Eodem hoc anno (1679) idem P. Thomas Pereyra moliri aggressus est aliud perquam elegans organum pro Pekinensi nostra ecclesia ... Sed quia incertum ad hoc est a regiis occupationibus otium, de ejus ultima manu mentio fiet anno sequenti siquidem Dominus vitam ac vires dederit." ARSI, *Jap. Sin.* 117, f. 188r.

opinion, could be situated in the first part of 1680[45].

All in all, on the basis of this evidence, there is the impression that Verbiest was occupied by this manuscript from the beginning of 1679 up to the first months of 1680 included, i.e., a period of more than one year! This seems quite a long time for a booklet like the *Astronomia Europaea*, which in its printed form contains only a hundred pages in quarto. For a correct and right understanding of this point, we must remember the numerous and varied tasks of the author, both inside and outside the Board of Astronomy, and his repeated complaints about a lack of time, for instance in the preface of the *Astronomia* itself[46]! On the other hand, we might expect that he worked on it at different moments with many interruptions, according to the time available. The last datable references in the text, thus, refer to the beginning of 1680 and, by their direct relation to the moment of writing, they place the last identifiable layer of the AE in early 1680. Because they appear in four distinct chapters (9; 23; 25-26), this layer may be a kind of general and last revision of the whole text. This would also better explain the contradictory indications within Chapter 25 (*Musica*), of which the beginning refers to a period shortly after January 1679, and the end, mentioning the nearly finished organ of Pereira, rather reflects a situation about the beginning of 1680. It is, then, quite tempting to relate this revision of early 1680 to the imminent departure of Couplet, who left his parish - as we know on independent grounds - at the very beginning of August 1680, and who most probably was responsible for bringing the manuscript of the AE to Macao, and from there to Europe.

Such circumstances of working harbor - as is the general experience - many occasions for errors of overlapping, and all kinds of inconsistency[47]. This is also the case with the *Astronomia*. Indeed,

---

45 "... Hoc ipso anno 1680 rex Tumkinensis ... accusavit Regem Cochinchinae ... jam Tartaro palam rebellem esse ... Nunc autem temporis quo haec scribo." AE, p. 28. See N. Golvers, *The* Astronomia Europaea *of Ferdinand Verbiest, S.J. (Dillingen, 1687). Text, Translation, Notes and Commentaries* (Monumenta Serica Monograph Series, XXVIII. Sankt Augustin - Nettetal, 1993), ch. 9, n. 28.

46 "Atque haec quidem multo ante in lucem Europaeam edidissem, si et otium scribendi et occasionem scripta mittendi nactus fuissem." AE, praefatio (p. 1). See also Josson and Willaert, *op. cit.*, pp. 160, 167 and especially p. 225, note 13 (in a letter of 15.8.1678!): "Cum autem hujus epistolae longioris saepius describendae mihi otium non sit, rogo etiam R.V. ut ..."

47 Other inaccuracies and errors (e.g. in chronological and numerical indications), cf. N. Golvers, *op. cit.*

referring thrice to some events of 1680, Verbiest apparently forgot his original aim to describe one significant period, the first decade of the restored European astronomy in Beijing 1668/1669-1678/1679. Further, it was Verbiest's intention to add the text of the *Compendium Latinum* to the AE, with some adaptations, none of them substantial. Originally, he thought of this CL as a kind of appendix at the end of the *Astronomia Europaea*, as we learn from a passage on p. 16 concerning the CLObs.: "... which I will append at the end of this treatise"[48]; he apparently afterwards changed his mind, and inserted the CL in the middle of the *Astronomia*, making it its Chapter 12, without adjusting his cross-reference on p. 16! This secondary transposition from the end to the center of the book also explains the bizarre and incorrect reference on p. 59 to a passage of the CLOrg. "Further on"[49], for the passage involved is found before p. 59 (actually on pp. 53-55[50]).

At the end of these considerations on Verbiest's treatises on European astronomy in China in the early Kangxi period, I offer my opinions and speculations on the question on the synoptic table, which will show how the Dillingen edition of the *Astronomia* was only the final stage in a process that took a decade. Meanwhile, I hope I have been able to afford some convincing new evidence, and to infer some correct specifications from it, that is, corrections on the conventional views. There are still more intriguing aspects in the history of the publication of the *Astronomia Europaea* which cannot be treated here due to lack of space.

---

48 See citation in note 8.

49 "Quam benigno pariter ac laeto vultu Imperator acceperit munuscula ... infra in Compendio octo priorum figurarum libri mei organici ad explicationem figurae octavae breviter indicabo!" AE, p. 59.

50 It cannot be totally excluded that this transposition is the work of Father Couplet (the "editor" of the *Astronomia* in Europe), and is to be seen in connection with the appending of his own *Catalogus* at the end of the *Astronomia*!

Table 1

Synoptic Table of the Development of the Dillingen Edition of the *Astronomia Europaea*

Ceyan ji lüe (1669/1670)		*Yixiang tu*	+ *Yixiang zhi* (1674)
Chinese text	+ 12 drawings	117 drawings in 2 vol.	+ 14 vol. text/tables
Compendium Sinicum	*Liber Observationum*	*Liber Organicus*	
Compendium Sinicum	+ Latin introduction = *Compendium Observationum* (after 1671)	8 or 57 or 117 drawings	+ Latin introduction = *Compendium Libri Organici* = 1676[1] 1678[2]

Compendium Latinum proponens etc.
1678

- with minor adaptations
- without drawings

Astronomia Europaea = preface + Ch. 1-11, 12, 13, 14-17, 28.
Draft: 1678/1679 - first half 1680
Printed edition: Dillingen, 1687.

THE RESEMBLANCES AND DIFFERENCES OF THE CONSTRUCTION OF FERDINAND VERBIEST'S ASTRONOMICAL INSTRUMENTS, AS COMPARED WITH THOSE OF TYCHO BRAHE

A Study Based On Their Writings

NICOLE HALSBERGHE

Nicole Halsberghe studied graphic arts at the Ghent (Belgium) St. Lucas Higher Institute, in Tokyo (Japan) and Ljubljana (Yugoslavia). She was assistant of Chinese studies at the Catholic University of Louvain and is currently professor at the St. Lucas Higher Institute. She completed a doctoral thesis on Verbiest's "Xinzhi lingtai yixiang zhi", in the spring of 1992.

On the island Hven, placed at his disposal by the Danish King Frederic II, the Dane Tycho Brahe founded the astronomical observatory Uraniborg, where he made observations and designed new instruments from 1576 to 1597. Between 1669 and 1673, Ferdinand Verbiest (Nan Huairen) designed and constructed six new bronze instruments by order of the Kangxi emperor. They were meant to equip the Imperial Observatory in Beijing, where Verbiest would stay until his death in 1688.

Both Brahe and Verbiest observed the celestial phenomena during a period of twenty-one years, from observatories they themselves had equipped with astronomical instruments. They were both working with the financial help of a sovereign, and they both constructed new appliances aiming at greater precision. What is more, they both wrote a book on their methods of construction, and their observations on these instruments. Brahe described his construction methods in his *Astronomiae instauratae mechanica* (written in 1598), and Verbiest did the same in his *Xinzhi lingtai yixiang zhi* (Essay on the newly constructed instruments for the Observatory) in 1674.

With the help of some examples from their instrument descriptions, I shall try to make clear in what way, and to what extent, Verbiest copied

- or deviated from - the Tychonean instruments.

In view of space limitations, I shall restrict myself to the description of the ecliptic armillary instrument, and Brahe's and Verbiest's explanations of graduation and subdivision systems, which were so extremely important for their intentions.

The reference to the ecliptic armillary instrument is important here, because Brahe - though he made a simplified model of his *Armillae zodiacales* by using four rings instead of the usual five or six - did not recommend this instrument, since its center of gravity shifts according to the position of the equinoctial junctions.

Why then, we can ask, did Verbiest construct an ecliptic armillary sphere (Ill. 1) modelled on Brahe's, though the maker himself did not speak very positively about this instrument? It must have been important for Verbiest, who made use of European astronomical catalogues and tables to define the ecliptic coordinates directly with an ecliptic armillary. Moreover, he wanted to confront his armillary with the old Mongolian instruments, to which he strongly reacted. His construction also meant a manifest change as compared with the instruments of his day, which were constructed with double split-rings.

Ill. 1: Ecliptic armillary instrument
(F. Verbiest, *Xinzhi yixiang tu* [Beijing, 1674], fig. no. 1).

Yet Verbiest did take account of Brahe's objections, which were mainly about the instrument's precision: though it was provided with a steel meridian ring, a very strong, solid basis, and braces, and though the wooden rings suspended inside were very light, the instrument could still bend easily by the changing positions of the rings, and this brought about a decline of the instrument's precision. As a result of this, Verbiest especially stresses the solidity of the construction and the maximum stability of his instruments. In other chapters of his *Yixiang zhi*, he illustrates this more precisely with principles from mechanics and statics.

If we compare both scientists' written descriptions of the instruments, we will come to the conclusion that Verbiest goes further into the matter of technical details of instrument construction. In mounting the meridian ring, he stresses, in the first place, the use of a supporting ring at its base to increase the instrument's stability, a ring that Brahe's instrument did not have. In indicating the equatorial poles, he uses a different method from Brahe's, who worked with the complement of the altitude, which he defined from the zenith. Verbiest, on the other hand, first points out the horizon, and then fixes his altitude of the place of observation from there. He may have based this way of thinking on the work of Chinese astronomers, for whom the zenith played no part.

He also technically explains the system of assembling the poles, which are details missing in Brahe's description. Verbiest also explains the assembling of the colure ring to the equatorial axes with the meridian ring. He indicates systematically the ecliptic poles (like Brahe on 23°31'30"), and from these points defines the ecliptic ring that is connected to the colure ring in the solstice points. The ecliptic latitude ring, which is attached to the ecliptic poles within the colure ring, contains in its poles a diameter axis. In Verbiest's instrument, this axis can be dismounted, so that its structure differs from Brahe's and rather resembles the system of the dismountable sighting tube of the Chinese armillary instrument (*jianyi*) of Guo Shoujing.

The structure of both explanations is also different. Verbiest describes the directions for the use of each instrument individually, after the technical explanation of the construction methods, and not maintaining the same order. Brahe keeps it within the individual explanation of the instruments. He describes together the construction method and the directions for the use of each instrument. If we compare the graduation systems of Verbiest and Brahe, we see that Verbiest uses the method of transversals only, and not the Nonnius system, which

Brahe used for some of his instruments, but then put aside as incomplete (Ill. 2).

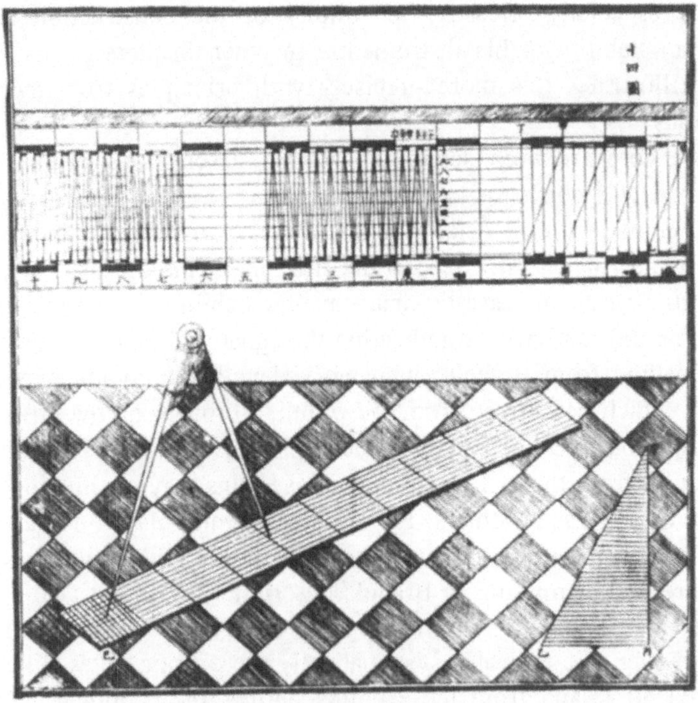

Ill. 2: Method of calibration
(F. Verbiest, *Xinzhi yixiang tu* [Beijing, 1674], fig. no. 14).

The use of transversal dotted lines was introduced by the British mechanician Richard Chancellor (1552), who is supposed to have used this system for the first time in the graduation of a Jacob's staff. Thomas Digges mentions this in his work *Alae seu scalae mathematicae*, published in London in 1573. The German astronomer Johann Ludwig Emil Dreyer, on the other hand, ascribes the introduction of transversals in precise graduations on limbs of astronomical instruments to Brahe, in his study on Brahe (Karlsruhe, 1894, p. 348). This was a variant of the zig-zag line system that was introduced by John Hommel (1518-1562). Brahe himself is supposed to have learnt this system from his fellow student Bartholomaeus Scultetus (1540-1614), who was one of Hommel's pupils, and who explained it in his *Gnomonice de solaris* (1572).

This system was eventually worked out on instruments in an effective way by Joost Bürgi in Kassel: instead of the dotted lines, he started using full lines, and added a scale to the alidad, the section of which - with a transversal line - indicated the number of minutes; in Brahe's system, each point of the dotted line corresponded to one minute. Brahe explained his recommended method of transversal dotted lines in an appendix of his *Astronomiae instauratae mechanica* (p. 107): "Supplementum de subdivisione et dioptris instrumentorum." Verbiest explained his method in three individual items: in Chapter One (pp. 31-32), under the title *kuibiao* (Observation pinnules), in Chapter Two (pp. 16-20), under two different items: *Xin yi ti ju ji fen miao zhi mingxi* (Detailed explanation of the extremely refined calibration of the new instruments), and *Xin yi fen fa zhi xiwei* (Precision of the calibration of the new instruments). In the first chapter, Verbiest confines himself to quoting two sighting systems with reading-off possibilities: the observation yoke, or beam (*kuiheng*), provided with oblique threads (which he will use for his horizontal azimuth instrument), and the alidad system with two pinnules, of which the remoter of the two can be a little cylinder placed in the center of the instrument or on a central axis. For both, the cylinder and the round axes have the special advantage that they can be used not only by one, but by two, observers at the same time. Verbiest seems to have borrowed the first of these two systems directly from the Chinese practice, probably from the *jianyi* instrument, or the simplified instrument of Guo Shoujing. The second system was a European one. We again find the construction with little cylinders and axes in Brahe's instruments. It was generally used till the end of the seventeenth century, when the Vernier system was to replace the method of transversals.

In his detailed explanation, Verbiest gives a geometrical exposition based on proposition 20 of the third chapter of *Jihe yuanben* (*Euclidis elementae*, translated by Ricci) which states: "In a circle, the angle at the center is double the angle at the circumference, when the angles have the same circumference as base" (Ill. 3). This shows that, when using this system with a pinnule on the edge of the limb of a quadrant or a sextant, one can double the size of one degree. He illustrates this from trigonometry by means of a working-sketch. This argumentation is, above all, meant to compare the proportions of his new instruments with those of the Chinese instruments and to sing the praises of his own instruments.

In the item on the precision of the calibration system, he explains his graduation system in a more detailed way, giving an exact method of working it out, and also expressing mathematically the larger accuracy of his instruments, as compared to the old Chinese ones. Here we can see that Verbiest used the principle of transversals, like Brahe, but that he worked it out in the spirit of Bürgi, attaching a graduation scale to the pinnule, which made the reading-off more accurate. The dotted lines for indicating the minutes are replaced by the intersection points of the transversals, with the concentrically placed circles on the graduation points.

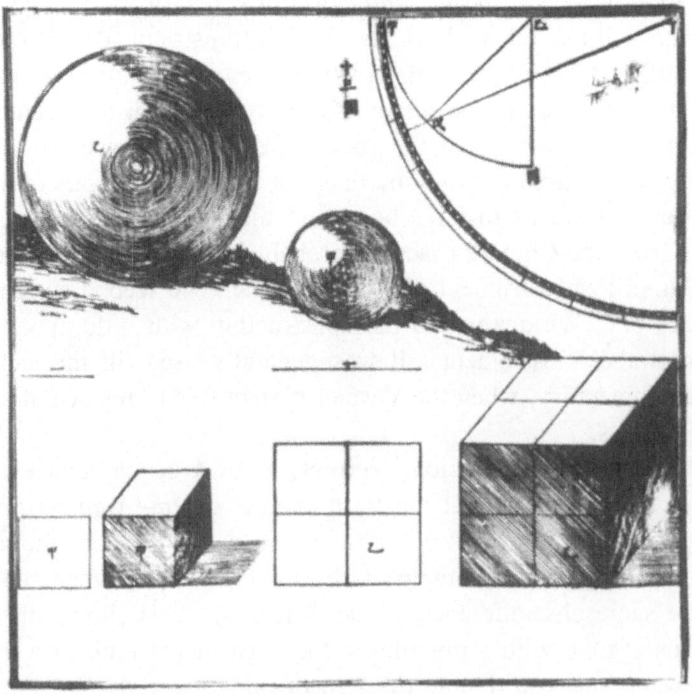

Ill. 3: Proportions
(F. Verbiest, *Xinzhi yixiang tu* [Beijing, 1674], fig. no. 13).

Comparing Verbiest's argumentation with that of Brahe's, we see a totally different intention. Brahe only tries to prove the precision of his method by calculating the smallest deviation by means of trigonometry. Verbiest gives no explanation about the construction of the pinnules. We can only reconstruct them with the help of his working-sketch in the *Yixiang tu*. Illustration 4 shows different types of pinnules, and in illustration 5 we can see a detached, movable pinnule, and a pinnule attached to an alidad. Brahe, on the other hand, does add an explanation, but his pinnules are more complex in their construction than Verbiest's. The pinnule closest to the eyes has slits on all four sides, and has small springs which are mounted on three sides, so that they can be pressed towards the pinnule or removed a little from it. This type cannot be found in Verbiest's instruments. In the ecliptic armillary instruments of both Brahe and Verbiest, the lower part of the base is pierced at four equidistant points by screws. With the aid of these screws, the whole instrument can be raised or lowered. On the other hand, the armillary instrument of Guo Shoujing has a water channel that is cut round the knee supports for levelling the instruments.

Ill. 4: Illustration of the different methods of observation (F. Verbiest, *Xinzhi yixiang tu* [Beijing, 1674], fig. no. 12).

From these examples, we can conclude that Verbiest's concept of instrument-building was an assimilation of Chinese instrumental design, but that he followed the alignment of European practice. He chose those methods that he thought most suitable for himself, and so we can also find Chinese sighting methods in his constructions. By no means can we say that Verbiest's instruments were exact copies of Brahe's instruments. His explanation and description methods were not new in China either. The *Yuanshi*, for example, gives a detailed description of the instruments of Guo Shoujing, while in Europe detailed descriptions like Brahe's were rather new. Verbiest's description of the instruments in his *Yixiang zhi* was more original because it was amplified and documented with elements from physics, mechanics, statics and optics to explain and illustrate the whole concept of his instruments.

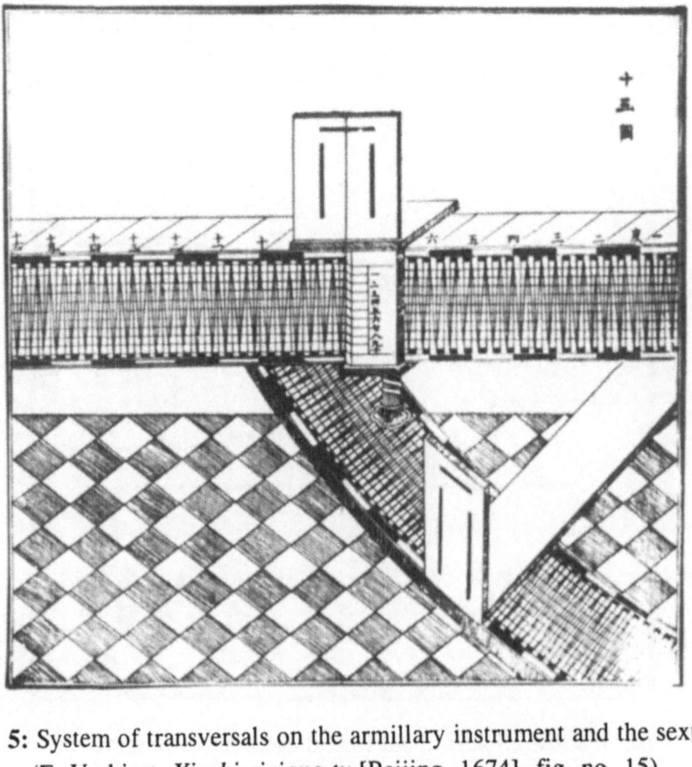

Ill. 5: System of transversals on the armillary instrument and the sextant (F. Verbiest, *Xinzhi yixiang tu* [Beijing, 1674], fig. no. 15).

SYNCRETISM BETWEEN EUROPEAN AND CHINESE CULTURE IN THE ASTRONOMICAL INSTRUMENTS OF FERDINAND VERBIEST IN THE OLD BEIJING OBSERVATORY

ISAIA IANNACCONE

Prof. Isaia Iannaccone holds a degree in industrial chemistry and in Chinese language and literature. At present he is contract-professor in History of Chinese Science and Technology at the "Istituto Universitario Orientale" of Naples. He is the author of several papers published in the proceedings of congresses or in scientific magazines.

I. The Preface of the *Yixiang tu*

With this work, we propose to undertake a primary study of the astronomical instruments built for the Chinese emperor Kangxi, according to the plans of the Jesuit scientist Ferdinand Verbiest, and under his direction. He himself installed them in 1674 on the terrace of the old astronomical Observatory of Beijing, where they can be admired to this day.

Two texts published by Verbiest in Beijing in 1674 are indispensable for the study of these instruments: *Yixiang tu* (Instrument Design) and *Yixiang zhi* (Instrument Description). Even if the two texts are strictly connected, we will refer exclusively to the first in this work, reserving an analysis of the translation and a technical scientific examination of the second for another place.

The *Yixiang tu* contains a preface in Chinese, 105 prints which illustrate the building and use of the instruments numbered from 1 to 117, utilizing Chinese numerals, and one unnumbered illustration showing the by now famous complete view of the instruments on the observatory terrace (Ill. 6)[1]. This text was also edited with the Latin

1 We studied a copy of this work in the Vatican Library in Rome, and it consists of two booklets; the complete title of this copy is *Xinzhi yixiang tu*.

title *Liber organicus astronomiae Europaea apud Sinas restitutae sub imperatore Sino-Tartarico Cam Hy appellato, autore P. Ferdinando Verbiest, Flandro-Belga Brugensi et Societate Jesu Academiae Astronomiae in regia Pekinensi Praefecto, anno salutis MDCLXVIII*[2].

We hold that the works *Yixiang tu* and *Liber organicus* ... are formally identical, apart from the arrangement of the Chinese preface itself which, instead of coming at the beginning of the text, is placed at the end in the case of the Latin edition. The only substantial difference between the two is the dating. If we refer to the common preface in the Chinese edition, we find that it indicates the year *jia yin* (the fifty-first) according to the traditional numeration, which uses the combinations of the duodecimal cycle (the twelve terrestrial branches) and the decimal cycle (the ten celestial stems) which, without a doubt, would be the thirteenth year of the reign of the Kangxi emperor, therefore 1674[3].

Why, then, is the Latin title-page of the *Liber organicus* ... dated 1668, and its Chinese preface 1674? The hypothesis which would attribute this to simple backward dating of the title-page of the *Liber organicus* ... does not seem very plausible[4]; more probably, the iconographic work of Verbiest began in 1668 and ended in 1674, and this would also explain why the Chinese preface in the *Liber organicus* ... is placed at the end: it could be an afterword. We can complete this hypothesis by adding that the *Yixiang tu* was published for the exclusive use of the emperor and the functionaries of the astronomical observatory, while the *Liber organicus* ... was probably supposed to serve for "internal use" among the Jesuits. This last supposition is supported by the affirmation of Louis Pfister, who claims the *Liber organicus* ... contained a nine-page Latin preface which, however, he himself had never seen[5]. Neither does it exist in the copies we have studied at the Vatican Library, nor in the specimen notes kept at the British Museum and the Museum of the History of Science at Oxford[6]. The preface, or ap-

2 We also studied a copy of this work in the Vatican Library.

3 P. Hoang, *Concordance des chronologies néoméniques chinoises et européennes* (Shanghai, 1910), p. 316; see also Tung Tso-pin, *Chronological Tables of Chinese History* (Hong Kong, 1960), 2:210.

4 Allan Chapman, "Tycho Brahe in China: The Jesuit Mission to Peking and the Iconography of the European Instrument-Making Process", *Annals of Science* (1984), 41:417-443.

5 Louis Pfister, *Notices biographiques et bibliographiques sur les Jésuites de l'ancienne mission de Chine (1552-1773)* (Shanghai, 1932), 1:359.

6 Chapman, "Tycho Brahe", *op. cit.*, p. 419, n. 4.

pendix, in Chinese common to the *Yixiang tu* and the *Liber organicus* ..., is of great interest because it reveals a rational working method that permits us to utilize the illustrations depicting the instruments and machinery in a practical way. They are not bound together, and can therefore be extracted singly in order to follow the description contained in another text to which the author refers, the *Yixiang zhi*. Furthermore, from this text we can deduce that Verbiest did not have only astronomical instruments, but also, for example, thermometers ("... to measure ... the heat and the cold ...") or hygrometers ("... to measure ... the dry and the moist ...")[7].

In proposing here the translation of the preface of the *Yixiang tu*, we must specify that we have chosen to present it in a nearly literal way, maintaining locutions such as "measure the heavens" (*ce tian*), since we hold that no other expression could render the scientific complexity and poetic synthesis which are understood. Terms such as *yi*, *qi*, and *hefa* have been rendered as "instrument," "device," and "law of science," which lend themselves well to the economy of the discourse. On first reading it may seem a bit stiff; but in reality, it reflects the polished prose of the time. The preface (reproduced in Ill. 7) is signed by Verbiest with his name in Chinese Nan Huairen. It can be translated as follows:

> Each instrument has its own method of construction, use, and installations, (and) there is also a reason for its being light (or) heavy (or) robust. There is a great number of instruments; therefore, even if the tablets used to write on were all exhausted, the subject would not be so. Therefore, if in order to shed light on the subject, drawings are not made and, what is more, circulated, how can we clarify the mystery (concerning the instruments)? Now each instrument has been clearly explained, and for this reason its design has also been included[8]. We are making (a book of) appendices (with the illustrations). Therefore, those who wish to discuss the subject, in seeing it (illustrated) can take notice of this. Drawings explain better than words can, and for this reason, they supply all that has not yet been explained; in following this procedure, we can understand that which was not clear before, and therefore clear up every mystery.

---

7 In 1671 Verbiest printed a book on thermometry: *Yanqi tushuo*. See I. Iannaccone, G. Merchionne, "La termometria europea rinascimentale nella Cina del XVII secolo: lo *Yanqi tushuo* di Ferdinand Verbiest", in F. Bevilacqua, ed., *Atti del X Congresso Nazionale di Storia della Fisica* (Milano, 1989), 239-251.

8 Verbiest makes reference to explanations in *Yixiang zhi*.

However, in all books, most of the drawings are placed under the explanation in order to read the text and see the image without having to look for it (on other pages). But not everyone knows that the description can be either complicated or simple, the illustrations also large or small. If (however) one acts in such a way that (text and drawing) are found on the same page, then certainly they might overlap each other, which would not be too favorable to the reading[9]. Therefore, we have made this book thus: the descriptions and drawings are in separate sections[10] so that, if the reader does not understand the description fully, he can clarify (everything) by looking at the drawing. What is more, the beginning and the end (of the book) are not bound, and[11], besides the six instruments, we have enlarged it with the explanations of other devices. First, for the purpose of clarifying the possibilities of the use of each instrument, we have had to cite in the previous text the laws of gravity; second, we have tried a number of times to explain that each instrument is in accord with the laws of science, so that they can be used at any place and at any time.

So, to measure the heavens, instruments exist which are fixed in a place, and which cannot be moved (for example, those which stand on [the terrace of] the Observatory). There are also instruments that can be brought along with oneself for use (for example, those which are in every province, and have been utilized to measure the changing of the seasons, the rising and setting of the sun, and the lengths of the day and night). Those instruments which are fixed to the ground serve to (make measurements of) *terra firma*; instead, instruments that are carried are used for navigation. There exist instruments to measure the heavens, the earth, the water, steam, the height of the mountains, the proximity or distance of the clouds, the heavy and the light, the cold and the heat, the dry and the moist, and so on. Each category has its instruments, and for this reason, there is a method for the building and use of (each) instrument.

II. The Instruments of Verbiest

Let us ask ourselves along with Gilles Gaston Granger: "Is a science possible without instruments? ... Absolutely not. One cannot, however, but underline the role that pure theatrical illusion often plays by

9 Literally: "is not pleasant for our eyes."

10 Literally: "they are not mixed between them."

11 Practically, *Yixiang tu* was formed by the drawings collected in a folder.

artificial instrumentation in every type of counterfeit knowledge"[12]. Now then, the instruments planned by Verbiest, and built under his direction for the Beijing Observatory, could be used to back up the accusation of "pure theatrical illusion" because of their wonderful artistic workmanship, which seems to take precedence of their technical use. In order to confute this impression, let us observe the instruments at closer hand.

The prototypes of the six instruments of Verbiest are traceable in those illustrated and described by Tycho Brahe in his *Astronomiae instauratae mechanica* of 1598[13]; these are :

- ecliptic armillary sphere (Ill. 8 and 9);
- equatorial armillary sphere (Ill. 10 and 11);
- horizon circle (Ill. 12 and 13);
- quadrant (Ill. 14 and 15);
- sextant (Ill. 16 and 17);
- celestial globe (Ill. 18 and 19).

A first summary technical comparison between Verbiest's instruments and those of Brahe was carried out by Allan Chapman[14], but this comparison - although of great interest for our subject - does not take into consideration the strict relation which exists between the illustrations (of the *Yixiang tu*) and the explanatory text (*Yixiang zhi*), nor does the author refer directly to any Chinese source. As has been indicated above, we prefer to await undertaking a detailed technical examination of the instruments until we have a full translation of the *Yixiang zhi*. At that time it will be meaningful to compare what has been written by Verbiest with the technical characteristics which Brahe gives in his work, as well as to reconstruct the sequences of the development of the instruments from Brahe to Verbiest, tracing both the Chinese technical influences and western ones lent by other famous European builders from which Verbiest might have drawn as, for example, Gemma Frisius, Thomas Gemini, Walter Arsenio, Gerardus Mercator, Girolamo

---

12 Gilles G. Granger, "E' possibile una Scienza senza Strumenti?", in *Gli strumenti nella Storia della Filosofia della Scienza* (Bologna, 1985).

13 T. Brahe, *Astronomiae instauratae mechanica* (Wandesburgi, 1598), translated and edited by H. Raeder, E. and B. Strömgren (Copenhagen, 1946). On Brahe's instruments, see also V.E. Thoren, "New Light on Tycho's Instruments", *Journal of History of Astronomy*, 4 (1973), 25-45; see also W.G. Wesley, "Tycho Brahe's Solar Observations", *Journal of History of Astronomy*, 10 (1979), 96-101.

14 Chapman, "Tycho Brahe", *op. cit.*

della Volpaia, Johannes Hevelius, and still others.

From elements that have emerged so far, some of the technical characteristics relative to some of the instruments, and their representation as regards the tradition of Brahe can be summarized in the following points:

- Verbiest, unlike Brahe and the other European designer-scientists, could avail himself of the Chinese refined bronze technology, and so realized more compact and steady carrying structures.
- Both the instruments and the relative construction processes can be illustrated in a very accurate fashion, and follow the European models - even if some of the technical solutions adopted by Verbiest turn out to be innovative; this demonstrates that the Jesuit was knowledgeable in the developments in instrumentalization from the time of Brahe until he left Europe (1657).
- The ecliptic armillary sphere built by Verbiest to measure the positions of the planets[15], was more elaborate than that of Brahe; but both illustrate the graduated scale with alternating black and white notches, as was the custom of the time. Both use, also, a screw-set in order to align the instruments. Furthermore, Brahe shows his sphere in a small well, while Verbiest, even if he illustrates it on a chess board-type pavement (used by Brahe on other occasions), accompanies the *Yixiang tu* (and therefore the *Liber organicus ...*) with tables which explain a construction of the well which will then be built on the terrace of the Beijing Observatory.
- The equatorial armillary sphere which Verbiest built to observe the equinoxes[16] is completely similar to that of Brahe, but differs sensibly from those already used by the Chinese, which were more complex in that they adopted a greater number of rings: those of the equator and of the horizon in the outer nest, and the ecliptical ring in the inner nest[17], as well as double rings.
- Among the innovations introduced by the Jesuit scientist was the use of the equinoctial colure of the semicircular bracing ring, which sustains the equatorial circle only in the southern hemisphere. This sharply reduced the difficulties in observing the heavens near the

---

15 Henri Bosmans, "Ferdinand Verbiest, directeur de l'Observatoire de Pékin", *Revue des questions scientifiques*, 21 (1912), 273.

16 *Ibid.*

17 J. Needham, *Science and Civilisation in China* (Cambridge, 1959), 3:339-382; see also H. Maspero, "Les instruments astronomiques des Chinois au temps des Han", *Mélanges chinois et bouddhiques*, 6 (1939), 183 *et seq.*

equinox.

- Verbiest's graduated scales, besides the problem indicated concerning the division into 360 degrees, was a very efficacious system of reading that used the method of diagonal lines. It represented the most advanced system of reading of a scale with the naked eye before the use of telescopic sights.
- Verbiest's sextant is mounted on the azimuth, and its geared setting renders the instrument more stable, as compared to the ball mounting used by Brahe (and by Hevelius); furthermore, the pulley and tackle permit one to follow celestial objects more delicately than with the stabilizing spikes that Brahe had used in his *Sextans astronomicus trigonicus pro distantiis* (Ill. 20). The technical solution of Verbiest is very interesting, and may have been suggested by the bipartite arch of Brahe (Ill. 16), which has a semicircular clamp-mounting which, however, cannot do the work of a gear.
- We also note that the circles on Verbiest's instruments are graduated 360 degrees, unlike the Chinese, which are graduated 360¼ degrees; the division by sixtieths of degrees and minutes operated by Verbiest was less precise than the decimal system in tenths and hundredths of the Chinese.

In reviewing the points above, one may notice that Verbiest's instruments were not built merely for show. There is no need, therefore, to let oneself be taken in by the decorative motifs of the instruments, which surely represent the most outstanding feature for whoever approaches them for the first time. Let us remember again, that in China the custom of decorating scientific instruments with symbolic figures (dragons, animals, clouds, etc.) was already in use with instruments much older than Verbiest's as, for example, the splendid equatorial armillary sphere built by Guo Shoujing in 1276, his equatorial torquetum (1270), and the ancient armillary sphere of Su Song (1090), which was also endowed with clockworks[18]. Verbiest, then, did nothing but resort to a congenial decorative mode in keeping with Chinese tastes which had already been codified for centuries, and was necessary for objects used under the direct supervision of the emperor. Certainly, we do not hold that the artistic aspect is wholly secondary in the scientific instruments of Verbiest. In fact, they prefigure as true art objects for both the complexity of the ornaments and

---

18 On Su Song's astronomical clock-tower and instruments, see J. Needham, Wang Ling, D.J. De Solla Price, *Heavenly Clockwork, The Astronomical Clocks of Medieval China* (Cambridge, 1960), reprinted in 1986.

the balance of spatial relationships. Of course, only the ability of the Chinese in bronze work permitted the practical realization of these instruments which, we are sorry to note, do not appear in any manual of oriental art history.

III. Conclusion

In reevaluating the course of Chinese science, it is emphasized that Verbiest and other previous Jesuit scientists introduced in China an astronomy that was profitable for calendrical science, but less advanced than the Chinese astronomy itself of the same period. This ancient Chinese astronomy was certainly more innovative than European astronomy with regard to intuition (for example, in the concept of the "infinite"), theoretical models (for example, in the precocious use of equatorial coordinates), and instrumentation (for example, in the equatorial armillary sphere). However, we cannot generalize this reasoning, and it would be more suitable to carry out a systematic study of the scientific body of writing produced by the Jesuits in Chinese. Let us remember that only a critical analysis of these texts would permit an objective evaluation of the work undertaken in China by the members of the Society of Jesus. First of all, the weight of scientific titles compared to those of religious divulgence for each author must be considered. From the studies that we have already undertaken in this direction, historians of science and sinologists have uncovered personalities such as Johann Terrenz Schreck[19] and Giacomo Rho[20], alongside Johann Adam Schall von Bell, who seems, inexplicably, to have been granted the merits of the others in his brotherhood, in addition to his own, which were great[21].

One cannot deny that these scientists were knowing participants in a magic moment in the history of science: the encounter (or confrontation) between two scientific systems. And it is just as undeniable - as much as one might wish to prove the contrary - that they were the representatives of a modern way of doing science, even though it was not free from contradictions. One of the accusations made against these scientists is

---

19 I. Iannaccone, "Johann Schreck (Terrentius): Scienziato, Linceo, Gesuita e Missionario nell'Impero dei Ming", *Asia orientale*, 5:6 (1987), 49-85.

20 I. Iannaccone, G. Arpaia, "Giacomo Rho: un Astronomo Italiano del '600 in Cina", in *Atti del VIII Congresso Nazionale di Storia della Fisica* (Naples, 1987), pp. 241-245.

21 *Ibid.*

that they propagandized the astronomic system of Brahe, and not that of Nicolaus Copernicus, in China. But, one can reply, Brahe's view - even if less advanced than the Copernican one - was more recently formulated. On the other hand, even in Europe, in spite of the fact that Galileo Galilei and Johann Kepler had solidified the Copernican theory, it was necessary to wait for the *Principia* of Isaac Newton (1687) in order to recast physics substituting Galileo's law of circular inertia with the law of linear inertia, or to enunciate a clear conception of force as the cause of acceleration more than of motion or, what is more, in order to introduce the concept of the law of gravitation, with which they were able to explain the laws of planetary motion and those of the fall of heavy bodies. It is in light of this evaluation, which can never be determined a priori, that one must also examine the work of Verbiest. For example, instead of underlining the fact that the building of the ecliptic armillary sphere was a step backwards, compared to the equatorial astronomy already adopted by the Chinese, one might remember that at present there is a re-evaluation of the ecliptic astronomy to determine the orbits of heavenly bodies.

We emphasize that it is precisely in the building of instruments that Verbiest can be classified without a doubt among the forementioned scientists who paved the way in the history of instrument development. With these, Verbiest has in common the rigorous planning-method and the yearning (ambition?) for divulgence, as his sophisticated theoretical and iconographic manuals - which explain the construction, installation and use in detail - attest. Furthermore, in the instruments of Verbiest, as in those of the major instrument builders operating at that time in Europe, the knowing search for precision is revealed, which manifested itself in the attempt to resolve the three principal problems connatural with astronomical devices:

- the accurate sight of the object studied,
- the accurate alignment and division of graduated circles,
- the accurate reading of graduations.

The diligence with which the French and German troops seized them after the sack of Beijing in 1900, is witness to the fact that Verbiest's instruments represented a significant step in the history of science and art, and still do today. The French kept the instruments which they had appropriated (the quadrant, the ecliptic and equatorial armillary spheres, the quadrant altazimuth of Kilian Stumpf built in 1714, and the famous equatorial torquetum of Guo Shoujing of 1270) in their embassy, and returned them only in 1902. To get back the instruments taken by the

Germans and brought to Berlin (the celestial globe, the sextant, and the horizon circle of Verbiest, in addition to the equatorial armillary sphere built by Ignaz Koegler in 1744, and a Chinese equatorial armillary sphere), the Chinese had to wait until the end of the First World War, when the Treaty of Paris (Article 1131) imposed the restitution of the instruments taken by the Germans. They were finally returned to Beijing in 1921.

The old Beijing Observatory, closed to all but a few scholars for decades, was reopened on April 1, 1983. The Chinese press has recently given great prominence to Verbiest's instruments (Ill. 21) and this, in addition to underlining the renovated interests of the Chinese in the study of the history of astronomy, has confirmed the fact that the instruments themselves represent, in a crucial manner, the cultural link between Europe and China (Ill. 22, 23, 24, 25, 26).

Ill. 6: Peking Observatory in the unnumbered woodcut of the *Xinzhi yixiang tu* of F. Verbiest (Copy of the Vatican Library).

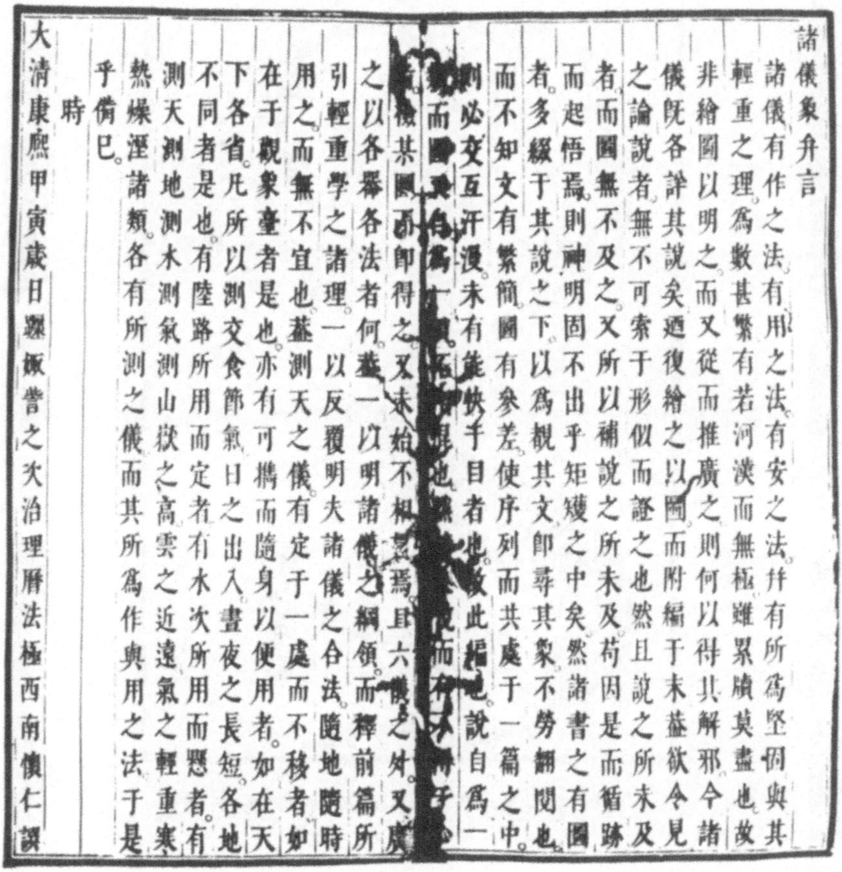

Ill. 7: Foreword of *Yixiang tu* of F. Verbiest (copy of Vatican Library).

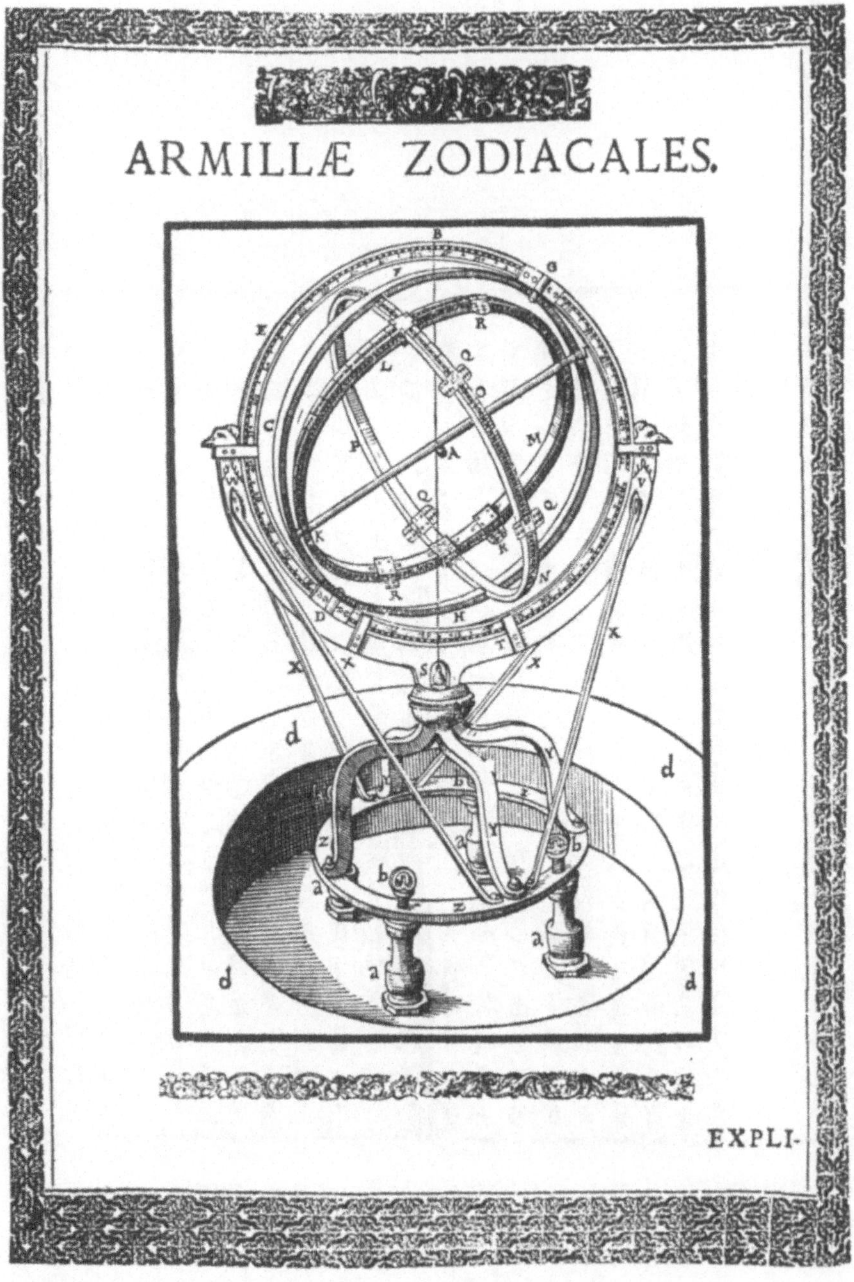

Ill. 8: Simple ecliptic armillary sphere from *Astronomiae instauratae mechanica* of Tycho Brahe (Wandesburg, 1598).

Ill. 9: Simple ecliptic armillary sphere from *Xinzhi yixiang tu* of F. Verbiest (Beijing, 1674), fig. no. 1.

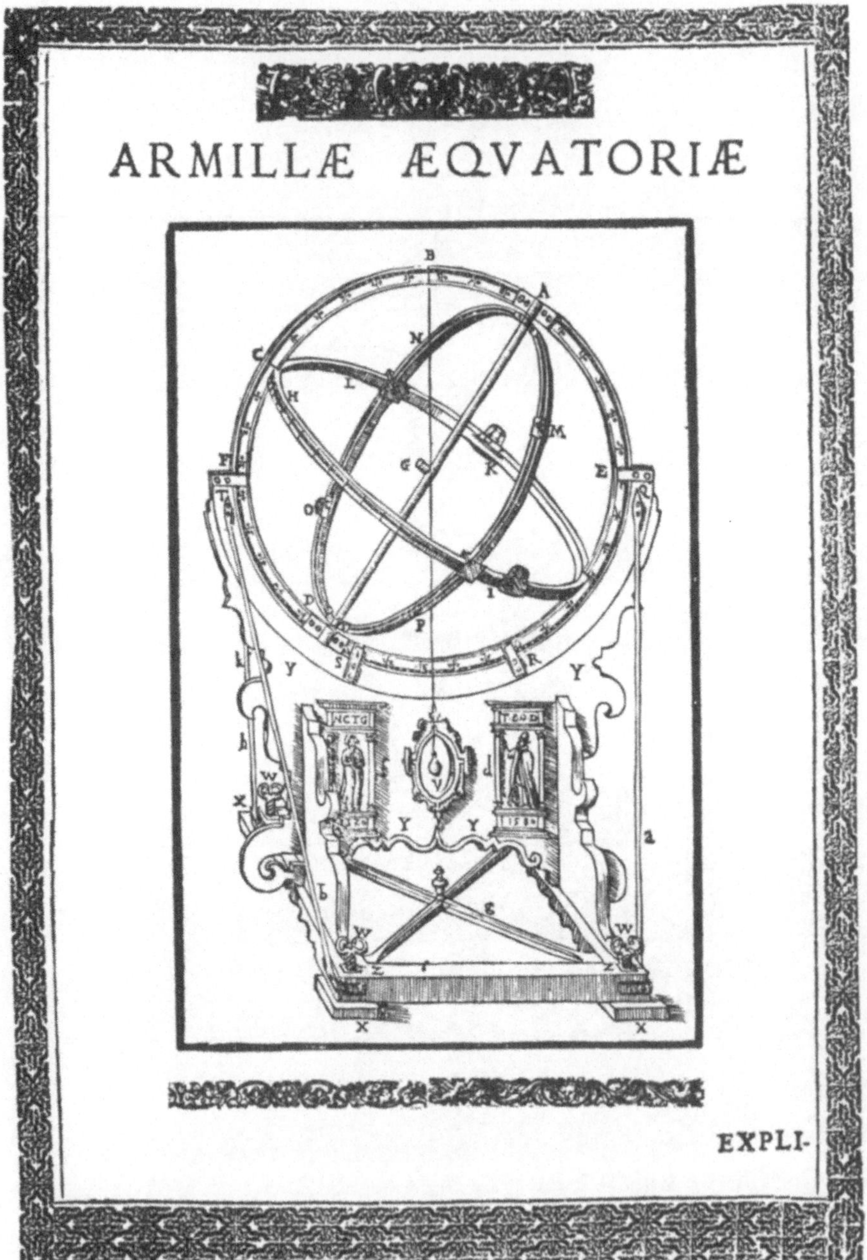

Ill. 10: Simple equatorial armillary sphere from *Astronomiae instauratae mechanica* of Tycho Brahe (Wandesburg, 1598).

Ill. 11: Simple equatorial armillary sphere from *Xinzhi yixiang tu* of F. Verbiest (Beijing, 1674), fig. no. 2.

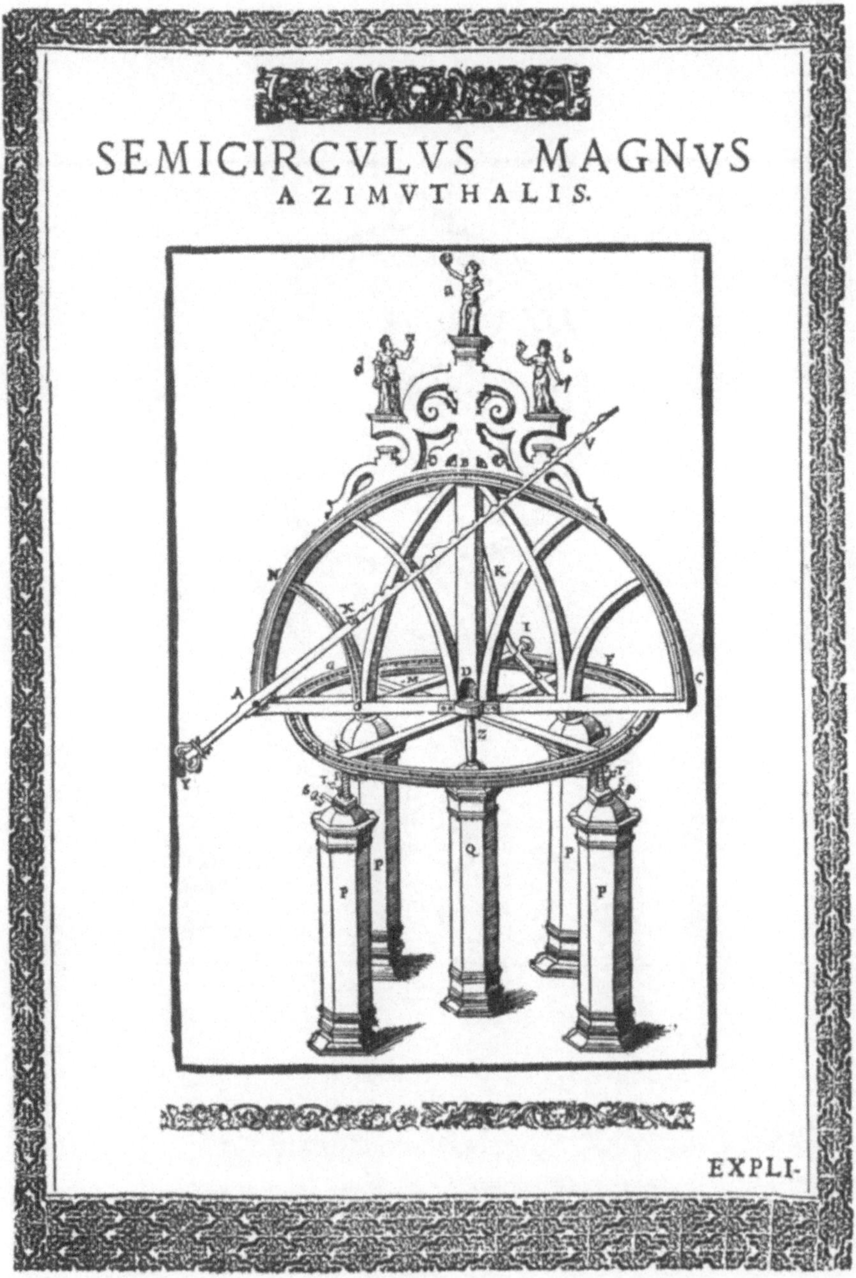

Ill. 12: Azimuth semicircle from *Astronomiae instauratae mechanica* of Tycho Brahe (Wandesburg, 1598).

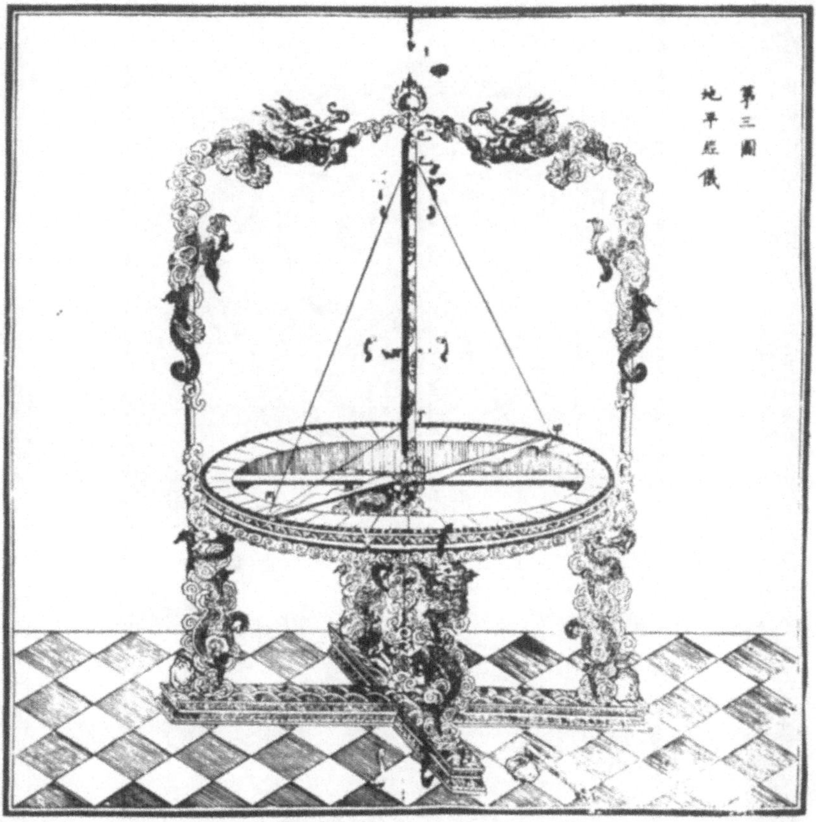

Ill. 13: Horizon circle from *Xinzhi yixiang tu* of F. Verbiest (Beijing, 1674), fig. no. 3.

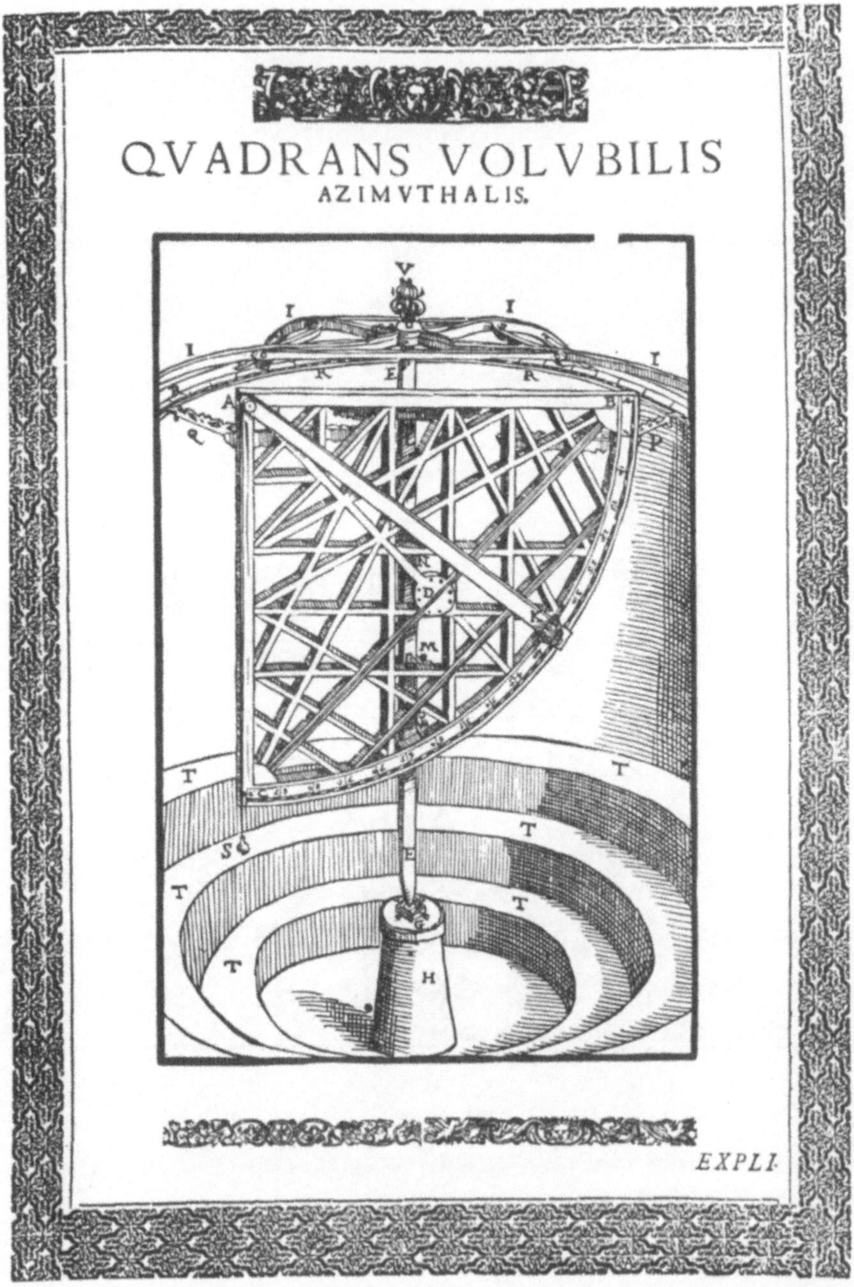

Ill. 14: Quadrant from *Astronomiae instauratae mechanica* of Tycho Brahe (Wandesburg, 1598).

Ill. 15: Quadrant from *Xinzhi yixiang tu* of F. Verbiest
(Beijing, 1674), fig. no. 4.

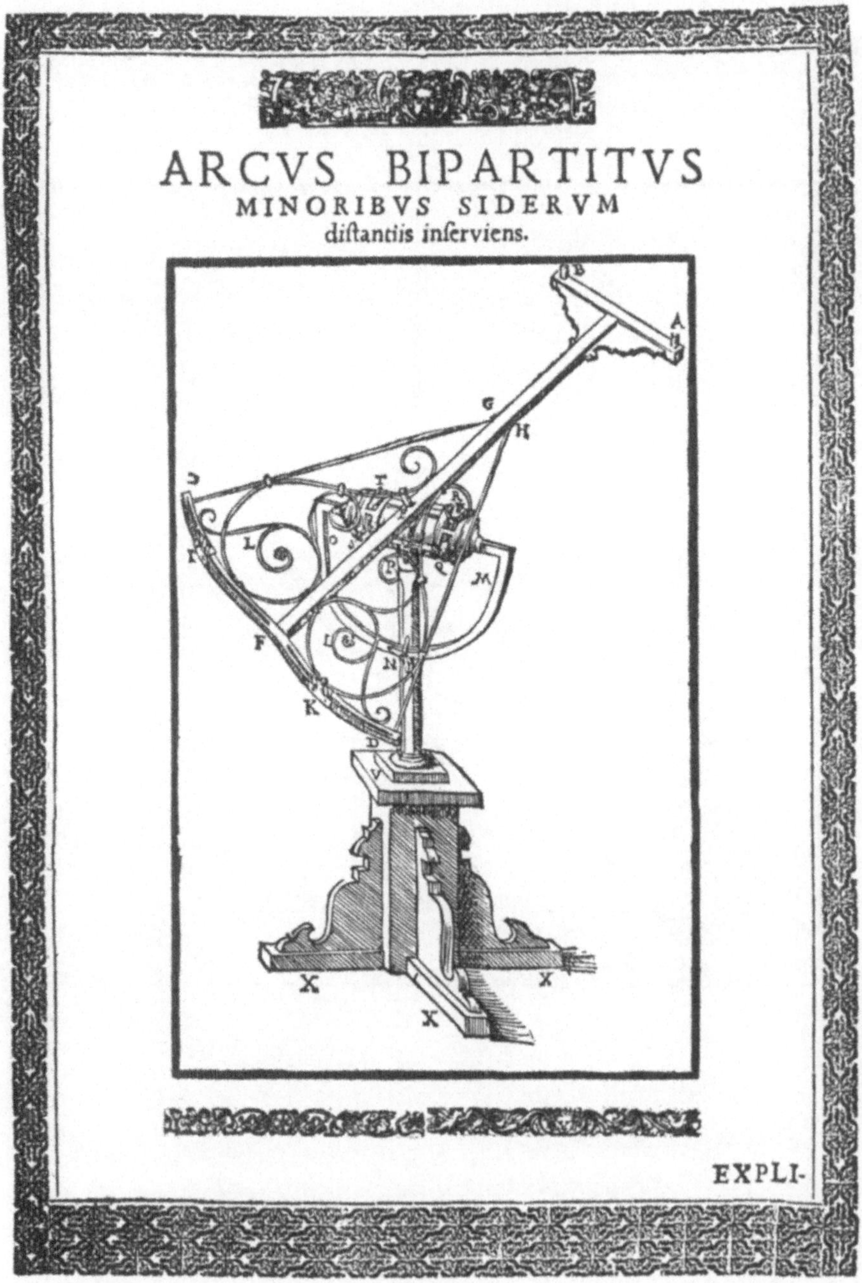

ARCVS BIPARTITVS
MINORIBVS SIDERVM
diſtantiis inſerviens.

EXPLI-

Ill. 16: Bipartite arc from *Astronomiae instauratae mechanica* of Tycho Brahe (Wandesburg, 1598).

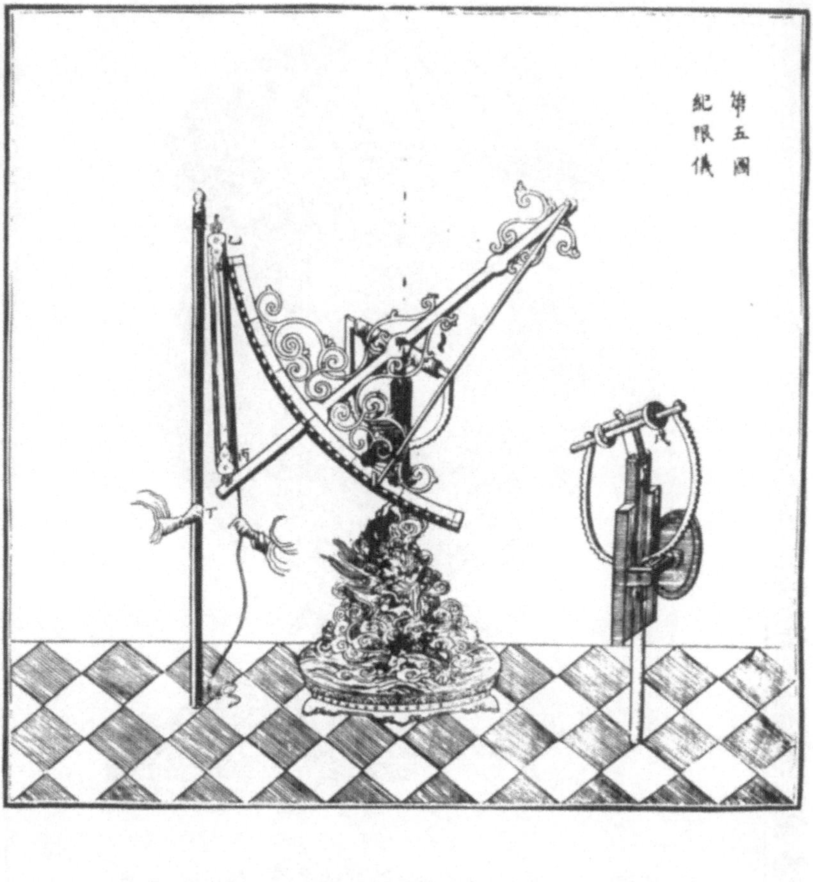

Ill. 17: Bipartite arc from *Xinzhi yixiang tu* of F. Verbiest (Beijing, 1674), fig. no. 5.

Ill. 18: Celestial globe from *Stronomiae instauratae mechanica* of Tycho Brahe (Wandesburg, 1598).

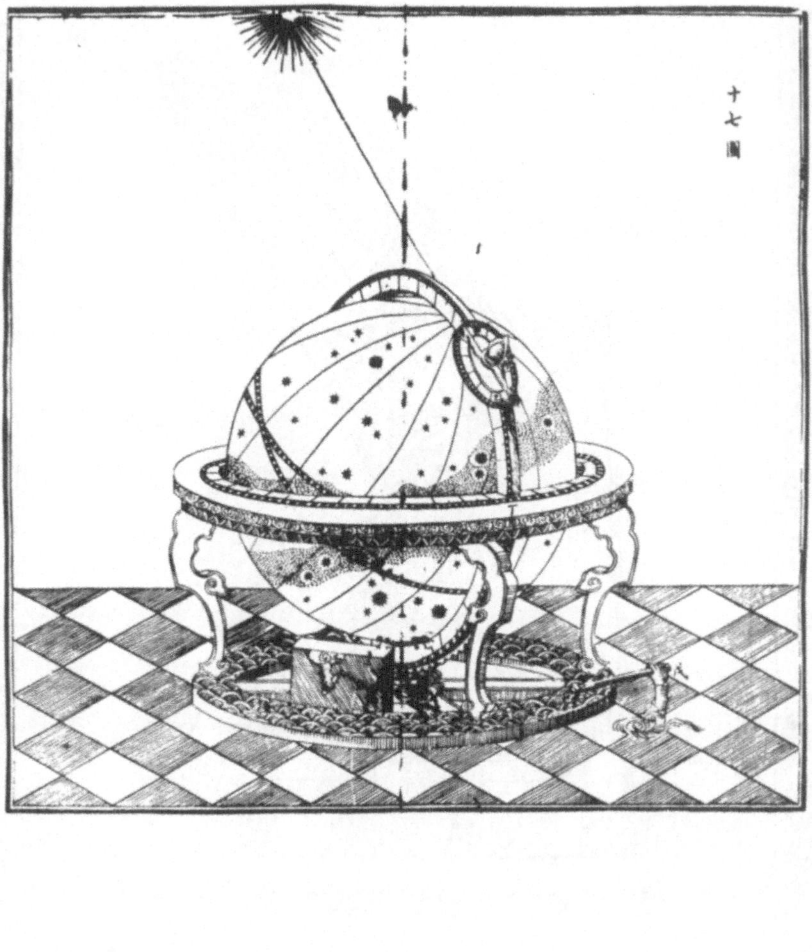

Ill. 19: Celestial globe from *Xinzhi yixiang tu* of F. Verbiest (Beijing, 1674), fig. no. 6.

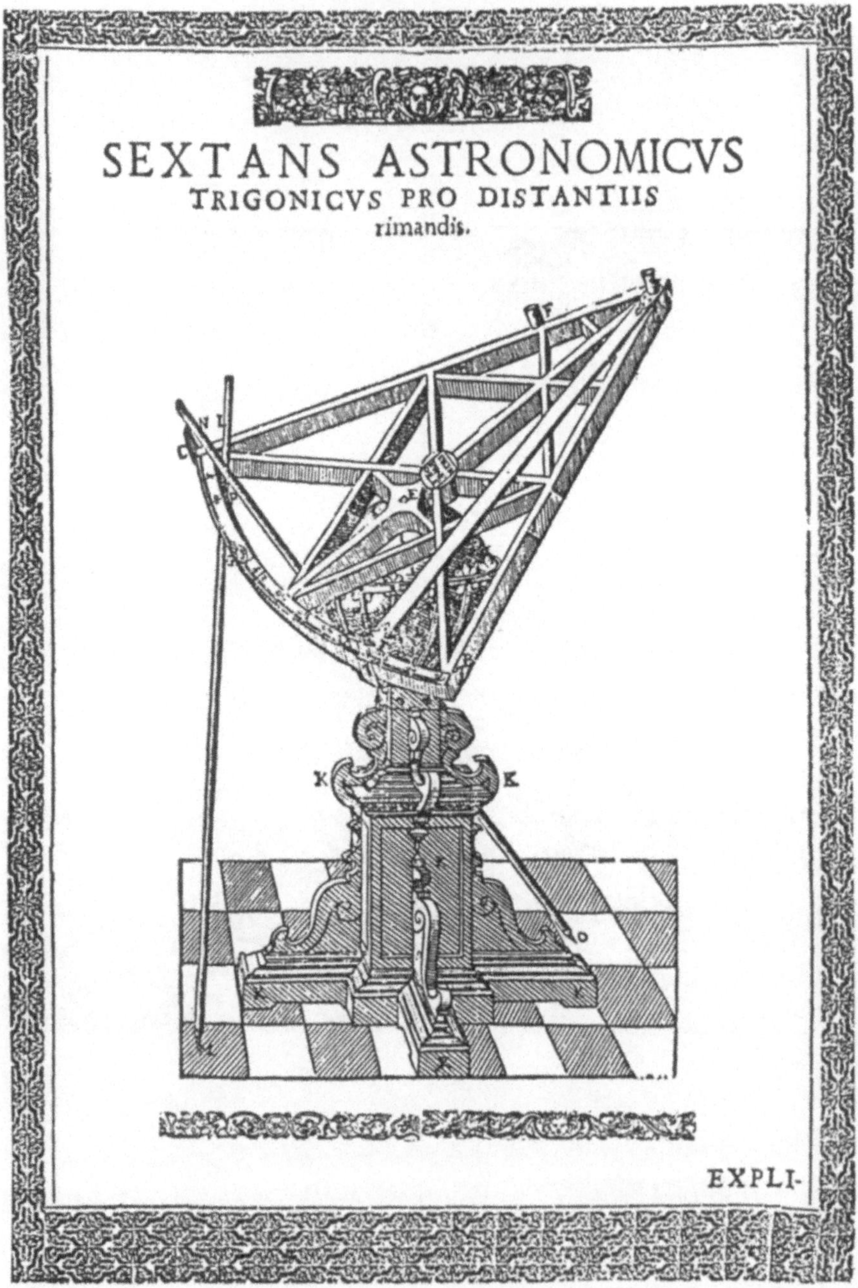

Ill. 20: Six-foot sextant from *Astronomiae instauratae mechanica* of Tycho Brahe (Wandesburg, 1598). Note the stabilizing spikes and the ball mounts.

Ill. 21: One of the editorials dedicated to Ancient Peking Observatory and to Verbiest's instruments (*Shenghuo canmou bao*, Beijing, March 12, 1987).

Ill. 22: The Ancient Peking Observatory. Note the instruments on the roof. (Photo 1987 by I. Iannaccone).

Ill. 23: Verbiest's sextant and equatorial armillary sphere on the roof of the Ancient Peking Observatory (Photo 1987 by I. Iannaccone).

Ill. 24: Verbiest's equatorial armillary sphere (1674) on the roof of the Ancient Peking Observatory (Photo 1987 by I. Iannaccone).

Ill. 25: Detail of the decoration of Verbiest's equatorial armillary sphere (Photo 1987 by I. Iannaccone).

Ill. 26: Verbiest's quadrant (1674) on the roof of the Ancient Peking Observatory (Photo 1987 by I. Iannaccone).

FERDINAND VERBIEST AND THE GEOGRAPHICAL WORKS BY JESUITS IN CHINESE 1584-1674

CHEN MINSUN

Chen Minsun is a professor emeritus and former chairman (1978-1984) of the Department of History, Lakehead University, Thunder Bay, Ontario, Canada. He received his B.A. degree from Qinghua University and his M.A. and Ph.D. degrees from the University of Chicago. His research interests include Sino-Western contacts and Canada-China relations. He is one of the co-editors of "China Insight" (1985), "East Asia Insight" (1988), and "East Asia Inquiry" (1991), all published by the Canadian Asian Studies Association in Montreal.

I. Introduction

In late Ming and early Qing China, a number of geographical works written in Chinese were produced by the Jesuit missionaries. These works included maps, books, and atlases. Ferdinand Verbiest (1623-1688) is the acknowledged author of three of these significant works[1]. This paper shall try to ascertain the contributions made by Verbiest to the authorship and publication of these geographical works by Jesuits in Chinese, which covered a time span from 1584 to 1674. It shall also discuss the common characteristics of these Jesuit geographical works and the Chinese reaction to the Jesuit missionaries[2].

---

1 Louis Pfister, *Notices biographiques et bibliographiques sur les Jésuites de l'ancienne mission de Chine, 1552-1773* (Shanghai, 1932-1934), 1:355-356; Fang Hao, *Zhongguo tianzhujiaoshi renwu zhuan* (Hong Kong, 1970), 2:87, 178-179.

2 This paper is partially based on an M.A. thesis written by this writer under the supervision of Professor Earl H. Pritchard at the University of Chicago. The full reference is: Chen Minsun, "Geographical Works by Jesuits in Chinese, 1584-1672", M.A. thesis, University of Chicago, 1959. This writer wishes to express again his gratitude to Professor Pritchard and to the

First we shall begin with a brief historical account. Matteo Ricci (1552-1610), the pioneer Jesuit missionary in the history of the Catholic mission in China, was the author of the first of these geographical works. After establishing good relations with the Chinese, Ricci went to Zhaoqing, the residence of the Guangdong and Guangxi viceroy, in 1583. Ricci's friends, mostly Chinese scholar-officials, saw a western world map hung in his residence. They liked the map, and encouraged Ricci to have it translated and made available in the Chinese language. Ricci thought that he could use the map to spread Christianity and to ease the Chinese fear of strangers, so he happily consented. Thereupon Ricci, using his western map as a model, translated the names of places and geographical terms into Chinese, and with some added notations, produced *Shanhai yudi quantu*, a world map[3]. Ricci's map was immediately engraved and printed, and it soon became popular among Chinese scholar-officials. Wherever Ricci went, as he moved his missionary post from the south to the north, he was requested to make new editions of his map. From 1584 to 1608, two years before his death in Beijing, no less than eight editions of Ricci's world map were printed and copied. While only three of these editions were authored by Ricci himself, the other editions were either reprinted or copied by Ricci's Chinese admirers without Ricci's prior consent. Ricci's map was circulated in many provinces of China, and copies of it even found their

---

United Board for Christian Higher Education in Asia (New York) for a fellowship which enabled him to write the thesis. In preparing this paper, this writer also received generous help in locating Chinese sources from Mr. Dai Wenbei, assistant librarian for the Chinese collection in the University of Chicago Library. To him, this writer wishes to extend his sincere thanks.

3 For a clear photographic reproduction of Ricci's world map of 1602, see Pasquale M. d'Elia, *Il Mappamondo Cinese del P. Matteo Ricci* (Vatican City: Biblioteca Apostolica Vaticana, 1938). This work is the most comprehensive account of Ricci's map, and a monumental work about Ricci's map-making activities. For English translations of the notes on Ricci's Chinese map, see Lionel Giles, "Translations from the Chinese World Map of Father Ricci", *The Geographical Journal*, 52 (1918), 367-385; 53 (1919), 19-30; and Kenneth Chen, "Matteo Ricci's Contribution to, and Influence on Geographical Knowledge in China", *Journal of the American Oriental Society*, 54 (1939), 325-359. For a translation of the part relating to the Americas, see L. Carrington Goodrich, "China's First Knowledge of the Americas", *The Geographical Review*, 28 (1939), 400-411. For a thorough account of the editions of Ricci's map, see Hong Weilian (William Hung), "*Kao Li Madou di shijie ditu*" (A Study of the World Map of Matteo Ricci), *Yougong*, 5, n° 3-4 (April 11, 1936), 1-50. See also J.F. Baddeley, "Father Matteo Ricci's Chinese World Map, 1584-1608", *The Geographical Journal*, 50 (October 1917), 254-270.

way to Japan[4].

Through his world map, Ricci introduced to China: a) a unified conception of the entire world - the idea that the world consists of five continents, b) the idea of the sphericity of the earth, and c) the division of the world into climate zones. Ricci also brought into China the most recent discoveries in western geography, the use of meridians, and the translation of geographical terms. All these ideas and information were new to the Chinese, and they immediately attracted attention. People with curiosity and a desire for knowledge hailed the publication of Ricci's maps, while the skeptics and conservatives wondered and doubted[5].

With the arrival of more Jesuit missionaries, Chinese interest in world geography in the seventeenth century became heightened. In 1623, a five *juan* work written in Chinese on the geography of the world - the *Zhifang waiji*, prepared by Giulio Aleni (1582-1649) - was printed by his Chinese friends[6]. According to Aleni's preface to the *Zhifang waiji*, after Ricci made his maps, Diego de Pantoja (1571-1618) was requested to translate a western printed map. On the basis of his own observations and knowledge, Pantoja completed some *Notes*. People in the city of Beijing were delighted to read the *Notes*, but they were not printed. Fearful of the disappearance of that work, Aleni revised and enlarged Pantoja's manuscript into a book and called it the *Zhifang waiji* - an account of the countries which are not included in the Chinese official records[7]. With the *Zhifang waiji*, world geography, as introduced by Ricci, became more detailed and systematized. Aleni's work included a complete map of the world and separate maps for Asia, Europe, Africa, and North and South America. It included a brief account of the countries in each continent and a general account of the four seas[8].

---

4 Hong, "*Kao Li Madou*", *op. cit.*, pp. 12, 28; Chen Minsun, *op.cit.*, pp. 30-33.

5 Ssu-yü Teng and John K. Fairbank, *China's Response to the West* (Cambridge, 1954), pp. 12-17.

6 Chen Minsun, *op. cit.*, pp. 51-52.

7 *Ibid.*; see also Bernard Hung-Kay Luk, "A Study of Giulio Aleni's *Chih-fang wai-chi*", *Bulletin of the School of Oriental and African Studies*, 40 (1977), 60-61.

8 The *Zhifang waiji*, which was reprinted in the various collections published in later centuries, contained varying arrangements of the maps. Early editions put one of the maps at the beginning of each *juan*. Later editions grouped all five maps at the beginning of the book.

Fourteen years after the publication of the *Zhifang waiji*, Aleni wrote and published in 1637 another vivid account of Europe entitled *Xifang dawen* (Answers to questions concerning the West). This work probably was not widely circulated outside of Fujian province, the author's missionary station. Aleni's two books were the most complete accounts on Europe available in Chinese at that time[9]. If Ricci's *World Map* was a pioneer achievement, Aleni's *Zhifang waiji* and *Xifang dawen* were truly milestones, because the *Xifang yao ji* and the *Kunyu tushuo* published in the second half of the seventeenth century, after an interval of thirty years, were the condensations and improvements on the style and content of Aleni's works[10]. Often in many Chinese geographical works, references were made to the *Zhifang waiji*. Seven years after the publication of Aleni's *Xifang dawen*, the Ming dynasty was overthrown. At first the new and vigorous Qing dynasty was busy consolidating its control in China, and paid little attention to the outside world. Interest in that world, however, was soon to become evident. In 1669, the *Xifang yao ji*, a concise account of the West, prepared by the Jesuits Ludovico Buglio (1606-1682), Gabriel de Magalhaes (1610-1677), and Verbiest was presented to the Kangxi emperor. In response to the emperor's queries, the Jesuits compiled a concise account of western civilization, and pledged themselves to do their utmost to serve the Chinese emperor[11].

Three years later, in 1672, Verbiest's *Kunyu tushuo* was published. In his preface, Verbiest stated that the *Kunyu tushuo* was a book which discussed important features of the world, such as topography, earthquakes, mountains, tides, movement of the sea currents, rivers and streams, peoples, customs and various products. His learned colleagues Ricci, Aleni, Alfonso Vagnoni, and Sabatino De Ursis had written about all this before. Verbiest, therefore, selected from their material, and added some new knowledge in order to further expound the truths which

---

9 John L. Mish, "Creating an Image of Europe for China: Aleni's *Hsi-fang ta-wen*", *Monumenta Serica*, 23 (1964), 2-3. Fang Hao, *Zhongxi jiaotong shi* (Taibei, 1959), 4:166-170.

10 The *Xifang yao ji* is basically a digest of the *Xifang dawen*, while the *Kunyu tushuo* is essentially based on the *Zhifang waiji*.

11 According to the preface as contained in the *Qingchaotang congshu*, the *Xifang yao ji* was presented to the emperor soon after he inquired about the conditions of Europe on the twenty-third day of the eleventh moon in the seventh year of the Kangxi reign, i.e., December 26, 1668. The work is also entitled *Yulan xifang yao ji*, a clear indication that it was presented to the emperor for his survey. See also Fang Hao, *Zhongxi jiaotong shi*, 4:170.

had been made known before by them[12]. In or about 1674, Verbiest published the *Kunyu quantu* (Map of the World), to complement the *Kunyu tushuo*[13]. Thus, from 1584 to 1674, a total of at least seven geographical works in Chinese were produced by the Jesuit missionaries. Among these seven different works mentioned above, only the manuscript by Pantoja was not published[14]. For the six published works, only the *Xifang dawen* was not widely known or circulated[15].

II. Characteristics of the Jesuit Geographical Works

There are several common characteristics shared by these geographical works. First, they were all produced or written in response to requests or demands. Ricci's maps and Aleni's books were produced at the urging of Chinese scholar-officials. Several of these scholar-officials even polished the Jesuit texts and wrote prefaces. The *Xifang yao ji* prepared by Buglio, Magalhaes, and Verbiest was prompted by the Kangxi emperor's curiosity. Verbiest's book and maps, evidently, were also produced to satisfy the interests of the Manchu scholar-officials who wanted to gain some knowledge about Europe and the world.

A second common feature is the high esteem and praise which the Jesuits expressed towards China, its rulers and scholar-officials. In Ricci's map of 1602, China is described as "most famous for her culture and products"[16]. In the *Zhifang waiji*, Aleni notes that since ancient times, China has continuously produced great emperors and sages. The beauty of her civilization, ritual, music, and customs, and the richness of China have been admired by far and near nations[17]. In the *Xifang yao ji*, the Jesuits praised the Chinese emperors for giving the Jesuits protection and stipends, and pledged their utmost to serve the Kangxi emperor[18]. The *Kunyu tushuo* by Verbiest, however, does not contain such laudatory remarks. It reads like a pure geographical work.

A third common characteristic is that each Jesuit writer presented his

---

12 *Kunyu tushuo* in the *Congshu jicheng*, p. 5.

13 Fang Hao, *Zhongguo tianzhujiaoshi*, 2:178.

14 See note 6.

15 *Zhifang waiji* in the *Tianxue chuhan* (Hanzhou, 1629; reprint, Taibei, 1965), 3:1306; Fang Hao, *Zhongguo tianzhujiaoshi*, 1:207.

16 K. Chen, "Matteo Ricci's Contribution", p. 335.

17 *Tianxue chuhan*, 3:1320.

18 *Xifang yao ji* in the *Zhaodai congshu*, 27/12a-12b.

work, not as the writing of a single author, but as the collective work of himself and his fellow Jesuits. Each work is essentially a collective product which represents the efforts of a number of Jesuits. Not only the names of the Jesuit revisers and the *imprimatur* were prominently given at the beginning of each work, but direct quotations from previous Jesuit writings were also liberally incorporated into the new text without specific notation. While Aleni evidently quoted a great deal from the manuscript of Pantoja, Buglio, Magalhaes, and Verbiest utilized Aleni's books rather extensively. In the *Xifang yao ji*, the Jesuit trio actually condensed the *Xifang dawen* without giving any indication that that was the case. In the *Kunyu tushuo*, Verbiest acknowledged that he selected materials from the writings of other Jesuits, especially from Vagnoni's *Kongji gezhi* (Treatise on the Material Composition of the Universe), Ursis's *Biaodu shuo* (On the Compass) and Aleni's *Zhifang waiji*. A major part of the *Kunyu tushuo* contains essentially direct quotation from the *Zhifang waiji*[19]. As explained above that each Jesuit work should be considered as a collective work by several Jesuits, one should therefore not charge any particular Jesuit with plagiarism. Verbiest was essentially the compiler, but not the author of the *Xifang yao ji* and the *Kunyu tushuo*. Verbiest's contribution, thus, involved only the preservation and expansion of the geographical works produced previously by his fellow Jesuits. It is a contribution which, though less important in Verbiest's colorful life and manifold career, should nevertheless deserve our recognition.

A fourth common characteristic shows that all Jesuit writers and editors were aware that they represented not only the Society of Jesus as missionaries in an alien country, but were also representatives of the European community, which had a different and highly developed civilization. Ricci called himself "a native of Europe" rather than a native of Italy. The Chinese records of that time also referred to Ricci as a person of Europe (*Ouluobaren*)[20]. Other Jesuit writers, on the other hand, stated that they went to China from the West or "Great West." They consciously acted as interpreters of Western or European civilization to the Chinese. Two of these Jesuit works exclusively contain accounts of Europe, the "West"[21].

The fifth common characteristic is the consistency in many of the

---

19 Chen Minsun, *op. cit.*, pp. 121-129.

20 *Ibid.*, pp. 131-132.

21 *Xifang yao ji*, 27/1a; Mish, "Creating an Image", p. 5.

statements which run through these Jesuit geographical works. For example, the Jesuit writers supported each other on such matters as the distance between Europe and China. Ever since Ricci reported that he traveled 90,000 *li* from Europe to China, and that it took him three years to reach China from Europe, other Jesuits repeated the same mileage and duration of time in their writings[22]. There were some elaborations made by Aleni, however, in saying that it took one year each to reach Portugal from the interior part of Europe, and from Portugal to India, and that a third year was needed to reach China from India[23]. On the whole, the 90,000 *li* distance was mentioned again and again. On the Jesuit missionaries themselves, an affirmation was made consistently that, though they were born in a European country they were willing to live and to work till death in the country to which they were sent[24]. While the distance between Europe and China was somewhat overestimated, the fact that European Jesuit missionaries were willing to labor till old age, and to spend the rest of their life in China, gave credence to their pronouncements[25]. The consistency in Jesuit accounts was probably based on the recognition that any discrepancy would create suspicion. Professor Jacques Gernet's point is well made when he says that in seventeenth-century China "one had to be extremely careful not to change positions on what had earlier been presented as eternal verity"[26].

III. Idealization and the "Mythistory" of Europe

The sixth, and indeed the most important common characteristic of Jesuit geographical works in Chinese, is the idealization of Europe. Europe is depicted as the Christendom where peace had prevailed from the beginning of the Christian era. The Jesuits used these geographical works not only to preach the Christian gospel, but also to inform the people of China that an ideal society already existed in Europe. This

---

22 *Tianxue chuhan*, 3:1320, 1495; *Xifang yao ji*, 27/12a.

23 *Xifang yao ji*, 27/2a; Mish, "Creating an Image", p. 7.

24 See Matteo Ricci's memorial to the Wanli emperor dated 1601 in Henri Bernard, *Le Père Matthieu Ricci et la société chinoise de son temps* (Tientsin, 1937), 2:8; *Tianxue chuhan*, 3:1936; *Xifang yao ji*, 27/11b.

25 Most of the four hundred Jesuit missionaries who were in China from 1538 to 1773 ended their career in China. See Pfister, *op. cit.*

26 Jacques Gernet, "Christian and Chinese Version of the World in the Seventeenth Century", *Chinese Science*, 4 (Philadelphia, 1980), 5.

firm belief had much to do with the strong and deep Christian faith which the Jesuit missionaries possessed. To borrow terminology from Professor William H. McNeill, it seems that the Jesuit missionaries also actually conveyed a "myth" or "mythistory" of Europe to the people of China. McNeill has indeed provided us with new insight into the study of cultural interaction when he points out that "myth lies at the basis of human society," and that "myths, moreover, are based on faith more than on fact"[27]. The relevance of this new insight to the study of seventeenth-century Sino-western cultural contact is particularly evident. For this contact involved the confrontation of the Sinocentric myth of the world with the Jesuit myth of Europe, and the rejection of the Jesuit myth of Europe by the majority of Chinese scholar-officials.

To ease the fear and distrust of the Chinese scholar-officials for the Jesuit missionaries, the Jesuits tried to idealize European political history. In his Chinese *Kunyu wanguo quantu* published in 1602, Ricci described Europe in the following words:

> The continent of Europe contains some (thirty) states, all of which enjoy a monarchical system of government. They follow no heterodox doctrines, but are reverent adherents of the holy Christian religion (Catholicism), which recognizes one Supreme Deity ... The workers are skillful and clever, while the people are well-versed in astronomy and philosophy. In their daily activities they are solid and honest, and have a high regard for the five relationships. Modes of production are plentiful. Princes and ministers are prosperous and healthy. Communications are kept with foreign countries at all times, while her merchants roam over the entire earth[28].

Again, in a letter which Ricci wrote to a Chinese scholar-official who championed Buddhism, Ricci further asserted:

> In my home (continent of Europe), ever since we adhered to the Catholic religion one thousand six hundred years ago, ... there has been no war and no strife among the thirty states of Europe for one thousand six hundred years[29].

---

27 William H. McNeill, "The Care and Repair of Public Myth", *Foreign Affairs*, Sixtieth Anniversary Issue (Fall 1982), 1; *idem*, "Mythistory, or Truth, Myth, History, and Historians", *American Historical Review*, 91 (February 1986), 1-10.

28 Giles, "Translations", *op. cit.*, p. 377; K. Chen, "Matteo Ricci's Contribution", *op. cit.*, p. 329.

29 *Tianxue chuhan*, 2:647. A similar view is expressed in the *Xifang yao ji*, 27/11b.

This idealized version of Europe was also contained in Aleni's *Zhifang waiji*:

> All European states, large or small, from their kings to the common people, adhere to Catholicism, no heterodox doctrine being allowed in their midst. The kings bind themselves by royal marriages and live peacefully generation after generation. Trade and goods move freely, and they never use selfish measures to covet what belongs to the public ...
> All government officials are well-provided with large salaries to allow them to have ample means for support, and also enough for spending on charities. There never have been any corrupt practices ... Taxes in European countries amounted to one-tenth of the income. Taxes are reported by the people themselves, and there is no hastening measure. The judicial process is simple ... All cases are to be decided according to evidence. No punishment is allowed before conviction. The judges cannot ill-treat a defendant. Since laws are well-prescribed, officials can neither exploit the people, nor can they use their positions to take advantage of the common people[30].

In the *Xifang yao ji*, there is also the same idealization: "European states, large or small, from their kings to the commoners, adhere only to the Catholic Religion; there is no other religious teaching"[31]. The probable reason, as mentioned above, that the Jesuit missionaries indulged in such an idealization of Europe, was that they were both imbued with Christian faith and were confronted in China with a Sinocentric myth. The Sinocentric myth consisted of the idea that China was the center of the world-civilization and the institution of the tribute system. The tribute system involved the recognition by the rulers of foreign countries the special position occupied by the Chinese emperor as the Son of Heaven, and the acceptance of vassalage on the part of these foreign rulers. Detailed regulations concerning the sending of periodic tributary missions to China were further stipulated[32].

IV. Chinese Reaction to the Jesuit Writings

Ricci's editions of his world maps and Aleni's *Zhifang waiji* were warmly received by some Chinese scholar-officials and those Chinese

---

30 *Tianxue chuhan*, 3:1356, 1362, 1370-1371.

31 *Xifang yao ji*, 27/11a.

32 For the development of the tributary system, see John K. Fairbank and S.Y. Teng, "On the Ch'ing Tributary System", *Harvard Journal of Asiatic Studies*, 6 (June 1941), 135-246.

who were newly converted to the Catholic faith, but a large majority of the Chinese scholar-officials looked on the geographical works written by the Jesuits with skepticism and suspicion. That was the situation in the late Ming period. After the establishment of the Qing dynasty, and in spite of the interest shown to the Jesuits by the Shunzhi and Kangxi emperors, a large number of Chinese scholar-officials maintained their skepticism and suspicion toward the new religion and new scientific knowledge, including geographical knowledge, which the Jesuits brought to China. The publication of the *Xifang yao ji* in 1669 and the *Kunyu tushuo* in 1672 did not alter the general attitude of distrust. The attitude of the compilers of the *Mingshi* is a typical example. The *Mingshi* argued that, since none of the new geographical knowledge brought to China by the Jesuits could be verified by Chinese historical records, none could be accepted. Benefited by hindsight, we may say that the Chinese scholar-officials of the seventeenth and eighteenth centuries showed little interest in verifying the new geographical knowledge. One must nevertheless add that an important change in Chinese cartography that did take place after the publishing of Ricci's maps, was the disappearance of the method of "using squares to indicate distance"[33]. The division of latitude and longitude was more accurate, and probably that was why the old method was discarded, although the new method was not widely utilized at once. Yet, as late as 1848, when Xu Jiyu, a well-known scholar-official, was making finishing touches to his *Yinghuan zhilüe* (General Account of the Maritime World), he was still very skeptical of the Jesuit writers.

Westerners like Ricci, Aleni, and Verbiest all lived in the capital (Beijing) for a long time, and became well-versed in the Chinese language. Consequently, the style of their books is quite clear and agreeable, but there is not just a little boastfulness and craftiness in their accounts[34].

Other evidence of the dominance of the Sinocentric myth occurred during the reign of the Qianlong emperor (1736-1796), when the *Siku quanshu* (The Complete Library of the Four Treasuries) was compiled. The editors discarded all the writings on western religion, philosophy, and history done by the Jesuits and their Chinese converts. Only the Jesuit writings on geography, geometry, mathematics, astronomy, and

33 Chen Minsun, *op. cit.*, p. 139.

34 Francis W. Drake, *China Charts the World: Hsü Chi-yü and his Geography of 1848* (Cambridge, 1975), p. 54.

technology were accepted into the *Siku quanshu*[35].

While the Jesuit "mythistory" of Europe was eventually discarded when the Christian missionaries reentered China in the early nineteenth century, the Sinocentric myth also finally gave way by the early twentieth century to a more realistic appreciation of the actual world situation. Needless to say - just like people in all nations of the world - the Chinese have also been busy since then in reestablishing a credible "mythistory" for their own nation. We would hope that all the "mythistories" of the various nations could somehow accommodate each other in an increasingly shrinking and interdependent world[36].

The writing of geographical texts occupied only a small dimension of Verbiest's illustrious career. Our admiration for Verbiest should by no means be diminished on account of some defects found in his geographical works in Chinese, which were produced by the needs of the time. These works were produced when Verbiest had just regained the trust of the Manchu imperial court. Busy with the tasks in the Board of Astronomy, of which he was the newly appointed supervisor, Verbiest most likely did not have the time or the inclination to break new ground in geographical writings. However, the *Kunyu tushuo* does contain an album of pictures of twenty-three strange animals and a large sea-going ship, the pictures of the seven wonders of the world as well as the arena (Colosseum) of Rome. It made Verbiest's book more interesting to read, and it probably amused more Chinese readers than other Jesuit geographical works. This could be one of the reasons that this album of pictures was later published separately by an enterprising printer as an independent volume under the title of *Kunyu waiji*[37].

---

35 Fang Hao, *Li Zhizao yanjiu* (Taibei, 1966), pp. 139-141.

36 McNeill, "The Care and Repair", pp. 12-13.

37 The greatness of Verbiest could be seen from the fact that he was "the only westerner on whom a posthumous title was conferred by the Chinese court," according to Fu Lo-shu, *A Documentary Chronicle of Sino-Western Relations (1644-1820)* (Tucson, 1966), pp. 99, 476. See also Fang Hao, *Zhongguo tianzhujiaoshi*, 2:163-167. A recent account of the contribution westerners made to China mentioned Verbiest's development of Chinese astronomy, but did not touch upon the geographical works written by him. See Jonathan Spence, *To Change China: Western Advisers in China 1620-1960* (Harmondsworth, 1980), pp. 23-33. On the *Kunyu waiji*, see Pfister, *op. cit.*, p. 356; Fang Hao, *Zhongxi jiaotong shi*, 4:186.

FERDINAND VERBIEST'S CONTRIBUTION TO CHINESE GEOGRAPHY AND CARTOGRAPHY

LIN TONGYANG

Lin Tongyang studied at the Université Catholique de l'Ouest (Angers) and at the Université de Haute-Bretagne (Rennes). He received a doctorate in history in 1982 from the Sorbonne University, Paris. At present he is a professor of history at Tunghai University in Taiwan.

I. Ferdinand Verbiest's Works in the Field of Geography and Cartography

The golden age of European geography and cartography in China had its beginnings in 1584, when Matteo Ricci made his first edition of the Chinese world map engraved on blocks at Zhaoqing, but it reached its peak in 1674, when the Flemish Jesuit Ferdinand Verbiest published his celebrated world map under the title of *Kunyu quantu* and compiled his world geography at Beijing. It is certain that Verbiest, with his keen intelligence and enthusiasm for science, not only made a great contribution to the development of Chinese astronomy but also created a wide vision of the world for the Chinese. He was probably the only Jesuit to compile and print such numerous works in the field of geography and cartography in the course of the seventeenth and eighteenth centuries in China. Here we may begin with an outline of Verbiest's works.

1. The *Yulan xifang yaoji* (1669, 1 *juan*)

This little book, relating various aspects of the social and cultural life, as well as the general geography of Europe, was a collective work of Verbiest, Gabriel de Magalhaes and Ludovico Buglio. Previously it was difficult to fix the date of its publication[1], because the earlier reproductions (Zhaodai, ed. and Xuehai, ed.) did not give any valuable indications and evidence. According to the report of Adrien Grelon (1618-1696), we are convinced that this book was presented for the first time to the Kangxi emperor on May 7-8, 1669 in order to satisfy the curiosity of the emperor[2]. Moreover, in 1901, Henri Cordier gave the date of its publication as 1669[3]. It is said that this little book of the three Jesuits of Beijing was translated later into Manchu[4]. After a careful examination, we believe that this book was based on the *Xifang dawen*[5] (Questions and Answers regarding the West) of Giulio Aleni. On the whole, all the text except two final sections (*jiaofa* and *xishi*) is copied from the first *juan* of Aleni's work mentioned above. As to the editions of the *Xifang yaoji*, at least three exist: the Zhaodai (1697), the Xuehai (1831) and the Qingzhaotang.

2. The *Kunyu quantu* (1674)

The Verbiest map entitled the *Kunyu quantu* (A Complete Geographical Map of the Earth) consists of two hemispheres, each measuring 154 cm. in diameter. It is engraved on wood-blocks and divided into eight scrolls; each measures approximately 179 cm. by 54 cm.[6], very similar

---

1 In 1982, I tentatively fixed the date between 1669 and 1672, see Lin Tongyang, "Études sur l'introduction de méthodes et de connaissances géographiques européennes en Chine (1583-1718)" (Thèse de l'Université de Paris-Sorbonne, 1982), 2:407, n. 6.

2 Henri Bosmans, "Documents relatifs à Ferdinand Verbiest: les lettres annuelles de la Vice-Province de la Compagnie de Jésus en Chine Année 1669 par Adrien Grelon", *Annales de la Société d'Emulation*, n. 2 (1912), 41-42.

3 Henri Cordier, *L'imprimerie sino-européenne en Chine. Bibliographie des ouvrages publiés en Chine par les Européens au XVIIe et au XVIIIe siècles* (Paris, 1901), p. 8.

4 Louis Pfister, *Notices biographiques et bibliographiques sur les Jésuites de l'ancienne mission de Chine 1552-1773* (Shanghai, 1932-1934), 1:242; Henri Bernard, "Les adaptations chinoises d'ouvrages européens. Bibliographie chronologique depuis la venue des Portugais à Canton jusqu'à la mission française de Pékin 1514-1688", *Monumenta Serica*, 10 (1945), 370.

5 This book is divided into two *juan* and was published at the Catholic Church of Jinjiang in Fujian in 1637. John L. Mish has provided a facsimile reproduction of this book in his study entitled: "Creating an Image of Europe for China: Aleni's *Hsi-Fang Ta-wen*", *Monumenta Serica*, 23 (1964),4-30.

6 This measurement is made according to the original copy preserved at the National Palace Museum of Taibei, Taiwan, n° 06-462.

to that of the Ricci map of 1602[7]. Each hemisphere consists of three scrolls, and the two outer scrolls are reserved for cartouches giving general information on physical geography. It was in 1674 that the Flemish Jesuit drew his famous world map by stereographic projection. This specific term was given by the Jesuit François d'Aguillon in 1613, but the stereographic projection, invented by the Greek Hipparchos (about 180 B.C.), had prevailed in Europe among makers of maps for geographical purposes in the last quarter of the sixteenth century[8]. Using this scientific method, Rumold Mercator, son of the great Renaissance cartographer Gerardus Mercator (1512-1594), drew his world map of 1587[9] *(Orbis terrae compendiosa descriptio)*.

The Chinese title of the Verbiest map appears at the top of the third and sixth scrolls, and at the bottom of the second and seventh scrolls, the date (*Kangxi jiayin*, 1674) and the author's name, with the indication of his official rank (*zhili lifa*, administrator of calendar-making), are clearly given. Evidently, the drawing of this Verbiest map was made for the interest of the imperial palace, especially of the Kangxi emperor.

Owing to its enormous dimensions (179 cm. x 432 cm.), the Verbiest map has enough space for miscellaneous things to be added and described. The contents of the map can be divided into four categories: (1) the fourteen cartouches explaining natural and meteorological phenomena, such as the four elements, the poles, the earth, the spherical shape of the earth, rain and clouds, wind, the movement of the air, the effect of the tides, the current of the sea waters, earthquakes, human beings, rivers, mountains, etc. ...; (2) the five continents (Asia, Europe, Libya, America and Magellanica) constituting the main part of the map; (3) the toponyms with notes and description; (4) the illustrations of

---

7 Each scroll of the Ricci map of 1602 measures 179 cm. by 69 cm. See J.F. Baddeley, "Father Matteo Ricci's Chinese World Maps 1584-1608", *The Geographical Journal*, 1:4 (October, 1917), 264, n. 1; Pasquale M. d'Elia, *Il Mappamondo cinese del P. Matteo Ricci, S.I.* (Citta del Vaticano. Biblioteca Apostolica, Vaticana, 1938), p. 109 and "Recent Discoveries and New Studies (1938-1960) on the World Map in Chinese of Father Matteo Ricci S.J.", *Monumenta Serica*, 20 (1961), 84.

8 Johannes Keuning, "The History of Geographical Map Projections until 1600", *Imago Mundi*, 12 (1955), 7.

9 The world map of Rumold Mercator consists of two hemispheres, each having a diameter of about 25 cm. See Gerard Mercator, *Atlas sive cosmographicae meditationes de fabrica mundi et fabricati figura*, 4th ed. (Amsterdam, 1619), p. 29.

European vessels and the different kinds of animals scattered on the land, as well as in the sea. Indeed, the Verbiest map, being rich in ornament and in content, is more striking than the world map of Ricci (1602 ed. or 1603 ed.).

In respect to the sources of the Verbiest map, we have tried to present two possibilities: (1) the world map entitled *Diversi globi terraquei* drawn by Matthew Seutter of Augsburg in the middle of the seventeenth century and (2) the world map of Rumold Mercator of 1587[10]. It is remarkable that Verbiest's map, the *Kunyu quantu*, is quite similar to Rumold Mercator's world map, not only in being drawn by the same stereographic projection, but also in certain characteristic features of the Flemish maps[11] (for example, the filling of the Arctic regions by four large islands separated by narrow passages; the presentation of the supposed southern continent, unknown, and not yet discovered; the gulf running from the north to the middle of North America, and then forming smoothly as a great river to be connected with a somewhat inland lake). According to the study of Christine Vertente, the world map of the Dutch cartographer Johannes Blaeu (*Nova totius terrarum orbis tabula*, Amsterdam, 1648) probably served as the main source of the Verbiest map[12]. In 1974 Marcel Destombes also mentioned briefly this possibility, but he considered the revision of the Blaeu map of 1648 (this revision was on twelve scrolls and appeared in 1660) as the real source which Verbiest used to draw his Chinese map engraved on wood-blocks[13].

In order to accomplish his world map, Verbiest was obliged to make the best use of the previous works in Chinese of his predecessors of the Society of Jesus. He utilized, thus, a great number of published works of Ricci, Aleni, Niccolò Longobardo, Sabatino de Ursis and Alfonso Vagnoni. However, we conclude that the Ricci map of 1602 (the *Kunyu*

---

10 See Lin Tongyang, *op. cit.*, pp. 145-146.

11 For the characteristic features of Flemish maps, see E. Heawood, "The Relationships of the Ricci Maps", *The Geographical Journal*, 50:4 (October, 1917), 271; André Libault, *Histoire de la cartographie* (Paris, 1960), 38-39; Johannes Keuning, "XVIth Century Cartography in the Netherlands", *Imago Mundi*, 9 (1967), 53-63.

12 Christine Vertente, "Nan Huairen's Maps of the World", in *Conference Papers of the International Conference in Honor of Ferdinand Verbiest Commemoration of 300th Anniversary of his Death (1688-1988)* (Taipei, December 17-18, 1987), pp. 226-227.

13 Marcel Destombes, "Une carte chinoise du XVIe siècle découverte à la Bibliothèque Nationale", *Journal Asiatique*, 262 (1974), 210-211.

wanguo quantu) and the *Zhifang waiji* (Notes Supplementary to the Geography of the Ancients) of Aleni were used for the purpose of describing the five continents, and the *Biaodu shuo* (The Explication of the Gnomon) of de Ursis, as well as the *Kongji gezhi* (A Treatise on the Material Composition of the Universe) of Vagnoni were cited simply for the cartouches. It is difficult to know whether Verbiest profited by the world map of Francisco Sambiasi[14], although the Sambiasi map drawn early in 1633 at Nanjing had the same title as Verbiest's map.

If we accept the report of the Japanese historian Ayuzawa Shintaro, we have to admit the existence of two reproductions of the Verbiest map - the Guangdong edition (1856) and the Korean edition (1860)[15]. Nevertheless, Shintaro did not give any further information about the form and the contents of these reproductions. He insisted that the copy preserved in the Library of the Imperial University of Seoul was an example of the Korean edition, and that this example had only six basic scrolls, without the two outer scrolls (the first and the eighth) of the original Beijing edition of 1674[16]. Recently, Vertente pointed out that the Verbiest map of 1674 appeared in three different forms: (1) the *Kunyu quantu* of 1674 in eight scrolls; (2) the *Kunyu quantu* on one page dated 1674; and (3) the *Kunyu tushuo* on one page with the same date[17]. Apparently it is the first time that anyone has tried to classify the Chinese world map bearing Verbiest's name without making the distinction between the original production and the reproductions. According to the opinions of Vertente, the original edition of the Verbiest map exists in black and white, as well as in colored versions, and these colored copies are preserved in the National Library of Australia, the Municipal Library of Kobe, and the University Library of Seoul[18]. The copy belonging to the last-named institution must presumably be the same one declared and described by Shintaro in 1931. If this is true, the copy preserved in Seoul must be an example of the Korean edition of 1860. The single colored copy on one page (85 cm. x 110 cm.), which is preserved in the National Library of Paris

---

14 This map is reported by Bernard-Maitre, *op. cit.*, p. 347.

15 See Ayuzawa Shintaro, "Nan Kaijin ga Shina ni shokai shita chirisho ni tsuite" (The book of world geography introduced by Ferdinand Verbiest to China), *Chikyû* (The Globe), 24:5 (1931), 387.

16 *Ibid.*, p. 389, n. 1.

17 Vertente, "Nan Huairen", p. 225.

18 *Ibid.*, pp. 225, 229, n. 6.

(Rev. Ge. C5358), has been reproduced and inserted in the *Découvreurs et conquérants. La grande aventure des explorations* edited by Éditions Atlas under the direction of the French scholar Alain Bombard[19]. It is surprising that the third map, presented by Vertente, is also called the *Kunyu tushuo* - the original title of the world geography of Verbiest[20]. It is unlikely that Verbiest would confuse things in this disorderly way. In such circumstances, we can assume that the second and third maps mentioned above might only be the reproductions posterior to the appearance of Verbiest's map in 1674, because there was no reason for Verbiest to engrave simultaneously two world maps (even in color) on a reduced scale, in addition to his original world map in eight scrolls. In a word, it is almost impossible to see and compare all the copies (original production and reproductions) preserved in different institutions[21] (at least sixteen) all over the world.

3. The *Diqiu quantu* (A Map of the Terrestrial Globe)

It was Henri Bernard-Maitre (1889-1975) who reported the existence of a Verbiest map printed on a reduced scale in 1674[22]. This map is said to be made up of six scrolls, each measuring 55 cm. by 175 cm. From its title in Chinese (*Diqiu quantu*), it must be a world map consisting of two hemispheres drawn by stereographic projection. Until the present time, nobody has seen and studied it. However, if the deduction of Father Louis Pfister is correct, then a Verbiest world map existed before the drawing of the celebrated original map of 1674[23]. Still it is impossible to know whether this map prior to 1674 had any connection with the *Diqiu quantu* reported by Bernard-Maitre.

---

19 See *Découvreurs et conquérants. La grande aventure des explorations, 4:54 (1981), 1079.*

20 According to Christine Vertente, two specimens of this map entitled the *Kunyu tushuo* (black and white, 63.5 x 158 cm.) are preserved in Paris and Venice (see Vertente, "Nan Huairen", p. 226). The specimen at the Biblioteca Marciana in Venice has been reported by Professor Amelia Pezzali. See Joseph Dehergne, *Répertoire des Jésuites en Chine de 1552 à 1800* (Rome-Paris, 1973), p. 290.

21 I have presented my study under the title: "Aperçu sur la mappemonde de Ferdinand Verbiest le K'un-yü-ch'üan-t'u" to the 5th *Colloque International de Sinologie*, Chantilly, 15-18 September 1986. And in note 4, I have given a list of the different institutions or places where the maps of Verbiest are preserved.

22 Bernard-Maitre, *op. cit.*, p. 375.

23 Pfister, *op. cit.*, Vol. 1, p. 355.

4. The *Kunyu tushuo* (1674, 2 *juan*)

According to the reports of Pfister and Bernard-Maitre, this geographical book was published in 1672 in Beijing with a preface by Aleni[24]. This preface (without indicating the date) has been edited by Father Joseph Xu Zongze (1886-1947) in his catalogue of the Jesuits' Chinese works[25]. Aleni died early in 1649 at Yanping in Fujian, and therefore he could not have written a preface for Verbiest's work. In fact, the so-called preface of Aleni is copied summarily from Aleni's preface to his world geography (the *Zhifang waiji*). Ayuzawa Shintaro and Akioka Takeziro have hesitated to believe in the existence of such an earlier publication of Verbiest's work[26]. Besides, no person has spoken of its existence since the reports of the two Jesuit scholars. In this situation, we are obliged to accept the edition of 1674 as the original edition of Verbiest's world geography.

The format of this edition of 1674 measures 25.5 cm. by 13.5 cm.[27]. It was in Beijing that this book, divided into two *juan*, was engraved and printed. The author's name, with the title of his rank and the date of printing, are clearly marked on the inset prior to page six. It is remarkable that one can find three extra pages inserted in the first *juan* of this book. Two of these extra pages are devoted to an exposé concerning the cartographic methods for the world map. It seems, thus, that the compilation of this book was made immediately after the completion of the Jesuit's monumental world map. At the beginning of the first *juan*, Verbiest acknowledged having benefited from the works of his predecessors (Ricci, Aleni, de Ursis and Vagnoni)[28]. The so-called preface of Aleni is inserted in the second *juan*, just after the list of contents. If Verbiest added this reduced preface to his book, it was simply to show his respect to Aleni, because he obtained a great advantage from the latter's world geography.

---

24 *Ibid.*; Bernard-Maitre, *op. cit.* p. 374.

25 Xu Zongze, *Ming Qing jian Yesu huishi yizhu tiyao* (Taibei, 1958), pp. 320-321.

26 Ayuzawa Shintaro, "Nan Kaijin no *Konyo zusetsu* to *Konyo gaiki* ni tsuite" (A sudy of Verbiest's *Kunyu tushuo* and *Kunyu waiji*), *Chikyu* (The Globe), 27:6 (1937), 429; Akioka Takeziro, "Nan Kaijin no *Konyo zusetsu* ni tsuite" (A study of Verbiest's *Kunyu tushuo*), *Chiri kiyô iku* (Geographical Education), 29:1 (October, 1938), 7.

27 This measurement is made according to the holding preserved in the library of Fu Sinian of the Academia Sinica, Taibei, n° Jen-916-312.

28 *Kunyu tushuo*, 1674 ed., *juan* 1, p. 1a.

5. The *Kunyu waiji* (1 *juan*)

This is an illustrated book dealing with various subjects such as animals, plants, the seven wonders of the world, and other rare objects. Shintaro considered it to be compiled from the *Kunyu tushuo* by one who was fond of exotic and strange things[29]. There are two editions of the book: the *Shuoling* and the *Longwei mishu*.

6. The *Kunyu gezhi lüe shuo* (1676, 1 *juan*)

It was in 1676 that this book was engraved and published in Beijing, with a preface by Xu Erjue (1603-1680), a devout Christian and the eldest grandson of Xu Guangqi. It seems to have nothing to do with the *Kunyu gezhi*[30] presented to the Chongzhen emperor by Johann Adam Schall von Bell in 1640. In fact, Schall's work is merely a treatise on the exploitation of mines. This last work of Verbiest was copied from the *Kunyu tushuo*, but the majority of the geographical descriptions were omitted. It contains only generalities about Asia, Europe, Libya, America and Magellanica. Today, the printed copies of this book are very rare, and there are no reproductions. That is the reason why it is not very known to the reading public.

II.
The Character of Verbiest's Works and Their Importance

On the whole, Verbiest's geographical works - excluding the *Xifang yaoji* - include almost all the contents of the fourteen cartouches inserted in the Verbiest map of 1674. Evidently the *Kunyu gezhi lüe shuo* (1676) was derived from the *Kunyu tushuo* (1674). The distinction between these books is that the latter is augmented by the addition of a second *juan*. Moreover, this extra chapter is devoted entirely to a discussion of the geography of the five continents, including the great southern continent of Magellanica. It seems that the sources of this chapter come uniquely from Aleni's *Zhifang waiji*, and not from the Verbiest map. In fact, the annotations and descriptions concerning the five continents in

29 Ayuzawa Shintaro, *op. cit.*, p. 430.

30 According to Henri Bernard-Maitre, this book was finished on July 20, 1640 and then presented to the Chongzhen emperor of the Ming dynasty. In compiling it, Adam Schall had adapted the work of Agricola, *De re metallica* (Bâle, 1556).

Verbiest's map are already summarized. For instance, the description of Magellanica in Aleni's world geography has 432 Chinese characters, while the Verbiest map and the *Kunyu tushuo* have 217 and 365 respectively. The fact that Verbiest added thirty-two illustrations in woodcut to this world geography (the *Kunyu tushuo*) is quite remarkable. Still, we need to note that most of these illustrations, with their annotations, are taken primarily from Verbiest's map, and afterwards constitute the essential contents of the *Kunyu waiji*. In short, all the geographical works of Verbiest, with the exception of his first one (the *Xifang yaoji*), are the fruits of his celebrated world map.

At times in Beijing, certain scholars repeatedly asked the Flemish Jesuit: "As the territory of China is so vast and immense, how can you place her on the world map in such a strict way?"[31]. In order to make it easy for the Chinese to understand the evaluation of the geographical location of China and its measurements on the world map, Verbiest wrote a short exposé about cartographic methods. He tried to explain the principle of the parallels (longitude and latitude), as well as the correlation between the parallels and the celestial body. We have to consider that the introduction of these notions to Chinese society was very important, because the application of survey methods to cartography in China (for example, the method of *zhunwang*, *fenlü*, etc...) lacked astronomical and mathematical foundations[32]. In fact, the method of determination of latitudes existed early in the Tang dynasty, and the conception of the sphericity of the earth was not unfamiliar to the Chinese since the Han astronomer-mathematician Zhang Heng (78-139) invented the ancient *huntian* theory[33]. Nevertheless, it is quite strange that Chinese cosmology had affected Chinese map-making so little.

Because of the difficulties of converting the Chinese to western beliefs, Verbiest emphasized in his works the importance of European civilization in order to make an impression which would improve, in some degree, the image of Europe, and thus obtain the respect of his contemporaries for Christianity. At the end of his work (the *Xifang yaoji*), Verbiest spoke voluntarily of the *jiaofa* (religious teachings) and the *xishi* (western scholars). According to his own description, Catholicism had dominated Europe for more than 1600 years, creating

---

31 *Kunyu tushuo*, 1674 ed., *juan* 1, p. 5a *bis* - the page prior to p. 6.

32 For this question, see Joseph Needham, *Science and Civilisation in China* (Cambridge, 1959), 3:438, note a, 543-552, 576.

33 Zhen Zungui, *Zhongguo gudai tianwenxue jianshi* (Shanghai, 1955), pp. 171-175.

an admirable, prosperous western world where peace and political stability reigned. He insisted, also, that the Jesuits had already devoted their lives to mission work in the Chinese empire since Francis Xavier sailed over the South China Sea and died near Guangzhou[34]. Shortly after the incident of Yang Guangxian in 1669, Verbiest presented his first geographical work to the Kangxi emperor for the "imperial reading" (*yulan*).

Just like Ricci's map of 1602 (oval projection), the *Kunyu quantu* possesses several characteristic features of the maps of the Flemish school (*école flandrienne*), which was influential in the sphere of Renaissance cartography. The Ricci map is based upon Ortelius's planisphere of 1570[35] (*Theatrum orbis terrarum*), but Verbiest's map follows the example of the world map of Rumold Mercator (1587). The maps of both Verbiest and Rumold Mercator are divided into two parts, one for each hemisphere. In both maps there are, within the Arctic regions, four large islands separated by narrow passages. Moreover, on both oriental hemispheres of the maps mentioned, there are two notes attached to the large Arctic islands. One of them states that, owing to the rapidity of the tidal current, the water does not freeze in winter. Here, however, we have to mention that this note, in Chinese on the Verbiest map, is derived from Ricci's map of 1602. Finally, on both maps of Rumold Mercator and Verbiest, the Nova Zemlia (*Novaya Zemlya*) explored by the Dutch navigator Willem Barentz (1550-1597) in 1596[36] is presented similarly in the Arctic Circle.

By following the precedents of Ricci, Verbiest performed his obligation to respect the traditional concept of *tianxia* (that what is under heaven) when he drew his own map. In order to place China and its surrounding territories toward the center of the world map, Ricci moved the prime meridian of the Fortunate Islands (Canary Islands) from the center of the traditional European planispheres to the extreme left. As a result, Ricci centers his map on the meridian of 170° E., which passes east of New Guinea. Because of the method of stereo-

---

[34] *Yulan xifang yaoji* (Xuehai leibian ed., 1831), pp. 10b-11b.

[35] This world map of Ortelius is an oval projection measuring 23.5 cm. by 49.5 cm. See Ortelius, *Theatrum orbis terrarum*. Photolithographic Edition according to the original preserved at the University Library of Leiden (Lausanne, 1964), p. 1.

[36] For the exploration of Barentz, see Mercator, *Atlas sive cosmographicae meditationes* ..., pp. 41-42; Jean Amsler, *Histoire universelle des explorations* (Paris, 1955), 2:378-385.

graphic projection used for drawing the oriental and occidental hemispheres, Verbiest had to find another solution. Therefore, he made the prime meridian run through Beijing instead of the Canary Islands.

If certain resemblances exist between both maps of Ricci and Verbiest, it would be chiefly the absence of geographical knowledge relating to the regions in the vicinity of North China. In 1602 Ricci had extracted and utilized Ma Duanlin's *Wenxian tongkao* (1322) as the main source for his account of seven notices, which have been discussed by Kenneth Chen (Chen Guansheng)[37]. It appears that Verbiest had no interest in adopting these geographical legends into his own map, due to their incorrectness. Thus he kept only the legend of a certain primitive tribe named *Wajiezi*, and placed it on the east coast of Siberia (50°-60°). Considering the political situation that had evolved in China since the fall of the Ming dynasty, Verbiest had to efface an annotation concerning the Nu'ergan (or Nurgan, a commandery set up by Yongle in 1404 on the side of the lower Amur River), placed previously in North Manchuria on Ricci's map of 1602. Simultaneously, with the effacement of this out-of-date note, Verbiest named the East China Sea "Da Qinghai", instead of "Da Minghai".

Between 1682 and 1683, Verbiest was chosen by the Kangxi emperor to accompany him on the imperial tour to Manchuria and Mongolia, and recorded nearly every important detail of these voyages[38]. During his first tour to Manchuria (March 23 - June 6, 1682), Verbiest brought with him all the necessary mathematical instruments and, in the emperor's presence, made celestial observations for determining, bit by bit, the state of the heavens and the elevation of the pole of every town and city along the way. It was during May 11-16, 1682[39] that Verbiest measured the latitude of the city of Shenyang (Mukden) and placed it at

---

37 Kenneth Chen, "A Possible Source for Ricci's Notices on Regions near China", *T'oung Pao*, 34:3 (1938), 182, 190.

38 For the descriptions of these two voyages, see Du Halde, *Description de l'empire de la Chine* (La Haye, 1736), 4:85-102. The Chinese translation of Verbiest's voyage in Manchuria is included in the *Qingdai xiren jianwen lu* (ed. by Du Wenkai) (Beijing, 1985), pp. 69-84. Henri Cordier has emphasized the importance of these two voyages: "Au point de vue historique, le plus intéressant de ces volumes (Du Halde, *Description de l'empire de la Chine*) est le quatrième, qui renferme les voyages des P.P. Gerbillon et Verbiest en Tartarie" - Henri Cordier, *Mélanges d'histoire et de géographie orientales* (Paris, 1914), 1:83.

39 This date is based on the notes of voyage of Gao Shiji, *Hucong dongxun rilu*, the facsimile reproduction of Guangwen (Taibei, 1968), pp. 32-37.

41°56' N.[40] - surpassing Beijing by about two degrees[41]. On concluding his survey of the geographical location and situation of Manchuria, he was convinced that it would be possible, for missionary purposes, to reach Korea and Japan by this overland route and by sea[42]. However, all this geographical information that he observed in the land of the Tartars was gathered eight years after the publication of his world map. As to the presentation of Siberia, the river systems of the Ob, Yenisei and Lena running north into the Arctic Sea are well-shown, but Verbiest did not give further details about them. The first military encounters between China and Russia along the Amur River took place in 1652[43], just twenty-two years prior to the presentation of Verbiest's map to the Kangxi emperor. When the Russian envoy Nikolai Govrilovich Milescu Spathary arrived at the court of Beijing on May 15, 1676, Verbiest was appointed by the emperor as interpreter, and thus rendered Spathary valuable services[44]. Verbiest seems to have been the first one to dream of searching for a trans-Siberian route from China to Europe[45]. Of course, in 1674 Verbiest could not gather any important geographical information about Siberia for his own map. In fact, the first scientific map of Siberia, *New Geographical Map of the Northern and Eastern Regions of Asia and Europe* (*Nieuwe Landkaarte van het Noorder en Ooster deel van Asie en Europa*), drawn by the Dutch cartographer Nicolaas van Witsen, appeared thirteen years later

---

40 See Du Halde, *op. cit.*, 4:90; Ruan Yuan, *Chouren zhuan* (Biographies of the Chinese and Western Mathematicians), 1799, *juan* 45, p. 20a; Zhang Qin, *Kangxi zhengyao* (Kuangxu ed.), *juan* 18, p. 19A.

41 Some years later, the emperor mistook the latitude of Beijing (39°56' N.) for that of Shenyang. See "Lettre du Père Chalier au Révérend Père Verchère", *Lettres édifiantes et curieuses* ... (Paris, 1843), 3:763; Antoine Gaubil, *Correspondance de Pékin 1722-1759* (Genève, 1970), p. 541.

42 Gaubil, *op. cit.*, p. 222.

43 Cf. Vladimir Miasnikov, *L'Empire des Qing et l'état russe au XVIIe siècle* (Moscow, 1985), for the French translation, pp. 94-95, 102.

44 Miasnikov, *op. cit.*, p. 185; Gaston Cahen, *Histoire des relations de la Russie avec la Chine sous Pierre le Grand (1689-1730)* (Thèse de l'Université de Paris, 1911), pp. 20-25; Joseph Sebes S.J., *The Jesuits and the Sino-Russian Treaty of Nerchinsk (1689)* (Rome, 1961), pp. 65-67.

45 H. Josson, S.J. et L. Willaert, S.J., *Correspondance de Ferdinand Verbiest de la Compagnie de Jésus (1623-1688) Directeur de l'Observatoire de Pékin* (Brussels, 1938), Introduction, p. 14; Sebes, *op. cit.*, pp. 95-96; Jonathan Spence, *To Change China: Western Advisers in China* (Harmondsworth, 1980), p. 30.

(1687)[46]. It is certain that Verbiest wanted to persuade the Chinese that the earth is spherical in shape, and for this reason he drew his world map in two large hemispheres, one showing Eurasia and Africa, and the other, America. In the days of Ricci, the Chinese were, indeed, impressed by hearing the truth of terrestrial sphericity. For example, in 1602 the brilliant scholar Li Zhizao (1566-1630) wrote: "It is said that the water of the seas is attached to the earth, and the earth is shaped like a sphere. Thus I learn with amazement that mankind lives on this spherical earth"[47]. And between 1664 and 1669, Yang Guangxian (1597-1669), a *buyi* (civilian) of Jiangnan, fought the Jesuits in Beijing (Schall, Verbiest, Magalhaes and Buglio), and criticized severely their astronomical methods of calculation for the Chinese calendar devised by Schall at the end of the reign of the Shunzhi emperor. Yang, without knowing anything of western science, sought to make an attack upon the conviction of terrestrial sphericity expressed in the *Yudi tu*[48], a world map drawn by Schall in 1636. He wrote: "The fault of western methods originated from the map of its religion, indicating that the earth is shaped in the form of a sphere which is in the midst of space"[49]. Although Ricci had drawn "two small maps"[50] (*doi mappamondi piccoli*) - the oriental and occidental hemispheres made by orthographic projection in Beijing towards 1604 - he did not have them engraved on wooden blocks. It seems that copies of this world map were scarcely known in the intellectual circles of the seventeenth century. In this

46 Cf. Gaston Cahen, "Les cartes de la Sibérie au XVIIIe siècle", *Nouvelles Archives des missions scientifiques et littéraires* (1911) 19:5-6, 11, 57; Pierre-Jacques Charliat, *Histoire universelle des explorations* (Paris, 1955), 3:82-83.

47 Preface of Li Zhizao to the Ricci map of 1602, see d'Elia, "Il Mappamondo cinese del P. Matteo Ricci", Tavola XI.

48 This world map, drawn as a conic projection, is included in the *Huntianyi shuo* (Construction and Usage of Celestial and Terrestrial Spheres), *juan* 5. For the details of it, see Kazutaka Unno, "To-Jakubo oyobi sho-Yujin no sekaizu ni tsuite" (J. Adam Schall von Bell's and Michel Benoît's Maps of the World), *Jimbun-chirigaku no shomondai* (Problems of Human Geography), edited by the Committee for the Commemoration of Professor S. Komaki's 70th Birthday, s.l., 1968, pp. 84-87.

49 Yang Guangxian, *Budeyi* (I Cannot Do Otherwise), photolithographic edition of Zhongshe, 1929, *juan* 2, p. 63a.

50 See Pasquale M. d'Elia, *Fonti Ricciane* (Rome, 1942-1949), 2:165-166. The reproduction of this world map, consisting of two hemispheres (each measuring 26.5 cm.) is inserted in the *Fangyu shenglüe* (1610 ed.), *Waiyi dili* (The Geography of the Outer Barbarians), *juan* 1, pp. 3b-4a, 4b-5a.

ambience, Verbiest might have made up his mind to draw a world map using stereographic projection in order to show the Chinese the sphericity of the earth.

The differentiation between the maps of both Ricci (1602 or 1603) and Verbiest is decided not only by the separate methods of projection, but also by the attribute of possession. In the maps of Ricci, there are always certain characteristic seals of the Society of Jesus[51] (oval and square), and his own signature: *Ouluobaren Li Madou* (The European Matteo Ricci), *Yesuhui zhongren Li Madou* (Matteo Ricci of the Society of Jesus). In contrast, Verbiest has marked only the title of his rank and his Chinese name on the bottom of the exterior circle of his oriental hemisphere. Thus it appears that the Verbiest map possesses a certain official character. Somehow, the Jesuit map still lacks an author's preface, as well as the possible prefaces of Chinese scholars. We know that all the various editions of Ricci's maps were to be engraved and published under the direction of Wang Pan, Wu Zuohai, Li Zhizao, and Li Yingshi. These men bore individually the cost of each publication, and wrote separately their graceful prefaces to Ricci's maps (Wang Pan did not write his preface). Since Verbiest worked in the Bureau of Astronomy and participated in other imperial services at the will of the emperor, he would have been prevented from enjoying contacts with the elite of Chinese society. Indeed, his scientific contribution to the court had nothing to do with the intellectual circles of his day. This is the reason why Verbiest did not profit from the geographical information (especially of East Asia) known to his Chinese contemporaries. The Flemish Jesuit gave up the correct contours of Japan, and ignored completely the toponyms inserted in the Ricci map of 1602. Then, for the geography of Japan, Verbiest transliterated certain western toponyms into Chinese characters, which are very strange to read. According to Vertente, these strange place-names in Chinese are derived from the map of Blaeu[52]. It can be argued that Verbiest might have been ignorant of the existence of the Chinese sources concerning Japan. The Australian historian W.Z. Mulder has also insisted that the Jesuit did not make the best use of the Chinese material available on the Philippine Islands[53]. In fact, since the establishment of Manila in 1571, the

---

51 Cf. d'Elia, *"Recent Discoveries and New Studies..."*, pp. 113-114.

52 Vertente, "Nan Huairen", *op. cit.*, p. 227.

53 W.Z. Mulder, "The Philippine Islands in the Chinese World Map of 1674", *Oriens Extremus*, 25:2 (1978), 219.

Chinese merchants of Fujian put on shore at Luzon frequently. At the beginning of the seventeenth century, Zhang Xie (1574-1640), author of a maritime geography of South Asia, described vividly the activities of the Chinese immigrants in Luzon, and the miserable incident of the massacre of Manila in 1603[54]. The Jesuits in China would not talk about the situation of this insular country colonized by the Spanish, on account of the ardent xenophobia of the Chinese.

Verbiest has not engraved on his map, as Ricci did, the diagram representing the notion of the nine celestial spheres based on the traditional Ptolemaic astronomy, the table giving the declination of the sun in degrees for any day of the year, the explanation of eclipses of the moon, the short discourse on the Four Elements (earth, air, fire and water), etc. ... Nevertheless, in the cartouche entitled *Diqiu nanbei liangji* (The Poles of the Earth), Verbiest wrote at the beginning: "The center of the earth is equally the center of the celestial body." The Jesuits in China - just as their contemporaries in Europe - did not dare to proclaim openly the new theory of the solar system established by Nicolaus Copernicus (1473-1543). In fact, the conservative attitude of the Catholic Church, rigorously interpreted to the disadvantage of the Copernican theory, was brought to China by the Jesuits with the arrival of Ricci in Guangdong. The Copernican quarrel had, perhaps, an even more tragic history in China than in Europe, because it went on until the second half of the eighteenth century, when the French missionary Michel Benoît spoke freely of the Copernican heliocentric system in his world map of 1761[55].

On the whole, Verbiest's map presents, to a certain degree, an artistic beauty because of the equilibrium of its two hemispheres, as well as the fourteen cartouches decorated in both Chinese and western styles. We see various kinds of aquatic and terrestrial animals illustrated everywhere, and ships sailing over the high seas. The illustrations of these animals, unknown to the Chinese, are undoubtedly destined to satisfy their curiosity - but the presentation of the three-masted sailing-ships must have a certain definite significance. Thus we see a fleet composed of three barques going on a voyage through the South Seas, and an isolated one navigating out of the east coast of North America - as if it was on its way back to Europe. It seems that the presentation of

---

54 Zhang Xie, *Dongxiyang kao* (1888), *juan* 5, 1a-7b.

55 Boleslaw Szczesniak, "Notes on the Penetration of Copernican Theory into China", *Journal of the Royal Asiatic Society* (1945), 31-32; Lin Tongyang, *op. cit.*, 2:415-416.

these Spanish sailing-ships is to indicate two of the most important inter-continental trade-routes of the seventeenth century: the galleon trade between Acapulco and Manila[56], and the shipping-route between Europe and the east coast of North America (the Dutch, English and French colonies)[57]. In order to match the illustrations of the vessels, Verbiest also wrote a description entitled *Haibo*[58] (The Vessels) dealing with ship-building, the organization of the voyages, the art of navigation, etc... With regard to the nautical methods we read: "The astronomers occupy themselves with observing the celestial phenomena. They calculate the sun by day and the stars by night. Making use of the sea-chart, they determine the exact degrees in order to see the dangers and the distances."

Regarding the introduction of natural geography, Verbiest contributed greatly to its development in China. He gave systematically, in his fourteen cartouches, an explanatory description of such natural phenomena as the shape of the earth being a spherical body, the periodic oscillation of the tides due to the effect of the moon which moves round the earth, the formation of rain and clouds caused by the condensed moisture of the atmosphere, the formation of the winds and monsoons which influence navigation on the high seas, the regional diversities of culture determined by different weather conditions, the origin of the rivers which give life to mankind, the formation of the mountains, the movements of the air, etc... The Chinese certainly had previous knowledge of some of these matters. For example, in the eighth century Dou Shumeng, a native of Zhejiang, compiled a work entitled *Haitao zhi* (A Brief Description of the Tides), in which he explained the change of the tides in connection with the movements of the moon[59]. Having realized the value of the climatic division into five zones described by Ricci in his map of 1602, Li Zhizao and Fang Yizhi (d. 1671) demonstrated the fact that the geographical location of China in the temperate zone (19°-

---

56 For this subject, see especially Pierre Chaunu, "Les Philippines et le Pacifique des Ibériques (XVIe, XVIIe, XVIIIe siècles)", *Introduction méthodologique et indices d'activité* (Paris, 1960), 241-269; "Le galion de Manille. Grandeur et décadence d'une route de la soie", *Annales E.S.C.*, 4 (October-December, 1951), 447-462.

57 Cf. Fréderic Mauro, *L'Expansion européenne (1600-1870)* (Paris, 1967), pp. 147, 153, 156-159; Amiral de Brossard, *Histoire maritime du monde* (Paris, 1974), 2:66-79.

58 This description is derived from the *Haibo* of the *Zhifang waiji*, *juan* 5.

59 *Zhongguo kexue wenming shi* (Taibei, 1983), pp. 340-341.

42°) was very beneficial to Chinese cultural progress[60].

It is remarkable that Verbiest criticized without hesitation the Chinese superstitious attitude towards natural phenomena. He added a diagram of the lunar eclipses (*yueshi tu*) to one of his fourteen cartouches, so as to contradict the faulty legends concerning the occurrence of eclipses of the moon. Moreover, Verbiest attacked the fatalism vis-à-vis earthquakes being caused by the earth-dragon, or the huge sea-turtle which was said to support the balance of the earth according to ancient tradition. The Jesuit offered then a dynamic theory of earthquakes: "Due to the effect of the air condensed and squeezed in the interior of the earth, violent movements of this inner force take place in the earth and, as a result, earthquakes happen normally."

We need to pay attention to the presence of the so-called five continents in Verbiest's map. We will briefly discuss certain particularities, and try to point out their importance. The general outline of East Asia, especially the curved figure of the Chinese east coast, seems more exact than that of Ricci's map of 1602. The Gulf of Bohai together with the Liaodong and Shandong Peninsulas, are well presented, and correspond to geographical reality. In the Ricci map, the Liaodong Peninsula appears as a tiny promontory attached to southern Manchuria, and the oblong figure of the peninsula of Korea is exceedingly close to the waters down near the mouth of the Yangzi River. In this regard, Ricci based his map upon Luo Hongxian's general map of China[61] in the four-volume edition of the *Guang yutu* of 1579. The Flemish Jesuit might have used Martino Martini's map of China as edited and inserted in the revision of Blaeu's world map of 1660[62]. It is almost impossible to make a distinction between Verbiest's map and the maps of Aleni (the *Wanguo quantu* and the map of Asia) as far as the outline of the East China coast is concerned. We assume that Verbiest might have profited from his ease in consulting the details of these maps of Aleni. Obviously, Aleni himself might have used a certain European map for his map of Asia, because he transliterated a single geographical

---

60 Li Zhizao, *Hungai tongxian tushuo* (A Treatise on the Stereographic Projection of the Celestial Sphere), *Tianxue chuhan*, photolithographic edition of the Xuesheng shuju (Taibei, 1965), 3:1741-1742; Fang Yizhi, *Wuli xiaozhi*, 1664, *juan* 1, pp. 3a-3b.

61 This map of China is entitled *Yudi zongtu* and measures 26 cm. by 20 cm., see Luo Hongxian, *Guang yutu* (Qiandai ed.), 1579, *juan* 1, pp. 1b-2a.

62 Destombes, *op. cit.*, pp. 210-211; Boleslaw Szczesniak, "The Seventeenth-Century Maps of China: An Inquiry into the Compilation of European Cartographers", *Imago Mundi*, 13 (1956), 116.

term into Chinese (Miyake, placed in Japan) from a certain western writing. Although Aleni's way of transliterating is similar to that of Verbiest, they did not use identical cartographic sources. According to Helen M. Wallis and E.D. Grinstead, the presentation of Eastern Asia on the David globe (painted in lacquer on wood, 59 cm. in diameter) is notably improved. Consequently, it is better than that of Ricci's map of 1602[63]. Comparing the presentation of the three great river-systems (the Yellow River, the Yangzi and the West River), and particularly the Great Wall in two maps of Ricci and Verbiest, we note that these maps do not have any striking differences. It is also important to note that Taiwan, as conquered and colonized by the Manchus in 1683, is already presented in the Verbiest map with the Tropic of Cancer passing through the middle of this fertile island. This is the first time that Taiwan is correctly shown in a Chinese world map. In contrast, Ricci named the island Da Liuqiu (the Great Ryukyu Island) in his famous map of 1602, and he did not give its outline so well as Verbiest. Of course, Verbiest left off the inscription *Da Ming yitong* (which is engraved on the interior of China in the Ricci map of 1602), as well as a short introduction of the Ming empire (45 Chinese characters), without writing a new one to replace it.

In spite of the fact that the west coast of New Holland[64] (*Hollandia Nova*), discovered by the Dutch captain Willem Janszoon in 1605, has been clearly presented on the oriental hemisphere of the Verbiest map, we still see the supposed Great Southern Continent (*Terra Australis*), which provoked a large number of discussions, especially during the epoch of the great trans-Pacific voyages (1550-1800)[65]. Moreover, just like his contemporary cartographers in Europe, Verbiest connected New Holland (*Xin alandiya*) to New Guinea (*Xin xinieya*), to which he added a notice saying: "This country is called New Guinea because it is similar in shape and appearance to Guinea in Africa. Europeans have only recently arrived in this country; hence, it is not yet known for

---

63 Helen M. Wallis and E.D. Grinstead, "A Chinese Terrestrial Globe A.D. 1623", *British Museum Quarterly*, 25 (June, 1962), 87.

64 Verbiest transliterated *Hollandia Nova* into Chinese as *Xin alandiya* and placed it between 18°-42° S. He put the Chinese name *Demensi* (Diemens Landt ?) on the northern coast, and the name *Neidesidi* (Terre de Diemens) on the southern coast. For the map of *Hollandia Nova* of the seventeenth century, see Melchisédec Thévenot, *Relations de voyages curieux* ... (Paris, 1696), 1:50 (the map is placed under page 50).

65 For this interesting subject, see Armand Rainaud, *Continent Austral. Hypothèses et découvertes* (Paris, 1893).

certain whether it is a piece of the continent or an island"[66]. Indeed, it is strange that Verbiest was not fully acquainted with the fact of the insular character of New Guinea, as revealed by his senior Aleni early in 1623[67]. To our surprise, Aleni located this equatorial island in its exact position from 1° to 10° S., and from 165° to 190° longitude east of the Canary Islands on his world map of 1623 (the *Wanguo quantu*). It was between March 31 and May 28, 1623 in Beijing, that Longobardo (1559-1654) and Manuel Dias Jr. (1574-1659) completed the construction of the so-called David globe in the form of a wooden ball, on which these two Jesuits primarily revealed the striking discovery of the Torres Strait by the Spanish navigator Luis Vaez de Torres in 1606[68]. But it was not until 1762 that the insular character of New Guinea was known to the public. At that time the English attacked and seized Manila, where the nautical and geographical documents written by Torres himself were conserved. Verbiest seems not to have recognized the value of this important geographical discovery. Moreover, the fact that Verbiest drew at will New Holland as an enormous terrestrial mass is radically contrary to the historical and cartographical information available in the middle of the seventeenth century. In doing so, he probably wanted to give a possible contour of this newly-discovered land. On the occidental hemisphere of Verbiest's map, we notice the presence of New Zealand (*Xin selandiya*, *Zeelandia Nova*), found by the Dutch mariner Abel Tasman (1603-1659) in 1642. In a word, the Flemish Jesuit Verbiest was the first one to introduce the discovery of New Zealand and Australia to China.

III.
The Influence of Verbiest's Works

It was his belief in the improvement of Chinese understanding of western culture and religion that led Verbiest to undertake his geographic and cartographic works. Early in 1669, after the incident of Yang Guangxian, Verbiest started compiling his first work (the *Xifang yaoji*) dealing with diverse subjects about Europe. Five years later he

---

66 This translation is cited by Lionel Giles, "Translations from the Chinese World Map of Father Ricci", *The Geographical Journal*, 53:1 (January, 1919), 22.

67 Giulio Aleni, *Zhifang waiji*, in *Tianxue chuhan*, photolithographic edition, 3:1467-1468; Lin Tongyang, *op. cit.*, 1:106-107.

68 Wallis and Grinstead, "A Chinese Terrestrial Globe", *op. cit.*, pp. 87-88.

drew his famous world map, which is still a splendid piece of cartography in China. From this map, he issued three additional geographical works: the *Kunyu tushuo*, the *Kunyu gezhi lüe shuo*, and the *Kunyu waiji*. Before the fall of the Qing dynasty, each of Verbiest's works was reproduced and printed two or three times, except the *Kunyu gezhi lüe shuo* and the *Diqiu quantu*. As far as we know at present, the total of these wood-cut reproductions comes to at least ten. The quantity of these reproductions reflected, to some degree, the popularity of the Flemish Jesuit's works in China before the end of the nineteenth century. Moreover, it is worth noting that Verbiest's *Kunyu tushuo* was compiled twice in the imperial collections: the *Gujin tushu jicheng* (1726) and the *Siku quanshu* (1773-1782). Bernard-Maitre considered Verbiest's works (not including the works on religion) as virtually "productions of the imperial printer"[69]. This is mostly true for Verbiest's astronomical studies, but it was not always true for his geographic works. Among all of Verbiest's geographic and cartographic works, only two original editions (the *Xifang yaoji* and the *Kunyu quantu*) could have been engraved and printed at the imperial palace. There is no evidence to prove whether or not the others were engraved.

Just like Ricci and Schall, Verbiest was well-versed in science and technology. He won the respect and admiration of the Chinese by his extensive works on astronomy, geography, and cartography, and also by his production of cannons. Doubtlessly, the prestige of the Jesuits in China was indispensable to support and commend the western religion. About fifty years after the death of Verbiest, the French Jesuit Antoine Gaubil insisted in one of his letters from Beijing to Étienne Souciet (August 31, 1734), that the interest in the Church from the earlier missionaries to Verbiest was "planted and conserved by the favor of astronomy"[70]. It is true that after Verbiest started occupying a post in the Imperial Board of Astronomy (*qintianjian*) in the spring of 1669, that he began to receive particular esteem from the Kangxi emperor "by his virtue and science"[71]. During the years of religious persecutions incited by the hostility of Yang Guangxian, the four Jesuits in Beijing were arrested, and the other twenty-five missionaries in the provinces were

---

69 Bernard-Maitre, *op. cit.*, p. 369.

70 Gaubil, *op. cit.*, p. 389.

71 Charles Le Gobien (1653-1708), *Histoire de l'édit de l'empereur de la Chine en faveur de la religion chrétienne* (Paris, 1698), pp. 10-11.

exiled to the city of Guangzhou[72]. How could they solve this intolerable situation? The Belgian historian Henri Bosmans stated that there was a unique issue for the Jesuits at that time: how to persuade the emperor of the superiority of western science, to show him its advantages and to present him the offering of services in exchange for the liberty to preach the Christian religion[73]. On this condition, Verbiest became an intimate tutor of the emperor, and taught the sovereign European science, including geography and cartography[74].

For a long time historians have neglected Verbiest's influence on the formation of the Kangxi emperor's vision of the world. At the middle of the seventeenth century, China was inevitably compelled to face the unexpected agitation caused by the coming of western powers to its gates, particularly the Russians on the northern border and the Dutch in the eastern seas. In May 1669, when the Yang Guangxian incident had come to an end, Verbiest and his companions de Magalhaes and Buglio presented their collective geographic work (the *Yulan xifang yaoji*) to the emperor. This young sovereign, who had ruled in person since 1667, was anxious to know about foreign affairs as well as the domains of his own realm. Attracted by the book offered by the three Beijing Jesuits, he summoned them twice (May 20 and May 30, 1669) to the imperial palace, and asked them "numerous questions regarding the things of Europe"[75]. Between the court receptions of the Portuguese ambassador Don Manoel de Saldanha in 1670 and the Russian envoy Spathary in 1676, Verbiest drew his world map in order to satisfy the emperor's needs. Thus it was Verbiest who had the opportunity to create a certain vision of the world for the sovereign of the Chinese empire. After the Flemish Jesuit passed away in 1688, the emperor kept in constant contact with other missionaries, especially the French Jesuits, so that his understanding of the world could be improved and perfected. Towards the end of his rule, the emperor already possessed a wide knowledge of geography, and often discussed the maritime intercourse between China and Europe; he even predicted: "Beyond the seas,

---

72 Bosmans, *op. cit.*, pp. 20, 55-56; Huang Bolu, *Zhengjiao fengbao* (Public praise of the true religion) (Shanghai: Cimutang, 1884), 1:53.

73 Henri Bosmans, "La notice nécrologique de Ferdinand Verbiest par son secrétaire Antoine Thomas de Namur ...", *Annales de la société d'émulation*, 2 (1914), 103.

74 Joachim Bouvet, *Histoire de l'empereur de la Chine* (La Haye, 1699, reprinted Tianjin, 1940), p. 83.

75 Bosmans, *op. cit.*, p. 46.

the western countries will some day constitute a real menace to China. This is what we say in advance"[76].

It was perhaps natural that Verbiest, having decided to lay the foundations of western science in China, should have certain influence on the attitudes of the emperor towards geographical methods. When the Jesuit was guided to Manchuria in the spring of 1682, he carried out a number of astronomical observations in the emperor's presence, so as to determine the position of certain cities (Mukden, Kirin, etc.)[77]. This scientific experiment gave the emperor an example of the relation between practical measurements and the geographical reality. The emperor thus began to pay attention to geographical methods, and asked the Flemish Jesuit Verbiest to show him how to use astronomical instruments.

According to an account by Jean de Fontaney, the emperor, in 1689, visited the city of Nanjing, where the French Jesuit was staying, and personally made an astronomical observation of the south star (*laojinxing*, *le Canope*) from the top of Guanxingtai (the old Nanjing Observatory).[78] Later on the emperor gave an imperial order to his third son (the hereditary prince Yinzhi) to establish an imperial academy of mathematics[79] (*suanxueguan*) at the summer palace in order to offer scientific instruction to men chosen from distinguished banner families, and to Chinese astronomer-mathematicians (e.g., Chen Houyao, Mei Gucheng, He Guozong). The emperor's generous support of the study of western science stimulated, without doubt, a revival of Chinese mathematics in the eighteenth century. Besides this, a few of the men trained in the Academy of Mathematics were to take part in the full cartographic survey of China. Among them were two lamas who would be commissioned by Kangxi to map out Tibet in 1711[80]. The imperial taste for western geographic measurements can be thoroughly seen from the dialogues between the emperor and the mathematician Chen Houyao (a native of Taizhou, *jinshi* of 1654) during their voyage to Jehol in the

---

76 *Donghua lu* (Donghua records), comp. Wang Xianqian (1887), *juan* 19, p. 27a; *Kangxi zhengyao*, *juan* 18, pp. 5b-6a.

77 Du Halde, *op. cit.*, 4:87.

78 De Mailla, *Histoire générale de la Chine ou Annales de cet empire, traduites du Tong-Kien-Kang-Mou* (Paris, 1780), 11:317, n. 2.

79 Fang Hao assumed the *suanxueguan* (Academy of Mathematics) would be incorporated into the Mengyangzhai. See Fang Hao, *Zhongxi jiaotong shi* (Taibei, 1959), 4:202.

80 Du Halde, *op. cit.*, 1:571.

summer of 1709. The emperor asked the eminent scholar some questions about determining latitudes and the diameter of the earth[81].

The presentation of Verbiest's map of 1674 to the throne was the beginning of the emperor's contact with scientific cartography. After that, the emperor took notice of the value of maps as essential to the governance of the state. He said:

> From the time of our youth, we have paid attention to geography. By studying maps and records, and widely investigating local opinions, we have made a point of ascertaining the names of all hills and streams from antiquity until now, even those in the frontier regions and remote deserts ...[82].

In February 1694, he declared to the secretaries of the Imperial Chancery and others: "We have often made inquiries about regional situations, and examined the maps attentively. That is why we know every important place and town on the frontier in detail"[83]. In fact, the Kangxi emperor was not only vigorous in maintaining his authority within his empire and in protecting and expanding the Chinese frontiers, but he also dispatched the French Jesuits to undertake a great geodetic survey of China from 1708 to 1717. It was because of his understanding of the accuracy of western scientific methods that the emperor decided to draw a general map of the Chinese empire, under the direction of Pierre Jartoux. In a letter dated June 15, 1717, by the French Jesuit Joseph de Moyriac de Mailla (1669-1748), we read:

> Another time, the great prince (Kangxi) ordered Father Verbiest and the other Europeans to renew the calendar .... Then, having perceived that the European maps which were presented to the throne by the missionaries could be clearly distinguished by the degree to which they corresponded perfectly with celestial phenomena, and that the Chinese maps were far from this perfection, he ordered the Europeans and the Tartars, by a decree of the Supreme Board, to survey all the empire at public expense, and draw maps of all parts of it[84].

---

81 Ruan Yuan, *op. cit.*, *juan* 41, pp. 1b-2a; Zhang Shen, *op. cit.*, *juan* 18, p. 19a.

82 Zhang Shen, *op. cit. juan* 18, pp. 8a-8b. Another translation has been made by W.R. Carles, see "The Emperor Kangxi's Edict on Mountains and Rivers of China", *The Geographical Journal*, 109 (January 1922), 264.

83 *Donghua lu*, *juan* 12, p. 12a.

84 *Lettres édifiantes et curieuses* ..., M.L. Aimé-Martin (Paris, 1843), 3:283.

The brilliant Jesuit Gaubil also insisted that the services of Verbiest and other Beijing Jesuits (Schall, Gerbillon, Fontaney, Bouvet, Antoine Thomas) to the emperor in the promotion of Chinese map-making, resulted in making up his mind to survey his immense domain[85].

In general, Verbiest's works served to meet the emperor's urgent needs. Nearly all of the scientific activities and accomplishments of the Flemish Jesuit Verbiest were realized within the imperial palace. He did not have enough opportunities to communicate with his contemporaries and open a wider vision for the intellectual community. In this he differed considerably from Ricci, who had played an important role in the introduction and popularization of modern geography and cartography in China. Just after the Yang Guangxian incident, Verbiest not only taught the Kangxi emperor about the general culture and geography of Europe, but also explained various subjects about the West to a certain prince, who has remained anonymous (in August 1670)[86]. It seems that Verbiest scarcely discussed geographical problems with the Chinese scholars because of his major preoccupations with imperial service. Unfortunately, this course of action which he initiated was also adopted by his successors, in particular, the French Jesuits. The geodetic survey of China and Benoît's world map entitled *Kunyu quantu*[87] (1761) both possessed an official character, and hence lost their possible influence on Chinese society. At the turn of the seventeenth and eighteenth centuries, the Jesuits in China preferred to introduce the Europeans to the varied and diverse aspects of Chinese things, rather than to communicate with the native scholars. None of Verbiest's works has a preface written by a Chinese scholar, except the *Kunyu gezhi lüe shuo*. Here we see only a short preface by Xu Erjue. In fact, that Verbiest's map of 1674 did not deeply penetrate Chinese society and become known to the majority of the intellectual circles, was due both to the limited quantity of its original printing, on the one hand, and to the highest priority of meeting palace necessities, on the other hand. Even though Verbiest neglected to keep in touch with the elite of Chinese society, he did argue in 1669 with his adversary Yang Guangxian about the spherical shape of the earth and the long distance between

---

[85] Gaubil, *op. cit.*, p. 542.

[86] See H. Bosmans, "Ferdinand Verbiest directeur de l'Observatoire de Pékin (1623-1688)", *Revue des questions scientifiques*, 21 (January, 1912), 267.

[87] This world map of Michel Benoît is divided into two hemispheres, each measuring 5 French pieds (165 cm.) in diameter. For the details, see Lin Tongyang, *op. cit.*, 2:409-416.

China and Europe[88]. In this respect, Verbiest emphasized how positively the spherical shape of the earth had been proved[89]. The substantial proofs that he presented were later included in one of the fourteen cartouches (*diyuan*) in his world map of 1674.

Verbiest's map eventually formed the basis for the world map of Benoît, who gave it the same name as that of Verbiest's map. It was in 1761 that the French Jesuit Benoît offered his map, still in manuscript form, to Qianlong to celebrate the fiftieth birthday of the emperor. Even though the original manuscript is lost, we can determine its outline and contents from certain copies of the same period. Just like Verbiest, Benoît ran the prime meridian through the capital of the Chinese empire, in order to respect the so-called *tianxia* concept. It is evident that Benoît used the majority of Verbiest's map for the geography of the five continents. However, the contrast between these two Jesuit maps is quite remarkable. Benoît added two important subjects from the domain of modern science to his map: (1) the results of the geodetic surveys in Peru and Lapland by the French *Académie des Sciences*, and (2) the importance of the Copernican heliocentric system.

To judge the response of the Chinese to Verbiest's geographic and cartographic achievements, we have to begin with Xu Erjue, who wrote a significant preface to Verbiest's last work (the *Kunyu gezhi lüe shuo*) in 1676. Having inherited his grandfather Xu Guangqi's spirit of criticism and realism, Xu noted the ignorance of his compatriots:

> The Chinese remain in a corner (of the world), just like the frogs living in the bottom of a well, without knowing the immensity of the great rivers. ... They know only the domain of China, the capital, and the provinces, but nothing special outside of China[90].

Xu thought that the Chinese did not dare circumnavigate the earth, as the "western scholars" had done. As a result, the Chinese could not discover the marvelous natural phenomena of the entire world: the mountains, the rivers, the sea-currents, the earthquakes, etc. ... Verbiest, "knowing the profound interest of the emperor," compiled his splendid works. For this reason, Xu esteemed the Flemish Jesuit very

---

88 See Lin Tongyang, *op. cit.*, 2:584-592.

89 Cf. F. Verbiest, *Budeyi bian* (The Dispute over the *Budeyi*), *Tianzhujiao dongchuan wenxian*, photolithographic edition of the Xuesheng shuju (Taibei, 1965), pp. 51b-52b.

90 Preface of Xu Erjue to the *Kunyu gezhi lüe shuo*, p. 1a. We use the manuscript copy in the library of Fu Sinian of the Academia Sinica, Taibei, n° Jen-916-321-9.

highly: "Mr. Verbiest has gotten knowledge of the secrets of the Creator and proved the perfection of the ancient saints' teaching. Is what he has contributed to learning for posterity not a great deal?"[91] Although Xu's impression and opinion is only one specific case, it might represent those who were baptized, and so apt to accept western science.

The Verbiest map was copied at least two times by Chinese geographers, first by Zhuang Tingfu, a native of Wujin, Jiangsu, and second by the distinguished scholar Wei Yuan (1794-1856). It was in 1788 that Zhuang drew a small-scale copy of the Jesuit map and had it engraved and inserted in his book entitled *Haiyang waiguo tubian* (An Illustrated Record of the Foreign Countries of the Seas). Apparently this small-scale copy was also printed separately. Although Wei Yuan had the chance to recognize the existence of such a single reproduction, he never got Zhuang's book[92]. Nevertheless, a short exposé written by Zhuang Tingfu under the title *Ditu shuo* (A Treatise on Maps)[93] has been included in Wei Yuan's splendid work *Haiguo tuzhi* (An Illustrated Gazetteer of the Maritime Countries, first ed. in 1844). Without knowing the principles of western cartography, Zhuang Tingfu accused Verbiest of having reduced the proportion of China, and having enlarged that of the foreign countries in Verbiest's own map. He ignored and opposed the value of the ornamental elements in Verbiest's map: "In this world map we see the presentation of reptiles, fishes, and even monsters which have no connection with terrestrial phenomena"[94]. Wei Yuan's *Haiguo tuzhi* also has a copy of the Verbiest map called *Yuantu* (The Round Map)[95]. This was engraved in a poor manner, and some of the older place-names had already been replaced by the new ones used freely in the mid-nineteenth century. Wei Yuan also copied the majority of the descriptions of the mountains, the rivers, the tides, the movements of the sea-waters, and the famous rivers of the world from Verbiest. Owing to his ingrained prejudices and deeply-fixed convictions of Buddhist cosmogony, Wei Yuan strongly attacked the Jesuit map's divisions into four or five continents. He concluded:

---

91 *Ibid.*, pp. 2a-2b.

92 Wei Yuan, *Haiguo tuzhi* (Guweitang ed.), 1849, *juan* 47, p. 26a.

93 This treatise is included in the *Haiguo tuzhi*, *juan* 47, pp. 26a-32a, and the *Xiao fang hu zhai yudi congchao zai bu bian* (edited by Wang Xiqi), 1897.

94 *Haiguo tuzhi*, *juan* 47, pp. 26a-27a.

95 This so-called Round Map is divided into two hemispheres, each measuring 23.5 cm. in diameter. See *Haiguo tuzhi*, *juan* 2, pp. 2a-2b, pp. 3a-3b.

"Among the four large continents revealed by the Sacred Books of Buddhism, the Europeans have hardly found two of them, and what remains is not yet discovered and explored"[96].

There are two other geographers (Chen Lunjian and Xu Jiyu) who were probably influenced by Verbiest's geographic and cartographic achievements. In 1744, Chen Lunjian, a native of Tong'an, Fujian, compiled a book entitled *Haiguo wenjian lu* (Record of Things Seen and Heard among the Maritime Nations), in which the author presented a traditional flat map representing the whole world. This flat map is formed in a round map, and named the *Sihai zongtu*[97]. The fact that Australia is described with a short sentence (saying that the human beings scarcely reach this land) in the map, tells us that the author might have consulted the Verbiest map of 1674. It was in 1848 that Xu Jiyu (1795-1873) finished compiling a geography of the world entitled *Ying-huan zhilüe* (A Brief Description of the Oceans' Circuits). Evidently Xu collected and utilized some geographical works printed by the Jesuits after the arrival of Ricci in China. But he did not hesitate to express his suspicions: "Having stayed one by one in the capital for a long time, the Europeans Matteo Ricci, Giulio Aleni and Ferdinand Verbiest mastered individually the Chinese language. That is the reason why they stated the things clearly and correctly in their works. Still, the things that they have described were fabulous and exaggerated in every way"[98]. Xu's world map (*Diqiu tu*) is divided into two hemispheres, but does not have any relation to Verbiest's map, because Australia's proper geographical name in Chinese (*Aodaliya*) was perfectly presented in Xu's map.

The attitude of the highly-placed Chinese scholar-officials vis-à-vis western science might reflect, to some degree, the tendencies of the orthodox and conservative scholarship in the seventeenth and eighteenth centuries. In fact, they almost neglected the existence of a real world due to their ignorance of circumstances beyond the seas of China. For instance, Ji Yun (1724-1805) participated in the compilation of the great Imperial Manuscript Library (*Siku quanshu*) as one of the chief editors. He might have taken advantage of his position to open and read the Jesuits' works printed before 1773. Nevertheless, he kept his suspicions

---

96 *Haiguo tuzhi*, *juan* 46, pp. 1a-10a.

97 See Chen Lunjian, *Haiguo wenjian lu* (included in the *Siku quanshu wuji*, Zhenben 102), *juan* 2, pp. 1b-2a.

98 Xu Jiyu, *Yinghuan zhilüe*, photolithographic edition, Taibei 1986, p. 28.

and skepticism toward the authenticity of Verbiest's world geography: "I suspect he (Verbiest) read the ancient Chinese books after his arrrival in China, and then copied and renewed some of them without respecting the truth"[99]. This attitude can be found equally among the chief editors of the *Da Qing yitong zhi* (Comprehensive Geography of the Chinese Empire, compiled in 1764, 500 *juan*): "In the ninth year of Wanli (1581) of the Ming dynasty, Matteo Ricci arrived at the district of Xiangshan near Guangzhou... Later on a large number of his adherents came to China. All of them adopted his ideas and instructions concerning the geography of five continents, and thus exaggerated things out of vanity"[100]. In spite of their doubts about the credibility of the Jesuits' works, these scholar-officials used and copied Verbiest's world geography to an enormous extent when they were obliged to describe the geographical facts about *Xiyang* (Europe). It is important to note that the chief-editors of the imperial collections could scarcely have realized and differentiated western things. For example, the official-editors of the *Qingchao wenxian tongkao* (Comprehensive Study of the Documents of the Qing Dynasty) mistook the Portuguese ambassador Don de Saldanha for an Italian. As a result, they placed the narration concerning the reception of the Portuguese embassy at the court in 1670 (Verbiest acting as interpreter) into the section called *Yidaliya* (Italia or Italy) of their collective work[101].

Among the works of other Chinese scholars, there is also evidence of Verbiest's influence on Chinese geography. In his work entitled *Bahong yishi* (1683, 4 *juan*), Lu Ciyun, a native of Qiantang, Zhejiang, mentioned briefly the arrival of Russian envoy Spathary at Beijing in 1676 and "the western scholar Verbiest serving as interpreter"[102]. He pointed out that *Xibailiya* (Siberia) was a cold and merciless region used for the transportation of Russian criminals[103]. Moreover, Lu cited extensively the text of Verbiest's world geography to support his way of dividing the world into four parts (the East, West, South and North), with its supposed center in China. Apparently it opposed the geography of five continents introduced by the Jesuits, and it simply tended to lead

---

99 *Siku quanshu zongmu tiyao* (Shanghai, 1930), 13:6.

100 *Da Qing yitong zhi* (Shanghai [Baoshanzhai ed.], 1902), *juan* 423-424, p. 1a.

101 *Qingchao wenxian tongkao* (Shanghai, 1936), *juan* 298, p. 7468.

102 Lu Ciyun, *Bahong yishi* (Longwei mishu ed.), *juan* 2, p. 2a.

103 *Ibid.*

to great confusion. Wang Shizhen (1634-1711) limited his interest to the illustrations of seven wonders inserted in the second *juan* of Verbiest's world geography. He cited, thus, the interesting story of the bronze statue of Rhodes (work of Chares, 32 m., 280 B.C.): "From the European Verbiest, director of the *qintianjian* and vice-president of the Board of Works, I have heard that the voyage from Europe to China measured 100,000 *li*. The vessels were very large. At the passage of a certain port, there was a statue in bronze standing tall. It was too tall to be calculated ..."[104]. The two authors of the *Aomen jilüe* (A Brief Record of Macao, 1751, 2 *juan*), Zhang Rulin and Yin Guangren, were merely interested in citing the description of the exotic and curious animals (turkey, ostrich, African civet, Italian taratola ... etc.). The fact that they place the names of these animals, with the annotations, under the section called *Yidaliya* in their book[105], demonstrates their ignorance of the divisions of the five continents and the geographical distribution of the animals. The eminent scholar Ruan Yuan (1764-1849) also showed his appreciation for the descriptions of animals. In 1822, he compiled a local gazetteer of Guangdong, to which he added some comments about the birds, beasts and reptiles of the "barbarian countries"[106]. All of his descriptions are certainly derived from Verbiest's *Kunyu waiji*. It seems that the Chinese admired peculiar things more than geographical knowledge. Finally, at the end of the eighteenth century, Fu Yunlong, a native of Deqing, Zhejiang, probably saw the Verbiest map of 1674. In his precise treatise entitled *Ditu jing-wei shuo* (A Study on the Latitude and Longitude of Maps), Fu wrote: "The meridian runs from the north to the south. In drawing a Chinese map, we have to take the meridian running through the observatory of Beijing as the prime meridian. In contrast, to draw a foreign map, one also takes the meridian passing through its own capital"[107].

Apparently, the Chinese travelers of the seventeenth and eighteenth centuries - Yu Yonghe, Wang Dahai and Xie Qinggao - knew nothing of Verbiest's works, for they made no use of its material in the notes of their voyages. A somewhat important work is the *Shenjianlu*[108] by

---

104 Wang Shizhen, *Chibei outan* (Shanghai, 1691), *juan* 24, 10a.

105 *Aomen jilüe* (*Zhaodai congshu* ed.), *juan* 2, pp. 54b-56b.

106 See *Guangdong tongzhi* (Taibei [*Zhonghua congshu* ed.], 1959), 2:1900-1901.

107 *Xiao fang hu zhai yudi congchao* (1897), 1:1a.

108 The text in Chinese can be found in Fang Hao, *op. cit.*, 4:187-195.

Louis Fan Shouyi (1682-1753), who accompanied the Italian Jesuit Antonio Provana (1662-1720), an envoy of the Kangxi emperor, on a trip to Rome in 1707. During more than eleven years (1708-1719) in Europe, Fan visited many cities in Italy (Rome, Naples, Florence, Turin, etc.) and returned to China in 1720. On October 12 of that year he was received at the summer palace in Jehol. In Fan's work already mentioned, certain Chinese place-names (America, Judea, Loreto, Italia, Roma, etc.) are derived from the geographical works of Aleni and Verbiest.

THE KANGXI CELESTIAL GLOBE: A MILESTONE IN THE HISTORY OF SINO-WESTERN CULTURAL EXCHANGE[1]

YI SHITONG

Yi Shitong was an associate professor of the Old Beijing Observatory until April 1991. At present he is the director of The World Harmony Science and Culture Institute. He is a member of the Chinese Institute of Astronomy, Museology, Archaeology and of the Chinese History Research Institute of Science and Technology. He is also a member of the Tibet Science and Technology committee, in charge of astronomy. He published "Chinese-English Star Catalogue" and some sixty articles.

Between 1669 and 1673, from the eighth to the twelfth year of the reign of the Qing emperor Kangxi, Ferdinand Verbiest, a Belgian missionary, designed and built six large bronze astronomical instruments to supplant their outdated counterparts at the Beijing Observatory, thus opening a new chapter in the history of Chinese astronomy[2].

The celestial globe was known in ancient China as *hunxiang*, that is: a model of the celestial sphere[3]. Whenever a *hunxiang* was mentioned in historical documents, the reign title of the ruling emperor was attached to make its meaning more definite.

The celestial globe of the Kangxi emperor is a representative work among the six astronomical instruments created during the early Qing dynasty, and it surpasses all the other pieces in the complexity of technology, period of creation, material used, and range of application.

---

1 The first draft of this essay was completed in 1986 in conjunction with the Sino-Belgian project of producing a replica of the Kangxi celestial globe. Three additional drafts led to revisions that are incorporated in the current text.

2 Seven astronomical and meteorological instruments were set up and displayed during the Ming dynasty on the platform of the Beijing Old Observatory. However, only three of them were large bronze instruments, that is, *hunyi* (armillary sphere, now in Nanjing), *hunxiang* (celestial globe, now lost) and *jianyi* (abridged armilla, now in Nanjing).

3 *Hunxiang* (celestial globe) is also called *huntianxiang*; confusion arises when it is sometimes called *hunyi* (armillary sphere) and *huntianyi*.

The celestial globe not only can comprehensively display the spherical positions of stars obtained by other instruments, but it can also indicate the diurnal or annual movements of celestial bodies at different latitudes, seasons and times. Its sphere and graduation on its band provide direct readings of the spherical coordinates of different systems of celestial bodies, or make possible the conversion of coordinates. The celestial globe has a strong audio-visual appeal, and it can considerably relieve the astronomer's heavy labor in complicated calculation. In a word, it is useful in many ways.

China has a long tradition of worshipping the heavens, and an astronomical instrument is also a ritual vessel devoted to this purpose. It represents the principal star worshipped by the emperor. According to ancient rites, the emperor made sacrifices to the heavens in the south suburb and to the earth in the north suburb. That was why Verbiest, as shown in *Lingtai yixiang zhi. Tu* (Illustrations to records of astronomical instruments at the observatory)[4], had the celestial globe placed in the center at the south of the observatory, which shows the great importance attached to the instrument, since it was considered the "first of all the astronomical instruments"[5].

Structurally, the celestial globe consists of three parts.

A. The crisscrossing beams at the base, with a knob in the center supporting the sphere. West of the beams is a transmission gear box with a crank, by means of which the altitude of the north pole on the sphere can be adjusted to simulate apparent spherical motion at different latitudes.

B. The adjustable mounting. Its horizontal (upper) ring has a groove which can be filled with water to keep it level. The foundation ring is equipped with foot screws for adjusting height and level, and their protective covers. Four animal-adorned legs connect the horizontal and foundation rings. In appearance, the mounting and

---

4 *Lingtai yixiang zhi. Tu* is almost the same in contents as the illustrations in *Lingtai yixiang zhi*, for the preface states that since no additional explanation is added to the diagrams, readers should use the two books together. However, because of its different size, this book of diagrams looks quite like an independent publication, and it is difficult to find a good one among the very few existing copies. The edition that this author studied is kept at the Purple Mountain Observatory in Nanjing and is said to be one of the best copies in China. In fact, some diagrams in that book have nothing to do with the *Lingtai yixiang zhi*.

5 See the first few volumes of the *Lingtai yixiang zhi*. The text or terms concerning the celestial globe actually came from *Huntianyi shuo* and this idea was much in vogue at that time.

the beams are all of one piece, but actually they are detachable, with room for adjustment.

C. The celestial sphere and the meridian ring. At the north pole of the ring is a dial with its pointer, and the flame-shaped nut for tightening the pointer. At the zenith of the ring is the zenith gauge, with its floral design that can slide along. At its nadir to the west is a toothed arc which is connected to the notched groove of the gyro wheel at the knob, supporting the sphere and enabling it to rotate.

The Kangxi celestial globe is a large bronze instrument composed of more than 120 parts and accessories, and 1,888 decorative stars. It is 2,760 mm. high, 2,630 mm. at its widest, and weighs 3.85 tons. The diameter of the sphere is 1,860 mm., the outer diameter of the meridian ring, 2,095 mm., and that of the horizontal graduated disk, 2,340 mm. The center of the sphere is 1,370 mm. from the ground surface. The celestial globe is rationally designed and constructed, majestic in shape, and magnificently ornamented. It is a rare work of art now under state protection.

What are the similarities and differences between the celestial globe, typical of the first astronomical instruments forged with western technique, and the traditional native instrument, the *hunxiang* ?

First, the *hunxiang* of the Ming dynasty (or of an even earlier period) was principally designed as a model to represent the ancient astronomers' basic theory that the sky is circular and the earth square. The lower part of the instrument, which symbolizes the earth, is a square box engraved with animal patterns on the four sides. Above the box appears the half-hidden sphere. Since the time of ancient Greece, western astronomers had attempted to interpret celestial movements by means of geometry. Basing themselves on the proofs of geometry, they regarded the celestial globe as having a perfect shape, and would under no ordinary circumstances seal it up in a model. Among large *hunxiang* mentioned in history with lower part exposed to view, the Kangxi celestial globe was a product without parallel.

Secondly, the traditional *hunxiang* had always been mounted with double meridian rings copied from the armillary sphere. Astronomical observation in China always attached great importance to data on the upper culmination. The armillary sphere with its double meridian rings poses the least obstruction on the field of vision commanded by the sighting tube. Being more solid and maneuverable, it is easier to assemble. The western celestial globe has only a single ring, one side of which can be used as the meridian plane, from which the shadow thrown on the celestial globe is the meridian. Although the construction of such instruments and the difference of principle on which they are built, as contemporary standards go, are minor issues, they involve the historical traditions of the East and the West, as well as their cultural, scientific, and technical backgrounds.

Thirdly, the system in ancient China was based on the circumference of the sky divided into 365 1/4 degrees. This method was based on the daily shift of the sun on its apparent annual position as the natural standard. A day was reckoned in a hundred *ke* or 12 *chen* (12 standard double-hours). The trouble is that they cannot be divided, and although a number of adjustments were devised in successive dynasties, they were not ideal. The western countries, on the other hand, divided a day into 24 hours and the sky into 360 degrees, and further subdivided them by 60.

Fourthly, in ancient China, great importance was attached to the observation of celestial phenomena. The spherical position of a star was decided by its equatorial coordinates. Although the ecliptic had been determined, for a long time no clear geometric conception of the ecliptic coordinates was formed. The value of the ecliptic coordinates as referred to in documents was actually that of the projection of the ecliptic coordinates relative to the equatorial coordinates. The geometric pole of the ecliptic had never been determined or used. Differing from historical conditions in China, the western countries had long since introduced the system of ecliptic coordinates with a clear geometric conception. Even before the Renaissance, they had been marking the spherical positions of celestial bodies with ecliptic coordinates. For the first time in China, the Kangxi celestial globe indicated the spherical positions of stars with ecliptic coordinates.

Fifthly, on a traditional Chinese instrument, divisions below *du* (degree) were marked by *tai*, *ban*, and *xiao*. Because of technical limitations, ancient instruments were rather small, so that only three lines were marked for one degree, providing a reading of only one quarter of a degree. When a celestial body approaches or passes over an engraved line, it is still possible to read an approximate value equivalent to about a twelfth of a degree. Beginning from the Yuan dynasty, as the instruments became larger, the hundred-mark system was introduced, with each degree divided into ten *fen*, each providing a reading of a twentieth of a degree. Proof of this can be found in data from Yuan dynasty documents. On the six instruments of the early Qing dynasty, a degree was divided into six *fen*. On each *fen* a diagonal line was drawn across ten equal parts formed by concentric circles. By means of a caliper gauge sliding along, the point of intersection provided a fractional reading as minute as 1/240th of a degree. Because of uneven graduation owing to technical limitations, actual precision fell far short of the reading, but was nevertheless much more accurate than in earlier times. Such a method of graduation was used only later in the West.

Lastly, the system of transmission differs with the exception of those operated by hand, all ancient Chinese *hunxiang* were equipped with a mechanism for tracking the apparent diurnal motion of celestial bodies, not unlike the mechanism of the modern sidereal clock, while the height of the pole of the sphere was fixed. The system of gear transmission of the Kangxi celestial globe was intended for adjusting or changing the height of the pole, while the diurnal motion was directed by hand[6]. The equatorial disk with its pointer, and the zenith gauge, were also introduced from abroad.

The celestial phenomena on the Kangxi celestial globe incorporated the results of the star catalogue recorded in *Chongzhen lishu* (Chongzhen's almanac)[7], collectively compiled and translated by such eminent Chinese and foreign scholars as Xu Guangqi and Johann Adam Schall von Bell in the late Ming dynasty. Although *Chongzhen lishu* was revised by Schall and others, and published in the early Qing dynasty under the title of *Xiyang xinfa lishu* (Almanac based on the Western new system)[8], its contents underwent little change. When Verbiest was put in charge of astronomical work, he merely corrected the positions of stars listed in the catalogue by adding the precession of equinoxes and

6 Either in China or in other countries, the celestial globe is a great classic instrument in the history of the development of science, which involves the interacting processes of some related systems in East and West. See my essay, "Xu Guangqi he wan Ming yiqi" (Xu Guangqi and the instruments of the late Ming dynasty), in *Xu Guangqi yanjiu lunwen ji* (Collected essays on Xu Guangqi) (Shanghai, 1986), p. 76.

7 Today there is hardly a complete copy of *Chongzhen lishu* (Chongzhen's almanac).

8 During the Qing dynasty, the *Chongzhen lishu* had been revised by Johann Adam Schall von Bell and renamed *Xiyang xinfa lishu* (Almanac based on the Western new system). The deletions were mostly words or terms about memorials to the emperor or official titles of the previous Ming dynasty. Apart from this, the Ming dynasty edition remains intact with very few scientific revisions. However, this author refers here only to the part which he himself proofread in the Palace Museum in Beijing more than twenty years ago.

calculated up to the epoch used by the Kangxi celestial globe. He used the year 1672 as the epoch of the ecliptic star catalogue, and the year 1673 as the epoch of the equatorial star catalogue, and based the positions of stars on the ecliptic star catalogue. When the globe was commissioned, both the ecliptic and the equatorial star catalogue were included in the *Lingtai yixiang zhi*[9].

This book was compiled by Verbiest in 1674, the twelfth year of the Kangxi reign[10]. Some thirty-one persons took part in the work. The book included an account of the design and operation of the six instruments, a star catalogue, and tables for observation and calculation. There was also a set of more than a hundred illustrations, against which the text might be compared. Because of the haste with which the book was written, the different sources of information, and the lack of coordination in the process of writing, considerable errors occurred which users of the book should guard against.

The star catalogue in the *Lingtai yixiang zhi* was compiled specifically for the celestial globe of Kangxi. The two should be carefully verified against each other to compensate for the former scholars' oversight. And so, from 1987 to 1988, when the replica of the globe was being made, we again checked the star catalogue and read it through against the spherical positions of stars on the globe. We found there were a total of 1,895 stars in the catalogue and on the sphere of the celestial globe. Of this figure, nineteen stars were found on the globe but not listed in the catalogue, while seven stars were listed in the catalogue but not located on the globe. In other words,

1,895 - 19 = 1,876 (number of stars in the catalogue);
1,895 - 7 = 1,888 (number of stars on globe).

However, the actual number of stars on the globe would be 1,895 if the seven in the legend of the sphere were taken into account.

The celestial phenomena as indicated on the sphere of the Kangxi celestial globe belonged to the system outlined in *Bu tian ge* (The Song

---

9 The value of ecliptic coordinates in the star catalogue of the *Lingtai yixiang zhi* is corrected according to the ecliptic star catalogue with the yearly difference of the ecliptic longitudes of the *Chongzhen lishu*. As for the value of the ecliptic latitudes, both books are the same. The value of the equatorial coordinate is converted according to the revised ecliptic star catalogue. The relationship in this transmission can easily be proved.

10 This involves textual research. See the unpaginated section entitled *Houji* (Postscript) in Yi Shitong, *Zhongxi duizhao hengxing tubiao* (Comparison of Chinese and Western tables of fixed stars) (Beijing, 1981).

of the March of the Heavens), compiled after the Tang dynasty. The research on astronomy during the book translation period in the late Ming dynasty mainly concerned the comparison of the Chinese and western celestial phenomena. The western star catalogue used for this purpose was that compiled by the Danish astronomer Tycho Brahe (1546-1601). Here special mention should be made of the stars near the south celestial pole discovered as a result of the development of navigation in western countries. The Kangxi celestial globe displayed 144 stars (or 155, including the additional ones) of twenty-three constellations near the south celestial pole for the first time, serving as the basis for the naming of stars near the south celestial pole.

The role played by the Kangxi celestial globe in the history of the evolution of Chinese celestial phenomena included the following:

1. It preserved the general picture of the traditional celestial phenomena of the late Ming period and established a relationship for comparison with the classical western star catalogue, thus making the traditional Chinese celestial phenomena and star names more easily understood by world astronomers.
2. The large-sized Kangxi celestial globe, with its distinctly marked star positions and catalogue, is the one and only instrument for comparing the Chinese with the western celestial phenomena as they met at the point of intersection. It is also an object for research into the technical perfection reached in the Qing dynasty. For instance, comparing the actual with the theoretical precisions of graduation and the positions of celestial bodies on the sphere with those supplied by the star catalogue, we can track down former technical errors and analyze the causes leading to such errors.
3. Although the Kangxi celestial globe provided a basis for succession for the positions of stars and their nomenclature in the Qing dynasty, it had little influence on celestial phenomena, especially the numbering of stars, during the middle and late Qing period. In view of its direct link with the work of translation in the late Ming dynasty, it is equally correct to say that it represented the end of the celestial phenomena of the Ming dynasty or the beginning of those of the Qing dynasty. As the *hunxiang* built during the Zhengtong (1436-1450) and Chenghua (1465-1488) reigns of the Ming dynasty had been melted down and recast into a pair of bronze lions in 1771, it would be nearer in keeping with history to consider the celestial phenomena on the Kangxi celestial globe as the summation of the evolution of the traditional *hunxiang* up to the Ming period.

4. The loss of historical documents and cultural relics after several hundred years has caused certain matters, at one time commonly known, to appear more and more unfamiliar to people of today. Nevertheless, it is against such a background that the Kangxi celestial globe not only provides us with a concrete picture of the science and technology of that period, but also gives rise in our mind to more ancient associations. Naturally such associations or conjectures are inferences, but they will not stand without the support of solid historical evidence. Even certain faults that occurred during the manufacture of the celestial globe may serve as valuable historical messages to astronomers of today, for they not only provide standards for comparison but also impel us to investigate the various factors leading to existing problems and to search for answers more fitting to historical reality. Such problems will remain unanswered if we do not probe deeply enough into the contradictions or deficiencies existing in documents and cultural relics[11].
5. The Kangxi celestial globe used for the first time more definite signs and serial numbers for stars and visual stars, which is something worthy of commemoration in itself.

The illustrations to the *Lingtai yixiang zhi* provide a glimpse of the distribution of instruments on the observatory in those years, but they represent a second plan of arrangement after reconsideration[12]. In his replies to questions put by the Kangxi emperor (KH 8:8:3), as recorded in documents, Verbiest had previously considered putting the ecliptic armilla at the southeastern corner and the altazimuth at the southwestern corner. The equatorial armilla was to stand at the center of the south side, the quadrant due east, and the sextant due west, while the celestial globe would take up its position in the center of the northern side. Such an arrangement would be quite different from the actual plan that eventually took place as indicated in the illustration.

What caused Verbiest to change his plan is not known, but probably was due to several historical considerations. The illustrations to the *Lingtai yixiang zhi* show a small brick platform located due east on the observatory, which Verbiest intended for the site of the quadrant. It was

---

11 The related original documents were deposited in the archives with the approval of the Qianlong emperor in 1771 and currently are in the Number One Historical Archives in Beijing.

12 This is specified in the *Xichao ding'an* of Ferdinand Verbiest. The copy this author consulted is that of the manuscript kept in the Institute of the History of Science in the Chinese Academy of Sciences.

an observation tower where astronomers on duty scanned the sky for extraordinary aspects used for divination[13]. In ancient times, the observation tower played an even more important role than the instruments, for naked-eye observation was something no instrument could replace. The observation tower functioned as a command post, because with the exception of the pre-arranged observation with instruments, the general rule was for the observer to first discover the celestial bodies before instruments would be employed to determine their spherical coordinates. Since the principal duty of the astronomer was stipulated, the observation tower could not be dismantled or moved, and that was why Verbiest had to change the distribution of his new instruments.

Comparing it with the first plan, we also note that the ecliptic theodolite was moved to the southwestern corner, while its original position was taken up by the equatorial theodolite. This change of the leading role of the ecliptic theodolite reflects a change in Verbiest's traditional thinking that the ecliptic theodolite would have priority over the equatorial theodolite[14].

In both the first and second plans the celestial globe had maintained its absolutely leading role, but in actual arrangement its position was shifted from due north to due south, the direction in which sacrifices were offered to the heavens. Such arrangements were quite natural, considering that in Chinese traditions heaven has always been considered as located to the south and earth to the north.

Today every consideration other than convenience for work is nonessential in making the arrangements. In history, however, such change reflected the provisions or customs of traditional rites in a country with an ancient civilization, which can be easily glossed over by present-day researchers. We should pay serious attention to such problems in the research of history or the handling of historical incidents.

People generally appraise the value of an ancient object by comparing its former cost of manufacture with what it would cost to build today. This is rather difficult, for the concept of value in the past differs from that of the present. So do the technical process and various intermediate links. The price for the same object may differ considerably in the past and at present.

---

13 See Yi Shitong "Zuogengtai kao" (Investigation about the observation tower), *Wenwu* n. 1 (1991), 74-79.

14 Actually this can only be described as a nominal concession and compromise.

According to the *Xichao ding'an* (Records of cases decided during the reign of the Kangxi emperor) compiled by Verbiest[15], the manufacture of the six instruments (including the reconstruction of the foundations for the instruments and the casting of the protective iron fences) had cost 12,027.3 taels of silver for the hiring of craftsmen. But the figure did not cover the overall expenses. The Qing court had laid down regulations for the issuing of materials such as gold, silver, bronze, iron, tar, wax, and rosin, most of which were allotted by warehouses and then written off in receipts sent to the emperor for approval. Obviously, in considering the prices of the instruments, it is necessary to make an estimate of the worth of the materials. So far only the total amount of silver spent for the construction of the six instruments has been made available. We still have to find out the cost of the celestial globe. During a weighing in of the six instruments in 1983, when the observatory was given a major overhaul, the total weight of the instruments added up to 14,418 kilograms[16]. Since the weight of each piece to the total weight is roughly proportionate to its cost to the total expenditure, it is possible to assess the cost of each piece from the total expenditure according to the ratio among the different weights of the six instruments. However, the net weight of each piece and the material that went into its making are two different matters because, in casting, workmen had to take into account the casting heads, the patching up or recasting of certain parts, and the remainder after processing.

Other miscellaneous expenses should also be taken into account. For a giant project like the casting of astronomical instruments, the regular court workers were understaffed and would be considerably reinforced by temporary workers. The ratio between the regular and temporary workers was one to two. Generally, the regular workers received meal subsidies during the term of the project, while getting their pay from the department responsible for the project (which was not included in the account submitted to the emperor). The last included wages for the hired hands, miscellaneous expenses and subsidies for the regular workers.

In view of the above, and in order to make an estimate of the cost of instruments, it is necessary to find out the total amount of materials used and the pay for the regular workers, then add the expenses sub-

---

15 See note 12 above.

16 See Yi Shitong "Beijing gu guanxiangtai de kaocha yu yanjiu" (Inspection of and research on the Beijing Old Observatory), *Wenwu*, n. 8 (1983), 44-51.

mitted to the emperor, and we have the answer[17]. Referring to the balance sheet for the wages and materials used in the building of the new armilla during the Qianlong reign, we have calculated by ratio the number of workers employed, the materials used, and the expense of silver involved in the making of the Kangxi celestial globe as follows:

Total amount of bronze:	5,248 kg.
Net weight of instrument:	3,850 kg.
Total number of workers:	24,635.
Subsidies for meal:	360 taels of silver.
Pay for 9,990 regular workers:	1,538 taels of silver.
Pay for 14,110 temporary workers:	2,173 taels of silver.
Pay for 535 odd hands:	43 taels of silver.
Miscellaneous expenses:	636 taels of silver.

The figures show that six workers were needed to refine each kilogram of bronze. Putting the cost of materials and the average pay of workers in terms of present-day prices we can compare them with the corresponding values in the year when the celestial globe was manufactured, and find out the difference. The results obtained by this means are reliable, but they only represent approximate values[18].

As a testimonial to the cultural, especially astronomical, exchange between China and the West, the Kangxi celestial globe was incomparable. This is because it involved data that taxed the joint efforts of many Chinese and foreign scholars after the late Ming period. Apart from construction technology, the celestial globe also introduced from the West a new coordinate system and measurement with stricter geometrical conception[19].

---

17 See Tong Yan, Situ Tong, and Yi Shitong "Jiheng fuchenyi" (Equatorial armillary sphere), *Gugong bowuyuan yuankan*, n. 1 (1987), 28-35, 48 (footnotes).

18 The value of silver varies from one historical period to another. Therefore one cannot estimate the cost of the instruments according to the current price of silver. It is more reasonable to convert the price of the instruments with the wages and the expenditure of material used.

19 The result of cooperative translation by the Chinese and the missionaries in the late Ming can chiefly be seen in the academic activities of Schall in his later years. These include: promulgating western (new) methods, issuing a "current calendar" (*shixianli*), and the gradual development of the final shape of the astronomical instruments that were improved several times through experiments. (In Verbiest's day they were finalized into six large instruments). It seems safe to view the early Qing achievement of astronomical instruments as the continuation of the late Ming legacy of the Bureau of Astronomy. As a direct successor of Schall, Verbiest clearly

The western star catalogue that corresponded to the celestial phenomena of the Kangxi celestial globe was composed by the Danish astronomer Brahe. The data of star positions were obtained through field observations by Chinese astronomers between the late Ming and early Qing dynasties[20]. Brahe discovered a nova in the Constellation of Andromeda on November 11, 1572. Known popularly as the Tycho Nova, it was a supernova belonging to the galaxy. In ancient China a nova was a guest star. When the Tycho Nova followed Brahe's star catalogue to China, it was indicated on the Kangxi celestial globe as a guest star. As a rule, a celestial body like a nova that appeared suddenly, only to fade again, would not be marked on a celestial globe. So it was rather remarkable for the Kangxi celestial globe to give an indication of a guest star. The structure of early Qing astronomical instruments, including the design of the stone platform for observation, was subject to a great extent to Brahe's influence. A casual glance through the illustrations to Brahe's instruments will prove the point.

Although the designation of celestial phenomena of the Kangxi celestial globe followed the western practice, the contents of celestial phenomena on its sphere still belonged to the system of *Bu tian ge* after the Sui and Tang dynasties[21]. The twelve zodiacal signs were translated as twelve *ci* (twelve Jupiter stations) and twelve *chen*, both being methods for dividing the sky in ancient China by making it into twelve zodiacal signs along the equator. The twelve *ci* were arranged from west to east, and the twelve *chen* from east to west. The twelve *chen* in the sky were represented by the twelve branches on their horizontal positions, namely, *zi*, *chou*, *yin*, *mao*, *chen*, *si*, *wu*, *wei*, *shen*, *you*, *xu*, and *hai*. They were in correspondence with the twenty-eight lunar

---

showed the connections. However, from small instruments to six large instruments, we can detect a qualitative change. For this change, Verbiest's contribution was enormous.

20 There was little direct evidence from experimental investigation in the Ming and the Qing periods; some was only concluded from indirect deduction. Two points are worth mentioning here. First, the position of the "dark star" (*anxing*) of Chinese traditional astronomy could not be found in the western star catalogue which the missionaries brought to China. Second, there was an obvious error in the systematical position of such stars and it showed that some data resources were based on specific observations conducted in China.

21 *Bu tian ge* is a kind of formula in verse for learning astrology which was handed down over the ages from the Sui and Tang dynasties and exerted great influence on the beginners for almost a thousand years. Although the versions vary from time to time, that does not make a difference for our discussion at this point.

mansions. When the mansions were at the southern culmination, the twelve *chen* fell in accord with the twelve branches. The twelve *ci* were known as *xingji*, *xuanxiao*, *quzi*, *jianglou*, *daliang*, *shishen*, *chunshou*, *chunhuo*, *chunwei*, *shouxing*, *dahuo*, and *ximu*. The application of the twelve *ci* and twelve *chen* was related with the recording of the year by Jupiter.

In ancient China, the twelve *ci* and twelve *chen* were arranged in equal division along the equator. During the late Ming period these terms were used to represent the twelve zodiacal signs. As a result, in addition to the twelve ecliptic signs, there were also the twelve equatorial signs, and it is necessary to differentiate by indicating whether they were ecliptic or equatorial. In addition, the Kangxi celestial globe was carved with twenty-four fortnightly periods, a result of the prolonged evolution in the preparation of the Chinese calendar.

The twelve western zodiacal signs were introduced into China with the translation of Buddhist scriptures during the Sui and Tang dynasties, but they were not used, probably because of the guiding principle of "putting the Chinese learning as the basis and applying the western learning in practice" advocated by Xu Guangqi and other late Ming dynasty scholars[22].

The Kangxi celestial globe, which involved both Chinese and western academic traditions and documents, represented the results of cultural exchange through the combined efforts of Chinese and western scholars from the late Ming to early Qing dynasties. These included the work of former astronomers and also Verbiest's creations. They can hardly be separated[23].

From the point of view of scientific and technical development, the Kangxi celestial globe marked the formal adoption of the western, or new calendar, by Chinese astronomy. It also means, at least in the area of astronomy, that the Qing court acquiesced that China lagged behind the West. As a product of Sino-Western cultural exchange, the Kangxi celestial globe represented the union of Chinese and western culture, academically as well as in technology, style and decoration. It reflected the long traditions and glorious achievements in Chinese and western astronomy, and bore witness to the advance of western science and technology as traditional Chinese science and technology gradually declined.

---

22 See note 6 above.

23 From available sources, it seems that Verbiest might have had better knowledge about metal processing and manufacturing than his contemporaries.

With the arrival of the western missionaries in the late Ming period, the Beijing Observatory became one of their early stopovers because of various historical factors and certain accidental conditions. With astronomy as the medium, they helped to bring about Sino-Western cultural exchanges, albeit unwillingly, least of all ideally. However, we must not underestimate the difficulties encountered by the missionaries in China, the results of which should not be ignored from any angle because of their world influence. The "new learning" or "western learning", especially astronomy, mentioned in modern Chinese history, and the "sinology" to which western academic circles often referred, were for the most part related with the activities of missionaries and with the history of the Beijing Observatory. The Kangxi celestial globe was a mark in history showing the height and maturity of Sino-Western cultural exchange.

Verbiest arrived in China in 1659. He did missionary work first in Shaanxi, but was summoned to Beijing the next year to become an assistant to Schall, the director of the Bureau of Astronomy. He succeeded Schall in his writings and translations, which were completed in the early Qing period when the Translation Bureau was established to introduce western literature into China. This brought Verbiest into focus in the propagation of classical western astronomy, the change of the calendar, and large scale translation. However, we should note that soon after Verbiest came to Beijing, the persecution of the missionaries took place. Schall died in 1666. His coaching of Verbiest was short. Actually, through Schall, Verbiest had followed and succeeded the collective achievements of writer-translators in the late Ming period, but he had mastered the basic learning in astronomy and engineering while studying in his native country. In creating the Kangxi celestial globe, Verbiest not only learned from his predecessors, but was justifiably credited with many new and unique inventions. Though the basic principle of the globe was drawn from *Xinfa lishu* (New almanac) and *Huntianyi shuo* (Explanation of the armillary sphere), turning theory into practice was by no means an easy job.

The *Lingtai yixiang zhi* lists more than sixty uses of the celestial globe, but these cover mainly the conversion of the spherical positions, times, and angles of the equatorial, ecliptic, and horizontal coordinates on the sphere of the celestial globe. The following are a few examples taken from the *Qinding da Qing huidian* (Collective statutes of the Qing dynasty) to illustrate the uses of the globe.

1. How to find out the time when a star is in culmination at one of the twenty-four fortnightly periods. Align the period position as marked on the ecliptic and the eastern side of the meridian ring, and fix the disk pointer at the noontime position (the horizontal position due south). Turn the star to the eastern side of the meridian ring. The time indicated by the caliper gauge is the time when the star is in culmination.
2. How to find out the star in culmination at one of the twenty-four fortnightly periods. Move the period marked on the ecliptic to the eastern side of the meridian ring, and fix the caliper gauge at the noontime position. Rotate the sphere till the disk pointer is at the required time. The star under the meridian ring is the star in culmination.
3. How to find out the times of sunrise and sunset at one of the twenty-four fortnightly periods. Align the period position on the ecliptic and the eastern side of the meridian ring, and fix the disk pointer at noontime. Turn the period position to the eastern horizon. The time indicated by the caliper gauge is sunrise. Turn the period position to the western horizon, and the time indicated by the caliper gauge is sunset. You can also find out the positional azimuths of the sun when it rises or sets from the azimuth graduations marked on the horizontal ring.
4. How to find out the horizontal position and altitude of a celestial body at one of the twenty-four fortnightly periods. First find out the celestial phenomena of the period with the method mentioned above. Then measure the horizontal position (horizontal longitude) and altitude (horizontal latitude) of the celestial body, using the high-arc caliper gauge. Because the original gauge has been lost, it is rather difficult to make an exact copy of it, though the principle is clear.

More than three hundred years have passed since the completion of the Kangxi celestial globe. During this interval, because of the precession of equinoxes, the skew angle of the ecliptic and the equator, and the self-motion of stars, the positions of celestial bodies on the celestial globe now differ by about five degrees with their actual positions in the sky. Still, the celestial globe faithfully records the celestial phenomena more than three hundred years ago, and will do so for a long, long time. From the point of view of the history of cultural relics, science and technology, among the six early Qing instruments, the celestial globe not only occupies a very prominent place; it is also the only celestial globe linking even earlier history because all the traditional *hunxiang* have been lost. Among the specimens extant, the Kangxi celestial globe is the largest and earliest, and provides reliable data for verifying the changes of the celestial phenomena of the Ming dynasty and the star names in the early Qing dynasty. From the technical point of view, the Kangxi celestial globe is a first-rate instrument with regard to design and ornamentation. It still shines with the brilliance of history and wins the praises of every visitor to the observatory.

The year 1988 marked the three hundredth anniversary of the death of Verbiest, and it was named by the Belgian government as the Year of Verbiest to commemorate this precursor in the East-West cultural exchange. A replica of the Kangxi celestial globe has been made which was put on exhibition, and now stands permanently on the campus of the University of Louvain, Verbiest's alma mater.

The reproduction is of the same size as the original, but heavier, in order to minimize the change of form wrought by time. A number of lost parts have also been restored according to photos taken earlier in the century. Certain errors that occurred in the course of the designing and processing of the original have been corrected, as far as possible, while the celestial phenomena and graduations conform strictly to the original. By common consent, the unseen parts and accessories are replaced by those now in common use, but all other parts have been reproduced by hand according to the original technology. The replica (excluding preparation) was completed in October 1988, having taken about one year. It sailed from a Chinese port in November and arrived in Belgium at the end of 1988.

As the common wealth and historical heritage of human civilization, the Kangxi celestial globe is a symbol and milestone of Sino-Western cultural exchange, as well as a monument. It enables people to recall history and look forward to a more beautiful future.

Explanations of the Table

The *Hengxing zongji* (General record of fixed stars) gives comparisons between a number of star catalogues of the Qing dynasty and the names and number of stars listed in *Bu tian ge*. Among the stars listed are the three enclosures, twenty-eight lunar mansions, and the constellations near the south pole, as well as attached constellations and the proper names of stars. The long evolution of Chinese celestial phenomena has caused confusion among the names of stars with different origins, signs of which are still evident (e.g. repetitive names and official titles of different periods). The added constellations near the south pole were introduced from abroad during the late Ming period.

In the statistics about the number of stars in the *Hengxing zongji*, *ge* refers to *Bu tian ge*, *zhi* to *Lingtai yixiang zhi*, *cheng* to *Yixiang kaocheng*, and *xu* to *Yixiang kaocheng xupian*.

The figures marked with an asterisk in the *Hengxing zongji* indicate that they do not agree with those given in the star catalogue. In our table, the figures have been corrected according to those in the star catalogue.

The statistics of constellations in the *Huangdao xingbiao* (Ecliptic star catalogue), and in the *Lingtai yixiang zhi* have been compiled exclusively for the reproduction of the Kangxi celestial globe. For the convenience of verification, the numbers of stars have been counted according to different constellations, and each constellation has been subdivided into six different ecliptic longitudinal sections. In the column on stars, *qi* refers to blurred celestial bodies observed through the telescope, such as nebulae, star clusters, and galaxies, including also certain misrecognized stars.

Statistics of stars include not only those in the catalogue, but also those on the sphere of the globe after verification. Indications are given where the number of stars in the catalogue does not agree with the actual number of stars on the sphere of the globe. Thus figures marked with Φ indicate that there are stars on the sphere but not in the table, while figures marked with ▪ indicate that there are stars in the table but not on the sphere.

Apart from checking the original texts of the *Huangdao xingbiao*, and of the *Lingtai yixiang zhi*, we have also compared them against the Kangxi celestial globe. Division of the twenty-four fortnightly period zones is indicated in the catalogue for the convenience of verification. For easier indexing, the scope of the ecliptic which starts with the Spring Equinox is included. The calculation of coordinates in traditional Chinese astronomy began with the Winter Solstice. Using the Spring Equinox as the start was introduced from the West.

Table 2

12 Zodiacal Signs, 12 Jupiter-Stations, 28 Lunar Mansions, 12 Standard Double-Hours, 24 Fortnightly Periods, Ecliptic Longitudes and Equatorial Time Zones

12 zodiacal signs	Capricorn	Aquarius	Pisces	Aries	Taurus	Gemini	Cancer	Leo	Virgo	Libra	Scorpius	Sagittarius
12 Jupiter stations	Xingji 星纪	Xuanxiao 玄枵	Juzi 娵訾	Jianglou 降婁	Daliang 大梁	Shichen 實沉	Chunshou 鶉首	Chunhuo 鶉火	Chunwei 鶉尾	Shouxing 壽星	Dahuo 大火	Ximu 析木
28 lunar mansions	Dipper Ox	Girl Emptiness Rooftop	House Wall	Legs Bond	Stomach Stopping place Net	Turtle Invest-igator	Well Ghosts	Willow-star Extended Net	Wing Chariot Platform	Horn Neck	Root Room Heart	Tail Winnow-ing Basket
12 standard double hours	Chou 丑	Zi 子	Hai 亥	Xu 戌	You 酉	Shen 申	Wei 未	Wu 午	Si 巳	Chen 辰	Mao 卯	Yin 寅
24 fort-nightly periods	WS/LC	GC/BS	R/AC	VE/CaB	GR/BSu	LF/GE	SS/LS	GH/BA	EH/WD	AE/CD	DH/BW	LS/GS
Ecliptic longitudes	270°/285° 270°/-300°	300°/315° 300°/-330°	330°/345 330°/-0°	0°/15° 0°/-30°	30°/45° 30°/-60°	60°/75° 60°/-90°	90°/105° 90°/-120°	120°/135° 120°/-150°	150°/165° 150°/-180°	180°/195° 180°/-210°	210°/225° 240°/-270°	240°/250° 210°/240°
Equatorial time zones	18"/19"	20"/21"	22"/23"	0"/1"	2"/3"	4"/5"	6"/7"	8"/9"	10"/11"	12"/13"	14"/15"	16"/17"

Abbreviations:
WS - Winter Solstice; LS - Lesser Cold; GC - Greater Cold; BS - Beginning of Spring; R - Rains; AC - Awaken of Creatures; VE - Vernal Equinox; CaB - Clear and Bright; GR - Grain Rain; BSu - Beginning of Summer; LF - Lesser Fullness; GE - Grain in Ear; SS - Summer Solstice; LH - Lesser Heat; GH - Greater Heat; BA - Beginning of Autumn; EH - End of Heat; WD - White Dews; AE - Autumnal Equinox; CD - Cold Dews; DHF - Descent of Hoar Frost; BW - Beginning of Winter; LS - Lesser Snow; GS - Greater Snow.

FERDINAND VERBIEST'S CONTRIBUTIONS TO CHINESE SCIENCE

XI ZEZONG

Xi Zezong is a professor of the history of science and the former director of the Institute for the History of Natural Science, Academia Sinica, Beijing. He is also a corresponding member of the Academy of History of Science of Paris. His main field of research is the history of astronomy, especially Chinese astronomy.

Liang Qichao (1872-1927), a famous Chinese scholar and one of the leading spirits of the Reform Movement of 1898, said:

> The most important event worth writing about in the history of Chinese scholarship in the late Ming and early Qing dynasties (sixteenth and seventeenth centuries), was the introduction of European astronomy and mathematics into China. If the propagation of Buddhism during the Jin and Tang dynasties is regarded as the first understanding between Chinese and alien cultures, the introduction of western science in the sixteenth and seventeenth centuries is certainly the second. Though Islamic civilization was introduced in the Yuan dynasty, it was on a small scale and had only little influence on the Chinese. In these new historical circumstances the style of study was transformed. Many scholars became interested in the New Western Methods and enjoyed talking about practical problems[1].

According to statistics compiled by Professor Nathan Sivin, from the beginning of the Qing period (1644) to Wang Yinzhi (1766-1834), of thirty-six famous scholars, eighteen left a total of seventy-two books on astronomy and mathematics. Although the rest of them wrote no special book on these topics, they often dealt with their problems. Sivin also noticed that these works not only had a great influence on the development of Chinese science itself, but also brought about changes in

---

1 Liang Qichao, *Zhongguo jin sanbainian xueshu shi* (The History of Science in China during the last 300 years) (Shanghai 1936), pp. 8-9.

Chinese academic circles both in ways of putting problems and in methods of solving them[2].

Of those who first introduced western science into China, three Jesuits made a prominent contribution and are generally regarded as the pioneers. They are Matteo Ricci of Italy (1552-1610), who arrived in China in 1582; Johann Adam Schall of Germany (1592-1666), who arrived in China in 1622; and Ferdinand Verbiest of Belgium (1623-1688), who arrived in China in 1659. In honor of the transcendent contributions of the three Jesuits to Chinese science, the government of the People's Republic of China repaired their tombs at Beijing in 1978. Here I would like to deal with the contributions of Verbiest. But because of the limited documents to which I can refer, my paper cannot cover the whole subject, and is nothing more than an introduction.

I.
The Struggle for the Application of European Astronomy to the Chinese Calendar

In the latter part of the sixteenth century, Chinese agriculture and handicrafts were developing rapidly. This development threatened the idealist system of philosophy maintained since the Song and the Yuan dynasties. The result of the currents was an advance in approaching problems of science and technology. Many well-known scientific books were published in that time, such as *Bencao gangmu* (Great *materia medica*, 1596) by Li Shizhen, *Tiangong kai wu* (Encyclopedia of Technology, 1637) by Song Yingxing, *Youji* (Travel Notes, 1640) by Xu Xiake and so on. As for astronomy, calendar reform was needed because the *Da tong* calendar, which had been in use for over two centuries without improvement, was out of date. Predictions of eclipses and other celestial phenomena based on the *Da tong* calendar often did not accord with reality. Just at this time European missionaries arrived in China to promulgate Christianity. They realized that by introducing science and technology, the door closed to Christians could more easily be thrown open. They expected that their help in calendar reform would win them favor with the court, which would give them greater opportunities for preaching. This idea was clearly stated in Ricci's letter of May 12, 1605:

---

2 Nathan Sivin, "*Wang Hsi-shan*", in *Dictionary of Scientific Biography*, 14:159-166.

> If this astronomer were to come to China, after we had translated our tables into Chinese, we would undertake the task of correcting the calendar, and thanks to that, our reputation would go on increasing, our entry into China would be facilitated, our sojourn there would be more assured and we would enjoy greater liberty[3].

At the invitation of Ricci, missionaries who had a good command of astronomy arrived in China in 1620. These included Niccolò Longobardo (1565-1654), Johann Terrenz Schreck (1576-1630), Giacomo Rho (1592-1638) and Schall. In 1629, after a serious and long-fermenting controversy, the Imperial Bureau of Astronomy (*qintianjian*) again failed to predict an eclipse. Therefore the Qing court determined to adopt the methods of European astronomy to compile a new calendar, and appointed Xu Guangqi (1562-1633) to take charge of the task. Under the leadership of Xu Guangqi, the Calendrical Board, consisting of about one hundred members, was founded. Rho and Schall were asked to compile and translate astronomical books. By 1634 a great compendium containing one hundred and thirty-seven volumes was completed, under the title *Chongzhen lishu* (Chongzhen Reign-period treatise on calendrical science). However, many scholars who adhered to the old line vehemently opposed the new method and impeded the implementation of the new calendar. The dissension continued for a long time, so that by the time of the fall of the Ming dynasty the new calendar had still not been put into use.

In the fifth lunar month of 1644, Manchu troops entered Beijing and the Qing dynasty was founded. Thereafter regent Dorgon summoned the officials of the former Bureau of Astronomy, and also Schall, to debate the calendars that each had prepared for the fall of the year (1645). As a result, Schall pointed out seven mistakes in the old calendar prepared by the officers of the Bureau of Astronomy. On the contrary, in the new calendar prepared by the missionaries, there were almost no mistakes to be found. The officials could only say: "We learned our calculation methods from the sages of former times, and did not dare to renounce the resulting calendar and be obedient to the foreign one." On the first day of the eighth lunar month of the same year, a solar eclipse took place. Schall and the director of the Bureau of Astronomy, bringing with them the prediction tables for this eclipse as calculated by themselves, held a test in the old observatory. The court sent Feng Quan, an

3 Matteo Ricci, *Opere Storiche* (Tacchi Venturi, ed.) (Macerata 1910-1913), 2:284-285; Henri Bernard, *Matteo Ricci's Scientific Contribution to China*, translated by E.C. Werner (Peiping, 1935), p. 56.

advisor to the emperor, as the arbiter of the test. The new western calendar accorded with the celestial phenomenon very well, but the *Da tong* calendar showed a difference of two *ke* (30 minutes), and the Islamic calendar a difference of three *ke* (45 minutes). Thereupon the court appointed Schall director of the Bureau of Astronomy and adopted the new western method to calculate the official ephemerides for the next year. On the front cover of the ephemerides the five Chinese characters *Yi xiyang xinfa* (Based on the new western method) were printed. From then on, Schall was given increasingly full confidence and had a position of special favor with the court; in the fifteenth year of his reign the Shunzhi (1658) emperor graciously granted him the rank of *guanglu dafu* (an official of the first class) and bestowed honors on his ancestors for three generations.

These favors distinguished the achievements of the missionaries, but thereby the contradiction between them and the feudal idealist philosophers became more acute. As Verbiest was summoned to Beijing to assist Schall in the compilation of the calendar in 1660, a storm over this new calendar was breaking out. In 1657 Wu Mingxuan, in control of the Moslem Calendar Section, first accused Schall in court of miscalculating the positions of Mercury. The accusation was based on the supposition that Schall's prediction that Mercury would not appear in the sky in the second month and the eighth month was wrong. The emperor thereupon appointed high officials to go to the observatory three times to examine the case (on the tenth day of the fourth month, on the twentieth day of the eighth month and on the fifth day of the ninth month, respectively). Mercury could not be seen on any of these occasions. The accusation thus proved false, and Wu Mingxuan was sentenced to death, though the execution was postponed.

When Wu Mingxuan attacked the calendar compiled by Schall, the controversy was also taken up by a scholar from Anhui province named Yang Guangxian, who knew about calendrical science to some extent. In the twelfth month of 1660 Yang presented to the Board of Rites a report entitled *Zheng guoti cheng* (To Rectify the State System), accusing Schall of usurping state power by printing "Based on the New Western Method" on the front cover of the official ephemerides, and of computing errors in computing the calendar, but his report was ignored. Then in the seventh month of the third year of the Kangxi reign (1664), Yang presented another report called *Qing zhu xiejiao zhuang* (Challenging the Heretics) in which he wrote:

> Adam Schall dodged and hid in the court for the purpose of stealing secret information under the cover of compiling the calendar. The missionaries ganged up and were hatching a sinister plot. That is why they are building churches in the capital and other important cities and promulgate their doctrines to tempt the people. Their activities have violated our law and must be condemned.

This alarming report caused consternation in the Qing court. However, because of this report the regent Oboi was placed in an ambivalent position toward the westerners. So, on the sixth day of the eighth month the Qing court brought Schall, Verbiest and other missionaries in Beijing to a public trial. At that time Schall was seventy-three years old, and had recently become ill, so that he was not able to say a single word in his defense. To undertake the cause of the missionaries, Verbiest replied as their spokesman, but he was stopped by the judge. Consequently, on the first day of the fourth month of 1665, the Boards of Rites and of Punishments drew up a proposal according to which Schall would be put to death by dismemberment, all of the missionaries in Beijing would be banished, others in every province would be deported under escort from Guangzhou, seven Chinese officials in the Bureau of Astronomy would be put to death by dismemberment, five others would be condemned to death, and Christianity would be proscribed. The charges against them were absolutely ridiculous, namely that the Bureau of Astronomy failed to choose the correct day for holding a ceremony for the youngest son of the Shunzhi emperor, and that the missionaries' so-called "Ten Thousand Years Calendar" only contained two hundred years. All of these charges were quoted from essays entitled *Xuanze yi* (On choice) and *Zhai miu shi lun* (On ten mistakes), written by Yang Guangxian. On the next day, while the regents were holding a meeting to ratify the proposal, they had to flee in alarm from a sudden earthquake. Thereafter, earthquakes continued from time to time and a comet appeared in the sky. The superstitious Qing court regarded these phenomena as manifestations of the anger and discontent of Heaven. According to the established custom of preceding dynasties, offenders in such cases must have their penalties reduced. Hence, only the five officials of the Bureau of Astronomy were executed, and Schall, Verbiest and the other missionaries were released from prison and allowed to stay in the Beijing church temporarily.

After Schall, Verbiest and the others were discharged, the Qing court appointed Yang Guangxian as director of the Bureau of Astronomy and Wu Mingxuan as the vice-director. Yang Guangxian tried five times to resign from the position because he only knew calendrical principles in outline and was not familiar with the methods of calculation. His resignation was not approved and he was obliged to be in charge of the Bureau of Astronomy. Later, he collected his *Qing zhu xiejiao zhuang* (Challenging the heretics) and other articles into a book, which was published under the title *Budeyi* (I cannot do otherwise). Yang Guangxian first adopted the *Da tong* calendar and then used the Moslem calendar, but whichever method he used, the results calculated often did not accord with the celestial phenomena. Therefore in the eleventh month of 1668, the Kangxi emperor, who was only fourteen years old, eventually decided to test by practice what was correct in the new and old methods. Many high officials were appointed as witnesses. In the observatory, Verbiest (by that time Schall had died), Yang and Wu were asked to predict the noonday shadow length of the gnomon. Verbiest calculated that if the gnomon was 8.49 *chi* (Chinese feet) in height, its shadow would be 16.66 *chi*. At the next noonday the shadow coincided exactly with the length that he predicted. Though Yang and Wu did not know how to calculate the shadow length, they still did not admit defeat. Wu said the shadow would be 9 *fen* (Chinese inches) longer, and Yang that it would be 6 *fen* longer, than that predicted by Verbiest. On the twenty-fifth of the same month the second test was held in front of the Wumen Gate of the palace. Verbiest correctly said that the 2.2 *chi* high gnomon would have a 4.345 *chi* shadow. On the next day the third test was in the observatory again. Verbiest correctly predicted that an 8.55 *chi* high gnomon would have a 15.83 *chi* shadow. These three test results proved that Verbiest was right in each case and that on the contrary, Yang and Wu were always wrong. Knowing these results, the emperor decided that the civilian calendar and the ephemerides of the sun, the moon and the five planets, compiled by Wu Mingxuan should be passed on to Verbiest to examine. Yang Guangxian was utterly discomfited and argued:

> The calendar calculated by your subjects is based on that of the emperors Yao and Shun; the throne and the state system inherited and carried on by your majesty are all those of the emperors Yao and Shun. Your majesty accepted everything inherited from emperors Yao and Shun, but why only reject their calendar?[4]

In order to handle the problem with more care, the Qing court in the eighth year of the Kangxi reign, sent twenty high officials together with Verbiest, Yang, Wu and others to the observatory to test the mistakes

4 Huang Bolu, *Zhengjiao fengbao* (Shanghai, 1903), p. 23.

pointed out by Verbiest. The initial moment of the (fifteen-day solar period) beginning of spring, the beginning moment of the (fifteen-day solar period) rain water, and the positions of the moon, Mars and Jupiter all accorded with those predicted by Verbiest; Yang and Wu were wrong in every case. With such results, Wu and Yang were discharged, and the calendar of the next year was to be compiled by Verbiest. Though he declined an appointment as vice-director of the Bureau of Astronomy, he was put in charge of astronomy as general editor of the calendar with a salary of two hundred taels of silver and twenty-five measures of rice per annum, the same as the vice-director. Verbiest held the office until his death and was succeeded by Claudio Filippo Grimaldi of Italy (1638-1712, arrived in China in 1669). From generation to generation until the sixth year of the Daoguang reign (1826) this position was held by a missionary. In that year Verissimo Monteiro da Serra went back to Portugal for treatment of an illness.

After Verbiest took charge of the Bureau of Astronomy, he wrote *Budeyi bian* (I cannot do otherwise refuted), and other books. In these books the fallacies presented by Yang Guangxian in *Budeyi* were refuted one by one. He also presented a report to the court demanding that Yang Guangxian be punished for having framed a case against upright persons, in that Li Zubai and four other officials had been executed, and for having trumped up a charge against Schall of plotting a rebellion. The emperor set aside the previous judgment. Schall, Li Zubai and the other officials were rehabilitated. A tomb and monument were built in memory of Schall and the emperor sent a high official to manage a memorial ceremony for him. All the persons previously absolved in the case were recharged, and Yang Guangxian was sentenced to death. On account of his advanced age he was excused from punishment and sent to his hometown. On his way, however, a pestilential disease killed him. This long-term struggle ended with the complete triumph of Verbiest and the thorough defeat of Yang and Wu.

Verbiest's triumph in the struggle was a victory for science. Yang Guangxian was blind in his opposition to everything foreign, and clung to traditional methods in order to assert feudal interests. His suspicion that Schall was plotting a rebellion was nothing more than a misunderstanding of Christianity. His knowledge about astronomy was so poor that, as Ruan Yuan, the author of *Chouren zhuan* (Biographies of mathematicians and astronomers), said, Yang was "forging records by

Ill. 27: The Ancient Peking Observatory.
Unumbered woodcut from the *Xinzhi yixiang tu* of F. Verbiest (Beijing, 1674).

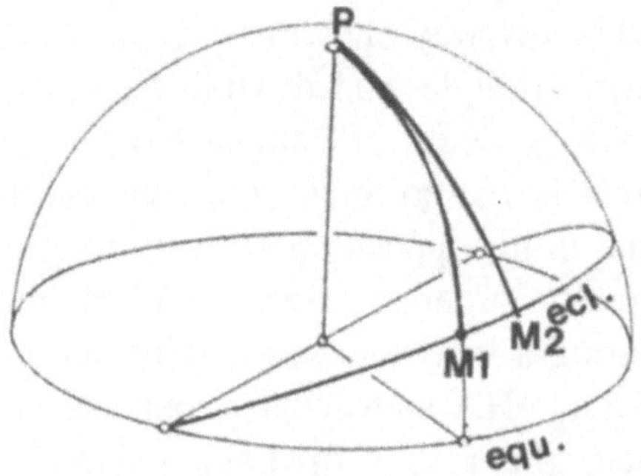

Ill. 28: The ecliptic extensions of the lunar lodges obtained by measuring the ecliptic arc between right ascension circles PM1 and PM2.

the old method, forcing the celestial bodies to be obedient to the imagination of man, misapplying the scientific apparatus, and disturbing the harmony of the course of the heavens"[5]. Also Lu Xun, the chief leader of the new Chinese cultural movement, and a great Chinese writer and thinker, satirized Yang's notorious xenophobic idea, that "preferred to use the old calendar in China rather than to have it improved by a westerner." Yang's attitude was laughed at by later scholars. Such a clownish stand against the development of science cannot be considered typical of the Chinese people. Actually, at that time, many scholars such as Xue Fengzuo (1600-1680), Huang Zongxi (1610-1695), Wang Xishan (1628-1682) and Mei Wending (1633-1721) had a correct attitude towards foreign things. These scholars studied European astronomy, analyzed it, and even more or less developed it. But their work is outside the scope of this paper.

II.
Making Six Astronomical Instruments

Verbiest considered that instruments are most important in compiling a calendar, because only "with the help of instruments could the theory and methods of the calendar become more and more precise and accurate, and it is impossible for the compiler to get a good calendar without making precise apparatus"[6]. After he had taken charge of the Bureau of Astronomy, he at once began to design and to make instruments. With hard work, for four years, six new bronze instruments were made in 1673, and are now still standing at the site of the old Beijing Observatory. They are the ecliptic armillary sphere (*huangdao jingweiyi*), the equatorial armillary sphere (*chidao jingweiyi*), the horizon circle and azimuth (*diping jingyi*), the quadrant (*diping weiyi*), the sextant (*jixianyi*) and the celestial globe (*tiantiyi*). Of them the ecliptic armillary sphere and the sextant were unknown before in China (Ill. 27).

Earlier Chinese bronze ecliptic instruments (*huangdao tongyi*) were simply made from the equatorial armillary sphere with an ecliptic circle. The so-called ecliptic extensions of the lunar lodges then were obtained by measuring the ecliptic arc between right ascension circles PM1 and

---

5 Ruan Yuan, *Chouren zhuan* (Biographies of Mathematicians and Astronomers), *juan* 45, Nan Huairen (Shanghai, 1935), p. 590.

6 Nan Huairen, *Lingtai yixiang zhi* (Record of Newly-made Instruments in the Observatory) (Beijing, 1674).

PM2, which separately pass the two controlling stars of neighboring lunar lodges (Ill. 28). This ecliptic extension is different from the longitude difference based on considering the ecliptic pole and circle as the cardinal point and circle. The inner and outer degree of the ecliptic is also different from the present latitude, because the former is measured along the right ascension circle rather than the longitude circle. Thus some scholars referred to ancient Chinese ecliptic data as polar longitude and polar latitude[7]. But using the ecliptic armillary made by Verbiest, we can directly get the longitude and latitude of celestial bodies. The structure of Verbiest's ecliptic armillary is shown in illustrations 29 and 30.

Ill. 29: Ecliptic armillary
(F. Verbiest, *Xinzhi yixiang tu* [Beijing, 1674], fig. no. 1).

7 Yabuuchi Kiyoshi, *Chûgoku no temmon rekihô* (Astronomical Science in China) (Tokyo, 1969), p. 55.

On an oblique cross there stand two clambering dragons which support a half circle with cloud ornaments, on which a meridian circle with a diameter of 6 *chi* is fixed. The uppermost point of the meridian circle is the zenith. For Beijing ($\Phi = 40°$), the north pole is located northward 50° from the zenith; the point opposite is the south pole. Through the south and north pole is installed a right ascension circle joined to the

Ill. 30: Verbiest's ecliptic armillary sphere
on the roof of the Ancient Peking Observatory (Photo: courtesy of Xi Zezong).

Ill. 31: Equatorial armillary
(F. Verbiest, *Xinzhi yixiang tu* [Beijing, 1674], fig. no. 2).

Ill. 32: Verbiest's equatorial armillary on the roof of the Ancient Peking Observatory (Photo: courtesy of Xi Zezong).

meridian circle by two short steel axles. The ecliptic poles are two points on right ascension, the angular distances from which were 23°21'30" at that time. The ecliptic circle is perpendicular to the ecliptic poles. It intersects with the right ascension circles at two points: the one near the north pole is the summer solstice and the other near the south pole is the winter solstice. So this particular right ascension circle is called the "circle through poles and solstices." It is joined to the ecliptic axis through the ecliptic poles. The longitude circle is a circle rotating around the ecliptic axis and with a smaller diameter. At the middle of the ecliptic axis there is a small cylinder perpendicular to the longitude circle.

For use in observations, several verniers could be fixed on the ecliptic circle or longitude circle. These verniers are now absent. Professor Chang Fuyuan has inferred that it was necessary for determining the longitude and latitude of a star to select another star, the position of which was already known, as controlling star. First one would locate the latter on the ecliptic circle by a vernier and then observe the former by another vernier. The longitude of the star under observation would be determined by comparing the positional difference of both stars. In the same way, by rotating the longitude circle and using verniers, one could get its latitude[8].

The mounting of the equatorial armillary (Ill. 31 and 32) is the same as that of the ecliptic armillary, but its structure is rather simple, having only three circles: equatorial, right ascension and meridian. Instead of an ecliptic axis, there is a steel axis through the north and south poles and from the latter stretch two quadrant arcs to support the equatorial circle. In the equatorial circle there is a right ascension circle which can rotate around the axis. The equatorial circle is combined with the meridian circle and both of them are fixed. In principle this instrument is the same as ancient Chinese armillary spheres, the only exception being that the observation is by means of a vernier rather than a sighting tube.

The horizon circle and azimuth (Ill. 33 and 34) have only one circle. It is six *chi* in diameter and four *chi* in height, and is supported by a vertical pillar and four bronze clambering dragons. On two of the clambering dragons (the eastern and the western ones) there stand two columns decorated with spiralling dragons, four *chi* high. Each dragon at the top of a column stretches a paw to support a "fire ball." This fire

8 Chang Fuyuan, *Tianwen yiqi zhi lüe* (Beijing, 1930).

ball corresponds to the zenith. Between the fire ball and the center of the horizon circle there is a rotating long axis, the top of which enters the fire ball, and the lower part of which enters the center of the central vertical pillar. When the standing axis rotates, an alidade perpendicular to and combined with it above the horizon circle rotates simultaneously. A wire stretched in a line from both ends of the alidade and attached to the fire ball forms two right triangles. Rotating the alidade and making the two wires coincide with a star, we can get the azimuth of the star by reading the graduation pointed out by the alidade on the horizon circle. This method is the same as that used in *liyunyi* (the standing-rotating instrument) invented by Guo Shoujing (1231-1316). The only difference is that the *liyunyi* could fix both the azimuth and the altitude of a celestial body, but this instrument could only find the azimuth. It must be admitted that this is a shortcoming.

Ill. 33: Horizon circle
(F. Verbiest, *Xinzhi yixiang tu* [Beijing, 1674], fig. no. 3).

Ill. 34: Verbiest's azimuth on the roof of the Ancient Peking Observatory (Photo: courtesy of Xi Zezong).

Ill. 35: Horizontal altitude instrument (quadrant)
(F. Verbiest, *Xinzhi yixiang tu* [Beijing, 1674], fig. no. 4).

Ill. 36: Verbiest's horizontal altitude instrument on the roof of the Ancient Peking Observatory (Photo: courtesy of Xi Zezong).

Ill. 37: *Jixianyi* (sextant)
(F. Verbiest, *Xinzhi yixiang tu* [Beijing, 1674], fig. no. 5).

Ill. 38: Verbiest's *jixianyi* (sextant) on the roof of the Ancient Peking Observatory (Photo: courtesy of Xi Zezong).

The apparatus made by Verbiest for determining the altitude or zenith distance of a heavenly body is the horizontal altitude instrument. It is also called the quadrant because it has only a quarter of a circumference (Ill. 35 and 36). There are two graduation scales on the quadrant. The graduation on the inside represents the altitude from the upper 0 to the lower 90 and that on the outside represents zenith distance from the lower 0 to the upper 90. The center of the circle is connected with both ends of the arc by two radial rods six *chi* in length. The plane formed by the arc and two radii is decorated with dragons and cloud ornaments. The quadrant is mounted in a frame set on a pedestal. The pedestal has a long base and two cross-pieces. At the intersection of each cross-piece, a pillar is set up; the tops of the pillars are spanned by a cross beam. From the midpoint of the pedestal to the midpoint of the cross beam there is a rotating axle, 9.6 *chi* in length. The solid plane of the quadrant is attached to this axle. At the intersection of both radii of the quadrant there is a small cross axle perpendicular to the instrument plane. A sighting tube is set on the small cross axle close to the instrument plane. The length of the sighting tube is equal to the radius of the circle. Attached to the lower end of the sighting tube is a small leaf. To take observations, one rotates the instrument plane and elevates the sighting tube. When the sighting tube is aligned with a star, the point on the quadrant's circumferential scale with which the leaf on the end of the tube coincides gives the altitude and zenith distance of the observed star.

In addition to these instruments to make observations in the three celestial coordinates, Verbiest made an apparatus to measure the angular distance between two stars. This is the *jixianyi* (the recorder of limitations), also called the sextant, because its main part consists of one sixth of a circumference (Ill. 37 and 38). The graduation of its scale begins from the middle of the arc and extends 30° on each side. The midpoint of the arc and the center of the circle are connected by a bronze radial rod mounted on a transverse axle which itself is the chord of a semicircular gear. The gear is supported by a column mounted on a pedestal decorated with dragons. A handwheel is provided to rotate the gear, which thus tilts the transverse axle. Meanwhile the plane of the sextant can be rotated around the vertical column. On the plane of the sextant there are a sighting tube and two verniers. The sighting tube and the bronze radial rod are joined by a small axle at the center of the circle. Rotating the sextant to face two stars, one observes the first star by aligning the standing ear of the radial rod with the small central axle, and then observes the second star through the vernier of the sighting tube in the same way. The number of degrees between the sighting tube and the radial rod is the angular distance of both stars.

Apart from the five instruments mentioned above, Verbiest made a demonstrational celestial globe (Ill. 39). The shape and structure of the celestial globe are similar to that of ancient Chinese *hunxiang* (celestial globes), but there are differences as follows: (1) stellar magnitude is denoted by stellar size; (2) there are constellations in the region of the south pole; (3) the ecliptic is divided into twelve equal parts, the longitude of which intersects at the ecliptic pole; (4) a time disc is installed on the meridian circle, divided into twenty-four parts representing twenty-four hours; and (5) a quadrant of a gear is mounted under the meridian circle, meshing with a gear in the instrument's housing rotated by a handle in order to change the altitude of the North Pole to suit the geographic latitude of the observational position.

Ill. 39: Verbiest's demonstrational celestial globe
(F. Verbiest, *Xinzhi yixiang tu* [Beijing, 1674], fig. no. 6).

It is now known that any of the three coordinates of a system can be transferred to each other. So long as there are data of one coordinate of a star, the other two coordinates can be obtained by calculation. Once the coordinate data of two stars are measured, to calculate the angular distance between them is possible. Therefore, one instrument is sufficient for astronomical observation. Why did Verbiest make five pieces of observational apparatus? This was because calculating methods then were not so developed, so that the transformation of coordinates was inconvenient, and also because Verbiest considered that by using various apparatus to observe celestial bodies, he could compare the results of each observation thus heightening its precision. He said:

> For the observation of heavenly movements, the more accurate instruments that are used, the more precise results are obtained. Six instruments used simultaneously, can be mutually complementary, as an observation may be inconvenient for this instrument, but convenient for that instrument. If the instrument is made precisely, the installation of the instrument is correct and the observation is in accordance with the correct method, the results of observation by various instruments will have to agree with each other. If there are differences, it is necessary to seek why they are not in agreement, and to correct them. After doing so, if there are still differences, use the observational value shared by most instruments. By this means, it would not be possible for the results of observations to be in disagreement with the movements of the heavens![9]

Verbiest's statement is quite correct. Using these instruments made by Verbiest, Chinese astronomers carried out observations for more than two centuries, using the data obtained to compile two star catalogues: one is the star catalogue of the *Yixiang kaocheng* (Description of astronomical instruments), consisting of ecliptic and equatorial coordinates of 3,083 stars for the year 1744; another is the star catalogue of the *Yixiang kaocheng xupian* (Supplement to the description of astronomical instruments) consisting of ecliptic and equatorial coordinates of 3,240 stars for the year 1844. These catalogues serve as a link between past and future. The present Chinese names of the stars are mainly based on these catalogues. Verbiest's contribution to making instruments is immortal. As historical witnesses to cultural exchange between the West and the East, his instruments today are still standing on the site of the old Beijing Observatory.

---

9 Nan Huairen, *op. cit.* (Beijing, 1674).

III.
Introduction of Physical Knowledge to China

After making his six astronomical instruments, Verbiest wrote a great work entitled *Xinzhi lingtai yixiang zhi* (Record of newly-made instruments in the observatory), or *Lingtai yixiang zhi* or *Yixiang zhi*. It comprised sixteen volumes. The first fourteen volumes explain how to make and use the six pieces of apparatus; these were printed during the Kangxi reign. However, the last two volumes - the fifteenth and sixteenth - are extremely seldom seen; they contain 117 figures to illustrate the preceding volumes. Although in the *Gujin tushu jicheng lifa dian* (Calendrical section of the Qing imperial encyclopedia) all 117 figures were printed, some of them had their original quality impaired by printing one figure in separate halves on two pages. According to the H.P. Kraus catalogue, n° 137, the Columbia University East Asian Library in the United States, the School of Oriental and African Studies, London, England, the Bibliothèque Nationale, Paris, and the Belgian Royal Observatory preserve copies of these volumes. But all of them contain only 106 figures. Fortunately, in the rare book room of the Beijing Library, the author has seen all of the figures printed exquisitely in four volumes instead of two volumes. In the beginning of this set of figures there is a foreword written by Verbiest, and many signatures and commentaries, as shown in illustration 40, written by well-known Chinese scholars after they had read the book.

1. Li Guangdi (1642-1718) commented: "I got this book from Verbiest. I have for a long time appreciated his astronomy and mathematics. These figures are drawn so clearly and exquisitely that much hard work may be eliminated by studying them."
2. Mei Wending (1633-1721) borrowed it for fifty days and then commented: "This book employs new methods to revise the old ones, bringing the methods of the Ming period up to the present day. When I borrowed this book I was appointed to revise the *Lülü shu* (Collected basic principles of music) and had not time to read this carefully, so I asked Mr. Gaiyu to transcribe a copy. After fifty days I returned it."
3. Yan Ruoqu (1636-1704) read and revised it.
4. Zhu Yizun (1629-1709) and Wang Xishan (1628-1682) read it together.
5. Li Zhonglun (1663-1706) read and revised it.

Some European scholars also commented:

Ill. 40: Verbiest's foreword to *Xinzhi yixiang tu* (Beijing, 1674).

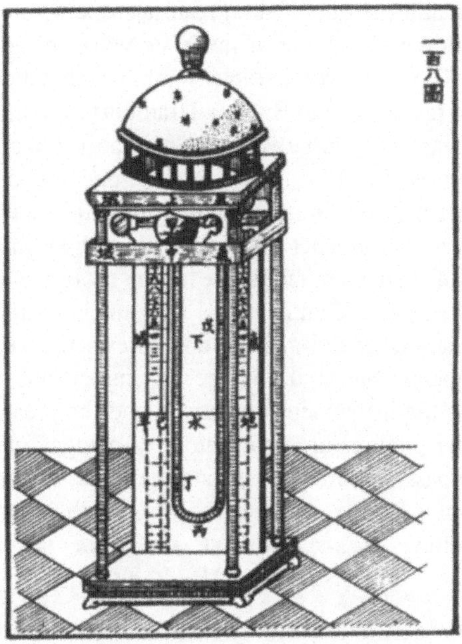

Ill. 41: Verbiest's air thermometer (thermograph)
(F. Verbiest, *Xinzhi yixiang tu* [Beijing, 1674], fig. no. 108).

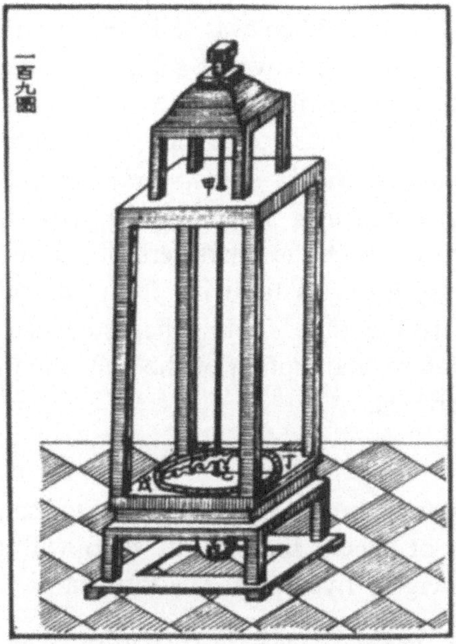

Ill. 42: Verbiest's hygrometer
(F. Verbiest, *Xinzhi yixiang tu* [Beijing, 1674], fig. no. 109).

> This publication is one of the greatest masterpieces of Sino-European printing and certainly one of the basic works of astronomical technology. ... The only comparable work which comes to mind is the group of a few of Tycho Brahe's Uraniborg buildings and instruments which appear in the large Blaeu Atlases at just about the same time as the present work. But the Chinese woodcuts must be considered vastly superior, both in intrinsic interest and in number. They display not only the instruments themselves, but show in amazing detail the processes of their manufacture, with the tools and implements used to produce them; the alignment and adjustment of their flat and curved surfaces; details of the gearing and screws used to adjust and direct the instruments; the civil engineering implements and processes used in building the instrument mountings and the great observatory tower itself. Other pictures are explanatory of navigational instruments such as the compass and cross staff, and their use; astronomical principles; and mechanical powers, such as that of the inclined plane, lever, screw, pulley, etc. (*Important Works in the Field of Science*, Catalogue No. 137 [New York: H. P. Kraus, n.d.], pp. 47-48).

In addition to the achievements mentioned by Kraus, we can find recorded in this book some other figures of instruments and physical concepts, which were advanced even by contemporary European standards.

1. Pendulum. This was called "The vertical string spherical instrument" by Verbiest. He said: "When was it invented? In recent decades a famous western scholar discovered some new characteristics of the vertical string spherical instrument." When he said "a famous western scholar," he referred to Galileo, and some of the contents of the fourth volume of *Xinzhi lingtai yixiang zhi* drew its materials from *Dialogues Concerning Two New Sciences* written by Galileo. Verbiest described two basic characters of the pendulum. He said: "The time the sphere takes to swing from one side to the other is equal, and so is the time from either side back to the original position." Therefore, the motion of the pendulum is isochronic. He also gave the formula:

$$l_1 : l_2 = T^2{}_1 : T^2{}_2 = n^2{}_2 : n^2{}_1$$

and gave the following examples: the string of sphere A is 1 *chi* in length and that of sphere B is 2 *chi*, so B will swing sixty times while A swings eighty-five times. The formula is:

$$1 : 2 = N_2{}^2 : 85, \text{ hence } N_2 = (1/\sqrt{2}) \times 85 = 60.8).$$

This result means that the length of the string is proportional to the square of the oscillation period, and inversely proportional to the square of the frequency.

2. The motion of freely falling bodies. When Verbiest discussed the pendulum, he incidentally discussed the motion of freely falling bodies and said: "The distance a body falls is proportional to the square of the time during which it is falling." This means that:

$$s = 1/2\ gt^2 \text{ and } s^1 : s_2 = t^2{}_1 : t^2{}_2.$$

He also said: "The proportion is a series of odd numbers, such as 1, 3, 5, ..." That is to say, if a body falls 1 *chi* in the first second, it then falls $2^2 - 1 = 3$ *chi* in the 2nd and $3^2 - (1 + 3) = 5$ *chi* in the 3rd second etc. That is, the ratio of the falling distance in each consecutive second is equal to the ratio of consecutive odd numbers. The formulation of the concept = agrees with that of theorem two, deduction one in proposition two, in the third day of Galileo's *Dialogues Concerning Two New Sciences*[10].

3. The thermometer. It is not enough to measure changes of the temperature only by the sense of touch. For accuracy temperatures must be measured by a thermometer. Verbiest said: "If the external temperature is equal to our body's, we cannot sense it; we can only do so in the case of higher or lower temperatures. To remedy this defect, I designed a special instrument with which we can easily detect a change of temperature not by touch but by vision, using the most sensitive sensory organ of the human body." The thermometer made by Verbiest, being an air themometer[11], shown in illustration 41, consists of a glass bulb, denoted A, which was attached to a stem with a fine bore, denoted BCDE. The CBD parts of the stem are filled with water. When the bulb becomes warmer, the gas in the bulb expands. The water in the left hand side of the stem will be pressed down and the water in the right hand side will rise. If the bulb becomes colder, the process will be the opposite. The amount of expansion is indicated on a scale, which can represent the temperature change.

---

10 Yan Ganjie, "Qie Lilei de gongzuo zaoqi zai Zhongguo de chuanbo" (The Early Propagation of Galileo's Work in China) in *Kexue shi jikan*, No. 7 (1964), 59-67.

11 Wang Bing, "Nan Huairen jieshao de wenduji he wenduji shi xi" (Analysis of the Thermograph and Tests with the Thermograph introduced by Ferdinand Verbiest), in *Kexue shi yanjiu* (Historical Research on Science), No. 5(1986), 76-83.

4. The hygrometer. Verbiest said: "Among all things in the world, the tendons and skin of beasts and birds are the most sensitive to changes in moisture and dryness." Hence he made a hygrometer of a fresh tendon of a deer, shown in illustration 42. The tendon is two *chi* in length, suspended from a bracket at its upper end and weighted at the bottom. The upper part of it was under tension and the lower part was wrapped around a rectangular tube behind which a plate was placed. The plate was divided into two parts, to the left and right of the tube. On both of them scales were engraved to indicate the change in humidity. The left-hand scale indicated dryness, and the right-hand scale indicated moisture. The tube was decorated with a dragon and a fish. A change of humidity will make the tendon become looser or tighter which gives rise to a rotary motion of the dragon tube. The dry air makes it turn to the left and the moisture makes it turn to the right. Recently, Wang Bing pointed out that Verbiest's hygrometer was very similar to that made by William Molyneux (1656-1698), but Verbiest's description was earlier than that of Molyneux[12].
5. The law of refraction. Verbiest said: "If the light of the sun, moon and stars travels from a body, through which light travels easy, into another body, through which light passes with difficulty, the light will bend towards the top of the latter. If the light travels conversely, it will bend away from the top." What he said is a statement of the law of refraction, invented by Christian Huygens in 1690. Verbiest knew that fact more than ten years earlier than Huyghens, but not with complete precision. Huyghens stated that $n_{21} = v_1/v_2$ (v_1, v_2 is the velocity of light in medium 1 and 2), that is, when a ray of light travels from an optically thinner medium into an optically denser medium ($n_{21} > 1$), the refractive light will be towards the normal line and when it travels conversely ($n_{21} < 1$), it will be away from normal.

IV.
Other Important Writings and Contributions

Apart from making the six instruments and writing the *Xinzhi lingtai yixiang zhi*, Verbiest wrote two other important works on astronomy: *Kangxi yongnian biao* (Perpetual tables compiled in the Kangxi period,

---

12 *Ibid.*

1678), and *Jiushidu biao* (Tables of ninety degrees, 1682). The former, together with the *Huangyu quanlan tu* (The imperial comprehensive map), were called the "two cultural treasures of the Kangxi period" by the Japanese scholar Inaba Kimiyama. The *Kangxi yongnian biao* consists of eight parts (solar motion, lunar motion, eclipses and the five planets), each comprising four volumes. Some fundamental data are given at the beginning of every part and then the ephemerides of a celestial body (such as Jupiter) is given for a period of 2000 years. Having these ephemerides, it is easy to compile a civilian calendar. The *Jiushidu biao* is a special table for calculating solar eclipses occurring at Shengjing (now Shenyang ($\Phi = 42°$), so it is also called "Ephemeris of Solar Eclipses." It lists the zenith distance, termed "Limit 90°." The "Limit 90°" means a point on the ecliptic, from which the distance to the intersection between the ecliptic and horizon circle is 90°.

However, the more important contribution of Verbiest to China is that, with his activities, he had a far-reaching influence on the life of the Kangxi emperor. Of all emperors in Chinese history, he was the most interested in natural science. When he talked with his ministers about the reasons for his study of natural science, he said:

> You only know that I have mastered mathematics, but you do not know why I have learned it. In my childhood, relations between the Chinese and western scholars of the observatory were not peaceful. They lodged many complaints against each other with the government. It almost led to some of them being killed. Yang Guangxian and Ferdinand Verbiest measured the noonday shadow length of the gnomon in the presence of the court ministers outside the Wumen gate of the palace, but none of the ministers understood the theory and method they used. I thought it was impossible to decide what is right if you did not know it yourself, so I made up my mind to learn natural science[13].

He took Verbiest as his teacher and learned mathematics, astronomy and mechanics. In 1682 Verbiest wrote to his confreres in Europe that Jesuits who mastered astronomy, optics and mechanics were welcome in China. Both the pay and the favorable conditions given to them by the Chinese emperor were more than that given to dukes or princes, and they always worked in the palace and talked with the emperor[14]. Among

13 "Kangxi huangdi yu ziran kexue" (The Kangxi emperor and Natural Science), in *Nankai xuebao* 1980, 59-67.

14 *Ibid.*

his teachers were Tomas Pereira (1645-1708), who arrived in Beijing in January, 1673 and also Jean-François Gerbillon (1654-1707) and Joachim Bouvet (1656-1730), both of whom arrived in Beijing in February, 1688. Among them, Pereira followed the emperor for thirty-six years until his death. Along with the minister Songgotu, Pereira and Gerbillon went to the Sino-Russian boundary to hold negotiations between China and Russia, with the result that the two countries signed the Treaty of Nerchinsk (Nibuchu), by which China recovered two thousand *li* of lost territory.

When the rebellion of Wu Sangui broke out, the emperor ordered Verbiest to make cannons. He made 350 pieces within a year. Once Verbiest held a cannon test at the Lugou Bridge, and Kangxi saw that the percentage of hits was very high. It is said that he was so happy that he took off his dragon costume and bestowed it upon Verbiest in gratitude for his services. To explain this technical knowledge Verbiest wrote a book entitled *Shenwei tushuo* (Illustrated book on martial prowess, 1681). Verbiest also painted a world map and wrote the books entitled *Kunyu tushuo* (Geography and maps) and *Kunyu waiji* (Appendix to geography and maps). The former was transmitted to Japan and, until the Meiji Restoration (1868), it was considered a definitive book for descriptions of the world.

Overall, the contributions of Verbiest to Chinese science were manifold. We do not have enough time to cover everything, but there is one more matter that we should deal with. It is that he made an experiment on the use of the steam engine in land and sea vehicles. In 1939 Leroy L. Thwing discovered that Verbiest had written a paper to describe his experiment in 1668. The English translation of the original text is as follows:

> Three years ago, to know the moving power of steam, they caused a wagon to be made of light wood about two feet long. It could be moved forward very easily. In the middle of it they placed a brazen vessel full of live coals, and upon that a boiler was placed. Through the hind axle a bronze gear was fixed, which was joined with a somewhat smaller gear through an upright shaft. If the upright shafts were turned around, the wagon would be driven forward. The vapor from the boiler came through a little pipe upon a sort of wheel made like the sails of a windmill; this little wheel turned another with an axle-tree, and by that means set the wagon in motion for two hours together; but, lest room should be wanting to proceed constantly forward, it was controlled to move circularly, in the following manner. To the axle-tree of the two hind wheels was fixed a small beam, and to the end of this beam another axle-tree, which

> went through the center of another wheel somewhat larger than the rest; and according as this wheel was nearer or farther from the wagon, it made a greater or lesser circle. This model indicated a principle of the moving power, and could be used in any other machine. They made a little ship as a present to the elder brother of the emperor. It had a boiler and could travel in a circle on the water. They also made use of it in a clock tower as the power source of a clock[15].

Verbiest commented on his experiment: "The moving power of steam being given, it is easy to make other applications of it." Verbiest's use of the word *they* is not a disavowal of his invention; it is probably a sort of editorial "we". Here Verbiest called his work an experiment rather than a discovery. His achievement may be based on Giovanni Branca's steam turbine. Branca was an Italian and his work *La Macchina* was published in Rome in 1629. Verbiest might have read this book, but it is of great significance that he then used a similar play-thing to drive a wagon or a little ship, improved them by adding steering gear, and suggested various uses for the steam turbine. In using steam as moving power, Verbiest's experiment was 123 years earlier than Symington's steamboat, 150 years earlier than George Stephensen's train, and 200 years earlier than Amédée Bollée's car. In using the steam turbine for moving power, Verbiest's experiment was 218 years earlier than Charles Parsons's steamboat and 243 years earlier than Ljungstom's train. Therefore, Verbiest should occupy an important position not only in the history of Chinese science, but also in the world's history of steam engines.

---

15 L.L. Thwing, *Technology Review*, 41:4 (1939), 169-170.

Verbiest

Engineer

THE "MANCHU CANNONS" CAST BY FERDINAND VERBIEST AND THE HITHERTO UNKNOWN TITLE OF HIS INSTRUCTIONS

GIOVANNI STARY

Giovanni Stary teaches Manchu language and literature at the University of Venice (Italy). His researches are devoted to Manchu literature, history and culture, with special reference to Manchu shamanism. He is editor-in-chief of the "Central Asiatic Journal" and published several monographs and articles.

Very few people are aware that seventeenth-century cannons were brought to Italy after the Boxer Rebellion. They are to be found in two Roman museums. Three of these cannons - quite small in size - are in the *Sala della Giustizia* of the *Castel Sant'Angelo* Museum. On the breech of each one there is a bilingual Chinese-Manchu inscription which is hard to read since it is considerably damaged, and thus fragmentary, and also, because of the none too happy installation of the cannons: vertically on a wall, and only a few decimeters off the floor. The inscriptions look very primitive; the ideograms measure six to eight millimeters, and the Manchu writing is equally small, with the word *baitalara*, for example, measuring fifteen millimeters. Despite this, however, I have been able to decipher the date of casting as the twentieth year of Kangxi, 1682, and also the name of the maker, Nan Huairen, in Manchu Nan Hûwai-jin. We are dealing, then, with the work of the Jesuit Ferdinand Verbiest, one of the most famous missionaries in Chinese missionary history.

There are another four cannons to be seen in the *Museo dell'Istituto Storico di Cultura dell'Arma del Genio*, at Lungotevere della Vittoria, 31. They differ from those already mentioned both in size and in the rich ornamentation on the barrel. They too have bilingual Manchu-Chinese inscriptions on the breech, more elegant in form and perfectly legible.

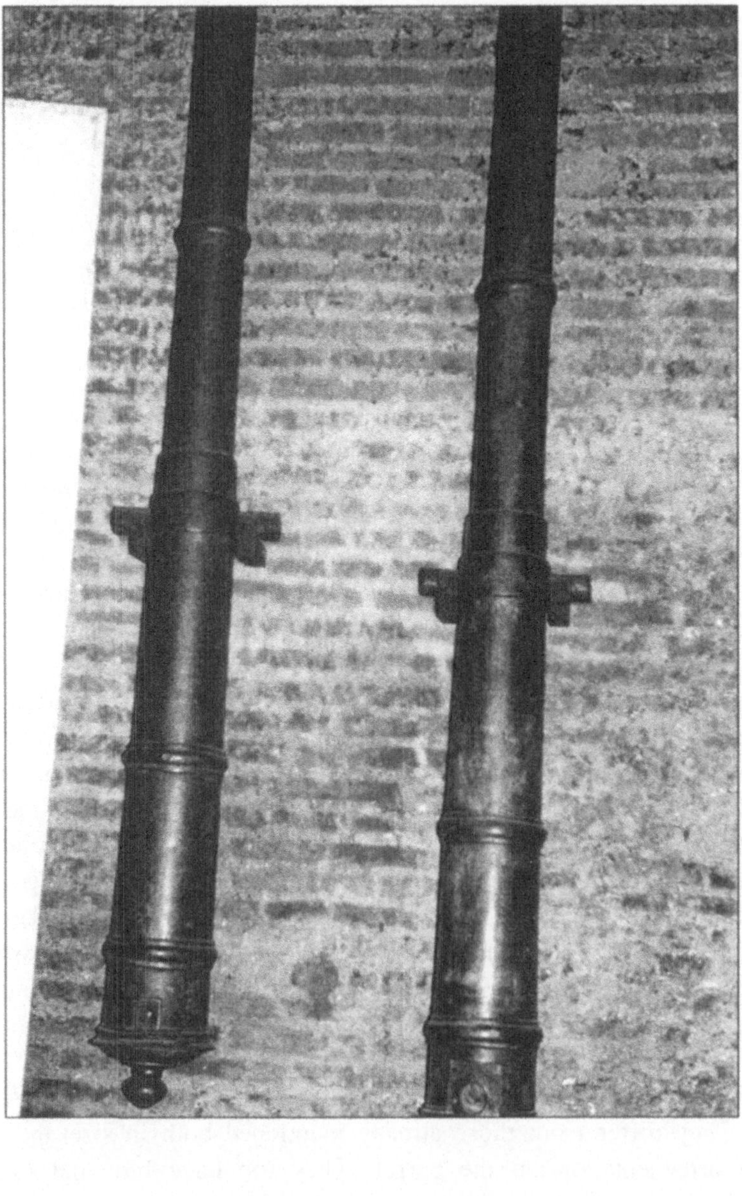

Ill. 43: Cannons of the 1st series, 1682, Rome, Castel Sant'Angelo (Photo by G. Stary).

Ill. 44: Cannons of the 2nd series, 1689. Cannon of Ingolstadt. Detail: dragon (Photo by G. Stary).

Ill. 45: Cannons of the 2nd series, 1689.
Cannon of Veste Coburg with inscription.

The Manchu text[1] reads as follows:

> *Daicing gurun-i Elhe Taifin-i orin jakûci aniya. hungkerehe horon sanggara enteheme akdun amba jiyanggiyûn. baitalara okto ninggun gin jakûn yan. eshun selei muhaliyan juwan ilan gin. sing ni den jakûn fun uyun li. durun ilibuha hafan Nan Hûwai-jin. tuwame weilehengge hafan Foboo. Sostai. zoguwan Wang J'i-cen: faksi da Li Wen-de: Yan S̱e.*
>
> In the twenty-eighth year of Kangxi of the Great Qing dynasty (1689) the (cannon named) "Great (and) Eternally Faithful Condottiere (and) Artificer of Imperial Power" was cast[2]. The powder necessary is 6 *gin* 8 *yan* (3879.304 g.)[3]. (With it is shot) a cast-iron ball of 13 *gin* (7758.609 g.). The rear-sight is 8 *fun* 9 *li* (28.48 mm.). The functionary (in charge of) designing the form was Nan Hûwai-jin. The functionaries (in charge of) supervising the work were Foboo and Sostai. The chief of the craftsmen was Wang Zhichen. The craftsmen were Li Wende and Yan Si.

The inscriptions on the other cannons are identical, except for their ballistic data. Two other cannons with different beautiful ornamentations are kept at the *Museum für Deutsche Geschichte*, Berlin (Unter den Linden). One cannon can also be seen at the *Bayerisches Armeemuseum*, Ingolstadt, *Altes Schloss* (cannon n° 18), and, also in Germany, at the

---

1 Parallel Chinese text, on the right side :

大清康熙二十八年
鑄造
武成永固大将軍用
藥六觔八两
生鐵砲子十三觔
星高八分九釐
製法官南懷仁
監造官佛保
碩思泰
作官王之臣
匠役李文德
顏四

2 The word *horon* is written above the text, as is *Daicing*, and thus justifies the translation "Imperial (power)" instead of the usual "terrible." It should be underlined here that the translation "...the name of the cannon is 'Generalissimo Wu Zhengyong'" (from the Chinese text) is absolutely wrong; "Wµ Zhengyong" corresponds to Manchu "horon sanggara enteheme akdun" and is not the name of a "generalissimo."

3 The units of measurements mentioned are: 1 *gin* (*jin*) = 596.815 g.; 1 *yan* (*liang*) = 37.301 g.; 1 *fun* (*fen*) = 3.2 cm.; 1 *li* (*li*) = 0.32 mm.; 1 *cun* (*cun*) = 3.2 cm.

Ill. 46: The cannon of Veste Coburg with inscription (detail). With permission of Kunstsammlungen Veste Coburg, Neg. Nr. 26254 and 26421.

Ill. 47: Cannon of Berlin. Detail (Photo by G. Stary).

Veste Coburg. A photo of the breech of this cannon has been published in 1988, together with a detailed description:

> Material: Bronze; Gewicht des Rohres: 1911 kg.; Gesamtlänge des Rohres: 315 cm.; Laufbohrung: 12.5 cm.; Laufbohrungslänge: 282 cm.; Kugelgewicht: 12-16 Pfund.
> Kunstsammlung der Veste Coburg. Inv. n° G9[4].

Two cannons can also be seen in London (The Armouries, Tower of London, Chelsea Royal Hospital)[5]. Two more cannons are kept in Vienna, *Heeresgeschichtliches Museum* (Arsenal, *Geschützhalle* 17, Kammer 2; inv. n° 81446 and 81447), and in Budapest. The two cannons in Budapest are carefully described by G. Kara:

> Les canons échouèrent en Europe au temps du soulèvement de 1900, comme butin des troupes interventionnistes européennes, et furent remis plus tard au Musée de l'Histoire Militaire de Budapest. ... Le canon n° 22 est plus grand; le calibre du tube est de 14 cm., la longueur de 330 cm. Celui qui porte le chiffre 23 a un tube d'un calibre de 12 cm. et d'une longueur de 300 cm. ... Le personnage mentionné sous le nom de Fo Pao par les inscriptions est probablement identique à Fo Lun qui vécut à la même époque et appartenait au drapeau blanc de la droite mandchoue; au temps de Kangxi il remplissait une haute charge militaire, cf. *Zhongguo renmin dazidian*, p. 303. De même on ne sait pareillement rien de certain sur *Cho Sseu-t'ai*, il est peut-être identique à Cho Tai qui s'était également rangé sous le drapeau blanc de la droite mandchoue ... cf. *op.cit.*, p. 1365[6].

These fourteen cannons were cast according to Verbiest's design, even after his death in 1688; the whole casting, of course, began earlier - as we can see from an annotation in the *Qijuzhu* under Kangxi twenty-sixth year, fourth month, *geng shen* (= thirteenth) day (May 23, 1687):

> *Nan Huairen tiqing chongtian pao keming zhi chu*
>
> 南懷仁題請衝天砲刻名之處
>
> "Nan Huairen submitted a memorial asking for a name to be engraved on the blustering cannons ..."[7].

---

4 E. Schmitt, Th. Schleich, Th. Beck (eds.), *Kaufleute als Kolonialherren: Die Handelswelt der Niederländer vom Kap der Guten Hoffnung bis Nagasaki 1600-1800* (Bamberg, 1988), pp. 213-215.

5 Cf. the catalogue of the exhibition *China, Hemel en Aarde. 5000 Jaar Uitvindingen en Ontdekkingen / Chine, Ciel et Terre. 5000 ans d'inventions et de découvertes* (Brussels, 1988), p. 418, n° 293.

6 G. Kara, "Sur deux canons chinois des Ts'ing", *Acta Orientalia Academiae Scientiarum Hungaricae*, X (1960), 249-253.

7 *Kangxi qiju zhu* (Beijing, 1984), 2:1617.

The circumstances that brought this missionary to the unlikely occupation of manufacturer of cannons are well known, being connected with the Sanfan Rebellion and the Sino-Manchu military expedition against the Russian fortress of Albazin (in Manchu: *Yagsa*), on the Amur, in the years 1685 and 1686. On September 10, 1674, Verbiest was given the pertinent order:

> The emperor sent orders to the Board of War: "When Our Grand Army marches to suppress the rebels, it will need firearms very badly. We order Nan Huairen, the Director of the Imperial Board of Astronomy, to cast light but effective cannons, convenient for transportation[8].

Everything leads us to believe that the cannons now in *Castel Sant'Angelo* are these "light but effective" cannons, thus suitable for transportation from Beijing to the northernmost point of the Amur. The casting of all the cannons in this first series must have been done by 1682. Those at *Castel Sant'Angelo*, in fact, carry this date, when Verbiest received this solemn encomium for his work:

> An additional title of junior vice-president of the Board of Works was conferred upon Nan Huairen, Director of the Imperial Board of Astronomy, Supervisor of the Administration of Calendar-calculation and Commissioner of the Office of Transmission (*tongzheng shisi*). He was thus promoted because the cannons, whose casting he had directed, were good and strong[9].

The *Museo del Genio* cannons and those found in other European museums, however, were cast according to instructions left by Verbiest. Indeed it has come to light that the Jesuit composed a small work in Chinese on the casting and use of western cannons, but no copy of it has been found to date.

---

8 *Kangxi shilu*, ch. 49, p. 6b, quoted in the translation of Fu Lo-shu, *A Documentary Chronicle of Sino-Western Relations (1644-1820)* (Tucson, 1966), 1:48.

9 *Kangxi shilu*, ch. 102, p. 3b, Fu Lo-shu, *op. cit.*, p. 58.

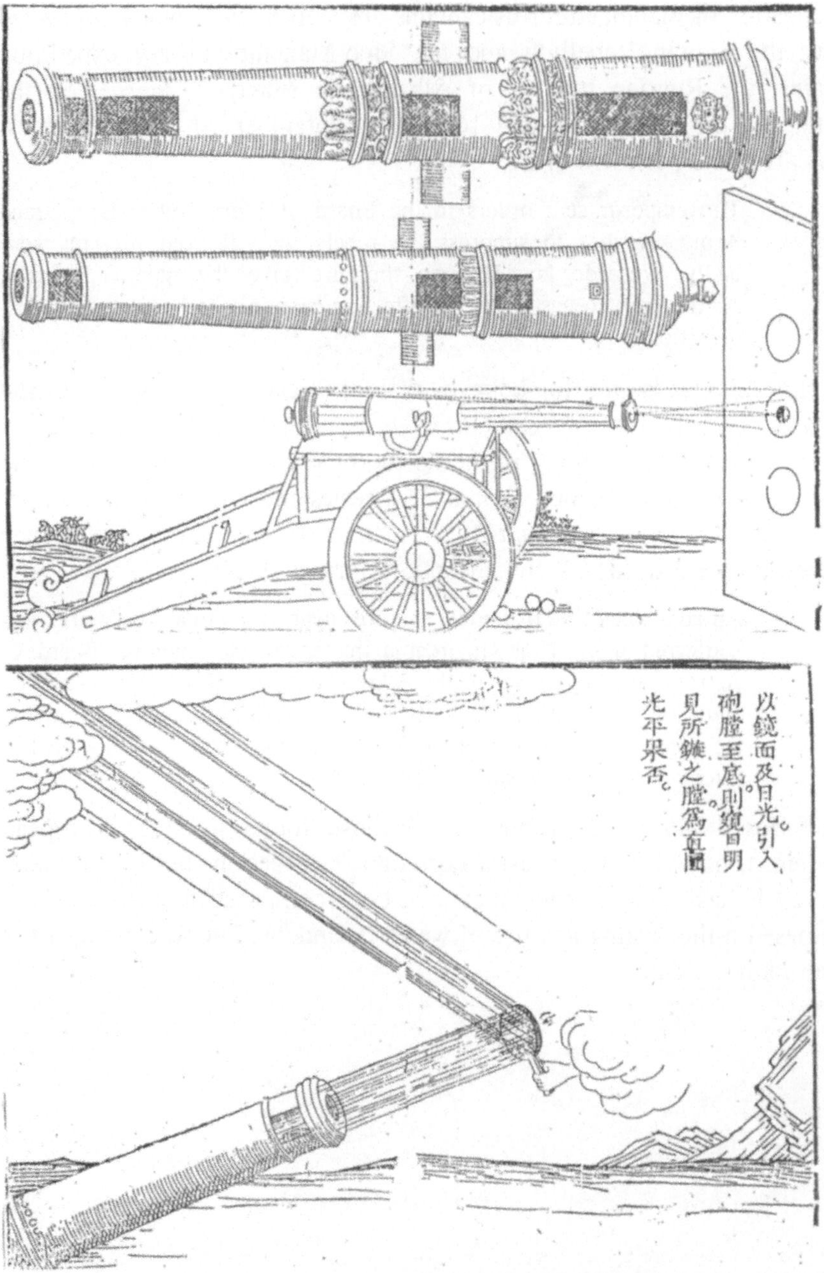

Ill. 48 and 49: Two pages from Verbiest's *Shenwei tushuo*
ARSI (Rome), Jap. Sin. II, 67 III. Reproduced with the kind permission of the late Father E. Lamalle, to whom this paper is dedicated.

Thus, even its title remained unknown for a long time[10]. All we have by way of evidence, is an observation of Fu Lo-shu, in a note saying that, "Verbiest, who had cast 440 cannons for the Qing government, wrote a book on how to cast cannons, entitled the *Shenwu tushuo*"[11]. This author, however, does not give us, in this case, the characters, which are an essential condition for identifying a Chinese title, so that any affirmation remains unproven.

During my research in this direction, I have analyzed numerous reports from Verbiest to the Manchu emperor, collected in booklets entitled *Xichao ding'an* in the *Biblioteca Nazionale* in Rome[12], the *Biblioteca Apostolica Vaticana*[13] and in the *Archivum Romanum Societatis Jesu* (ARSI)[14]. In the ARSI collection my researches met with success in that a report of Verbiest to the throne reads thus:

> *Nan Huairen jinzou wei jincheng shenwei tushuo qi zhun liu chuan ...*
> 南懷仁謹奏爲進呈神威圖説氣准留傳
> Nan Huairen has drawn up a reverent report to the throne, presenting the *Shenwei tushuo ...*

Thereafter, this phrase is repeated in a document of the Ministry of the Interior[15]:

> *Nan Huairen xian zhupao erbai sanshi er wei you shenweipao erbai sishi wei.*
> 南懷仁先鑄砲二百三十二位又神威砲二百四十位
> Nan Huairen, in the past, cast 232 cannons, and later, 240 *shenwei* cannons.

It is evident, then, from the above-mentioned report, that *shenwei* is the abbreviation of *shenweipao*, i.e. "wonderful, terrible cannon," also known by the Manchu name *Ferguwecuke horonggo boo*[16]. The title of

---

[10] Cf. Pfister, *Notices*, p. 359: "Je n'ai pas rencontré le titre chinois. Il renferme 44 feuilles de figures nécessaires à l'intelligence de cet art et des instruments propres à pointer les canons;" see also Blondeau, *Mandarijn*, p. 499: "Er is nog geen enkel eksemplaar van dit werk teruggevonden zodat de Chinese titel ontbreekt." This Chinese title, but without sources, was given by Zhang Li and Liu Jiantang in their book *Zhongguo jiao'an shi* (Chengdu, 1987), p. 103.

[11] Fu Lo-shu, *op. cit.*, 2:464, N. 83.

[12] "Sala manoscritti", n° 72.C.530.

[13] *Borgia Cinese*, 350, 27.

[14] Folder *Jap.-Sin. II*, 72, 73, 67/3, 67/5.

[15] Cf. p. 179a, and on pp. 183a and 184a-b of the same collection, i.e. copies of original documents not preserved.

[16] *Wuti qingwenjian* (Tamura Jitsuzô, Imanishi Shunjû and Satô Hisashi, eds.),

Verbiest's work on the casting of cannons is, therefore, *Shenwei tushuo* (and not *Shenwu tushuo*, as Lo-shu Fu thought): "Explanations and Illustrations (of the cannon named) Wonderful (and) Terrible." Since the above-mentioned report is dated the tenth day of the fourth month of the twenty-first year of Kangxi, i.e. May 16, 1682, we can deduce that the work was finished, and therefore presented to the throne, in the early months of 1682. As has been written above, in the same period (May 23, 1682) Verbiest asked the emperor for a name to be engraved on the breech. The work itself has not yet been found but, in an ARSI folder (*Jap.-Sin. II* 72, 73, 67/3, 67/5), there are, along with Verbiest's reports, two contemporary illustrations (here reproduced) on the use of cannons and, in all probability they are the only evidence to be found to date of that work of Verbiest entitled: *Shenwei tushuo*.

Conclusion

At present, seventeen cannons cast by Verbiest are found in some European museums as can be seen from the following table. All cannons, except those of the first series (1682, *Castel Sant'Angelo*) and those of London (second series, 1689), are characterized by beautiful, different ornamentations, but their inscriptions are identical in Manchu and Chinese. A difference exists only in their ballistic data.

---

2 vols. (Kyoto, 1966-1968), 1:n° 17418.

Table 3

Verbiest's Cannons in European Museums

Place	Year of casting	Powder	Iron ball	Rear-sight
FIRST SERIES				
Rome, Castel Sant'Angelo Rome, Castel Sant'Angelo Rome, Castel Sant'Angelo	1682 1682 1682			
SECOND SERIES				
Rome, Arma del Genio (*)	1689	6 gin 8 yan	13 gin	8 fun 9 li
Rome, Arma del Genio (*)	1689	6 gin 8 yan	13 gin	8 fun 2 li
Rome, Arma del Genio (*)	1689	6 gin 8 yan	13 gin	8 fun 2 li
Rome, Arma del Genio (*)	1689	6 gin 8 yan	13 gin	8 fun 2 li
Berlin	1689	10 gin	20 gin	9 fun 9 li
Berlin	1689	10 gin	20 gin	9 fun 9 li
Ingolstadt	1689	6 gin 8 yan	13 gin	9 fun 8 li
Coburg	1689	5 gin	10 gin	9 fun 8 li
London	1689			
London	1689			
Vienna	1689	6 gin 8 yan	13 gin	11 fun
Vienna	1689	5 gin	10 gin	8 fun 2 li
Budapest	1689	6 gin 8 yan	13 gin	9 fun 1 li
Budapest	1689	5 gin	10 gin	1 cun

(*) Photos in F. Bortone, S.J., *I gesuiti alla corte di Pechino* (Roma, 1969), between pp. 128-129.

FERDINAND VERBIEST AND THE CASTING OF CANNONS IN THE QING DYNASTY

SHU LIGUANG

Shu Liguang graduated in 1963 from the Department of Mechanics in Wuhan Engineering Institute. At present he is a senior engineer of Beijing Fork Lift Truck Foundry. From 1987 to 1988, he participated in making the replica of the Celestial Globe of Verbiest.

Ferdinand Verbiest was born in Belgium on October 9, 1623[1]. His Chinese name was Nan Huairen, his poetic name Xun Qing[2], and his courtesy name Dun Bo. He was baptized very young and was determined to devote all his life to God. He did a lot of scientific research, especially in the field of astronomy[3]. In 1657, together with the Italian missionary Martino Martini (Wei Kuangguo), he left for China. Having reached China in 1659, he at first did some missionary work in Shaanxi province. Already in 1660 he was ordered to come to Beijing as an assistant to Johann Adam Schall von Bell, the German missionary who was in charge of the Imperial Astronomy Board[4]. Later, in 1668, he was asked to treat personally the affairs of the Board. A year later, because he had made an accurate calculation of the calendar, he was promoted vice-president of the Board[5]. In 1674 he was appointed minister of the *taichang si*, the administration of the affairs of the court, for having made astronomical instruments. Based upon a

---

1 Xu Zongze, *Ming Qing jian Yesu huishi yizhu tiyao* (An Annotated Bibliography of Works Written or Translated by Jesuits during the Ming and Qing Dynasties), p. 390.

2 *Cihai*, vol. *zi*, 1980, p. 479.

3 Xu Risheng and An Duo, *Nan xiansheng*, p. 1.

4 Xu Zongze, *op. cit.*, p. 390.

5 Yin Luanzhang, *Qing jian*, p. 192.

calendar list for two hundred years drawn up by Schall, Verbiest compiled in 1678 the calendar of the Kangxi reign, the *Kangxi yongnian lifa* (32 vols.). He was then granted the new title of Head of the Office of Transmission, *tongzheng shisi*. In 1682 he was finally appointed head of the Board of Works, due to his brilliant achievement in casting cannons. When he died in Beijing on January 28, 1688, the emperor himself wrote the funeral oration and the inscription for the tablet[6]. Verbiest was even granted the posthumous title of *qinmin*, meaning "diligent and intelligent." The emperor's uncle and many other ministers were sent as part of a large ceremony in honor of Verbiest at his tomb.

Among the more than a thousand missionaries of the past four hundred years, Verbiest was the only one to be granted a posthumous title of honor[7]. One of the reasons for conferring this title lies without doubt in Verbiest's role in the casting of cannons during the Kangxi reign. The Kangxi emperor even said after his death: "Since the year of the rebellion of Wu Sangui had started, Verbiest realized a very important military exploit by casting cannons"[8]. Why did Nan Huairen, as a missionary, assist in the casting of cannons? How many and what type of cannons were cast? What were the characteristics of these cannons? How important is Verbiest's contribution to the history of cannon-casting in ancient China? All these questions will be dealt with in the following essay, in order to clarify the facts of Verbiest's cannon-casting.

I.
The Historical Background to Verbiest's Cannon-Casting

In 1673 Wu Sangui, who was in charge of the garrison of Yunnan province, rose in rebellion with the purpose of disrupting the vast Chinese empire. Geng Jingzhong, in charge of the Fujian garrison, and Shang Zhixin, son of Shang Kexi, who was in charge of the Guangdong garrison, as well as some other Han generals who had good relations with Wu Sangui, answered his call to rebellion. Half of China's territory was thus swept with warfare. In Chinese history this period is called "the Rebellion of the Three Feudatories."

Having been informed of the rebellion, the emperor decided to crush

6 Nan Huairen, *Xichao ding'an*.

7 Fang Hao, *Zhongguo tianzhujiaoshi renwu zhuan*, vol. 2, p. 164.

8 Nan Huairen, *op. cit.*, s.n.

it. Elite troops of the Qing army were sent to Jingzhou to cut off the north-south route from the rebellious troops. At the same time Qing troops were sent to Jiangxi province to prevent a possible reunification of Wu's and Geng's rebellious armies. The army stationed in the south of the Yangzi River was reinforced to protect the financial resources located in the southeast. Due to all these measures temporary stability was gained. The emperor, however, was still deeply worried that only a military confrontation between the Qing dynasty and the rebellious army would enable the Qing army to penetrate the battlefields. The reason was that the rebel troops consisted of elite troops of the Qing army. The strongest troops of the rebellions were those of Wu Sangui, who, stationed in Yunnan province, had trained his men and prepared armaments for more than ten years[9]. The regions occupied by the rebellious armies were mountainous, carved by rivers and inconvenient for transportation, and therefore extremely easy to defend and hard to attack.

Facing the facts, the emperor realized that the repression of the rebellion would require considerable effort to improve the artillery. The Qing army was particularly in need of mobile and nimble cannons, easy to be transported along mountains and across rivers without detaining the troops[10]. To meet this urgent need, the emperor ordered the Ministry of Defence to ask Verbiest to take part in the casting of light cannons[11].

II.
Verbiest's Important Achievements in the Casting of Cannons

Of the total 905 cannons cast during the Kangxi reign, more than half were cast under the supervision of Verbiest. On August 14, 1674 the emperor sent his chamberlain to the residence of Verbiest to deliver the following order: "You are asked to work out a good technique to cast cannons that meet the need of military activities in mountainous areas, carved by rivers"[12]. Carrying out the imperial order, Verbiest took up the task and soon presented a model of a wooden cannon to the emperor for examination. On March 14, 1675 the emperor ordered his minister

---

9 Wei Yuan, *Shengwu ji*, vol. 2, "Kangxi kanding sanfan ji", part 1, pp. 2-3.

10 *Ibid.*

11 *Qingchao wenxian tongkao*, vol. 194, p. 6587.

12 Nan Huairen, *op. cit.*, p. 166.

of the Imperial Guard, Da Deng, to accompany Verbiest to test the cannons. The testing met with success, in that over a hundred shots were fired without damaging the cannon itself. He was then ordered, on April 19, to cast the cannons following the model. On May 24, the Kangxi emperor was personally present at the final testing. He highly praised the success of Verbiest[13].

According to similar cannons collected in the Palace Museum in Beijing, the wooden cannon was in fact a type of cannon with an iron barrel, the mouth being equipped with a bronze ring, and to the rear part a bronze ball is added. The whole is covered by painted wood (Ill. 50). At the rear of the cannon a fire gate with cover is placed, and the cannon is loaded on a chariot. Behind the bronze ball there is an iron bar with a whorl, which serves to adjust the slant of the barrel. The degree of resistance of a wooden cannon against the pressure of the shot is said to be not as strong as that of bronze and iron cannons.

Ill. 50: Verbiest's wooden cannon
(Palace Museum, Beijing. Photo: courtesy of Shu Liguang).

[13] *Ibid.*, p. 166.

In November 1675 the commander of the local army of Shaanxi, Wang Fuchen, rose in rebellion, thus menacing Shanxi and Gansu. The Dutch cannons, imported in 1604 from Holland, were much needed to put down Wang's rebellion. Verbiest and his workers had cast twenty of these Dutch cannons within only twenty-eight days[14]. They were the most powerful cannons of that time. Also in November Prince An Yuele, who was preparing a counterattack from Jiangxi against the rebellions in Hunan, sent a letter to the emperor, stating: "No military base can be broken down without the Dutch cannons." Later he set forth another demand: "The Dutch cannons are too heavy to be transported from Guangdong to here. We therefore would like to have twenty light cannons, which are easier to move in our conquest of the rebellions." The emperor replied: "The cannons cast by Verbiest will be sent to the camp of Prince An in Jiangxi in the amount asked for"[15].

In the first month of 1676 Verbiest was ordered to cast some more Dutch cannons. In March he was asked to test two cannons at Haizi (Nanyuan, Beijing). The emperor himself attended the testing of the cannons. At a banquet, he praised Verbiest's work by awarding him with money[16]. From 1674 to 1676, Verbiest made 132 wooden cannons and Dutch cannons to put down the rebellions. In 1677 the emperor sent the Dutch cannons cast by Verbiest to the troops of Prince An[17]. At the beginning of 1679 government troops reconquered Yuezhou and, consequently, Hunan was generally under control. In August, the emperor prepared his army for the conquest of Taiwan province. All twenty cannons therefore were ordered to be transported from Guangdong to Fujian, because "They are powerful, light, and easy to move"[18].

In 1680 the rebellion of the three princes was crushed with success, and the Kangxi emperor, who was determined to unify the whole of China, began to pay close attention to the improvement of artillery. He ordered the old and damaged bronze cannons in Zhili to be collected, melted down, and cast into new cannons by Verbiest. All told 320

---

14 *Ibid.*, p. 181.

15 *Qing shengzu shilu*, 58:8.

16 Nan Huairen, *op. cit.*, p. 162.

17 *Qing shengzu shilu*, 66:13.

18 *Ibid.*, 83:1.

Ill. 51: Design of Verbiest's cannon presented to the emperor (Palace Museum, Beijing. Photo: courtesy of Shu Liguang).

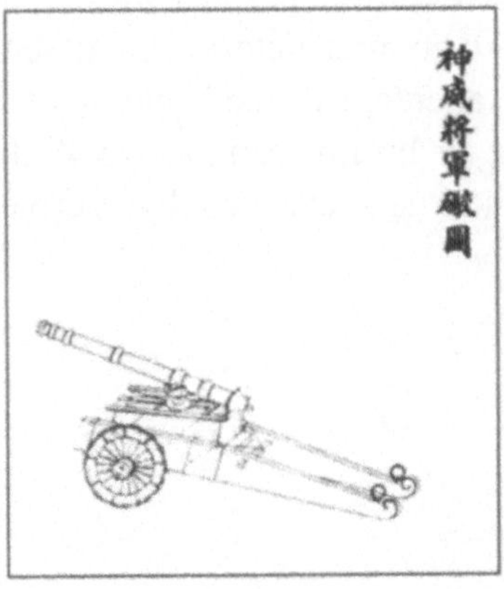

Ill. 52: The *shenwei* cannon (*Qinding da Qing huidian tu wubei*, vol. 98).

cannons were planned to be cast, so that each section of the bannermen (consisting of eight sections) would receive forty cannons. On January 23, 1681 Verbiest tested a new type of cannon in Qinghe (Beijing). The emperor once more personally attended the testing, and pointed out that the principle of aiming was based on aligning the sights and the target. Verbiest was asked to do another experiment with his newly-made cannon in Yuqianshan to show the technique of alignment. On August 11, another 240 cannons, named *shenwei*, were cast under the supervision of Verbiest. The emperor then ordered Verbiest to carry out an experiment in the presence of the head of the Board of Works, Dang Guli, at the cannon site Lugouqiao. At the same time many soldiers were sent to attend this experiment, so that they could learn more about the technique of alignment. During a training period of three months (August 16 - November 14), a total of 21,600 shots were fired by 240 soldiers. The manoeuvre held on October 19 at the Lugouqiao cannon site, and attended by the emperor and his princes, was successful. The emperor said: "The cannons that you have cast have proved to be useful in Hunan, Guangdong, Shanxi and Jiangxi in the past years. Up to now I have not seen such an accuracy shown by cannons ever before." Having said this, he offered Verbiest his own sable coat as a gift of honor[19].

At the beginning of 1682 the Qing government turned its attention to the northeast border and prepared military actions to drive away Russian aggressors. In 1683 a meeting of the princes in charge of the administration was held to inspect the reserve of old cannons cast at the end of the Ming dynasty and at the beginning of the Qing dynasty. They planned to cast fifty-three new cannons to bring the total amount of cannons to eighty, so that ten cannons could be distributed to each section of the mandarin troops[20]. All these cannons were cast by Verbiest with the exception of a kind of 500 kg. bronze cannon ordered in 1687.

On July 28, 1686, Chancellor Zhao Chang delivered an order of the emperor to Verbiest which said: "Verbiest is ordered him to design a type of cannon, that is able to shoot 30 *jin* of cannonballs each time, and that is loaded on a chariot and attached to the bottom by a plate"[21]. Verbiest soon finished his design and presented it to the emperor through Chancellor Zhao (Ill. 51). The emperor examined the design

---

19 Nan Huairen, *op. cit.*, pp. 169-170.

20 Xu Risheng, *op. cit.*, p. 3.

21 Nan Huairen, *op. cit.*, s.n.

and ordered to cast a model of it, as well as twelve or fifteen cannonballs. Verbiest was also asked to measure the distance of different degrees of slant, and of the time the cannonball took to hit its target[22]. This type of cannon, called *chongtian pao* and preserved in the Palace Museum of Beijing, is like a trench mortar with a large barrel. The cannon looks like an erected bell, injuring the enemy with a power through a sharply curved ballistic, just the same as a modern trench mortar does. It is loaded on a four-wheeled chariot, and differs from the other cannons in the barrel, which clearly consists of two parts: the muzzle and the bore. The cannon is put into operation by filling the bore with cannon powder, some wooden chips, and a small amount of moist soil. The cannonball is then put on the powder in the bore. The ball should be covered with powder, and the mouth of the barrel with moist soil, in order to avoid sparks. It is necessary to light the fuse of the cannonball first, and then quickly the fuse of the bore. The advantage - and at the same time, the strongest point of the cannon - is that the enemy is injured by the chips of the exploded cannonball[23].

Towards the end of his life Verbiest, having the title of engineer, cast sixty-one cannons named *wuchengyonggu*, and eighty cannons named *shengong*[24]. According to the *Xichao ding'an* and the *Qingchao wenxian tongkao* (General reference of documents of the Qing dynasty)[25], Verbiest had cast 513 cannons. In addition to the cannons recorded in the book *Nan xiansheng xingshu* (The Deeds of Master Verbiest), written by Xu Risheng (Tomas Pereira) and An Duo (Antoine Thomas), the total amount of cannons cast by Verbiest reached 566.

Three types of cannons designed by Verbiest are mentioned in the *Qinding da Qing huidian* (Great Qing code approved by the emperor). The code, divided into 1590 volumes, consists partly of drawings and partly of calculations. This important imperial code was edited during the Kangxi reign, and enriched during the next reigns: the Yongzheng, the Qianlong, the Jiaqing, and the Guangxu reigns. These three selected cannons are mentioned in the code as follows: *shenwei*, *wuchengyonggu* and *shengong*.

---

22 *Ibid.*

23 *Qinding da Qing huidian tu wubei*, vols. 98, 100, s.n.

24 *Qingchao wenxian tongkao*, p. 6588.

25 Cf. vol. 194 chapter about Military Affairs, p. 6588.

The *shenwei* cannon

This cannon is shaped like a bamboo cap of a Chinese peasant. Its mouth is narrow and its rear part is large (Ill. 52). It weighs 400 *jin*, and measures 6.7 *chi* (1 *chi* = 0.31715 m, according to the measurement completed by the author with the authentic ruler preserved in the Palace Museum). The cannon is undecorated, but reinforced by five bronze rings. Near the mouth of the barrel is a gun sight, and the central part of the barrel's body is equipped with two bars. It bears the following inscription in Chinese:

> *Shenwei*, cast in 1681 in the Great Qing dynasty. For every shot 8 or 9 *liang* of cannon powder is used; a lead cannonball weighs 18 *liang*; the height of the gun sight is 00.7 *chi*; the engineer is Verbiest; the supervisors are Fa Bao, Qian Qibu, Tao Santai, Ningguta, Wu Ladai; the craftsmen are Li Wende and Yan Si.

The cannon is carried by a two-wheeled chariot, whose shaft measures 9.5 *chi*. It is undecorated, but a couple of rings are attached to the end of the shafts. The other data are all identical to the golden dragon chariot cannon approved by the emperor[26]. A target at a distance of 100 *gong* (about 200 m.) could be hit with 8 *liang* of cannon powder; at 150 *gong* with 9 *liang* of cannonpowder[27].

In the collection of the Palace Museum in Beijing there is a *shenwei* cannon cast in the Daoguang reign, bearing the following inscription:

> *Shenwei* cast in the 22nd year of the Daoguang reign of the Great Qing dynasty. For every shot 8 or 9 *liang* of cannonpowder is used; the height of the gun sight is 10.11 *chi*; supervisors are Wu Fu, Guan Heng, Zeng Nian; the technician is Chen Laiming.

The data on this cannon are almost the same as those on the *shenwei* cannon designed by Verbiest, except for the height of the gun sight, which is 0.04 *chi* higher than the one of Verbiest (Ill. 53). Here are the measurements of the *shenwei* cannon cast during the Daoguang reign:

> Length: 213 cm. (= 6.7 *chi* in the Qing dynasty).
> The inner diameter of the barrel: 5.3 cm.
> The outer diameter of the barrel: 10.4 cm.
> The diameter at the end of the firegate: 19.2 cm.
> The length from the firegate to the mouth: 195.5 cm.
> The depth of the barrel: 196 cm.

---

26 *Qinding da Qing huidian tu wubei*, vol. 100, s.n.

27 *Qinding da Qing huidian shili. Gongbu Zhupao*, 894:1.

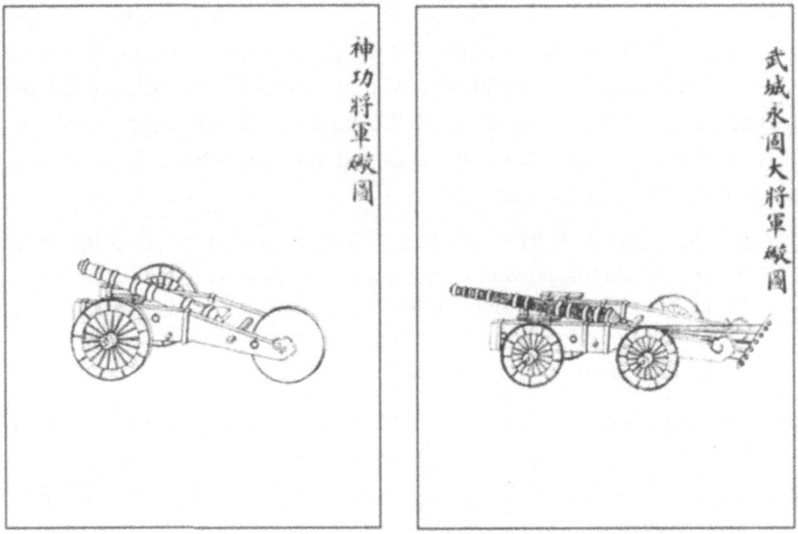

Ill. 53: The *shenwei* cannon (*Qinding da Qing huidian tu wubei*, vol. 98).

Ill. 54: The *wuchengyonggu* cannon
(Palace Museum, Beijing. Photo: courtesy of Shu Liguang).

These measurements are close to those of the *shenwei* cannon, described in *Da Qing huidian*[28]. Another proof that the Daoguang cannon is an imitation of Verbiest's cannon is based on a document of the manufacturing section of the Imperial Palace Bureau. During the Daoguang reign a *shenwei* cannon in the city of Qiqihar was to be replaced. The ministry of manufacturing set forth this demand to the emperor. Two months later the emperor replied: "I approve the casting of five cannons, following the model of the *shenwei* cannon of the Xianghuang troops, and by the data given in 'the Collection of the Designs of the Ceremonial Instruments'"[29]. This work was edited in 1759, and the cannon shown in this collection was cast under the supervision of Verbiest in 1681. We may therefore conclude that the *shenwei* cannon cast in the Daoguang reign is, indeed, an imitation of the one cast by Verbiest during the Kangxi reign. The cannon, made of bronze, is fine and beautifully shaped, reinforced by five rings at the mouth of the barrel to increase the resistance of the barrel against the pressure occurring by the explosion. When designing this cannon, Verbiest took into account the movement of the cannonball before it left the barrel, and the alteration of the inner pressure. The rear part is reinforced by two rings to concentrate the pressure of the explosion produced by the burning of the powder, and to prevent the rear part from breaking. The mouth of the barrel is also reinforced by two rings to prevent it from breaking at the moment the cannonball passes. The *shenwei* cannon cast by Verbiest has a record of firing off 300 to 400 shots without causing any damage to the cannon[30]. It is easy to imagine that these cannons were welcomed by the military. At the eve of the Yakesha War in 1683, there were twelve of these cannons in the city of Qiqihar alone[31].

The *wuchengyonggu* cannon

The rear part is a little wider than the mouth of the barrel. It weighs between 3600 and 7000 *jin* (3.5 ton); it measures between 9.6 and 11.1 *chi*. It is beautifully ornamented, and is reinforced by ten rings. The

---

28 Cf. vol. 100 in the part on armament.

29 *Qing neiwufu Yangxin dian zao banchu ge zuocheng zuohuoji Qingdang*, April 1840, n° 3013, s.n.

30 Nan Huairen, *op. cit.*, p. 168.

31 *Qing neiwufu*, *op. cit.*, s.n.

Ill. 55: The *wuchengyonggu* cannon
(Museum of Chinese History, Beijing. Photo: courtesy of Su Liguang).

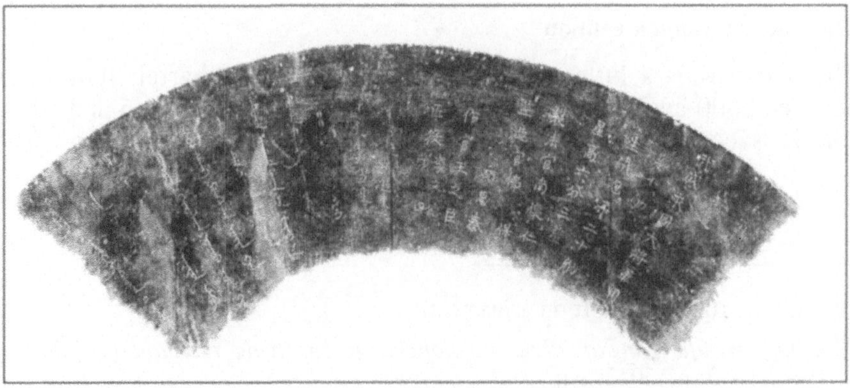

Ill. 56: Inscription on the *wuchengyonggu* cannon
(Museum of Chinese History, Beijing. Photo: courtesy of Su Liguang).

cannon sight is near to the mouth of the barrel (Ill. 54). On both sides of the barrel is engraved the name of the cannon, as well as ballistic data:

> *Wuchengyonggu* cannon cast in 1689. For every shot ten *jin* of cannon powder is used; a cannonball weighs twenty *jin*; the height of the cannon sight is 4.9 *fen*; the engineer is Verbiest; supervisors are Fa Bao and Shuo Sitai; the chief of the craftsmen is Wang Zhichen; the craftsmen are Li Wende and Yan Si.

The inscription is in both Manchu and Chinese. For a smaller type of this cannon, five *jin* of cannon powder is necessary, and an iron cannonball of ten *jin*. The cannon is loaded on a four-wheeled chariot; the length of the shaft is fifteen *chi*, reinforced by seven iron rings. The other data are identical to the *shenwei* cannon[32]. In the collection of the Chinese History Museum in Beijing, there is an authentic model of this type of cannon (Ill. 55) bearing the following inscription (Ill. 56):

> *Wuchengyonggu* cannon cast in 1689 in the Great Qing dynasty. For every shot ten *jin* of cannon powder is used; a cannonball weighs twenty *jin*; the height of the cannon sight is 6.3 *fen*; the engineer is Verbiest; the supervisors are Fa Bao and Shuo Sitai; the chief of the craftsmen is Wang Zhichen; the craftsmen are Li Wende and Yan Si.

The only difference between the two inscriptions lies in the height of the cannon sight (the height of the first cannon being 1.4 *fen* smaller than that of the cannon preserved in the History Museum of Beijing). Here are its measurements:

The outer diameter: 46.15 cm.
The diameter of the barrel of the rear part: 52.87 cm.
The diameter of the barrel: 15.5 cm.
The general length: 362 cm.
The weight of the cannon: 4 tons.
The length of the barrel: 330 cm.
The diameter of the wheel of the chariot: 120 cm.
The width of the wheel: 15 cm.

The remaining part of the chariot has been proved to be the rear part. The model was preserved in the court of the Embassy of Germany, located in Dongjiaominxiang Street in Beijing. In 1946 it was collected by the Museum of History, the present Museum of Chinese History.

The casting technique of this huge and beautifully ornamented cannon, which bears the name of Verbiest, proved to be of the highest level

---

32 *Qinding da Qing huidian tu*, *op. cit.*, s.n.

in the history of ancient China's cannon-casting. This combination of Verbiest's knowledge and Chinese craftsmanship is a historical testimony of the scientific and technological exchange between China and the West. It has been selected to belong to the most important national collection of antique objects.

The *shengong* cannon

The mouth of the barrel is a little narrower than the rear part, which is shaped in the form of a bamboo cap of a peasant in the south of China (Ill. 57). Its weight is 1000 *jin*, and its length is seven *chi*. It is not ornamented, but has five rings as a means of reinforcement. The cannon sight is near to the mouth of the barrel, and bears the following inscription in Chinese:

> *Shengong* cannon cast in 1689 in the Great Qing. For every shot one *jin* twelve *liang* of cannon powder is used; a cannonball weighs 3.5 *jin*; the height of the cannon sight is 4 *fen*; the engineer is Verbiest; the supervisors are Fa Bao and Shuo Sitai; the chief of the craftsmen is Wang Zhichen; the craftsmen are Li Wende and Yan Si.

The cannon (Ill. 54, left side) is loaded on a three-wheeled chariot, supported by iron chains, and the length of the shaft, between which a board is located, is 10.2 *chi*. The other data are identical to those of the *shenweiwudi* cannon[33].

Ill. 57: The *shengong* cannon
(Museum of Chinese History, Beijing. Photo: courtesy of Su Liguang).

33 *Ibid.*

All of these three cannons belong to the type of front loaded cannons, although differences among them do occur. The cannon powder and the cannonballs are put into the cannon directly through the mouth of the barrel and lit by a fuse. The cannons were always cast in bronze. Towards and under the central part of the barrel are situated two axles which support, balance and adjust the slant, and at the same time enlarge the cannon's power and its movability. From the inscriptions we learn that all these cannons were cast by the craftsman Li Wende and his co-workers. The author also found his name in an inscription on a *weiyuan* cannon cast in 1690 and in 1718, and preserved in the Palace Museum in Beijing. Besides the name of Li appearing in the inscription, the name of the cannon-casting factory in Jingshan, Beijing, where Li worked, was also mentioned. We therefore are convinced that these cannons were all cast in this factory.

III.
Verbiest's Important Work on Cannons: *Shenwei tushuo*

On January 27, 1682 Verbiest dedicated his book *Shenwei tushuo* to the emperor. This book, described by Verbiest in another work entitled *Xichao ding'an*, explains the theory and the method of firing cannons. The work is divided into twenty-six subheads and contains forty-four figures, as well as some formulas of calculations. Dr. John Lee pointed out in his famous book *A History of Science and Technology in China*[34], that Verbiest's work seemed to have been destroyed, for no trace at all could be found of it, neither in China nor anywhere else in the world. The only reference we have is the book *Xing xing zhi litui* (Ratiocination of Figures and Characters), written by Verbiest himself in 1683, twenty-one pages long and divided into sixteen small subheads. In it Verbiest set forth his method of correcting the slant of the cannonball by bringing the sights in line with the target in order to fix the position of the barrel. Other works, such as *Paodan gaodu biao* (A list of the height of the cannonballs), *Paodan qizhi suo xing qingke miao wei zhi biao* (A comparative list of distances, heights and the time it takes a cannonball to cross them), and *Paodan yuandu bili biao* (The ratio of the cannonball and its distance), *Tuizhong wu dao yuanjin gaodi zhi yi* (An Instrument Describing the Trajectories and Ballistics of Cannon-

---

34 John Lee, *A History of Science and Technology in China*, vol. 7, pp. 398, 598.

balls), and *Sanlu fa* (The Method of Three Ratios), all offered important research material for Verbiest's *Zhunpao zhi fa* (Methods for Sighting).

The *Zhunpao zhi fa* is composed of two kinds of contents. A greater accuracy of shooting became possible by correcting the slant of the cannonball by means of aligning the sights and the target. The front cannon sight, called the "star," is in alignment with the backsight, called the *dou*. The frontsight is attached to the ring near the mouth of the barrel, and the backsight, to the rear part of the barrel[35]. Restricted by the time in which Verbiest lived, it was very difficult to cast barrels of an equal thickness. If the left side of the barrel would be thicker and heavier than the right side, the cannonball would surely turn to the left[36]. The cannons should therefore be corrected before being delivered, by simply adjusting the position of the sights. The position is corrected as follows: first attach a couple of sights with wax to the barrel, then aim and fire at the target[37]. Measure the distance between the ballistic point and the target, thus modifying the position of the sights according to the measurements[38]. The position of the sights can also be corrected by using the method of the three ratios, e.g., take a hundred *chi* from the backsight to the target as the first ratio, the wrong rate (ranging from four *chi* left from the target) of the ballistic point as ratio 2, and a distance of six *chi* between the sights as ratio 3, then calculate the correct position using the formula of Verbiest given in his *Sanlu fa* (Method of Three Ratios)[39]:

$$\frac{\text{ratio 2 x ratio 3}}{\text{ratio 1}}$$

The result is the correct position of the sight. Move the cannon sight in wax according to the result mentioned above to the left side, and try again in a different site. The theory is explained in his work *Jihe yuanben* (The Theory of Geometry), exercise 4 and 2, volume 6[40].

In order to aim at a further target (within the maximum range of shooting), under the condition that the weight of the cannonball, as well

---

35 Nan Huairen, *Xing xing zhi litui* (Ratiocination on Figures and Characters), p. 19.

36 *Ibid.*, p. 18.

37 *Ibid.*, pp. 18-19.

38 *Ibid.*, p. 20.

39 *Ibid.*, p. 20.

40 *Ibid.*, p. 19.

as the weight of the cannon powder and the ballistic time of the cannonball are constant, and the degree of the slant is forty-five, the distance from the point of impact can then either be found by checking the list drawn by Verbiest: *Paodan qizhi suo xing qingke miao wei zhi biao* (A comparative list of distances and the time it takes cannonballs to cross them), or by calculating according to the formula of the *Sanlu fa* (Method of three ratios), ranging from the degree of the slant 1-45, or 45-89. In practice, the soldiers could easily obtain a greater accuracy when they knew the distance between the target and the position of the cannon and the slant of the cannon. The slant of the cannon could be adjusted with an instrument invented by Verbiest. Undoubtedly *Paodan gaodu biao* (A list of the height of cannonballs) and *Paodan qizhi suo xing qingke miao wei zhi biao* must be familiar.

IV.
The Position of Verbiest in Chinese History and the Development of the Cannons

Gunpowder was invented in China before 808 A.D. At the beginning of the tenth century gunpowder had already been used for military purposes. Soon after the casting of the first primitive gun (*tuhuoqiang*) in 1259, which could fire bullets, metal gun barrels appeared. In the Song dynasty burning firearms and explosive firearms, as well as barrel shaped shooting weapons, appeared. Even primitive missiles pushed by a jet with burning powder came into use. The power of firearms had been raised greatly by using bronze to cast barrels. In this way, the use of heavier cannonballs and more powder became possible. In the Yuan and Ming dynasties firearms (*huotong*) were cast and used in large quantities.

Chinese gunpowder and firearms found their way to European countries via the Arabic countries in the twelfth and thirteenth centuries. Guns developed much faster in Europe than in China, but western firearms and casting techniques were introduced in China only from the beginning of the sixteenth century. Among them, the French mechanic gun, the *niaotong* (bird shooting gun), and the Dutch cannon had known a great influence. In the last years of the Ming dynasty Xu Guangqi and Li Zhizao had been ordered by the government to cast a large quantity of Dutch cannons to be used in the war against the Houjin invaders. The introduction of western firearms and casting techniques had pushed the development of the technology of cannon-casting and the military industry to a high level at the end of the Ming dynasty. The latest development of early Chinese cannons was during the Qing dynasty. The Qing cannon-casting industry passed successively the stages of creation (1616-1661), growth (1662-1795), and decay (after 1795). By the middle of the nineteenth century early Chinese cannons had all been replaced by modern ones.

The casting of ancient cannons reached its peak in the Kangxi reign. In meeting the need to crush the rebellion, as well as to resist the invaders and to unify the country, cannons had been cast on a large scale. Not only was the amount of cast cannons enormous, but also the level of casting techniques, the variety of the cannons, and the quality of the cannons had been proved to the utmost. Over a period of fifty years of the Kangxi reign, 905 cannons had been cast[41]. More than half of them were designed by, and cast under the supervision of Verbiest. Their perfect technique, their beautiful shape, and their solid body were never to be equalled by the cannons of later reigns. The most impressive cannon preserved is the *wuchengyonggu* cannon, which bears the engraved name of Verbiest. Of these 905 cannons, only 201 cannons weigh more than 250 kg., compared to the lighter cannons, which weigh less than 250 kg. Of the 704 light cannons, one-third were designed by Verbiest and cast under his supervision. The data above show that Verbiest was creatively devoted to the casting of low-weight and easily transportable cannons, thus meeting the practical needs of warfare of the Qing dynasty.

Besides having contributed to the improvement of the Qing cannons, Verbiest also carried out systematic research, and wrote various essays on methods for sighting. The influence of his book, *Shenwei tushuo*, was also proved by an investigation done by the author of the files of the manufacturing section of the imperial court, in which a commentary on aiming and aligning said the following: "The keys to accurate shooting are the slant of the cannon and the right position of sights; only then does the occurrence of mistakes become impossible." These words reflect the influence of Verbiest's writings[42]. We can say, therefore, that Verbiest was beyond doubt a very important figure in the history of the development of ancient China's cannons, and also in the history of the transfer of western science and technology to China.

---

41 *Qingchao wenxian tongkao*, pp. 6587-6589.

42 *Qing neiwufu Yangxin dian zao banchu ge zuocheng zuohuoji Qingdang, qiangpao chu* (September 1767), n° 3284, s.n.

BEIJING PRECURSOR[1]

J. DITLEV SCHEEL (1918-1992)

Jorgen Ditlev Scheel entered the Foreign Service of Denmark in 1943 and was posted to Bern and London. Between 1961 and 1979 he was consul general in Canada and ambassador in Tunisia, Algeria, Libya, Colombia, Ecuador, Panama, Costa Rica and Nicaragua. He contributed to general and specialized publications on automotive history and automobile technology. Since 1988 he was a contributing editor of "Automobile Quarterly." He died on February 14, 1992 in Green Valley, Ontario, Canada.

The first, documented, automotive machine was built and run in Beijing 300 years ago. Great care has been taken with the phrasing of this statement, and the choice of words is stressed intentionally.

First is defined by reliable dictionaries as "being number one in a countable series; earliest in time." *Documented* should exclude all the splendid, fiery chariots and similar conveyances met in myths and prophecies, secular or religious. Countless inventive tales in folklore or literature are also left out. *Automotive* means, according to Webster's[2], "containing within itself the means of propulsion." That excludes machines propelled by outside forces such as wind, beasts of burden, or hidden muscular power. *Machine*, a word of joint Greek (*mechane*) and Latin (*machina*) roots, originally described "a ... constructed thing ... conveyance, vehicle ... and assemblage of parts ... that transmit forces, motion ... one to another in some pre-determined manner."

---

1 Excerpts of my 1984 monograph on the Verbiest vehicle were first published in "Precursor from Peking. Meet Ferdinand Verbiest, Seventeenth-century Missionary, Diplomat, Astronomer and Automotive Engineer", *Automobile Quarterly*, 23:3 (1985), 268-275.

2 *Webster's Third New International Dictionary* (Boston, 1966), s.v. "automotive."

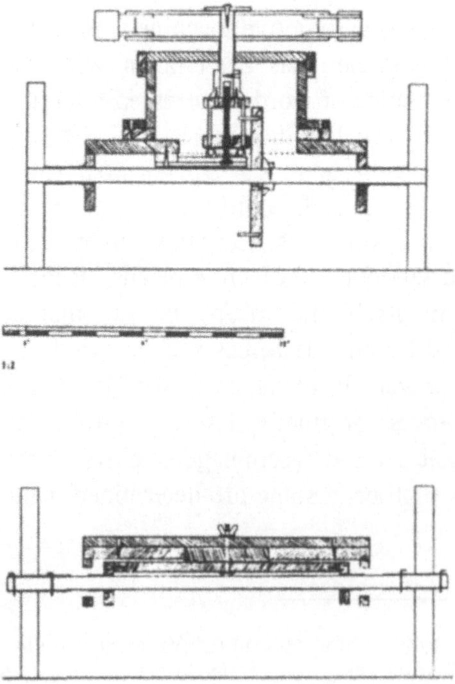

Ill. 58: Reconstruction and diagram of F. Verbiest's automotive machine (by J. Ditlev Scheel).

Again Webster's provides the helpful definition. However, Oxford[3] has an even better one "an apparatus for applying mechanical power, consisting of a number of parts each having a definite function, 1673" which is right at the epoch of the subject matter dealt with here. A further Oxford definition reads, "vehicle of any kind (usu. wheeled)" and that also hits the mark nicely.

In other words, about one hundred years before Nicolas-Joseph Cugnot's (1725-1804) first, experimental fardier[4] a self-propelled conveyance ran in Beijing. Its existence is well documented - in fact, so fully described by the man who built it that a reconstruction of this ancient automotive machine is possible, following the outline printed in 1687.

The Machine

The many writings by Ferdinand Verbiest, mainly in Chinese, have fortunately been conserved for the most part. They deal principally with astronomy and the making of astronomical and other instruments. Theological questions also fill many pages. Mathematics and physics are other prominent subjects. As far as it has been possible to establish, none of the Chinese texts mention or illustrate the automotive machine. That epochal achievement is recorded only, and so to say *en passant*, in *Astronomia Europaea* published in 1687 by Johann Kaspar Bencard at Dillingen-an-der-Donau in Bavaria[5].

Bencard was an astute businessman, son of a bookseller and publisher in Würzburg[6]. In 1675 he bought the printshop run by the Jesuit university at Dillingen and employed some of the leading skilled printers of the day. Verbiest's book was beautifully set and printed by Johann Federle. The book runs to 126 pages plus a two page index. The manuscript received the mandatory Papal imprimatur, on November 26, 1685.

---

3 *The Shorter Oxford English Dictionary* (Oxford, 1939), s.v. "machine."

4 Griffith Borgeson, "Cugnot Up to Date: Locomotion by Steam Pressure and the Power of the Printed Word," *Automobile Quarterly*, 20 (1982), 4-11.

5 Ferdinandus Verbiest, *Astronomia Europaea sub Imperatore tartaro-sinico Cam-hy appellato ex umbra in lucem revocata.*

6 Josef Benzing, *Die Buchdrucker des 16. und 17. Jahrhunderts im deutschen Sprachgebiet* (Wiesbaden, 1963), p. 80.

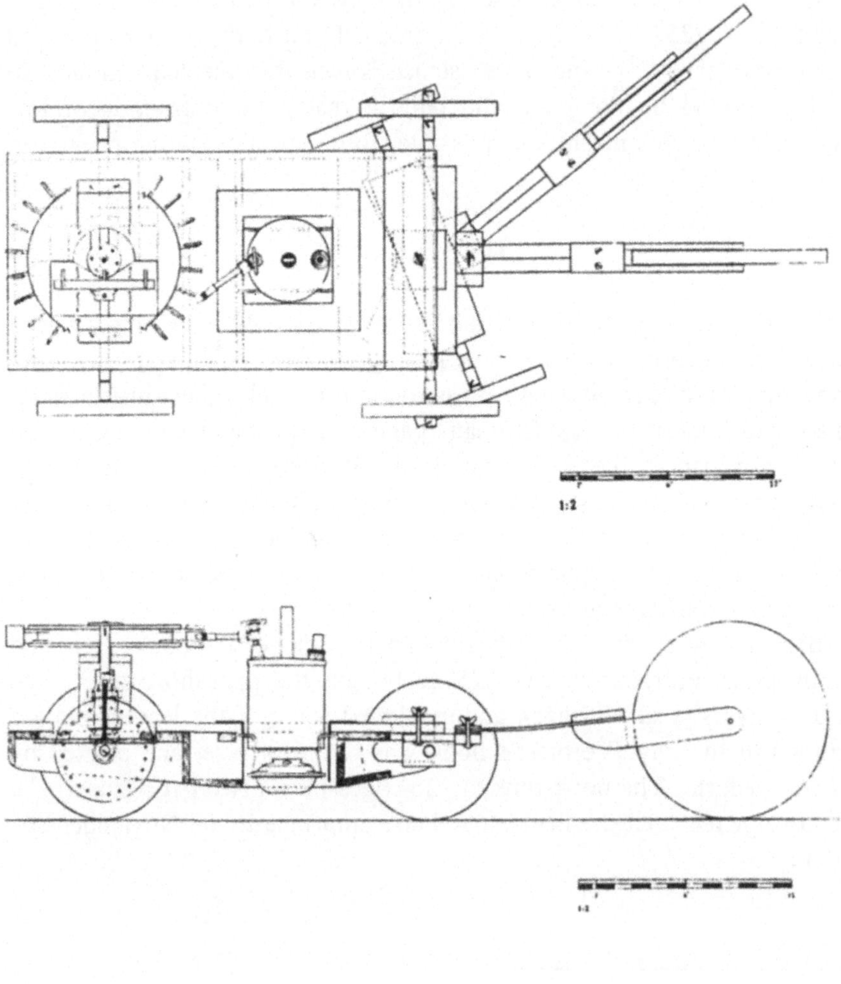

Ill. 59: Diagrams of F. Verbiest's automotive machine (by J. Ditlev Scheel).

In spite of the title, this work is more of a memoir about the scientific and technological activities of Verbiest during a quarter-century in China. Under the heading "Pneumatica" (pp. 87-88), he describes his automotive machine. The headline is a word of Greek origin meaning wind, air, breath or spirit. At the time of writing, steam was included.

Considering the supreme historical importance of this Latin text, it is translated here *in extenso*:

> Then three years ago, when I investigated the powers of the aeolipyle, I built a vehicle of light wood, two feet in length and easily movable on four wheels. On its center I put a small container filled with burning coal and placed an aeolipyle in the container. To the front axle I fitted a wheel with metal, transverse, and horizontal teeth that engaged a similar wheel mounted on a vertical axle, which thus moved the vehicle when rotated. On this axle I then fitted another, horizontal wheel one foot in diameter, and on the periphery of this wheel I placed small double bars like wings. The air forcibly expelled from the aeolipyle nozzle hit these wings, pushing the wheel round and thus propelled the vehicle, which would maintain a considerable speed during an hour or more (that is, as long as the air was expelled with force from the aeolipyle). In order that the distance covered by the machine should not be too long, I mounted a tiller, movable in all directions, at the center of the rear axle, and in the forked end of this tiller I put an easily rotating wheel of a larger diameter. When I turned this tiller right or left and locked it in position with a screw, the machine, propelled by the air from the aeolipyle, would run steadily in a larger or smaller circle, adjusted to the size of the courtyard or hall in which it ran, all according to the larger or smaller angle of adjustment for the tiller.
>
> This machine is a method of movement which can easily be applied to another vehicle, for instance, a paperboat with sails seeming to belly before the wind, continually sailing in a circle such as I demonstrated only for the elder brother of the emperor, but hiding the machinery so that only the hiss of the air expelled from the aeolipyle was heard, just like the real wind or the water rushing around the boat.
>
> Later I split the force of the air expelled by soldering a second jet-tube to the aeolipyle. When I fashioned the tip of this tube as a whistle, I could truly reproduce the song of a nightingale.
>
> Then I made a carillon play its tunes activated by this mechanism; once familiar with this method of movement many other diverting applications are easily contemplated[7].

---

7 As mentioned, for "air" read "steam" throughout the text.

Before analyzing the various elements in Verbiest's technological breakthrough, it seems appropriate to return briefly to the opening statement: first, documented, automotive machine - this time taking the wording in reverse order.

As described, Verbiest's vehicle clearly "consisted of a number of parts, each with a definite function." It also applied mechanical power. That is: a *machine*. Taking the constructor's word as evidence, so far neither corroborated nor contradicted, his machine also contained "within itself the means of propulsion." Hence: *automotive*. The *Astronomia Europaea* text, standing alone but unrefuted, supplies: *documentation*. And as no similar experiment is recorded as having been conducted at any earlier moment, it may be deemed: *first*. But when did it actually happen?

Beginning with the imprimatur of 1685 and going back the three years mentioned by Verbiest, 1682 is reached. That leaves, however, no time for transmission of the manuscript from Beijing to Rome. Communications over such distances were both slow and uncertain in the seventeenth century. It is, therefore, reasonable to believe that the manuscript was carried as safely as possible. Philippe Couplet (1623-1693), a close friend and fellow missionary who sailed from Macao for Europe in 1681, seems to have brought it with him, as he was instrumental in the eventual publication. Verbiest evidently worked on the text to the last moment possible, so the model-year of the little steam-carriage may safely be pushed back to 1678. It could be even earlier, but fully conclusive evidence is lacking. During 1670 and 1671 Verbiest demonstrated various "scientific toys" for the young emperor. His letters mention these intriguing toys and the instruments he made for the Observatory, but unfortunately say nothing about the automotive machine. Nor do, apparently, his surviving Chinese works - whether text or illustrations - contain anything on that subject. It may, however, safely be assumed that the automotive machine first ran some time between 1670 and 1678. Anyway, the priority of a full century before Cugnot is indisputable.

Turning to the detail components of the Verbiest machine, it is possible to trace back even further the origin of the parts he brought together with such elegance. Verbiest writes that he was investigating the powers of the aeolipyle. That word is of Greek/Latin origin: Aiolos was the Greek god having dominion over the winds (in Latin: Aeolus), and *pila* means a ball or sphere. In other words, a sphere of the winds, or more prosaically, a reaction wheel driven by air which, at the time, also meant steam. Oxford[8] says, "instrument for showing force of steam escaping through narrow aperture."

Ill. 60: Reconstruction of steam-turbine carriage in the Museo Nazionale della Scienza e della Tecnica, Milan. Earlier reproductions in J. Needham, *Science and Civilisation in China*, 4/2, fig. 472, opposite p. 226; R. Blondeau, *Ferdinand Verbiest als Oost en West elkaar ontmoeten*, p. 10.

8 *The Shorter Oxford English Dictionary,* s.v. "aeolipyle."

Power units of that kind date right back to antiquity. They were described by Heron of Alexandria in the first century of our calendar. The most widely used type is a hollow globe or sphere that can rotate on two trunnions. One of these tubes leads into the sphere which has two jet-nozzles diametrically positioned and angled 90 degrees to the axis of rotation. Steam admitted to the sphere under pressure will be discharged through the nozzles and spin the sphere on the trunnions to make the mechanism a wind-wheel.

In the accompanying illustration[9], of later date, vessel p is the boiler, and tube o admits the steam to sphere x. When the steam is discharged through jets z and w the sphere rotates on trunnions q and o. The unmarked individual crouching under the boiler, intently looking at the fire, has a function closely linked to Verbiest's concept of the aeolipyle. That figure is a sufflator, or steam-jet fire-blower. Such a fire-blower is clearly embraced by the Oxford definition of an aeolipyle, being: a boiler with a pinhole opening. Through such an aperture a steam-jet can be directed towards a fire, and the stream will aid the combustion considerably.

Fire-blowers, or stationary aeolipyles, were used already two centuries before our calendar and, maybe, even earlier. In fact, some evidence points to Central Asian origin going back several centuries. Certain Tibetan sufflators were fashioned like birds, while early European models were sculpted like human heads, busts, or full figures as illustrated[10]. In the representation here shown, the steam passes from the boiler through the hollow, righthand, feathered lion-leg into the figure that so obviously blows at the fire.

Verbiest's merit was to make the stationary aeolipyle his power unit instead of a combustion-aid, and to direct the powerful jet of steam onto a turbine. He invented neither the boiler with a pinhole nor the turbine wheel and made no such claims. But he did take a major technological step forward by fashioning a working combination of the two.

Tracing the origin of the turbine leads again into obscure antiquity.

---

9 [Editor's note: The author did not send the illustration with the essay and efforts to obtain it after his death were not successful.]

10 Joseph Needham, *Science and Civilisation in China* (Cambridge, 1965), 4/2, 225-228; Joseph Needham, "The Prenatal History of the Steam Engine", *Transactions of the Newcomen Society*, 35 (1963) 3-58. [Editor's note: The illustration mentioned was not sent with the essay and efforts to obtain it after the author's death were not successful. However, see the Tibetan steam-jet fire-blower surmounted by a bird's head and illustrated in Needham, *Science*, 4/2, figure 473 opposite page 227.]

Waterwheels were a very early invention, probably derived from gear-wheels that were also thought of thousands of years ago. Gradually the conception was refined, and remarkable progress was made during the century preceding Verbiest. From that period several reliable descriptions and illustrations survive to show water-driven mills or crushers in some detail.

For waterwheels both external and internal vanes, or blades, were used. Horizontal waterwheels, probably the earliest form, were still used in the sixteenth century in Europe. Some, even with curved blades, were illustrated in 1579[11].

The transition from running water to a jet of steam as motive power may also have very deep roots. Leonardo da Vinci suggested the use of a sufflator to make rotary motion. The first clear illustration or a rudimentary steam turbine appeared, however, more than a hundred years later when Giovanni Branca published *Le Machine* ... in 1629[12]. He proposed a boiler/sufflator shaped as the traditional Aeolian Cherub puffing through a short pipe that pointed the steam jet towards vanes in a horizontal turbine wheel. Various applications of this power unit were proposed, one driving - through a fancy train of gears and a trip-mechanism - two pestles pulverizing drugs in mortars. This and other suggested Branca machines were probably never made. His book was seventeenth-century science fiction, though with some practical background, as the author was in charge of the workshop at the prominent Santa Casa di Loretto shrine. His suggested power mechanisms must have been studied by the young Verbiest when completing science courses some ten years after *Le Machine* was published. The mature Verbiest, on the other hand, showed his independent approach to technology by avoiding the traditional cherubic torso and, instead, directing the pressurized steam from an unadorned, stationary boiler onto vanes protruding from a turbine wheel. This was a notable improvement on Branca's suggestion.

The transmission between the turbine and the driven wheels employed, according to Verbiest, technology similar to water-driven mills, and also suggested by Branca. Toothed wheels and right-angle gearing (lantern wheel and pin wheel) were well known to ancient

---

11 Flynn White, Jr., *Medieval Technology and Social Change* (Oxford, 1962), pp. 80-81.

12 *De machine diverse del Signor Giovanni Branca, Cittadino Romano, Ingegniero, Architetto delle Sta. Casa de Loretto* (Rome, 1629), p. 25.

engineers and fully illustrated in works[13] that must have been read by Verbiest during his scientific studies. The state of the art was evident some hundred years before Verbiest decided on his transmission, and there is no reason to believe that his solution differed markedly from preceding examples. He says nothing about the gear ratio chosen, but that will be commented upon later. As clearly stated in *Astronomia Europaea*, Verbiest specified driven front wheels for his little carriage. So - with truly remarkable, albeit fortuitous anticipation· of present design trends - the first automotive machine had a turbo-frontwheel drive layout. Unable to use the front wheels for any other function than driving the vehicle, Verbiest had to employ rear wheel steering. According to the description, he chose a fifth, tiller-mounted rudder wheel for that function. He must also have made arrangements for the rear axle to swivel on a chassis pivot. Such turning trains (for horse-drawn vehicles) were fairly standard in the sixteenth century, and had been used already in the first century before our calendar, although no connection can be established between the early and later applications of this design feature[14]. The refinement of pivoted steering trains must have been known to Verbiest, and he may even have been aware that the brilliant Dutch engineer Simon Stevin employed a rear turning train on his sail-carriages in 1600. If not, it was quite within the compass of Verbiest's established ingenuity to put a turning train at the rear when the front was otherwise engaged. Further comment on this design detail will follow. Verbiest's layout of drive and steering has escaped his most recent biographer[15], who writes about steering front wheels.

About the chassis Verbiest says nothing beyond light wood and a length of two feet. It must have been some kind of platform in order to accommodate the fire-vessel, the boiler, and the turbine support, but no other chassis features, other than those just mentioned, are given. Nevertheless, Verbiest thus left a pretty adequate description of his little machine. But that has not prevented a surprising number of later commentators from making rather extraordinary statements regarding this vehicle. The distortions began early. When Jean-Baptiste du Halde,

---

[13] Georgius Agricola (Jürgen Bauer), *De re metallica* (Basel, 1556); Agostino Ramelli, *Le diverse e artificiose machine del capitano Agostino Ramelli* (Paris, 1588).

[14] Charles Singer et al., eds. *A History of Technology*, 8 vols. (Oxford, 1954-1984), 2:548.

[15] Roger A. Blondeau, *Ferdinand Verbiest als Oost en West elkaar ontmoeten* (Tielt, 1983), pp. 100-101.

S.J. (1674-1743) published his four-volume *Description ... de la Chine* ... in 1735[16], he carefully avoided Verbiest's name when mentioning the automotive "pneumatic machine." He translated the passage from *Astronomia Europaea* word for word with only the slightest alterations enabling him to infer that another Jesuit missionary, Claudio Filippo Grimaldi (1638-1712), was the originator. This is probably the first imprint on automotive history of nationalism's cloven hoof, a manifestation that would plague researchers ever since. The writer was French and wanted to glorify his country. Therefore he deftly eliminated Verbiest who was from Flanders, a Spanish domain. Instead, he promoted Grimaldi, who hailed from Piemonte, at that time part of the lands ruled by the Duke of Savoy. The Duke was under firm French tutelage, and his subjects thus qualified to represent France and prove French pre-eminence.

Other, later writers also contributed to the general confusion, by manifestly avoiding the authoritative Latin text in *Astronomia Europaea* and using or producing garbled versions. The prize for misinformation must go to Dr. Kenneth Richardson, senior lecturer in Social and Economic History at Lanchester Polytechnic, Coventry, England. In his otherwise weighty book *The British Motor Industry 1896-1939*[17], he makes the following statement:

> The first actual production of (a self-propelled) vehicle, however, appears to have been achieved in China during the seventeenth century ... It was the result of an interesting meeting of European and Chinese minds at the Imperial Court. Two French (sic) Jesuit missionaries, Ferdinand Verbeist (sic) ..., and Philippe-Marie Grimaldi ... both constructed powered vehicles which were anticipations of the steam turbine. We do not know how Verbeist (sic) made his carriage, or for how long it lasted, but apparently in 1665 (sic) it took him through the streets of Pekin ...

Richardson provides full notes on his sources, and that makes the passage quoted even more incomprehensible. As one source he mentions: "Huc, Abbé, *Astronomia Europaea*". But the Abbot Evariste-Régis Huc (1813-1860) who lived two centuries after that book was published was, of course, not the author. Huc's work was entitled *Le Christianisme en*

---

16 Jean-Baptiste du Halde, S.J., *Description géographique, historique, chronologique, politique et physique de l'empire de la Chine et de la Tartarie chinoise*, 4 vols. (Paris, 1735), 3:334.

17 Kenneth Richardson, *The British Motor Industry 1896-1939* (London, 1977), p. 1.

Chine ...[18] and came out in 1857. He mentions Verbiest and quotes from *Astronomia Europaea*. But Richardson seems, according to his own notes, to have relied on an English 1858 book about James Watt, in which Huc is quoted.

Richardson further gives Joseph Needham as a source[19]. Again the reference is incomprehensible. Needham reprints in full du Halde's French text mentioned earlier. That text is, in turn, an almost verbatim translation from *Astronomia Europaea*, and especially states that the vehicle was two feet long. The dimension has completely escaped Richardson, who errs also on Verbiest's nationality, cannot get the spelling of his name correct, and would have him ride a carriage 24 inches long.

The late Jacques Ickx thought that the American pioneer constructor Charles Edgar Duryea, shortly before his death in 1938, propounded a similar dimensional misconception. Contemporary evidence does not support Ickx, for in an article published in *The Motor* (GB) June 29, 1937 (p. 943), Duryea states "a little steamer," and contributes a drawing that basically follows the *Astronomia Europaea* description.

In 1956 the respectable Swiss weekly *Automobil Revue* printed a short article entitled "A Wooden 'Automobile' - Thousands of Years Ago"[20] by Alfred Kring. That writer explained how self-propelled carriages were built, tested, and described in China already during the Zhou dynasty, for which the period is given as 1122-225 before our calendar[21]. He further wrote that Verbiest had learned about such technical wonders from the emperor who "was well aware of this invention, and let the missionary be shown the annals in which the information was elaborately recorded." Kring then stated that "Verbiest relates how he found a description so detailed and clear that - with imperial approval - he could reconstruct the wonderful carriage." It is also mentioned that Duryea "rather long ago" had devoted himself to "this account," and whether that of the Chinese annals, or that ascribed to Verbiest, is not clear. A technical survey, in some detail, of the vehicle, based on "very old descriptions from China," is obviously gleaned from *Astronomia Europaea* or, rather from one of the later,

---

18 Evariste-Régis Huc, *Le Christianisme en Chine, en Tartarie et au Thibet*, 4 vols. (Paris, 1857-1858), 3:147-148.

19 Needham, *Science*, 4/2, 225-226.

20 "Ein hölzernes 'Automobil' vor Jahrtausenden," *Automobil Revue*, n. 21, May 2, 1956.

21 The period accepted by Needham is ca. 1030-221.

garbled versions. The latter is more probable, as Kring expressly stated: "How the transmission from this (i.e. turbine) axle was affected to one or both carriage axles - if through crown- or toothed wheels, strings, belts, or other means - is not evident."

Intrigued by many details in this article, I wrote to the author a few years after the publication for more information. The reaction was most interesting. First he complained about not having been sent an international mail coupon for his return postage. Then he explained that he was unable, at the moment, to check his extensive archive of Chinese source material for further documentation. If a copyright question was involved, he expected to hear further. Reasons totally unconnected with this correspondence prevented me from pursuing the matter then and my research came to rest for some twenty years.

Documentation[22] published in the meantime renders it less than plausible that any ancient Chinese steam vehicle ever existed and, consequently, that any Chinese annals could have recorded such an invention. Kring's failure to grasp the workings of the Verbiest vehicle's transmission further shows that he had read neither *Astronomia Europaea*, nor du Halde's *Description* ... The only evidence available about ancient Chinese "self-propelled" carriages is to be found in legend. Needham[23] quotes the traditional tale, going back to the fourth century before our calendar, of Kungshu Phan, who constructed a wooden chariot and horses with a wooden driver. Once this mechanical marvel was ready, his mother was taken for a drive. She sped away, never to return.

No technical details have survived of that remarkable conveyance; but the legend may have fostered the Kring version of seventeenth-century automotive happenings in China. It could also have originated Richardson's passage about the meeting of Chinese and European minds on such matters, an event for which no documentation has so far been produced by any scholar. Other erroneous statements by various authors regarding details of the little steam carriage will be mentioned later. The pictorial misinformation encountered is dealt with separately in "Note on Earlier Illustrations of the Verbiest Vehicle."

---

22 Needham, *Science*, 4/2, 228.

23 *Ibid.*, 574.

The Replica

Having now translated Verbiest's own description of the automotive machine, and accounted for the more persistent misinformation spread during the three centuries following the event, it is time to consider the verbs of the opening statement "was built and ran." That is, to answer the cardinal question: Did it really run?

There is absolutely no reason to doubt Verbiest's own words that "the vehicle ... would maintain a considerable speed ... as long as (steam) was expelled ... from the aeolipyle." In the absence of any corroborating evidence there is, however, no other way of proving the performance of the machine than to build a replica and try it out. That was precisely the intention that started this research project.

Ever since first finding oblique references to Verbiest and his machine, I have been intrigued by this seventeenth-century precursor of the motor car. Information was slowly gathered when a heavy professional schedule allowed time for such frivolous studies. During twenty years, that was not frequent, and some research material earlier obtained was lost in movements between four continents. But the desire to reconstruct and prove Verbiest's automotive machine persisted. When, finally, the necessary time became available, a lot of research remained to be done before any building could be considered. As already explained, the only authentic description is twenty-six lines in Latin. All other representations are purely conjectural. Therefore it was necessary from the outset to penetrate the mind of Verbiest in order to fill the textual gaps. Hence, I studied his background, education, activities, and correspondence, even if unconnected with the turbo-front wheel drive carriage.

Given the length of the vehicle proper (without the rudder wheel) at two feet, and the turbine wheel at one foot[24], it looked natural and pleasing to dimension the width of the chassis at one foot. The track would then be approximately sixteen inches, which would give nice stability. Sketches were made to find the other likely dimensions, leaving room for the fire and the boiler without getting the jet-tube too long and risking condensation. Verbiest specified "light wood," so it was decided to use balsa, and the first wood for the chassis was cut in mid-January 1983. Two weeks later the basic chassis was ready. It weighed 1 pound 3 ounces (500 grams) and was constructed as a 1/2-

---

24 *Pied du Roi* was the pre-French Revolution continental standard measure. One PdR was 30.5 cm, i.e. approximately 1.07 English foot.

inch thick platform resting on two side rails of 3/8 inch thickness, 3 inches deep, tapering towards the front and rear ends. Three vertical crossmembers, front, center, and rear stiffened the structure and provided support for the yet unspecified fire/boiler arrangement, vertical turbine axle bearings, etc. Six horizontal traverses had similar functions and served to hold the platform firmly together.

The structural design is seen in the side-elevation drawing. The assembly was done throughout with countersunk brass screws, both for cosmetic reasons and to facilitate disassembly for expected modifications.

No dimensions are given in the Verbiest text for the wheels, but a reasonable compromise between stability and ground clearance (in particular, for the still-to-be-determined crown wheel) indicated an approximate 8-inch diameter. So the next step was to make wheels. Having no willing Chinese assistants - like Verbiest at the Observatory - I reluctantly gave up any ambitions of delicate spoked wheels, as were prevalent in China when the machine was first built. Instead I settled for simple disc wheels made of three 9-inch balsa strips, 3 inches wide and 3/8 inch thick, glued together side by side, and then glued cross-ply onto three similar strips. This sandwich was then shaped roughly before being turned, and finally drilled at the center to receive a 1/2-inch wooden axle. The process sounds rather uncomplicated as here reported. It required, however, both trials and errors until four acceptable wheels were produced. In the interest of stability, both axles were placed at the extremes of the chassis, although the front axle had to be kept somewhat back to avoid too large an overhang of the turbine wheel. The resulting wheelbase was 18 1/2 inches. The 1/2-inch wooden axles were an expediency. Whether metal was used in the original is not known, but in the interest of weight-saving, wood seemed quite acceptable - and readily available, even pre-fabricated.

In order to avoid the friction of wooden axles in wooden bearings (through the side rails) suitable copper tubing - actually joining muffs for 1/2-inch pipe - was cut and glued into the chassis sides. This made satisfactory axle bearings, protruding from the side members in order to position the wheels. A bushing, one inch long, made from the same copper muffs by removing the inner ridge, was interposed between the chassis bearings and the front wheels. The actual front track was 17 1/4 inches measured to the outside of the wheels.

For the first experiments with the rolling chassis the rear steering was made exactly as described by Verbiest: a non-swiveling rear axle from the center of which a tiller with angular adjustment carried a fifth, larger (11-inch) wheel. It did not work. Though the tiller had been fashioned as a single-leaf spring to keep the steering fifth wheel in fairly close contact with the surface, it just skidded, while the four main wheels - running parallel to the chassis - firmly kept the vehicle in a straight line. Back to the drawing board.

The next step was to make a turning-train to pivot on the chassis centerline, cut the necessary slits in the chassis for the rear axle lateral movements, and attach the tiller to the turning train. This arrangement will be apparent from the rear axle drawing and overhead view. It worked like a charm, suitable relative angular adjustments to the rear axle and the tiller provided. Duryea had, seemingly, reached similar conclusions for his pictorial reconstruction. I only feel that his employment of a worm-screw to adjust the tiller relative to the turning-train was a little too sophisticated.

With a rolling and steering chassis, it was now time to think of the turbine, the transmission, and the boiler. Obviously, the dimensions and positions of the turbine and the boiler had to be laid out with some care in order to ensure maximum effect from the steam jet. The only given dimension was that of the turbine, so the point of departure suggested itself. In spite of the two specifications for the turbine laid down by Verbiest - one foot diameter and protruding blades - some strange statements and representations have been made about this item over the years. A presidential address to the highly respected Newcomen Society[25] maintained that the turbine wheel had four wings, and the president even quoted *Astronomia Europaea* specifically. Exactly the same statement - four wings - was made ten years earlier by Duncan[26], not the most reliable of sources.

It can be traced to an otherwise well-documented biographical article[27] that appeared in Belgium in 1839, and was repeated, slightly modified in 1845[28]. The author quotes from, but does not use, the *Astronomia Europaea* description. Nor does he explain the "four wings," but perhaps this number was reassuringly equal to that used by

---

25 C.F. Dendy Marshall, "The Germs and Development of Some Mechanical Inventions," *Transactions of the Newcomen Society*, 16 (1935-1936), 19-20.

26 H.O. Duncan, *The World on Wheels* (Paris, 1926), p. 81.

27 Charles-Louis Carton, "Notice biographique sur le Père Ferdinand Verbiest," *Annales de la Société d'Émulation*, 1 (Bruges, 1839), 83-156.

28 Charles-Louis Carton, "Le P. Verbiest," *Album biographique des Belges célèbres*, 1 (Brussels, 1845), 65-98.

more familiar windmills. The "four wings" reappear in Huc's 1857 work. Again the text quotes from *Astronomia Europaea*, but is, evidently, lifted practically word for word from the 1839 Belgian biography.

Others, like Lord Montagu and Anthony Bird[29] and myself fell into the trap of following Branca's drawing, that is assuming the use of internal vanes after the water wheel model, instead of external blades. Duryea, on the other hand, put in a multitude of curved turbine blades, semi sunken/protruding. Clearly, the turbine blades should be external and uncurved, as explained by Verbiest. And their number should, of course, be higher than four. But how many? The Duryea drawing with 48 blades looked too modern to me, and after some studies of sixteenth- and seventeenth-century water wheels with external blades, a drawing was made of a turbine with the blades spaced at 15 degrees, i.e., 24 blades. This layout gave an impulse angle of nearly 90 degrees with the jet hitting the tip of one blade and clearing the following.

Verbiest used "small double bars" protruding from the periphery "like wings." In order to have a lightweight construction and obtain a good flat surface for the jet stream to hit, I built the turbine in this way: the main wheel was a sandwich construction of two 3/8-inch layers of 3-inch wide balsa strips glued together cross-ply, just like the road wheels. The sandwich was then roughed, turned, and center-drilled. The final diameter was 7 3/4 inches. In both surfaces of that wheel I machined 24 radial grooves of 1/8 inches deep, each reaching 3/4 inches inwards from the periphery. In each groove one hardwood bar of 1/8-inch diameter and 2 1/2 inches long was glued, after the outer ends of the bars had been shaped to receive the blades. These were cut from 1/16-inch balsa strips of 1 inch width, each being 1 1/4 inches high. After the blades had been glued in place, two circular covers, 8 3/4 inches in diameter were made of 1/8-inch thick balsa strips, 3 inches wide, glued together side by side, roughed, turned, and then glued onto each face of the turbine wheel, partly covering the bars. The height of the blades thus corresponded exactly to the total thickness of the now four-layer sandwich of the turbine wheel. The bars carrying the blades were firmly in place, and the whole thing looked neat without weighing too much. The total diameter, blades included, was 11 1/4 inches - near enough to the prescribed one foot. As the front axle center was set back

---

29 Lord Montagu of Beaulieu and Anthony Bird, *Steam Cars, 1770-1970* (London, 1971), pp. 14-15 with a reference to Branca's book as *La Macchina*, thus missing the plural of the misspelt title.

5 1/2 inches from the leading edge of the chassis, the turbine overhang was only 1/8 inch, with the turbine shaft positioned just over the center of the front axle.

At first a boiler capacity of approximately one (US) pint or some 500 cc., seemed reasonable. It was also conveniently close to the contents of an empty tobacco tin made available by a friend. The tin was 4 1/4 inch high and of the same diameter. It had a well-fitting screw-lid that could be made steamtight with teflon tape, and gave easy filling. The jet-spout was cobbled by soldering a 3/4 inch copper tube onto the lid, having cut the necessary hole near the periphery. A copper-tube elbow fitted easily to the soldered-in tube, and the conical end of an aluminum baster, picked up in the kitchen corner of a hardware store, completed the first test assembly. Again, without any precise directions, it was arbitrarily decided to position the boiler 2 1/2 inches to the rear of the turbine periphery to let the jet impinge at nearly 90 degrees to the tip of the blades. An opening had been cut in the chassis deck to accommodate the boiler resting on two horizontal metal supports, fixed between the transverse chassis members. That put the boiler lower than probably positioned by Verbiest, but the deviation from the original description was made to lower the center of gravity and shorten the turbine shaft.

As yet, the heating of the boiler had not been decided upon. The glowing coals used by Verbiest seemed too precarious. An alcohol burner, as used in fondue sets, appeared less dangerous, but that option was first superseded by solid fuel tablets. They were found in a shop specializing in outdoors equipment, together with a neat little metal burner that fitted as tailor-made to the boiler supports. The burner held the small "hard alcohol" tablets securely while burning under the boiler. Problems like steam pressure and nozzle size were left for later investigations. Duryea faithfully drew "burning coals" as the heating element, and placed a cylindrical boiler lengthwise in the brazier. The arrangement looks a little incongruous, but does not deviate materially from the original concept.

Now that a general layout had been achieved, it was time to ponder the transmission of the power that, hopefully, could be generated. Again, information given in *Astronomia Europaea* about this particular subject is less than complete, although a useful guide. The right-angle transfer between the vertical turbine shaft and the horizontal front axle is clear in principle - but not so the design of two gearwheels or the ratio employed.

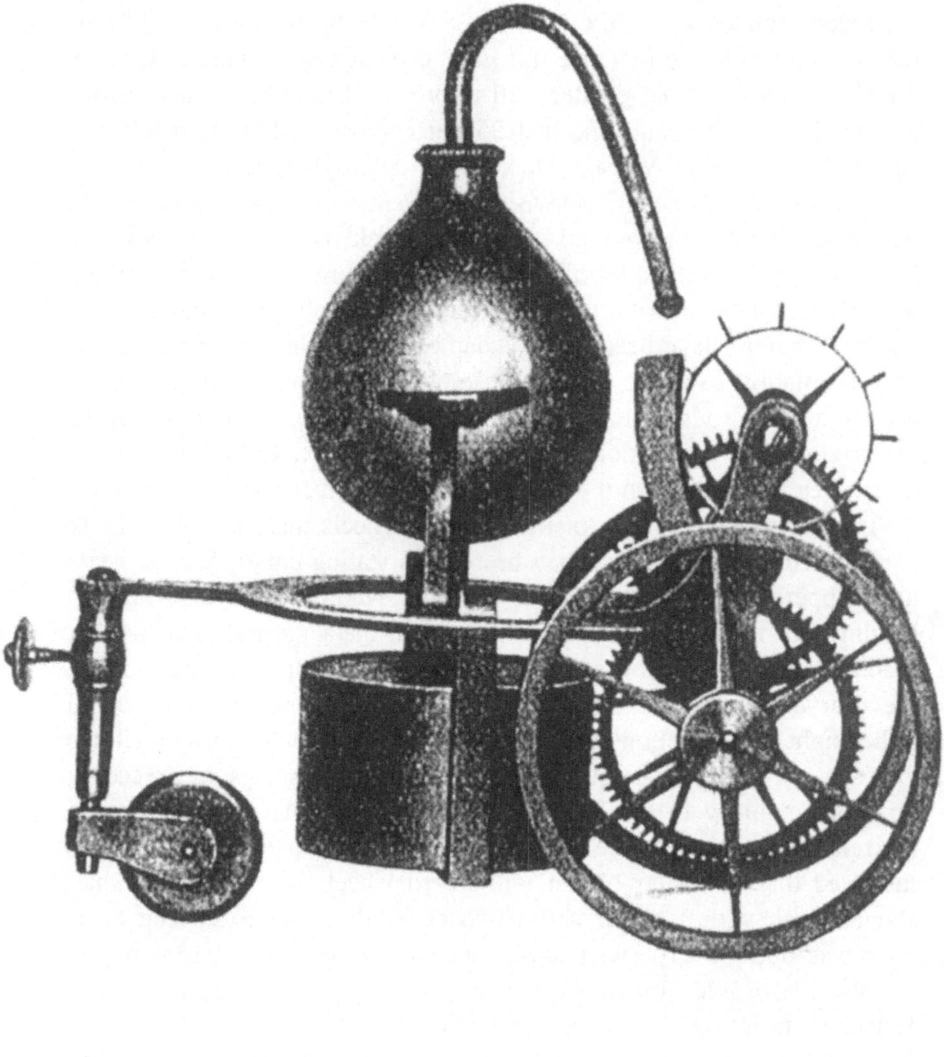

Ill. 61: Model of the 1775 "Karlsruhewagen," from: C. Dollfus - L. Beaudry de Saunier - E. de Geoffroy, *Histoire de la locomotion terrestre* (Paris, 1936), 2: 219.

Logic let the gear on the turbine shaft engage the front axle gear at the uppermost part. That would also keep the turbine shaft short. The Duryea drawing follows the same reasoning, but has the turbine rotating anti-clockwise. This may be a minor point, but in fact all early machinery rotated from left to right. Several explanations are offered for the undoubted preference for this direction of rotation. Charles Singer[30] points out that greater skill in one hand is linked to dominance of one side of the brain, and that 95 per cent of modern humanity are righthanded. Henry Winram Dickinson[31] states that the direction of rotation has always been clockwise, a convention possibly based on the way a winch handle is worked. Accepting clockwise rotation as primary I am, however, inclined to think that rotation from left to right is based on early human observation of the sun that seems always to move to the right. This must have been very fundamental for primitive humans, and so well established as the natural sense of direction that discovery of the earth's rotation round the sun could not break the convention. Whatever the origin of the convention, in the absence of directives to the contrary, the replica was built with the turbine spinning conventionally clockwise.

The material and shape of the two gearwheels must also be largely conjectured. Verbiest's text may be read as stating either that the gears had copper (or bronze) teeth, or that the whole gear was of metal. Having no facilities for fabricating such metal gears, and neither practical nor theoretical experience in that field either, I elected to make the gearwheels of balsa with metal teeth. This would fit the general lightweight construction of the machine. It would also facilitate experiments with various gear ratios. Studies of transmissions preceding and contemporary to the Verbiest machine, and general reflections on the relative speeds possible and desired for the turbine and road wheels indicated that a smaller pinion, or lantern wheel, on the turbine shaft should mesh with a larger crown wheel on the front axle. The exact ratio was open to experimentation, but something in the 1 to 4 region seemed reasonable. Initial experiments in a jig were not successful. Whatever ratio, radius, length of teeth, etc., the teeth on the crown-wheel clashed constantly with those of the lantern. After much head-scratching and exquisite fulmination, it dawned on me that the spacing of the teeth was critical and decided the radii. Once that was established, it was pretty plain sailing, and the revised gears worked

30 Singer, *History of Technology*, 1:24-25.

31 Henry Winram Dickinson, *A Short History of the Steam Engine* (Cambridge, 1939), p. 79.

well. Both in the jig and on the machine, the gear ratio was 1:4 (6 and 24 teeth, respectively, made of construction nails). Six months after cutting the chassis parts, I could run the machine with the turbine rotated by a blowing shop-vacuum.

To put the turbine wheel level with the boiler-spout, a supporting structure had been made. The vertical shaft ran in a copper bearing fitted into the upper cross-member of this support, and the lower end of the shaft, carrying the lantern wheel, had been drilled axially to fit around a steel axle fixed below (actually a 3 1/2 inch construction nail, duly modified). This arrangement would keep the turbine shaft with the pinion properly aligned.

In order to make the turbine/shaft/pinion assembly run as easily as possible, the lower face of the pinion was given a lightweight cover of thin metal sheeting, and a ball bearing (from a bicycle repair shop) put between this pinion face and the lower support for the steel axle that kept the turbine shaft in position. Before anyone cries foul, it should be pointed out that roller bearings were probably used in China before our calendar and known almost as early in Europe[32]. They were an established feature in seventeenth-century European technology. Ball bearings seem to have been employed by Italian craftsmen in the sixteenth century. So they are not entirely out of place in a seventeenth-century machine designed by someone fully versed in contemporary technology.

The machine now rolling and steering nicely under outside power, the next steps involved an effective boiler/nozzle aggregate and the fire arrangements. A good friend living nearby, and equally expert as producer of superb honey and mechanical marvels, volunteered to make a test boiler and nozzle. First a suitable can was subjected to stringent internal pressure tests, and duly reinforced. Then an elegant little safety valve, readily adjustable, was fashioned to function also as convenient filler. A small steam-cock was included in order to build up pressure before letting the steam discharge. A 3-inch length of 1/4-inch copper tube, annealed for careful bending, was fitted with a beautifully turned .035-inch nozzle. The safety valve set to blow off at 21 psi (per square inch), the steam jet whirled the turbine most convincingly in a test jig.

A full-scale test with this boiler in the chassis two weeks later was both promising and dramatic. The machine tried to move under its own power, but promptly caught fire in a rather impressive manner. The

---

32 Needham, *Science*, 4/2, 94.

conflagration was doused before any real damage resulted. Fortunately I had been provident enough to paint the immediate surroundings of the "fireplace" with heat-resistant aluminum paint. That prevented the flames from catching. The water dousing of the flames caused no harm, as the whole machine was carefully varnished with liquid plastic to give the balsa a harder surface and make the turbine blades impervious to steam.

Back again to the drawing board and the workbench for fabrication of a sheet metal firebox to protect the wooden chassis. As tested before, the installation of this firebox proved able to contain a fierce fire. Installed in the chassis, it marked a major improvement towards the safe operation of the Verbiest replica machine.

In between an .032 inch nozzle had been made for the next test. In spite of the firebox, this test was disappointing. The experimental boiler developed a leak that prevented pressure from building up. Repairs seemed pointless, and instead it was decided to make a proper copper boiler. The very first boiler had been discarded earlier. The expert produced a delightful upright, cylindrical boiler, with a content of 1000 cc (one quart). A "chimney" with two cross-tubes passed through the boiler to raise steam more quickly. The assembly was hydraulically pressure-tested, first at 50 psi, later at 75 psi. With the safety valve at 28/30 psi and the .032 inch nozzle in place the test was made. Solid fuel tablets flamed mightily, emitting evil smelling gasses. The safety valve hissed with powerful pressure. Steam rushed from the nozzle when admitted - and nothing else happened. The machine sat immovably on its wheels amidst all the sound and fury.

Much too quickly the fuel tablets burned out, and the first head of steam was gone in vain. We tried once more, and lifted the driven wheels to see if the turbine could turn them when unencumbered. It could, and although unable to move the machine, it seemed on the verge of so doing. After head-scratching and brave talk to hide our disappointment, a threefold remedial action was undertaken: a warmer, more sustained fire, higher boiler pressure, and a selection of nozzles to see which size would create the necessary torque at the turbine blades.

During the early planning, a liquid fuel burner had been considered, as mentioned. But such burners are not too easily obtainable, although one had actually been procured after long delays, and without buying a complete fondue set. While waiting for this burner, the fuel tablets had seemed an acceptable substitute, but in practice they showed limitations. So the firebox was modified to hold the liquid fuel burner. The expert produced an elegant set of interchangeable nozzles in three sizes: .050 inch, .060 inch, and .070 inch. The safety valve was re-adjusted to 35 psi, a temporary pressure gauge was fitted, and - to reduce rolling resistance - a slim rubberstand was stretched around each wheel. Thus prepared for the next test, optimism was tempered with some apprehension.

Again the test fizzled in clouds of steam. The new burner was an obvious and great improvement, but the 0.70-inch nozzle was too wide. Steam pressure got lost too quickly without generating sufficient force to move the machine.

In quiet desperation, the safety valve was readjusted to 40 psi blow-off, and the .060-inch nozzle was fitted. The fire was lit, the boiler simmered, pressure mounted rapidly, the steam-cock was opened and the machine moved forward at a steady, dignified speed.

It was a momentous experience to see the machine become automotive. I shook the house with a resounding yell of joyous relief, and felt a strangely intense contact across the span of centuries with the man who first saw his machine move so long ago. The time was 15:26 EST on February 20, 1984, some three hundred years after the original vehicle first ran in the Imperial Palace of Beijing.

Later tests, first with the .050-inch nozzle and blow-off at 45 psi, then with the .040-inch nozzle and 50/55 psi, were consistently successful. Now it was just a matter juggling pressure and nozzle sizes to obtain longer lasting performance, if so desired. Anyway, the point had been made.

Fourteen months after cutting the first wood for the chassis, I was able to state without any reservation that Verbiest's automotive machine did run as he reported. His imaginative *principium motus* was sound, and his concept really far grander than the *alteri machine* he so modestly claimed.

Summing Up

The automotive machine was only one of Verbiest's many technological triumphs. He possessed a rare versatility that went beyond science and mechanics, but his technical skill and imagination was, perhaps, the most astounding aspect of his brilliance.

He did, undoubtedly, construct the first documented automotive machine and, yet, he did not invent the automobile, as that conception is understood today. Verbiest's "fire-carriage" had no carrying capacity, and was too small to do more than divert the young emperor of China and Tartary. Verbiest did not think, either, that he had invented the automobile. In all humility he thought of his machine as "a method of movement" for which "other diverting applications are easily contemplated." In those contemplations he was, of course, several centuries ahead of the general technological and industrial development.

Research in depth about his life and times confirms that Verbiest was an extraordinarily gifted man, resourceful and ingenious. It may certainly be argued that his automotive machine did not influence the general development of the steam engine - neither the piston-, nor the turbine-variety - or of the automobile. That does, however, in no way diminish his stature in the "Pantheon of Technical History." A true pioneer is usually so far ahead of contemporaries that a breakthrough is neither understood nor even recorded. Fortunately, Verbiest did record most of his activities in some detail, and the records have even been preserved. Later commentators have distorted the evidence for various reasons ranging from inability to comprehend Latin, to nationalistic fervor.

The valuation of Verbiest's machine is, naturally, also dependent on the general level of technological development at the time the evaluation was made. Thus one serious writer classified Verbiest in 1839[33] as a steam engine pioneer. Six years later, in 1845, the same writer[34] emphasized Verbiest's experiments with marine steam-propulsion, and suggested that a Belgian tug-boat be named after Verbiest. It is not known if the suggestion was acted upon.

Sixty-seven years later, in 1912, another serious writer[35] mentions Verbiest's automotive machine in extremely well reasoned terms. He was not the inventor of the steam engine or the aeolipyle, but was a man who applied these known mechanical elements in a "new, very ingenious" way to "a small road locomotive." He goes on to say that Verbiest belongs to the precursors of the steam engine inventors. An irrefutable claim. The object of the present essay has been to clear the misunderstandings accumulated over three centuries, and to give Verbiest his proper place in automotive genealogy. The research brought me into contact with a multi-faceted, engaging personality of

---

[33] Carton, *Annales*, p. 130.

[34] Carton, *Album*, p. 67.

[35] Henri Bosmans, "Ferdinand Verbiest, directeur de l'observatoire de Pékin (1623-1688)," *Revue des questions scientifiques*, 71 (1912), 400.

great integrity and rare technical brilliance: a true mechanic who turned his unparalleled skill to the realization of new concepts. A visionary precursor.

Note on Earlier Illustrations of the Verbiest Vehicle

Research efforts have failed, so far, to turn up any contemporary illustration of the original Verbiest automotive machine. Later pictorial reconstructions are mostly conjectural, not to say hypothetical. The first came some 250 years after the event from the drawing board of the American pioneer Duryea shortly before he died in 1938. His drawing shows body-work decorated with a dragon. That is clearly apocryphal, but not entirely out of character. Verbiest liked the idea of hiding mechanisms (see his text about the steam boat). And some of his cast bronze astronomical instruments were, in fact, decorated with dragons. However, various of Duryea's technical details are questionable, and were commented upon in the main text.

The Duryea drawing somehow found its way to the Italian writer, Giovanni Canestrini, and appears in his 1938 book *L'Automobile*. Later it reappeared in an Italian history of technology from the Middle Ages to the present[36]. This book lists Canestrini as one of the contributors.

In 1956 the Swiss weekly *Automobil Revue* printed a story about a wooden "automobile" made in China "thousands of years ago"[37]. The illustrations published with that story were ascribed to Verbiest but correspondence with the author failed to produce any evidence for that claim. They were, as far as he could or would recall, a "graphic reconstruction." They seem to be awkward renditions of the Duryea design, with the turbine and transmission omitted. In the article the author stated his inability to explain the transmission. These illustrations are surely from a hand much later than Verbiest's.

A third representation of Verbiest's machine is to be found in the Museo Nazionale della Scienza e della Tecnica "Leonardo da Vinci" in Milan, Italy. There is displayed a small (20 x 20 cm. or appr. 8 x 8 inches) metal model built in 1956 for the exhibition celebrating the sixtieth anniversary of the Touring Club Italiano. As shown by the photo, the three-wheeled metal construction, with an elaborate train of

---

36 Giovanni Canestrini, *L'Automobile: il contributo italiano all'avvento e all'evoluzione dell'autoveicolo* (Rome, 1938), p. 423; Arturo Uccelli, *Storia della tecnica dal medio evo ai nostri giorni* (Milan, 1945), p. 637.

37 See the main text for comment about that story.

vertical gears between a small vertical turbine and the driven axle, has nothing to do with Verbiest's description. Enquiries to the Museum director promptly provided the information just quoted about the size and presentation of the model. It was further explained that the model was based on a picture published by *L'Illustration* in its history of terrestrial locomotion[38].

Unable at this writing to consult that sumptuously illustrated, but not too reliable volume, I can, however, state that the model in question is, in fact, of the German "Karlsruhewagen," a toy made in 1775 and thought still to exist in the city of its origin where, incidentally, Karl Benz grew up and was educated less than one hundred years later. How it came to be represented as the Verbiest machine is a mystery.

In the Italian history of technology mentioned before, the Milan model is illustrated, but in the text, not ascribed to Verbiest. Other works also reproduce the same photo as Verbiest's machine, which it is definitely not.

Concluding this review of pictorial misinformation regarding the Verbiest vehicle, I give a brief comment on the drawing that appeared in my book[39]. The artist, Verner Hancke, worked from a rough sketch made too hastily by me. The following errors are entirely mine: the turbine wheel should show protruding vanes, the gearing is not appropriate, the boiler is less than practical, and the rear wheel steering ineffective. Of this, more in the main text.

---

38 L. Beaudry de Saunier, *Histoire de la locomotion terrestre* (Paris, 1936; 2nd edition 1942). [Editor's note. The writer states he was unable to consult this work but that the Milan model is that of the German Karlsruhewagen of 1775. The following is additional data about this footnote reference: Charles Dollfus and Edgar de Geoffroy, *Histoire de la locomotion terrestre*, 2 vols. (Paris: L'Illustration, 1935-1936). Volume 2 was written by Louis Beaudry de Saunier, Charles Dollfus and Edgar de Geoffroy. For the account about Verbiest, the incorrect attribution by du Halde that Grimaldi copied Verbiest's design and an illustration of the 1775 model, see 2:219. The Milan 1956 model is thus based on the illustration of the 1775 German model as found in the 1936 book and as the illustrations of both models indicate. The writer concludes that this is "definitely not" Verbiest's machine.]

39 Jorgen Ditlev Scheel, *Cars of the World* (New York, 1963), p. 9. [Editor's note: The second edition of this work appeared in 1964 and the third in 1967. This is an English translation of his original study, *Beromte biler, tegnet af Verner Hancke* (Copenhagen: Politikens Forlag, 1962). The French translation is entitled *Les automobiles du monde entier* (Paris, 1964)].

Verbiest

Diplomat

FERDINAND VERBIEST AND HIS ROLE IN THE FORMATION OF SINO-RUSSIAN DIPLOMATIC RELATIONS

VLADIMIR S. MIASNIKOV

Professor Vladimir S. Miasnikov is deputy-director and corresponding member of the Russian Academy of Sciences. He published several monographs on Russian-Chinese relations, as well as articles on historical and contemporary aspects of the relations between Russia and China.

Three hundred years have passed since the death of Ferdinand Verbiest. A Flemish sinologist, a Jesuit missionary, a mathematician and a writer[1], a cartographer and a diplomat, he reached a high position at the court of the Kangxi emperor and played a prominent role in the establishment and development of diplomatic relations between Russia and China in the second half of the seventeenth century.

Verbiest was born on October 9, 1623 in Pittem near Kortrijk. He began his studies in September 1641 in Mechlin, and continued his theological education in Rome and Seville. Sent by the Society of Jesus as a missionary to China, Verbiest arrived there in 1659. He started to serve as a preacher in Shaanxi province, but then Johann Adam Schall called him to Beijing to join the research in the field of astronomy.

At the beginning of the Kangxi emperor's rule, when persecutions of the Christians were occurring, Verbiest, along with other Jesuits, was imprisoned. Schall, who had served on the Board of Astronomy, was replaced by Yang Guangxian, a Qing official, and a Chinese Muslim (who, unfortunately, did not have enough knowledge of astronomy and thus brought a lot of disorder into the emperor's calendar). Ten years after his arrival in China, Verbiest, having gained the deep respect of the emperor because of his knowledge in various fields, became head of the Board of Astronomy. In fact, he became a personal friend of the

---

1 The bibliography of F. Verbiest contains more than 35 titles.

emperor and one of his closest advisors. He took the Chinese name of Nan Huairen and his courtesy name was Dun Bo.

On May 15, 1676 the Russian embassy, headed by Nikolai Govrilovich Milescu Spathary, entered Beijing. The Russian ambassador was, by European standards, a prominent scholar and qualified diplomat. He spoke several languages, including Latin. The Qing government delegated several high officials to begin negotiations with the Russian mission, and Verbiest was appointed as interpreter. So, in the course of those negotiations, it was Latin that was used for the first time as the language of communication between the Russian and the Qing diplomats, and Jesuits served as mediators between the two states.

Acting in the interests of the Society of Jesus, for which he was seeking entry in Moscow, Verbiest - enjoying the emperor's trust and fully informed of Qing diplomatic activity - tried to give the Russian ambassador the right bearings, even at the risk of jeopardizing his own position. He conveyed the foreign policy outlook of the Qing court to Spathary, and told him of the situation in the country and at the court. He let Spathary know that the officials were misleading him when they said that they did not dare report his request to be received by the emperor, and to deliver the tsar's message to him. In fact, the emperor was continually inquiring about the Russian embassy. The information Verbiest passed on to the Russian diplomat should not be underestimated. In the course of those negotiations, the Qing diplomats actually rejected all Russian proposals aimed at developing relations between the two states.

At this time, Spathary learned the true contents of the emperor's message to the tsar delivered in 1675 by Ignatiy Milovanov, a Russian representative. (Verbiest had secretly translated it into Latin for him). "Henceforth, should you wish to live in peace," the emperor wrote to the Russian tsar, "you must return the fugitive Gantimur. ... If you do as I say, peace will prevail. That is expressly why this edict is being sent"[2]. Peter Gantimur, it must be mentioned, was a prince, belonging to a nomadic tribe in northern Manchuria; to protest the persecutions of the Qing authorities, he left for Russian domains. The Qing authorities viewed that development as a dangerous precedent, and demanded Gantimur's surrender. The Russian authorities, however, turned down their demands.

---

2 *Russko-kitaiskiye otnoshenia v XVII veke. Dokumenty i materialy* (Russian-Chinese relations in the seventeenth century), 2 vols. (Moscow, 1969), 1:411, n° 183.

Spathary and Verbiest were both scholars and diplomats. They felt a warm interest in one another. Their close relationship enabled the Russian diplomats to learn the true intentions of the Qing government. Verbiest told the envoy, under the oath of secrecy:

> ... that the Bogdykhan was contemplating war if His Majesty the Tsar did not surrender Gantimur. He also intended to attack the border forts of Albazin and Nerchinsk, which he now considered a source of danger, having come to learn that the people lived there with the knowledge of His Majesty the Tsar, (previously he had thought they were living there of their own free will, as people once lived on the Amur, and wreaked havoc whenever they wished). He also knew that the forts were poorly garrisoned, that they were far away from Moscow and near China, and he would prefer that the number of troops did not multiply on his border[3].

Verbiest was hoping to provide the Jesuits with a bridge for communication and travel between Europe and China via Russia, and that was why he rendered such services to the Russian diplomats. He let them understand that the demand to surrender Gantimur was nothing more than a diplomatic ploy. The Qing real interest did not lie with Gantimur, he said, for all they wanted to know was the tsar's intentions. Verbiest explained:

> They are a shrewd people, and they know that the forts were built for the protection of yasak-paying people. Should the tsar surrender Gantimur, the chief of all heathens, then the heathens will either follow him or drift away one by one. The tsar would then have no reason to keep his troops there, as there would be no yasak-paying people on the border[4].

Knowing the further course of events, we can see how accurate this piece of intelligence was. All the difficulties, setbacks, and losses of Spathary's embassy were worth this most valuable information about the emperor's true intentions. Moreover, Verbiest said that the Russian government still had time to fortify its Far Eastern outposts, because the Manchus would probably wait for Moscow's reaction to Spathary's embassy before taking action.

This confidential relationship between Verbiest and Spathary did not pass unnoticed; both of them were under suspicion. One day, on his way back from the palace, the envoy met Verbiest, who added to his

---

3 *Ibid.*, p. 433.

4 *Ibid.*

distress about the embassy's setbacks. He told him that someone in the Russian embassy "had done him, Verbiest, harm," by telling Amuhulang, a member of the Qing court, and Mala, a *de facto* director of the "Russian department" in the Qing imperial intelligence, "that the Jesuit had told the envoy all sorts of things about Chinese affairs"[5]. In the course of a lawsuit against Spathary in Moscow, upon his arrival home from the Qing capital, his contacts with the Jesuits in Beijing were used as incriminating evidence against him as well.

Still, Verbiest did not fear suspicion. A few days later he visited the Russian ambassador again, and explained that the demand to surrender Gantimur was being made so obdurately because the Qing "know you will not surrender him, thus giving them justifiable grounds to go to war and ravage Nerchinsk, Albazin and nearby areas"[6].

In March 1682, the emperor decided to visit Manchuria, the native realm of his dynasty. He wanted to see the places that would soon be the staging area for a system of measures aimed at changing the situation along the northeastern borders of his empire. It was officially announced that the emperor was going to offer sacrifices at the tombs of the first three Manchu khans near Mukden. On his journey, the emperor chose Verbiest as his closest companion. He wanted to verify the details of his scheme in conversation with this highly educated European who understood the Russians so well. Officially, Verbiest was to determine the astronomical coordinates of the bigger Manchurian cities. The emperor, too, liked to watch the nocturnal sky in the company of the chief of the Board of Astronomy to show his intimates that he was no less skilled than the European in recognizing constellations and calculating the time of the night by the stars[7].

The emperor's scheme was to establish outposts on the northeastern borders of the empire, so as to push the Russians out of the Amur region. Back in Beijing, Verbiest received a new appointment: he was made chief of the Board of Civil Works (*gongbu*) and charged with equipping the Qing army with European weapons.

In 1686, when the Qing army failed to break the resistance of the defenders of the Russian fort of Albazin, and the emperor was constrained to enter diplomatic contact with the Russian government

---

5 *Ibid.*, p. 443.

6 *Ibid.*, p. 449.

7 N. Cameron, *Barbarians and Mandarins. Thirteen Centuries of Western Travelers in China* (New York, 1970), pp. 245-247.

again, he chose an indirect approach. It is probably safe to assume that this approach was suggested by Verbiest, the emperor's advisor on "European affairs." The emperor picked the Dutch ambassador, Vincent Paats, as a mediator. By that time, Moscow also decided to resume diplomatic ties with the Qing court, and sent Nikifor Veniukov and Ivan Favorov, members of the former Spathary mission, to Beijing.

From their first visit to Beijing, Veniukov and Favorov already knew Verbiest, who had once applied to the Russian government (through Spathary) to be taken into its service, and was, indeed, the only one who could and would help the Russians. But the tsar's diplomatic service had not considered this possibility; none of the Russians who had come to Beijing knew Latin[8].

During Veniukov and Favorov's stay in Beijing, Verbiest acted with extreme caution. He managed to let the Russian representatives know that it was impossible to discuss anything in the presence of Qing officials, for this would inevitably create suspicion. He promised, however, to give them a letter for Spathary. He could trust no one, not even the Russian couriers, although they did not know Latin. Even so, he sent them his pupil Francisco, a Czech who had come to Beijing (with the Dutch embassy) to study Chinese. Questioned by the couriers about the plans of the Qing government, Francisco said the emperor was eager to "stop" the war, on the condition that the Russians surrender Albazin. His information about the Manchu army was of exceptional value. He said that there were no military men cognizant of modern science, "no engineers in the Chinese state," but that production of cannons and grenades had been started by Verbiest.

An interesting point is that, at the end of the Russian diplomatic visit to China, the envoys had received the emperor's order saying that the correspondence between the Russian and Qing court should be carried out not only in Latin, but in Mongolian as well. In his letter to Spathary, Verbiest wrote that the emperor had sent a message to the Russian tsar with the Dutch ambassador.[9]

In 1688, both states were engaged in the preparations for negotiations to take place in the Russian city of Selenginsk. In Beijing, the emperor was preparing a delegation to go to Russia. The experience with Spathary had shown that Latin was a most effective means of communication. Verbiest would, beyond question, have been the best

---

8 *Central State Archives of Ancient Acts. "Russia's Relations with China,"* Vol. 8, p. 86 r-v (in Russian).

9 *Russko-kitaiskiye otnoshenia v XVII veke* (Moscow, 1972), 2:789.

possible choice for the Selenginsk talks. But shortly before the appointments were made for the distant and difficult journey, Verbiest - astronomer and artillery expert, the emperor's tutor in the European sciences, his hunting companion, head of the Jesuit corps in Beijing who enjoyed the imperial favor - fell off a horse and died[10]. It is hard to say whose loss was greater: that of the Qing government or that of the Russian diplomats. Unexpected as it was, Verbiest's death gave another Jesuit, the Portuguese Tomas Pereira (Xu Risheng) the chance to be chief advisor of the Qing government in its diplomatic dealings with the Russian state.

Though Verbiest had passed away, his influence with Russian diplomats continued. In May 1689, on the eve of the Russian-Qing negotiations in Nerchinsk, a Russian representative, Ivan Loginov arrived in Beijing. He had brought a message from the Russian government to the Kangxi emperor. An interesting point is that Loginov asked for official permission to see the Jesuit missionaries. The Russian diplomatic service was unaware of Verbiest's death, and that is why Loginov, who did not know Verbiest personally, mistook Pereira for him. Since none of the Russians knew Latin - or any other West European language for that matter - the mistake was never rectified. The Jesuits were aware that Russia had entered an alliance with Rome and Poland against Turkey. The Russian courier was pleased to inform them that the allies had scored a series of victories[11].

A few days later the Jesuits gave Loginov a secret dispatch for ambassador Fedor A. Golovin, the head of the Russian delegation in Nerchinsk, in which Pereira introduced himself as a member of the Qing embassy. Pereira wrote of the previous year's journey to Selenginsk and, since he expected peace to be established between the Qing empire and the Russian state, hinted transparently that the Christian missionaries had a stake in such peaceful relations. He advised the Russian envoy to use the full titles of Songgotu and Tong Guogang, the heads of the Qing delegation in the negotiations, because prestige had an overriding importance for the Manchus. Hoping that Golovin's embassy would succeed, Pereira stressed that he had no intention of "going against Christian law" and aiding the heathens, though he was glad to serve the Qing emperor, so long as the latter did not hinder him from preaching the Gospel. The Jesuit said that he and his companions

---

10 Some authors suggest that Verbiest died after a short but difficult disease.

11 *Russko-kitaiskiye otnoshenia*, 2:489-490.

remembered "the many tokens of grace" shown to them by Spathary during his stay in Beijing[12]. Actually, in the course of the conference in Nerchinsk, Pereira's attitude differed from that expressed in the message.

In considering Verbiest's role in the Russian-Chinese contacts, we cannot but refer to another aspect of his activities in Beijing. In our view, he was an author of one of the maps of the northern borders of the Qing empire, and he produced that map under the influence of Russian cartography.

Spathary, who had obtained considerable information about China from the works of European missionaries and from his conversations with them in Beijing, augmented it with his own observations as traveler and explorer. As a result, his works were used for a whole series of European maps of East Asia in the late seventeenth century. Verbiest had been the first to learn from Spathary about the route to Europe through Siberia and "Muscovy". In fact, he paid for any future transit of missionaries across Russian territory by rendering services to the Moscow government via Spathary. Yet, he also needed the consent of the Qing government for such overland travel. It is safe to assume that Verbiest produced a drawing of the border between the Qing empire and the Russian state, with the adjoining zones, for his talks with the Kangxi emperor on this subject. A copy of the drawing was later used as a fragment for Antoine Thomas's map, and prior to that, the original could have been used by the emperor. That the emperor trusted Jesuit maps is clear from the fact that, after the initial version of *Da Qing yitong zhi* was completed, he instructed Jean-François Gerbillon on December 10, 1707 to draw a map of the entire country (completed by January 1, 1718)[13].

Even though the surmise that Verbiest played a role in producing a sufficiently detailed map that included the Russian Amur region calls for confirmation, it is obvious that the emperor did have such a map, whether Jesuit or Manchu, at his disposal.

---

12 *Ibid.*, pp. 490-491.

13 A. Hummel (ed.), *Eminent Chinese of the Ch'ing Period (1644-1912)* (Washington, 1943-1944; reprint, Taipei, 1964), p. 310.

FERDINAND VERBIEST AND SINO-RUSSIAN RELATIONS

HAO ZHENHUA (1927-1992)

Prof. Hao Zhenhua has been a research-fellow at the Institute of History, Chinese Academy of Social Sciences, and deputy secretary-general of the Chinese Society for Historians of China's Foreign Relations. He studied the history of the cultural exchange between China and Europe, and published many books and articles, among them "The History of Christianity in China before 1550." He died on April 30, 1992.

After its reintroduction into China by the Jesuit Matteo Ricci in 1582, Catholicism spread rather successfully for about 135 years. Because of the Rites Controversy, the Kangxi emperor decreed a prohibition of Catholicism in 1717, which seriously restricted the spread of the Catholic religion. In 1724 the Yongzheng emperor reaffirmed this decree, and almost all the missionaries were allowed to stay only in Guangzhou. What was the reason behind the continuous presence of the Jesuits at the Qing court during the reigns of the Kangxi, Yongzheng and Qianlong emperors? Generally speaking, the Qing government needed the Jesuits' knowledge of military affairs, the calendar and astronomy, as well as their scientific and technological expertise. But another, often neglected, reason was that the Qing government needed both the Jesuits' language and their diplomatic skills in its foreign relations[1].

In the sixteenth and the first part of the seventeenth century, Portuguese, Spanish, Dutch, English and French colonial forces expanded by sea into the southeastern coastal area of China. In the north, Russian colonialism invaded the Amur valley. After the Qing

---

1 In another article I concentrated on the relations between Verbiest and the Russian envoy, N.G. Spathary. See "Analysis of Spathary's Report of the Diplomatic Mission to the Qing Empire", in *China & Europe* (Louvain, 1986), pp. 110-139.

dynasty had come to power in 1644, foreign diplomatic envoys repeatedly came to China to negotiate trade. Examples are the Spanish envoy in 1646, the Dutch envoys in 1656, 1667 and 1686, the Portuguese envoys in 1673, 1676, and the Russian envoys in 1656, 1676 and 1686.

The Manchus were a minority group originating from the northern bank of the Amur river in the north of China. When they started to reign over central China, their socio-economic development was still in transition from slavery to feudalism, and their cultural level was very low. In dealing with such complex international affairs, they met with language and ideological problems. For its foreign relations, therefore, the Qing government had to rely on the European Jesuits who were in China at that time. These Jesuits understood not only many European languages, but also Chinese and Manchu, and were good at conducting diplomatic affairs. Ferdinand Verbiest, Dominique Parrenin, Kilian Stumpf, Ignaz Koegler and Antoine Gaubil, for example did diplomatic translation; Jean-François Gerbillon and Tomas Pereira participated in the Sino-Russian Treaty Conference of Nerchinsk (1689); Claudio Filippo Grimaldi and Joachim Bouvet were sent as diplomatic envoys to Europe. Among them, Ferdinand Verbiest was the man who aided Sino-Russian relations most effectively. This article discusses his activities in this respect.

I.
Bridge in the Foreign Relations Between China and Russia

The first prerequisite for two countries to be able to have diplomatic relations is the existence of a common language. Without it, any exchange of ideas is impossible. Before 1676 Sino-Russian relations lacked this common tongue.

Russia was originally a European country. Between Russia and China lies the vast Siberian region, with its many peoples who lived in various societies with a very low level of civilization. Along the upper reaches of the Yenesei lived the Eleuts, who were Mongolian-speaking nomads. Along the middle reaches of the Lena lived the Yakuts, who spoke Turkish and lived on hunting and fishing. In the region of Lake Baikal and along the upper reaches of the Lena river lived the Brati, who led a nomadic life and spoke Mongolian. All these peoples separated China and Russia completely.

In 1676 the Russian Tsar, Alexei Mikhailovich, sent Nikolai Govrilovich Spathary Milescu as a diplomatic envoy to China. Spathary's ancestors were Greeks from the Peloponesian peninsula, but he himself was born in Moldavia and educated in Constantinople, the capital of the Ottoman empire. There he learned Turkish, Greek and Arabic, and studied theology, philosophy and history. Later he went to Italy to study Latin and Italian, the natural sciences and mathematics. He worked for several decades in the service of the Danubian principalities, and had much political experience. He also had frequented the courts of large European countries and was well-versed in European diplomatic relations. In 1671 he was introduced to the tsar by the Jerusalem patriarch, Dositheus, as a Greek and Latin interpreter, and worked in the Posoliski Prikaz on the Moscow Razriad of Russia.

Spathary was, at that time, considered an excellent scholar in Russia. But for this diplomatic mission to China, his knowledge was by far insufficient. He did not understand any eastern language. On the way to China, while passing through Tobolsk and Nerchinsk, he looked for some interpreters. All of those people spoke Mongolian, and some of them understood a little Russian or Manchu. They were, however, all illiterate. None had ever held public office and none knew anything about diplomatic affairs. Spathary was, therefore, constantly concerned about translation problems. At the Qing government it must be said, none of the Manchu officials understood a western language either. So language was obviously a major stumbling block in the foreign relations between China and Russia.

Before Spathary came to Beijing, the Qing government had already considered the problem of translation. They knew that Verbiest could speak many western languages as well as Chinese and Manchu, and that he was erudite. They, therefore, decided to ask him first to try to talk with Spathary.

On May 15, 1676 Spathary arrived in Beijing. The minister (*shangshu*) of the *lifan yuan*, Amuhulang, came along with Verbiest to visit the Russian envoy. When they met, Verbiest and Spathary greeted each other in Latin and, after a first talk, were very happy to understand each other. Spathary said to the Amuhulang: "Thank the Bogdikhan (the Kangxi emperor) for having sent such a man who can not only orally explain all the affairs of the two sovereigns, but can translate correctly, word for word, their titles and letters"[2]. From then on Verbiest became the bridge in Sino-Russian relations, opening the door for the Jesuits to Chinese diplomatic circles. Latin became the formal language used for

2 John F. Baddeley, *Russia, Mongolia, China* (London, 1919), 2:333.

the diplomatic relations between China and Russia. The association between the two countries was unblocked.

Spathary arrived in Beijing in May and left for Russia on September 1. During this time, Verbiest did a lot of translation work. On June 5, Spathary handed over a letter[3] to the Kangxi emperor written by Tsar Alexei Mikhailovich on February 28, 1675 in Russian, Latin and Mongolian. The letter stated that the tsar's family descended from the Roman emperor Augustus Caesar, and that since nobody at the Russian court could understand Chinese, the contents of the four letters China had previously sent were unintelligible. The tsar also hoped that friendly and neighborly relations would be established between China and Russia. The Kangxi emperor ordered Verbiest first to translate the tsar's letter orally. Afterwards he said: "Neither my father, nor any other person, has ever received such a letter from the White Tsar. It must therefore be preserved in the Treasury as something very precious"[4]. Later Verbiest translated the letter from Latin into Manchu, and submitted it to the council chamber.

Following the Russian government's instructions, Spathary proposed, on June 13, twelve conditions of negotiation to the Qing government. On June 16, Verbiest was asked to translate them from Latin into Manchu. After reading the translated text, the Kangxi emperor ordered the Council of Elders to study it thoroughly and prepare a reply. The Qing government replied to every condition, but none of these were given to Spathary. Of the Qing government replies, the one concerning the second condition was most noteworthy. That condition said: "We beg the Chinese emperor to please write future letters in Manchu and Latin; our letters in the future will be written in Russian and Latin." The reply of the Qing government was: "We will act according to the requirement"[5]. From then on Latin became the common language used in the diplomatic relations of both China and Russia. This is proved by the following, yet incomplete statistics: there were one hundred and ten letters translated by Jesuits from Latin into Manchu or vice versa

3 The letter from the Russian Tsar Alexei Mikhailovich to the Kangxi emperor of China, February 1675, in *Qingdai Zhong-E guanxi dang'an shiliao xuanbian* (Selected Historical Materials of Russian-Chinese Relations during Qing Dynasty) (Beijing, 1981), 1:39.

4 Baddeley, *op. cit.*, p. 352.

5 The memorial to the Kangxi emperor by Fuquan about the claims of the Russian envoy, June 27, fifteenth year of the Kangxi reign, in *Qingdai Zhong-E guanxi*, 1:28.

between the fifteenth year (1676) of the Kangxi reign and the eighth year (1730) of the Yongzheng reign.

Spathary had also brought along four letters from the Chinese emperor to the Russian tsars: two of them written during the Ming dynasty and two during the Qing dynasty. According to Spathary, they had not been translated because nobody in Russia understood Chinese. The tsar asked the Chinese government to send them back after translation into Latin. The two letters dating from the Ming dynasty were written in Chinese, one[6] in September 1618 by the Yongle emperor to Tsar Mikhail Theodorovich, the other[7] in September 1642 by the Wanli emperor to the same tsar. The contents of the two letters were basically the same.

In them, the Chinese government asked the Russian merchants to trade with China. My personal view is that both letters are false, for reasons discussed in another article[8]. On June 17, the Kangxi emperor ordered Verbiest to translate both letters into Latin, and on June 23, Verbiest gave the translations to Spathary. Of the Qing dynasty letters, the first[9] was written by the Shunzhi emperor to Tsar Alexei Mikhailovich in Manchu in 1655. In it, he praised Russia for sending tribute to China. The second letter[10], by the Kangxi emperor to the same tsar in 1670, was in Manchu and Mongolian. It asked the Russian government to send the Chinese deserter Gantimur back, and to stop the plundering of the people at the Chinese border. Verbiest also translated these two letters into Latin. But in the end, the Chinese government did not give the two translations to Spathary because they realized that they showed an arrogant attitude as well as impolite language.

The difference in historical and cultural background between China and Russia led to divergent views on diplomatic relations. This resulted in heated arguments between the Chinese and Russian representatives: during negotiations, during the submission of letters or presents, and

---

6 Baddeley, *op. cit.*, p. 72.

7 *Ibid.*, p. 73.

8 Hao Zhenhua, "An Analysis of Two Letters from the Two Emperors of the Ming Dynasty of China to the Russian Tsar", *The Journal of World History*, n° 1 (Beijing, 1986), 22-29 (in Chinese).

9 The letter from the Shunzhi emperor of China to the Russian tsar, May 22, twelfth year of the Shunzhi reign, in *Qingdai Zhong-E guanxi*, 1:18.

10 The letter from the Kangxi emperor to the Russian tsar for asking him to send back Gantimur, May 13, ninth year of the Kangxi reign, in *Qingdai Zhong-E guanxi*, 1:21.

during the ceremony of receiving envoys.

After China had been united during the Qin and Han dynasties, the idea of *tianxia guo* (the unique kingdom in the world) arose, and it called itself *tian chao* (the supreme court). All the surrounding lands which fell within China's cultural sphere were considered as belonging to China. All the countries who obeyed China, or admired its culture were considered as its vassal states. That is why the notion arose in ancient China, that "no place on earth does not belong to the Chinese king, and every person on earth is his servant." This idea was carried out in Chinese foreign relations until the Qing dynasty. The tribute system was a special product of China's history and culture because, before modern times, China's culture was excellent and its economy prosperous. Only by admiring its culture and honoring its emperor could foreign countries establish relations with China. This admiration and honor were expressed in the paying of tribute, either regularly or irregularly and, in this way, China established its metropolitan position. The emperors of the Ming and Qing dynasties inherited this system and these ideas, and considered themselves the general rulers of the universe. When the sea routes opened and westerners gradually came to the East, China called the first Europeans "vassal people" and the western countries were considered "vassal states." The Dutch and English were called "red-haired barbarians." When western countries wanted to establish relations with China, they first had to send an envoy to pay tribute, so as to obtain the qualification of a tributary state. Up to the Daoguang reign of the Qing dynasty, China did not recognize the existence of the international community or the principle of equality between countries. The negotiations with Spathary were conducted in just this same spirit.

Before he left, Spathary asked the emperor to send a reply to the Russian tsar in Latin, treating him as an equal. In his answer to Spathary's request, the Grand Secretary of the Qing government, Mingzhu, made three suggestions:

> The first is: that every ambassador who comes to us here in China, must frame his speech as if he came "up" from a humble and inferior place to an exalted one, or throne; and we, in reporting to the Khan[11] the arrival of such ambassador, or other envoy, will likewise state that he has come from such and such a Sovereign, from a lower place to Your most exalted Throne, to strike his forehead on the ground. The second article is: that all gifts sent by any sovereign whosoever,

[11] Khan, i.e. Chinese emperor.

> through his ambassador, to the Khan, are called by us in our report, not gifts, but tribute, or yasak, declaring, "Such and such a ruler has sent Your Majesty, the Khan, tribute, or yasak," and not "presents." The third article is: when the Khan sends gifts or countergifts to such Ruler, they are not called "gifts," but it is stated that in return for services rendered, the Khan sends gratuities and presents to the Ruler, and to all his subjects. These three articles must be accepted by all ambassadors and other envoys who come to us, without reserve. In sum: I. When the Khan writes an answer to any other sovereign, he frames it as from the Most Exalted Throne to an inferior place. II. Of presents, it is said: "The tribute (or yasak) you sent us has been accepted." III. As to gifts or countergifts, it is written "Wherefore, we send to you and to all your country, for services rendered, a gratuity and presents"[12].

We know that from the end of the fifteenth century onwards, Europe entered a new period of international relations. Many states arose, and they were prepared to relinquish the idea of "our monarch is the most honorable" in favor of foreign relations, putting benefit to country and people in the first place. Accordingly, they adopted the principle of equality in foreign relations. Russia had adopted this way of dealing with diplomatic problems in Europe. By the middle of the seventeenth century, it had become one of the strong European countries. It was, therefore, understandably impossible for the Russian envoy to accept these humiliating regulations.

Spathary, upon hearing the proclamation of the Grand Secretary, Mingzhu, immediately criticized it. To the condition that China was a superior country, and that diplomatic envoys of other countries should strike their forehead on the ground in front of the Chinese emperor, he answered that in the whole world the tradition that sovereigns of large countries should respect each other had been established and accepted as a standard for a long time. Russia was not China's vassal state, so why should it call China a superior country? It was a general rule at that time, that diplomatic envoys bow in front of the sovereign of any country. The Russian envoy represents his sovereign. If he strikes his forehead on the ground in front of the Chinese emperor, this would be an insult towards the Russian tsar. To force a diplomatic envoy to perform etiquette that would insult his sovereign, would be something unheard of in the whole world. In answer to the point that China considered the presents of another country as tribute, Spathary argued: when China calls the gifts of another country tribute, it does much harm

12 Baddeley, *op. cit.*, p. 408.

to that country. The Russian tsar gave presents to China, hoping for friendship and good will. If China should call them tribute, it would consider Russia a vassal state. This would be an insult towards Russia. To accept this regulation would be the same as to sell the Russian tsar and his country; this is something he could never do.

This argument was the most important one during these negotiations between China and Russia. This was the first confrontation between eastern and western diplomatic ideas. After this debate, both countries recognized each other as sovereign states. This exerted a significant influence on the development of Sino-Russian relations.

This fierce debate was first translated as follows: the vice-president of the *libu* (the Board of Ceremonies) of the Chinese government, Mala, translated from Manchu into Mongolian, and then the Mongols at the Russian side translated from Mongolian into Russian. But because the Mongols proved to be not equal to the task, both sides agreed to ask Verbiest to do the translation. He did it directly from Manchu into Latin, and brought the negotiations to a good end. In this way, the exchange of ideas of foreign relations between the East and West had materialized.

II.
Recommendation to Send Grimaldi on a Diplomatic Mission to Russia

Ten years after Spathary went back to Russia the relations between China and Russia became more strained. This caused competition between the Jesuits and the Dutch for the favor of the Kangxi emperor. After 1676 the internal and external situation of China changed a lot.

In 1681, "The Rebellion of Three Feudatories," which had lasted for eight years, was finally suppressed. In 1683, Qing troops defeated the army of the Zheng clan that occupied Taiwan, and thus brought Taiwan again under the Qing government's supremacy. After these changes, the Qing government focused its attention on its northern border. There Russian aggression against the Amur valley had never stopped; it had rather intensified. The Russians did not confine themselves to the upper reaches of the Amur, but moved forward to its middle and lower reaches and up to the eastern bank of the Argun. In February 1685 Chinese troops defeated the Russian army at Albazin. In August of the same year, a large number of Russian troops occupied the place again. In June 1686 the Chinese government mustered a large army to reconquer this place.

It was under these circumstances that the Dutch governor-general of Batavia sent a diplomatic corps headed by Vincent Paats. It left Batavia on July 20, 1685, and arrived in Beijing on July 31, 1686. On August 3, the emperor held a banquet at which the Dutch envoy asked the Chinese government to allow the Dutch to engage in trade in Guangzhou, Changzhou, Fuzhou, Ningbo, Hangzhou, etc..., to be able to leave the ports without the Chinese government's permission, to trade with all the merchants, and to establish settlements in all ports mentioned above. The Chinese government, however, refused all these demands for trade privileges.

As already mentioned, the situation on the Chinese-Russian border had worsened, at this time, and a war could break out at any moment. Since 1681 China had sent six letters to Russia to discuss border problems, but they were like stones dropped into the sea; none of them received a reply. In spite of this, the Chinese government still wished to settle this problem in a peaceful way, and intended to send another letter to Russia. Officials were ordered to ask the Dutch envoy to communicate with Russia. The Dutch answered affirmatively, whereupon the officials submitted a memorial to the Kangxi emperor:

> At present, we have asked the assistance of the tributary ambassador of Holland, who reports that Holland's boundary is closely connected with Russia's and that officials of each country can understand each other's language. He has been informed of our problem several times, and understands all the details thoroughly. Now we have written a letter, using the seal of the Board of War, and given it to the Dutch ambassador to forward to the Chahan Khan[13], asking him to withdraw the Luochas[14] at Yakesa and Nibuchu[15], suggesting that we define the boundaries so that both sides will refrain from crossing. When he answers this letter, he will let his envoys come directly to Beijing by the overland route. If the overland route proves to be too difficult, let them send his memorial to Holland to be forwarded to us[16].

The emperor was, of course, very happy with these arrangements and commanded the Board of War to write immediately a letter to the Dutch

---

13 Chahan Khan, i.e., the Russian tsar.

14 Luocha, i.e., Russians on the Sino-Russian border.

15 Nibuchu, i.e., Nerchinsk.

16 The memorial to the Kangxi emperor by Ledehong asking the Russian tsar to withdraw the Russians at Yakesa, July 27, twenty-fifth year of the Kangxi reign, in *Qingdai Zhong-E guanxi*, 1:58.

envoys, who would pass it on to the Russian tsar. This letter[17], written in Manchu, Latin, and Russian, was dated September 17, 1686. It was also the day that the Dutch diplomatic corps left Beijing.

The Dutch diplomatic corps had taken its own interpreter along, a man named Frans Flettinger. Though he could speak the language of southern Fujian fluently, nobody at the Qing court understood him. As a result, the official interpreter of the Qing government, Verbiest, had been accepted as interpreter for the two sides.

Verbiest could not but feel worried when he saw that the Chinese government asked the Dutch to send a letter to the Russian tsar. As a Jesuit, Verbiest had to consider its consequences for Catholicism. He knew that the Netherlands was a Protestant country, where people did not believe in Catholicism. The dispute between Catholicism and Protestantism in Europe had brought about many bloody wars. In Indonesia and on the Indian coast the Dutch had invaded Catholic areas, and in Siam Catholicism was threatened by the Protestants. Verbiest feared that if the Dutch would succeed in delivering the letter, the Chinese government would probably bestow special favors upon them. If the Dutch would get a trading port in China, it would mean a decline of Portuguese influence in China, and Macao, the Jesuit's entry point to China, would be in a precarious state. If Protestants would emerge in Beijing, the Rites Controversy would intensify and endanger the existence of Catholicism in China.

Verbiest had to devise tactics to resolve this serious problem as quickly as possible. These tactics were very clearly explained in Verbiest's letter to the General of the Society of Jesus, Charles de Noyelle. Written on September 21, 1686, it said:

> Under these circumstances, while we were seriously considering how to support the inhabitants of Macao, and how to diminish their fear - and, at the same time, the hope of the Dutchmen - God inspired us with the thought that Father Philippus (Filippo) Grimaldi should be sent to Macao by the emperor under one or another pretext. At this favorable moment, the best pretext was undoubtedly that he could take along the letters of the emperor, directed to the Grand Duke of Muscovy, and that he could send them on via shorter ways in Europe, especially because the Dutchmen (when they leave here), will not be able to reach their ships taking off to Europe in the beginning of the next year. They will be forced to leave the letter for Muscovy in

17 The letter from the Board of War of the Qing government to the Russian tsar to withdraw Russian invaders and define a boundary line at Yaku, July 30, twenty-fifth year of the Kangxi reign, in *Qingdai Zhong-E guanxi*, 1:59.

> Batavia for a whole year. This detention (delay) did not please the emperor at all. If, for one or another reason, Father Grimaldi would not reach Rome, I beg Your Reverend Father to see to it that the letter for Muscovy should be delivered to the Grand Duke as fast as possible, and ask him to send an answer as soon as possible through several channels, so that we can check the efforts of the Dutchmen, and the favors which they consequently hope for. In this way, we can keep open our approach to China without difficulty or interruption[18].

Verbiest then approached the Kangxi emperor: "If His Majesty should send an envoy to Europe by way of Macao and Portugal, the letter would reach Muscovy much sooner." He also proposed that Grimaldi undertake this important task. The emperor accepted his idea gladly. His mandate: "Send Grimaldi with a letter of the Board of War by sea through Europe to Russia"[19]. This letter was written in Manchu, Latin and Russian, on September 22, 1686 (fifth day of the eighth month, twenty-fifth year of the Kangxi reign), and the contents were the same as those of the letter the Dutch envoy had taken along. Grimaldi left for Russia towards the end of November 1686.

On the eve of his departure for Russia, Verbiest asked the Russian envoy, Nikifor Veniukov, to bring Spathary a letter written by himself, hoping in this way to facilitate Grimaldi's mission. It read:

> Father Philip (Filippo) Grimaldi, our confrere, who frequently visited your honor here, was to be sent to open up communications between China and Rome. Taking advantage of this good opportunity, our Autocrat also gave him letters addressed to the Grand Prince concerning the same matters which had been treated in previous letters sent to the Grand Prince through various places. When no answer to them ever came, our Autocrat doubted their having reached the hands of the Grand Prince. We entrusted others, therefore, to Father Grimaldi, and one copy to the Dutch envoy as well, who was here, at that time, so as to be delivered by a safe hand. At present I beg your honor that if, by any chance, it should happen that Father Grimaldi (who intended to go through Austrian Vienna) was able to cross the Grand Prince's frontiers and reach your neighborhood, you would extend to our confrere the same good will you showed to all of us here[20].

---

18 Letter No. 72, Josson and Willaert, *Correspondance de Ferdinand Verbiest* (Brussels, 1938), p. 530 (according to the English version provided by Prof. Ulrich Libbrecht, Leuven University, Belgium).

19 Huang Bolu, *Zhengjiao fengbao* (Shanghai, 1904), p. 90.

20 Verbiest's letter to Spathary, November 24, 1686, in Baddeley, *op. cit.*, p. 436.

Grimaldi was not able to go to Russia. From Goa he sent the letter to Isphahan, where it was given to the Jesuit, Charles Maurice Vota, who brought the letter to Konrad Terpilowski in Astrakhan. The latter submitted it to Moscow on March 18, 1690[21]. Unfortunately, however, this letter arrived in Moscow two months later than that sent by the Dutch. Both letters, as it happened, fell behind Sino-Russian relations as they had by then developed, for China and Russia had already signed the Sino-Russian Treaty of Nerchinsk in 1689.

The Kangxi emperor was very much concerned about Grimaldi's diplomatic mission to Russia, and looked forward to his early return and report. On December 20, 1690, the emperor sent Joseph Suares to Guangzhou to find out if Grimaldi had already returned[22]. On July 23, 1692 the emperor figured that by now Grimaldi should be back in China, and sent Antoine Thomas to meet him in Macao. He was to tell Grimaldi that if there were westerners skilled in astronomy or in the calendar, they were to come along with him to Beijing to serve the emperor, while any others could settle wherever they wanted[23].

On his mission to Russia, Grimaldi went via Rome and France, and stayed a long while in Germany. While in Rome he met the famous German scholar, Gottfried Wilhelm Leibniz. Leibniz was much interested in ancient Chinese culture, and in July 1689 he sent a letter to Grimaldi in which he asked thirty questions about China. Among them were such topics as plants and agriculture, the production of paper, silk, porcelain, leather, medicine, geometry, astronomy, chemistry, metallurgy, machines and instruments for military and nautical uses[24]. In his discussions with Grimaldi, Leibniz developed ideas about promoting cultural exchange between East and West. Leibniz said: "In exchange for the great mathematical sciences of the Europeans, the Chinese could teach Europeans the natural arcana they have observed through their long history"[25]. He thought Russia should serve as the bridge in the Chinese-European cultural exchange. In 1716 he wrote a letter to the Russian tsar, Peter the Great, saying:

---

21 Nicolai Bantish-Kamensky, *A Collection of Diplomatic Documents between Russia and China, 1619-1792* (in Russian) (Kazan, 1882), p. 44, n. 3.

22 Huang Bolu, *op. cit.*, p. 109.

23 *Ibid.*, p. 113.

24 David E. Mungello, *Leibniz and Confucianism: The Search for Accord* (Honolulu, 1977), p. 33.

25 *Ibid.*

> He could, for mutual benefits, establish sound trade relations between Moscow and China, in order to transplant to Moscow and Europe the sciences and arts known in China, but unknown to us. We could thus increase the importance of our manufacturing and the commodities of life, and Moscow would also derive more than a little profit from it[26].

Grimaldi knew that Leibniz and Peter the Great were good friends, so before he left he asked Leibniz to write to the tsar in the hope that he would be allowed to return to China via Russia. The request, however, was refused by the tsar. In 1694, Grimaldi arrived by sea in Guangzhou and went at once to Beijing. He told the emperor all his experiences while serving as the envoy to Russia. The emperor was glad to receive his report, and bestowed many gifts on him. In the beginning of 1688 Verbiest had died, and Grimaldi was appointed administrator of calendar-making. While he was on the diplomatic mission to Russia, his position was filled by Thomas and Pereira. Upon his return, the emperor asked Grimaldi to take on full administrative responsibility.

III.
Opening an Overland Route Among Rome-Russia-China

It was extremely hazardous to navigate the seas during the sixteenth and seventeenth centuries. Epidemics and many diseases often occurred on board, not to mention the shortage of food and fresh water. Ships in distress were very common. There were also the dangers of being robbed by pirates or being captured by enemy ships. The possibility of finding a much safer, shorter overland route from Rome to China became, therefore, a frequently discussed topic among the Jesuits after Ricci had arrived in China. In 1622 Dutch ships began to attack Macao, through which the Jesuits used to enter China. Thereafter, the power of Holland in the Indian Ocean and the Chinese seas became stronger day by day, while the Portuguese force declined. Under these circumstances, it became even more pressing to find an overland route between Rome and China. In the time of Johann Adam Schall von Bell, the Jesuits were concerned most about the route from Persia through central Asia to Beijing. This was an old trade route, but by the sixteenth century it had been completely forgotten. In 1666, when Schall died, Verbiest became the leader of the Society of Jesus in China. From then on, he took the responsibility for finding the overland route.

---

26 Philip P. Wiener (ed.), *Leibniz Selection* (New York, 1951), p. 598.

In 1676 the Russian envoy, Spathary, came to Beijing. From him Verbiest learned of a route from Moscow to China via Siberia. The idea struck him to use this way. When Spathary decided to go back in early September of that year, Verbiest asked him to bring a letter to the tsar, in which he expressed his wish to be of service as regards the many languages he had mastered, and presented as a gift one of his works entitled *Astronomia Europaea sub Imperatore tartaro-sinico Cam-hy appellato ex umbra in lucem revocata.* Although he did not explicitly mention what he had in mind, his purpose was clear: he wanted beyond doubt to make a good impression on the Russian tsar.

During the decade following Spathary's departure from Beijing, relations between China and Russia had become so strained that war was inevitable. Verbiest realized that it would be impossible to open up a route through Russia to China; hence he placed his hope in the king of Poland. On November 17, 1685, he expressed this desire while writing to Nicolas Avancini. He said:

> I would also like to know if we have missionaries in Muscovy, and if they are able to penetrate the court of the Grand Duke. If there are none, it would be very desirable that Polish or German fathers open that way. This could easily be realized through the King of Poland. This is the reason why I write him this letter with pleasure. It is true that the power of Muscovy is widely spread over many eastern countries. It even reaches the borders of the Chinese empire, where it has established colonies, fortresses and posts of occupation. In this way, an enormous area is created in which to establish missions and to make the transition to the numerous countries which were never penetrated by the evangelical light[27].

In view of the increasing Turkish threat, Russia and Poland signed the Andrusov Cease-Fire Treaty in 1667, thus leading to the normalization of relations between both countries. Yet Russia refused to let anyone of any European country pass through its territories, for fear that these countries would then begin to compete for the oriental market.

In 1686 a new situation emerged in Sino-Russian relations. Just as Chinese troops were encircling Yakesa, Nikifor Veniukov and Ivan Favorov were sent by the Russian government to deliver to the Kangxi emperor credentials[28] in which the Russian tsar asked the Chinese

---

27 Letter n° 69, November 17, 1685, Verbiest to Nicolas Avancini, Josson and Willaert, *op. cit.*, pp. 521-522 (according to the English version provided by Prof. Libbrecht).

28 The letter from the Russian Tsar Peter the Great to the Kangxi emperor

government to stop attacking Yakesa, and claimed Russia would accept the Chinese government's peace proposals. Meanwhile, a delegation headed by Fedor Alexeevich Golovin had also been sent for boundary negotiations.

Verbiest had yearned for this cessation of fighting and the signing of a peace treaty between Russia and China for a long time. Naturally he now felt very pleased to see this new change in Sino-Russian relations. On October 1, 1687, Verbiest expressed his optimistic view on the future of Sino-Russian negotiations to Noyelle, the General of the Society of Jesus in a letter which was written by Thomas. It says:

> Après la conclusion de la paix, à laquelle les Russes travaillent activement, et quand ils auront obtenu l'autorisation de faire le commerce (avec la Chine), les lettres de Rome nous arriveront fort vite, en tout temps, par cette voie. Il y a mieux. Nos missionnaires pourront se répandre en Tartarie, voire même venir au secours de la Chine: cette route sera de toutes la plus sûre et la plus facile, pour les Polonais, les Allemands, les Belges, etc. Du fleuve Cobi jusqu'à Pékin, ou de Tobolsk, capitale de la Sibérie, l'ambassadeur (de Russie) comptait 600 lieues allemandes. Avec bagages, marchandises etc., il avait fait, disait-il, la route en quatre mois. Quand le commerce sera ouvert, cette voie sera très fréquentée et les courries semble-t-il, n'y feront pas défaut[29].

In February 1686, the Russian negotiating team left Moscow for Nerchinsk. At that moment the Chinese government was also busy preparing its negotiating team. Originally, the emperor expected to order Verbiest to hold the post of Latin interpreter for the Chinese delegation. But in 1687 Verbiest was seriously ill and could not go. Verbiest was very well aware of the background of the ongoing negotiations between China and Russia. He knew that Russia was exhausted because of the endless wars in Europe. It was thus impossible for Russia to undertake military ventures in the eastern countries; its only interest there was trade.

China, for its part, wanted neither war nor trade; it desired only to protect the territory it regarded as its own from the Russian sphere of influence, and to keep peace. China's main purpose was to keep Russia out of Mongolia and the Amur valley. The cessation of war and the

---

asking him to raise the siege of Yakesa and send an ambassador to negotiate the problems of the frontiers, December 20, 1685. *Qingdai Zhong-E guanxi*, 1:70-72.

29 Henri Bosmans, S.J., *Le problème des relations de Verbiest avec la cour de Russie* (Bruges, 1913), pp. 215-216.

pursuit of peace were thus common goals to both Russia and China. This also fitted in with the hopes of the Jesuits: the establishment of an overland route from Rome to China via Russia. So Verbiest thought that, supposing the Jesuits could mediate in promoting peace and the signing of a peace treaty between both countries during their border negotiations, they would be very grateful towards the Jesuits, and this would serve their own personal goals. Verbiest, unfortunately, died on January 28, 1688, but his vision was shared by his colleagues, Pereira and Gerbillon. According to Verbiest's recommendation before his death, the emperor told the *lifan yuan*[30] on April 13, 1688: "I believe that the westerners in our service were very trustworthy. I want to send Pereira to Russia, since he knows Latin, which is a very precise language. You can also send a letter to Russia and negotiate with Russia"[31].

In June 1689, Pereira and Gerbillon went to Nerchinsk with the Chinese diplomatic corps headed by Songgotu. During the Nerchinsk negotiations, they faithfully carried out the testament of Verbiest for peace between China and Russia. They used their tact and ability to conclude a successful peace treaty.

The Sino-Russian Treaty of Nerchinsk (1689), based on the principles of equality and the mutual benefits of international law, demarcated clearly the eastern part of the Sino-Russian boundary. China's northeastern boundary was to remain peaceful for 160 years, and Russia gained trading advantages: between 1689 and 1727 alone, official Russian caravans came to Beijing twelve times.

The triumph of the Jesuits, however, was bittersweet. On the one hand, they won the confidence of the Qing government. In 1692, the emperor issued a decree giving freedom for Catholics to do missionary work in China, thus realizing the Jesuits' dream. But not long after the signing of the Nerchinsk Treaty, Tsar Peter the Great declared that during the negotiations, the Jesuits had been partial to China, thereby betraying western interests. Because of this, he closed the Jesuits' residence in Moscow and expelled them altogether from Russia. The Jesuits' long hoped overland route from Rome to China via Russia became, at the same time, an impossibility.

---

30 *lifan yuan*, the Mongolian superintendency, was in charge of the tribes of Mongolia and the affairs of Tibet. It also conducted China's relations with Russia.

31 The Kangxi emperor's decree of April 13, 1688 in Ferdinand Verbiest, *Xichao ding'an*.

FERDINAND VERBIEST'S RELATIONS WITH KING JOHN III OF POLAND

EDWARD KAJDANSKI

Edward Kajdanski was born in Northeast China and graduated in 1951 from the North-Manchurian University. The same year he came to Poland. In 1963 he was sent to Beijing to work in the Polish Commercial Counsellor's Office till 1967. He spent seven more years in China, working in the Polish Embassy in Beijing and in the Polish Consulate General in Guangdong. He published several books on Far Eastern topics and more than 150 articles mainly concerning the cultural relations between China and Poland or between China and the West. His book "Michal Boym: the Last Envoy of the Ming Dynasty" has been translated into Chinese.

King John III Sobieski is recognized by historians as one of the most prominent Polish commanders, as well as a far-sighted politician. In 1673 Sobieski, who was at that time commander-in-chief of the Polish army, defended the Turkish army of Hussein Pasha near Chocim. While it was the biggest victory over the Turks carried off in Europe in overland operations, it did not remove the Turkish threat from Poland. After his election as King of Poland (1674-1696), Sobieski was anxious to regain the duchy of Prussia, which was lost in the war with Sweden. His Prussian-Baltic plans he sought to achieve with the help of France, with which he had signed a secret treaty of alliance a year after his election. But without terminating the war with Turkey, it was impossible to start any armed actions in the West, as the Turkish problem was closely connected with the question of relations with the second powerful neighbor in the East: Russia. In this regard King John III was seriously thinking of a Polish-Turkish-Tatar alliance against Russia, which was not completely groundless. In the spring of 1675 the Khan of Tatars, Selim Gerei, proposed to Sobieski "the confederacy of three swords" (that is, of the Poles, Turks, and Tatars) against, as it was clear, the common enemy: Russia[1]. It would seem that events favored such an alliance. In 1676, after King John III managed to stop the advance of a superior number of Turkish troops under Ibrahim Sheytan, the armistice of Zurawno was concluded. To transform this armistice, reached with the mediation of the French envoy François Gaston Marquis de Béthune, into a durable peace, the great Polish embassy of Jan Gninski was sent to Stambul. While it stayed there for two years, it failed to reach the aims intended by the king. This was due to the unyielding position of the new vizier Kara Mustapha, as well as to the internal opposition against the king's projects. The opposition, which included a part of the higher clergy, was interested in establishing an anti-Turkish League, instead of allying with "infidels" - and was encouraged from both Rome and Vienna.

In this particular situation, after being informed of the difficulties met by Polish diplomacy in the European capitals, the king sent an embassy to Moscow to negotiate a treaty with Russia. But Russia was uninterested in joining hands with Poland against Turkey at that time; it rather intended to conclude a separate treaty with that country itself (and, in fact, did so in 1681 at Bakhchesaray, for a period of twenty years)[2]. None of the great plans of the new king were realized: the Polish-Lithuanian state concluded neither the treaty with Turkey against Russia, nor that with Russia against Turkey. Poland finally allied itself with Austria, concluding an anti-Turkish treaty with that country in April, 1683. Carrying out his obligations as an ally, King John III was quick to come to the Austrian capital's aid, and in the battle of Vienna, he crushed the overwhelming Turkish forces. It was with just this victory that one of Ferdinand Verbiest's known letters is connected - although we do not know whether this letter, dated November 17, 1685, was the first of their correspondence[3]. Verbiest's letter was addressed to Nicolas Avancini, the German assistant to the General of the Jesuits in Rome. To the Avancini letter Verbiest attached a personal congratulatory letter to John III, and a eulogy from the Chinese Kangxi emperor to the king. Unfortunately we do not know the contents of this imperial

---

1 M. Bogucka, "Szlachta polska wobec Wschodu turecko-tatarskiego", *Miedzy fascynacja a przerazeniem (XVI-XVIIIw)* (Sobotka, 1982), n° 304, p. 189.

2 Z. Wojcik, *Rzeczypospolita wobec Turcji i Rosji 1674-1679* (Wroclaw, 1976), p. 75.

3 B. Szczesniak, "Diplomatic Relations between Emperor K'ang-hsi and King John III of Poland", *Journal of the American Oriental Society*, 89:1 (1969), 157-161.

calligraphy, nor anything else about it (whether it was delivered to the king, where it is located ...). The contents of the letter to Avancini proves, however, that the Chinese emperor knew about the existence of Russia's very powerful western neighbor, which was able to withstand even Russian pressure. The wars with Russia resulted in a number of Polish prisoners of war, who, since the time of Ivan the Terrible, were sent to Siberia.

There are four quite extensive documents about one of those prisoners of war, a certain Nikifor Jaxa (Jaksa) Czernichowski. The first document was found in the tsarist archives (and then lost for a second time), and was written by an anonymous agent in the year 1680 for Hildebrand von Horn, the secretary of the Danish king. It had fallen into the hands of Russians and was in the beginning of the last century turned to account by the Russian historian Gregory I. Spasski[4]. The second document was found in the work of the well-known Dutch diplomat and traveler, Nicolaas van Witsen. He lived in Russia for some time, and wrote a work on Siberia edited in 1692 in Amsterdam, and entitled *Noord en Oost Tartaryen*. In it we read:

> About two years ago a voivode of one of the cities situated on the river Jenisei dishonored the sister of a Polish captain in the service of Moscow. Taking revenge, this captain killed the voivode and ransacked the stores. He afterwards escaped in a south-easterly direction, together with other mutineers, and built there a town called Albazin, where he lived peacefully for ten years. When the time passed, he sent somebody to Moscow to request that his crime be forgiven. In return, he promised that from that time on, he and his men would pay annual taxes. The former act was forgiven, and the voivode of Jeniseisk was ordered to supply Albazin with all necessities[5].

The third document is found in the *Description of the Siberian Kingdom*, written by a Russianized German, Gerhard Friedrich Müller, and published in 1750 in Saint Petersburg[6]. And finally the fourth document, of a much later date, was the work of a Polish exile (of the

---

4 I have not seen this publication of George I. Spasski, but was informed of it by Mr. Stanislaw Lukas, Harbord, Australia.

5 Nicolaas van Witsen, *Noord en Oost Tartaryen* (Amsterdam, 1692; 2nd edition 1785), 1:113-115.

6 Gerhard F. Müller, *Opisanie Sibirskago Carstwa i wseh proizshedshih w nem del* (Saint Petersburg, 1750).

so-called 1863 January Uprising), whose name was Marian Dubiecki[7]. He was sentenced to lifelong banishment in Siberia, and years later was allowed to settle in Irkutsk, thus having the opportunity to use in his works the rich materials from the Irkutsk archives, which were completely burned in the great fire in 1879. From these four sources we know that the officer mentioned by Witsen was Czernichowski, a native of Zytomierz and a nobleman. He was exiled to Siberia after the Moscow war of 1632-1634, together with his father and sister. In 1650 he was the administrator of the agricultural colony at Chechensky Volok, and two years later became the administrator of the saltworks in Ust-Kutsk. In the neighboring city of Kirensk a big fair was held once a year where, in the summer of 1665, Czernichowski arrived with his sister. The city was also visited by Lawrence Obukhov, a voivode from Illimsk.

The fair had to be closed on August 6, but as early as the night of July 25, the voivode made haste to leave Kirensk. With all his suite, he boarded boats and, leaving behind his judicial and fiscal duties, sailed up the Lena river to Illimsk. Czernichowski assembled his party and followed in pursuit of the voivode. The anonymous author of the Latin text of 1680 says:

> Nikifor Czernichowski had a sister, and a heavy insult done to her resulted in his killing Obukhov. The brother sought revenge for the violence done to the honor of his sister[8].

Dubiecki writes that Czernichowski and his eighty-four companions, most of them Poles from Eastern Polesia, could not expect mercy from the part of the authorities. After a short deliberation, they resolved to follow the Lena river downstream to the mouth of the Olekma, and then to sail along the Olekma to Tugur. From there, it was a relatively small distance to the Amur river. What is especially worth noting is that an Orthodox priest named Hermogenes, an abbot of a monastery in Kirensk, voluntarily joined the refugees. During their escape, one group of Czernichowski's companions fell into the hands of Tungus aborigines

---

7 Marian Dubiecki, "Osadczy ziemi Mandzu", *Tygodnik Ilustrowany*, 14, n° 362 (1874), 341.

8 Earlier authors, including Witsen and Müller, wrote that the pretty woman dishonored by Obukhow was Czernichowski's sister. Among nineteenth- and twentieth-century Polish and Russian historians the opinion prevailed that Czernichowski's mutiny was, in fact, a result of the long-lasting dissatisfaction on social grounds, and that injustice coming to his sister could have been only the spark that caused the fire.

and was murdered. At last, only sixty-nine men of the original eighty-four reached the Amur river.

Why did Czernichowski and his companions decide to escape in the direction of China? Among the inhabitants of Siberian towns, so-called "ostrogs," there were ceaseless rumors about the riches of the Chinese state and about Chinese humanitarianism in respect to the prisoners of war. A number of exiles and Cossacks, it was said, were escaping to China. It was also known that they were allowed to settle down in a special district in Beijing, that they were not doing badly, and that some of them were reckoned among the privileged imperial guards. They were allowed to marry the local women and set up families. Franz Flettinger, a German sergeant who arrived in Beijing with the Dutch embassy under Vincent Paats, wrote that there was, at that time, a community of captive Russians numbering 564, of whom 300 were still alive early in 1687[9].

Czernichowski and his companions rebuilt the town, which was situated in the place where the Emuerhe river flows into the Heilongjiang or Amur river. They cultivated the ground around the town and made hunting expeditions, selling their sable and other furs to the Chinese and local buyers. It is worth mentioning that Czernichowski and his men simply wanted to live in peace with their neighbors and - unlike Khabarov and other rich merchants - did not need any riches themselves. During several years after their arrival on the banks of the Amur, there were no notices in either the Chinese, or the Russian records about any wars, clashes, or murders from their side[10]. Polish historians maintain that Czernichowski changed the old name of Albazin (after the Daurian prince Albaza) to his nickname Jaxa (Jaksa in modern Polish language), which was also the name of a legendary prince of the Slavs from the banks of the Elbe and Oder[11]. It is also a well-known

---

9 John E. Wills, "Some Dutch Sources on the Jesuit China Mission 1662-1687", *Archivum Historicum Societatis Iesu*, 54 (1985), 282.

10 Czernichowski's independence in Albazin-Yakesha lasted from 1665 until 12 February of the year 7182 (that is, 1674), as the minutes of the handing over of the town by Czernichowski to Russian authorities from Nerchinsk bear that date. From these minutes we know that there were altogether 109 armed men, a few families and several Tungus people in the town, and that the commissioner took over the property of the poor settlers: "ten sickles, several saws, two balances, one banner, one drum, one quire of paper." Dubiecki, "Osadczy", *op. cit.*, n° 365, p. 374.

11 During this conference Dr. Giovanni Stary informed me that Yakesha could be an old Manchurian word.

fact that this name is still used by the Chinese for Albazin, though they write it as Yakesha. There is another word in the Chinese language which is claimed by Polish historians to be of Polish origin: *luocha*. It was later broadened to indicate all Cossacks, or Russians, although it is probable that this word developed from the Polish *lowczy* meaning "hunter," as Czernichowski and his men called themselves hunters[12].

The Chinese authorities knew, of course, about the second establishment of Albazin in 1666, but for many years they did not resume attempts to destroy it. Similarly, it was known to the Russian government, and without any doubt also to the Jesuit missionaries in Beijing. It is certain that from 1669, at least, Czernichowski's Albazin had its own seal with an eagle, similar to the Polish eagle, but carrying a bow and arrow in its clutches[13]. This heraldry differed considerably from the blazons of Siberian cities. Ten years after Czernichowski's arrival on the banks of the Amur, the head of the Russian embassy to China (1676-1678), Nikolai Govrilovich Spathary Milescu, reported to the Russian authorities that there were many deserters living in Beijing who attended the Jesuit Catholic church, and were training Chinese cavalry in the use of firearms from the saddle. This kind of tactic was widely used in Poland by the light-horse detachments formed in the beginning of the seventeenth century by Alexander Lisowski[14].

There are many signs that King John III knew about the existence of this unusual Polish center on the Amur river much earlier than it was published by Witsen. He maintained close relations with many Jesuit and Franciscan scientists; from his youth he was fascinated by geo-

---

12 In the *Cihai* encyclopedia one can find the information that *luocha* is an abbreviated form of *Aluochasuo* and that it was a Chinese transcription of the Indian word Raksasa. At any rate, it came to stand for the Russians in the second part of the seventeenth century only, when Russian expansion in an easterly direction brought them to the vicinity of the Qing empire. I think that *lowczy*, being so close phonetically to *luocha* (Poles pronounce the word *lowczy* as *louchy*), was the real source of *luocha* and that the Chinese later identified this word with Raksasa, as it was the foreign devil who drank human blood and ate its flesh.

13 This seal is now in the Hermitage Museum in Leningrad. It was reproduced in *Russko-kitaiskiye otnoshenia v XVII veke. Dokumenty i materialy, Vol. 1 (1608-1683)* (Moscow, 1969), n° 178, p. 331.

14 He took part in the victorious battle of Kircholm in 1605 against the Swedes, was then in the military service of False Dmitri I and False Dmitri II, and at last in the service of Sigismond III of Poland. He died in 1616, but the irregular light horse detachments he created and known in Poland as *Lisowczycy*, took part in many battles in Europe as mercenaries until they were dissolved in 1636.

graphy and by the Orient, and this included China. In the royal library there could be found many works about China, as, for example, *Legatio Batavica* of Joannes Nieuhof, and *China Illustrata* of Athanasius Kircher. One of the editions of Kircher's *Mundus subterraneus* was dedicated to King John III of Poland. With the mediation of his librarian, the prominent Jesuit mathematician Adam Adamanty Kochanski, the king was corresponding with the son of the cartographer of Louis XIV of France, Nicholas Sanson, proposing his help in the proper representation of the territory of Poland on the maps drawn by Sanson. When the Italian Franciscan Vincent Coronelli, author of valuable geographical works, established his *Academia Cosmographica degli Argonauti* in Venice in the 1680s, considered to be the first geographical society in the world, King John III became its honorary member. According to Dubiecki, the above-mentioned Latin manuscript on Czernichowski belonged to the well-known and enlightened bishop of Smolensk, Alexander Kotowicz. Then it found its way to Vilno, from where it was carried away by the Russians. If so, a document of such importance should have been presented for the king's information.

Verbiest was at that time the superior of the Jesuits in Beijing. He administered the affairs of the Imperial Bureau of Astronomy, and was made a chief of the Board of Civil Works, being charged with equipping the army with European cannons. He was favored with the great confidence of the Kangxi emperor. This confidence can be confirmed by the fact that in the years 1682 and 1683, he was invited by the emperor to accompany him on his inspection journey to East Tartary (that is, the former Manchuria, or today's northeast China) and to West Tartary (that is, the eastern part of today's Chinese Inner Mongolia and the Mongolian People's Republic). In the course of this journey, Verbiest was able to collect important information concerning the distances between Beijing and many places to the northeast of it which were visited by the Russian embassies to the Chinese court. He was convinced that the land route to the Chinese capital through Siberia would be much more safe and convenient than the sea voyage around the Cape of Good Hope, or the combined sea and land route through Armenia, Persia and India to Goa, and then around Indo-China to Macao and on to South China. This was the reason why Verbiest was a devotee of all projects helpful to the opening of a land route through Russia for the Jesuits. Although this is not confirmed by documents, we can guess that the lost letters forming the correspondence between

Verbiest and King John III concerned this topic as well[15]. Anyhow, we know with certainty that Sobieski took several steps in the diplomatic field to get the access to the land route through the Russian territories for missionaries working in China.

In 1680 Verbiest had sent one of his missionaries, Philippe Couplet, to Europe with the aim of filling up the vacancies in the China mission. Among the newly-enlisted Jesuits was a Frenchman, Philippe Avril. It is evident that the preparations for sending him to Beijing, not by the ordinary routes, but using a new way, were made with the knowledge of both the French and the Polish authorities. King John III's wife, Maria Casimira, was a Frenchwoman, the daughter of Henri Marquis de La Grange d'Arquien, whose participation in the opening and financing of Jesuit missions in the Orient was remarkable[16]. The French envoy to the Polish court, Marquis François Gaston de Béthune was, incidentally, her relative. Avril left Paris and traveled to Lyons in the company of Couplet. He visited Rome and embarked on a ship in the Italian port of Livorno. Then he had to cross Armenia, and continued his journey to China through Astrakhan, Bukhara and Samarkand. When traveling through Russian territory he was detained, and after some troubles, sent back to the West, namely to Poland. A Polish historian, Father Xawery Liske - who investigated the circumstances of Avril's stay in Poland - says that King John III rendered accessible to Avril "the memorial which was left to him by the Russian legate Nikifor, who, just before his arrival in Poland, had returned from the diplomatic mission to Beijing." This memorial, describing the route to China and the map drawn

---

15 Prof. Hao Zhenhua wrote in his paper "Ferdinand Verbiest and Chinese-Russian Relations," which I received during this conference, that Verbiest realized that it would be impossible to open up the route through Russia from China, and hence he placed his hope in the king of Poland. There is a quotation from Verbiest's letter to Avancini, November 17, 1685: "I would also like to know if we have missionaries in Muscovia, and if they are able to penetrate the court of the Grand Duke. If there are none, it would be very desirable that Polish or German fathers opened that way. This could easily be realized through the king of Poland. This is the reason why I write him a letter with pleasure." It is evident from this extract that the lost letter of Verbiest to King John III concerned that question.

16 The Jesuit mission in Isphahan became an important place on the land route leading to and from China. It was established with the financial help of Maria Ludwika, the French wife of a Polish King, Wladyslaw IV. It was visited by Henry Roth and Francis Storer when they traveled from Europe to India, and by Michal Boym, who in 1651 carried on his mission as a Ming envoy to Venice and Rome.

according to it, was sent by Avril to his superiors in Paris[17].

The authenticity of this information was later confirmed in the correspondence of the king's librarian Kochanski with the great German philosopher Gottfried Wilhelm Leibniz. The original letters of Kochanski to Leibniz, and the drafts of Leibniz's letters to Kochanski, covering a period of thirty years and found in the library of Hannover, were published in Poland by Samuel Dickstein in 1901 and 1902. In his letter of January 8/18, 1692, Kochanski says that in the possession of "Rex Noster" (that is, King John III) there was "a map of the route through Siberia, Mongolia, etc., to China. This map was based on the account of Nestorius, a Muscovite who had previously taken part in the embassy to Beijing and then to Poland, but it did not contain any information concerning geographical longitudes, or latitudes"[18].

According to the prevailing opinion, shared also by Felix Alfred Plattner in his book about the activities of the Jesuits in China, this information came from the archives of the royal secret service[19]. We know, however, that the Russian envoy who was called Nestorius by Kochanski, and Nicefor by Liske, was actually Nikifor Veniukov who actually took part in Spathary's embassy to Beijing in 1676-1678. Several years earlier the Siberian Governor Peter I. Godunov ordered the preparation of a new map of Siberia. This map, used by Spathary and Veniukov, was drawn in 1667, and was for many years considered as lost. But in 1899 a certain Swedish scientist (the Soviet source does not give his name) found a collection of maps in the Stockholm archives, and among them were two maps of Siberia. He had them published in a Swedish journal. These maps belonged to the Swedish officer Eric Palmquist, who stayed in Russia in the years 1673-1674. The Soviet author Andrei I. Yefimov found that Palmquist's map was in reality the "lost map of Godunov"[20]. We know now that "Godunov's map" was copied by another Swedish traveler who accompanied the

---

17 Xawery Liske, "Filip Avril i jego pobyt w Polsce 1687-1689", *Niwa*, 11 (1877), 407.

18 Samuel Dickstein, "Korespondencja Kochanskiego i Leibniza", *Prace Matematyczno-Fizyczne*, 12 (1901), 225-278; 13 (1902), 237-284. See also Walter Drzewieniecki, "The Knowledge of China in XVIII century Poland as reflected in the Correspondence between Leibniz and Kochanski", *Polish Review*, 12:3 (1967), 53-66.

19 Felix A. Plattner, S.J., *Gdy Europa szukala Azji* (Krakow, 1975), p. 209.

20 Andrei V. Yefimov, *Iz istorii russkih ekspeditsyi na Tikhom Okeane. Pervaya polovina XVIII veka* (Moscow, 1948), pp. 60-61.

Swedish envoy to Russia on his journey around the country, and that the same map was used by Witsen[21].

There is still one more version concerning the way in which the map of Siberia finally found its way to the king's archives. Surprisingly, it is connected with Verbiest's journey to East Tartary in 1682. The report of this journey was known in London in 1684 or in 1685. In the third volume of *Miscellanea curiosa*, which was published by the Royal Society in London in 1702, Verbiest's account of his travels was given, together with an anonymous enclosure. The following extract is worth noting:

> Monsieur d'Arcy, who commands one of the king's ships in the fleet of Monsieur Le Marescal d'Estrées, recently informed us that, having served in Poland, and having been made governor of a place around Muscovy, the Muscovite ambassadors visited him on their return and were treated so well they were in an especially good mood. One of them showed him a chart of the countries between Muscovy and China, and told him that, from the three cities he showed him (Lopsla, Abasinko, and Nerchinsko) - all three under the government of the Great Dukes, and situated in the great Tartary - there was a way to Beijing, which was not more than 25 or 30 days' journey. This map, it seems, was to be kept very secret in Muscovy, for the next day the Muscovite was in despair for having shown it, saying that if it should be known, he would come to great harm. The officer, once back in France, gave a copy to the king, and another to Monsieur le Marquis de Signelay. To confirm this, it may be added, a Frenchman wrote from Muscovy that within the last two months they were actually raising troops to go to war with the Chinese[22].

Perhaps it was Monsieur d'Arcy who took the map from the hands of a drunken Russian legate, and perhaps there were many persons who took advantage of the occasion. At any rate, King Louis XIV of France and King John III of Poland were both fortunate to receive some knowledge about Siberia. The "Explanation" had to have been written before 1688, as the famous "Treaty of Nerchinsk" between Russia and China was concluded in 1689, being bargained for under the rattle of weapons from both sides. The two places mentioned by d'Arcy - Abasinko and Nerchinsko - are, of course, Albazin-Yakesha and Nerchinsk. It is more

---

21 All these maps are reproduced in A.V. Yefimov (ed.), *Atlas geograficheskih otkrytii v Sibiri i Severo-Zapadnoi Amerike* (Moscow, 1964).

22 "A Voyage of the Emperor of China into Eastern Tartary, Anno 1682", *Miscellanea Curiosa. Containing a collection of curious travels, voyages and natural histories of countries, as they have been delivered in the Royal Society*, 3 (1702), 179-184.

difficult to identify Lopsla, as there was no town of that name in either Russian or Chinese in that region. No such place is even mentioned in Witsen's *Noord en Oost Tartaryen*[23].

The map of Siberia was not very useful to Avril, despite the promising circumstances. King John III accepted Avril's joining the embassy sent at that time to Persia under Suleiman K. Count Syri, and King Louis XIV gave him a letter to the emperor of China. Sobieski, moreover, secured two Polish Jesuits as traveling companions, both speaking Russian, Turkish and Persian. They were Franciszek Ignacy Jampolski and Konrad Terpilowski. But, as we already know, Avril never reached China, despite the great hopes for this mission in France and Poland. A letter of Louis XIV's confessor, François d'Aix de La Chaise, dated in Paris, January 15, 1688, was found expressing the expectations of the French court that French missionaries would be able to pass through Russia and Siberia on their way to China[24].

The only known letter of King John III to Verbiest was written in Jaworowo, and is dated November 16, 1688. This letter never reached Verbiest, as the latter died in Beijing in January of the same year. His letter was sent to Rome by his successors. It was in the main Jesuit archives, ARSI, that it was found by Father Stanislaw Bednarski in 1933[25]. While there is no mention in the letter itself about the route through Siberia, it is evident that the king and Verbiest had exchanged several letters previously. We understand from this letter that Sobieski highly appreciated the Chinese emperor, whom he calls *Sinico-Tartaricus Monarcha*, and that he presented his portrait to the emperor. The king wrote:

> Moreover, we applaud the Most Powerful Emperor on land and sea, who governs such an illustrious people, and whose virtues are known over the whole world. By the will of Providence, may His Majesty rejoice in long years of glory, prosperity and happiness. Know that we eagerly offer him our faith and fraternity, as well as sincere friendship[26].

---

23 On Godunov's map are shown Selenginsky Ostrog and Nerchinsky Ostrog. To the east of them one can see the schematic drawing of the Lopsla castle. It is situated in the lower course of the Argun river. On the same river - but far to the south - the name Kalar is written, which I have identified as today's Hailaer, or Qing's Hulunbuir. See also note 35.

24 Szczesniak, "Diplomatic Relations", *op. cit.*, p. 160.

25 Stanislaw Bednarski, "Chinski list Krola Jana III", *Przeglad Powszechny*, n° 12 (1933), 31-34.

26 Szczesniak, "Diplomatic Relations", *op. cit.*, p. 159.

These were courteous, but totally empty words. We must see in them the reflections of the great statesman's political plans to find in the emperor an allied partner against the common enemy. Polish historians and orientalists are not in perfect agreement in this respect. Some maintain that Sobieski wanted to secure Chinese help against Turkey[27], but there is also another viewpoint. Walter Drzewieniecki writes: "They seem to have overlooked the important point that the pressure against Poland at that time was coming not so much from Turkey, but from the growing power of Russia"[28].

In 1686 the Jesuit missionary Claudio Filippo Grimaldi came to Europe from China, with the emperor's instructions to contact Russian authorities and to give them the Board of War's letter of September 22, 1686[29]. To this end he tried his best to get permission to travel through Russia as an imperial envoy. Grimaldi entered into correspondence with Leibniz through the mediation of Kochanski, as he thought that Peter the Great's extraordinary respect for Leibniz would help him to achieve this goal. We know that it was Leibniz who asked Kochanski for the recommendation of King John III to the shah of Persia, and that it was in fact granted. In the correspondence between Leibniz and Kochanski we also find their opinion on the reasons which prevented Russia's accepting missionaries to travel within its territory. In his letter of May 28, 1694 Kochanski supposed that "within the last few years Russians have annexed many Tartaro-Chinese provinces," and that "they do not want Poland (and perhaps France) to have knowledge about this"[30]. Leibniz, in his letter of August 10, 1696, wrote that it was known to him from some other sources that the reason for this Russian conduct lay in the clashes between them and the Chinese[31].

Finally, I would like to say some words about the description of the places lying to the north of the territories formerly belonging to the Ming state, as this was made by Verbiest. The most northern part of China under the Ming dynasty was known in the seventeenth century as Liaotum or Liaodong according to the maps of Michal Boym (1652),

---

27 B. Baranowski, *Znajomosc Wschodu w dawnej Polsce do XVIII wieku* (Lodz, 1950), p. 234.

28 Drzewieniecki, "The Knowledge of China", *op. cit.*, p. 59.

29 The letter was submitted to the Russian authorities by the above-mentioned Konrad Terpilowski in Moscow on March 18, 1690. See also the paper of Hao Zhenhua in the present volume.

30 Dickstein, "Korespondencja", *op. cit.*, p. 61.

31 *Ibid.*

Martino Martini (1655), and Kircher (1667). On Chinese maps from the Ming dynasty, this region was shown as located within the Great Wall (in fact, within the eastern part of it, being an earthen rampart built in the fourteenth century and already destroyed before the time of Verbiest's stay in China). The territories known to the European cartographers of that time were limited to those south of the Great Wall. Along the chain of the Chinese military posts that extended from Shanhaiguan in the west to the borders of Korea northward, the furthest was Shenyang - which in Verbiest's account is called Xin-Yam. The regions north of this line were not known in Europe at all, as these were not shown either on Chinese or on Russian maps.

In 1682 the Kangxi emperor carried out the inspection of the territories known in Europe as Eastern Tartary (later on Manchuria, and today North-East China or Dongbei). Verbiest was the first European to accompany the Chinese emperor on such a journey covering, as it did, the northern part of modern Liaoning province, Jilin province and even touching the southern borders of Heilongjiang province. The route of this expedition led from Beijing through Shanhaiguan, Ningyuan, Jinzhoufu, Shenyang, Tieling, Changtu, Yitunmen, Tudeng to Jilin. The furthest point of Verbiest's journey was the city of Ula which, according to him, was the ancient seat of the emperor's ancestors. He located the place at the distance of 32 *li* north of Jilin, and we know that the expedition reached it by the Songoro river, known to his companions as Sumhoa (this is modern Songhuajiang). I think that the identification of the city of Ula could be interesting to Verbiest biographers, as I found much controversy concerning its position in both foreign and Chinese sources. For example, the well-known Chinese historian Li Ung Bing says that the old seat of Manchu chiefs was Hartu Ula, known also as Xinjing[32]. But Xinjing was the military border-post of the Ming, and lies to the east of Shenyang, far south of Jilin. Moreover, it has no water communications with Jilin. In the *Cihai* encyclopedia Hartu Ula is identified as today's Laocheng in Liaoning province. In my opinion, the city of Ula visited and described by Verbiest is the modern Ulajie in Jilin province (at that time called Dashenula).

The emperor ordered Verbiest to take with him his mathematical instruments, and as a result of his observations, the correction of the latitude of Shenyang was made. He found that this was 41 degrees and

32 Li Ung Bing, *Outlines of Chinese History* (Shanghai, 1914), p. 275.

56 minutes - and not the 41 degrees as shown on European maps. Verbiest also delivered the first information on the latitude of the city of Ula, which was 44 degrees and 20 minutes. From his report, Europe realized more or less the true distance separating Beijing from the mouth of the Amur river. For Verbiest recorded the distance between Shenyang and the unknown Ula as 1,060 *li*, and from Ula to Nincrita (that is, Ningguta), today's Ningan as about 700 *li*. According to Verbiest, Nincrita was at that time a big river port, from which it was possible to sail by the Songoro and the Heleum (Heleum is, of course, Heilongjiang, the Chinese name for the Amur river) and to reach within forty days the Eastern Sea, which Verbiest thought was the Anian Strait, from European maps.

There is information of some importance in this Verbiest report concerning border questions at that time. He says that in the city of Kirin on the Songoro river, the inhabitants maintained a big fleet of barks - or junks - ready to repulse the Muscovites "who often came into this river to dispute the fishing of pearls"[33]. This could be simply the repercussion of the clashes at Albazin-Yakesha on the Amur river, but it is not impossible that the lower flow of the Songhuajiang also became a target of casual Russian expansion[34].

The aim of the emperor's inspection journey to Jilin and Ula was to assess the situation on the northern borders, and choose the means for their defence against the Russians. It was the emperor who ordered a hunting expedition to be sent across the Amur river in order to ascertain the strength of the Russian position at Albazin-Yakesha. As the report was encouraging, a fleet of wooden junks was built at Ningguta for service on the Amur river, and the post of military governor for Heilongjiang province with its residence in Aihun, was created. The towns of Qiqihaer and Mergen (modern Nenjiang) were both fortified, and ten postal stages were established along the main route to the Amur river in order to facilitate the transportation of troops and military stores. In 1683 a small party of Cossacks was captured in the vicinity of Aihun and taken back to Qiqihaer. Among these captives were also Poles from the party of refugees of Czernichowski. They were the

---

33 *Miscellanea Curiosa*, p. 189.

34 Near the mouth of the Songhuajiang there is a town called Tongjiang, which at that time was called Lahasusu. Maybe it was the Lopsla of the Russian cartographers. When I was a young boy, I was living in Harbin, in Heilongjiang province and I remember that there were some kinds of shells in the Songhuajiang where pearls were found.

authors of the Polish translation of the emperor's ultimatum to vacate Albazin-Yakesha. The news about a Chinese attack was reported by the then voivode Alexis Tolbuzin:

> In this year 193 (that is 7193 [our 1685] from the creation of the world, according to the Russian calendar prior to Tsar Peter's reform) there arrived in Albazinsky Ostrog the two Russian hunters Fiedka Tielsky (evidently a Pole: Tadeusz or Teodor Cielski) and Gavrilka Dmitriev Sotnik, Jr., and they brought with them three letters. The first was written in Russian, the second in Boghdoyan (that is, in Chinese), and the third in Polish...[35]

And the voivode continued:

> Boghdoyan voivodes and a multitude of Boghdoyan troops are approaching on ships and overland, with the intention of taking over Albazinsky Ostrog. On each ship are fifty men, and there are a thousand or more cavalry, and a hundred cannons...[36]

These were certainly cannons cast by Verbiest. Albazin was destroyed in the following year, and the peace talks followed in Nerchinsk, or Nibuchu. The Treaty of Nerchinsk, which regulated the border questions between the two countries, was negotiated during four years, and was concluded in 1689.

---

35 *Dela iz Albazinskago Archiva*, Russkaya Biblioteka (Moscow, 1832), p. 47.

36 *Ibid.*, p. 48.

FERDINAND VERBIEST IN THE *CIHAI* DICTIONARY

JEROOM HEYNDRICKX, C.I.C.M.

Father J. Heyndrickx holds a M.A. degree in Education from Loyola University of Chicago. After a six-year term as a provincial superior of the China province of the CICM fathers, he was from 1974 to 1981 vicar-general of the congregation. In 1983 he founded the Ferdinand Verbiest Foundation, K.U. Leuven. At present he is the director of the Foundation as well as the executive secretary of the Theofiel Verbist Study Center at Taibei. Since 1987 he is a lecturer at the department of Oriental Studies of the K.U. Leuven. He organized the International Conference on Couplet (1986), on Verbiest (1988) and on the Historiography of the Catholic Church in China (1990). He has published several articles on the Church in China.

The Great Dictionary *Cihai zidian* is, so far, the most important, comprehensive and largest reference book in China. It is both a dictionary and an encyclopedia, and exerts a tremendous influence on Chinese cultural circles. Ferdinand Verbiest is included in the *Cihai* for his proper historical status. Though the article on Verbiest is short, it summarizes the appraisal of Chinese academic circles of Verbiest. Since its first publication, about half a century ago, the *Cihai* has been frequently revised and enlarged. When one compares the items on Verbiest in different editions of the *Cihai*, one is able to see the changes of the appraisal by Chinese academic circles of Verbiest. At the occasion of the three hundredth anniversary of Verbiest's passing away, it is of academic value and practical significance to discuss the appraisal of Verbiest in the *Cihai*.

At present, the most popular edition of the *Cihai* is the one published by the Shanghai Dictionary Publishing House in 1979. I was told that three million copies of its one-volume edition, and ten million of its twenty-six-volume edition have been sold. It is estimated that about one hundred million people use the *Cihai*. Its influence is therefore obviously very great, especially on many mainland Chinese scholars who refer to this official text in their articles.

南怀仁(Ferdinand Verbiest, 1623—1688) 清初来中国的天主教耶稣会传教士。比利时人。顺治十六年(1659年)在葡萄牙殖民势力支持下来中国。初在陕西传教，次年到北京，成为*汤若望的重要助手。后以汤若望案下狱，释放后幽居北京。康熙七年(1668年)复被起用，管理钦天监监务，制造天文仪器，并为清政府监铸大炮。1675—1678年沙俄派使臣斯巴法里-米列斯库(Николай Гаврилович Спафарий-Милеску, 1636—1708)来中国时，他借任清政府译员之便，向斯巴法里-米列斯库递送有关清政府内部政治、财政、军事等机要情报多件。著作有《教要序论》、《康熙永年历法》等。

Ill. 62: "Nan Huairen", from: *Cihai* (Shanghai, 1979), 187.

【南懷仁】(Ferdinandus Verbiest, 1623—1688) 字勳卿，又字敦伯.比利時耶穌會之敎士.清順治間來華，與湯若望等同至北京傳敎.以明數理，任欽天監副.詔命改製觀象儀器，成新式者六種；復繪圖立說，用新器測爲諸表十六卷，名曰新製靈臺儀象誌；康熙中，擢監正，加太常寺少卿.并命督造大礮，試驗有效；繼湯若望恆年表，預推七政交食，成康熙永年曆法三十二卷；後進職右侍郎，隨駕盛京，測得其地高度.卒，謚勤敏.

Ill. 63: "Nan Huairen", from: *Cihai* (Beiping, 1936), 479.

The full text of the item on Verbiest in the 1979 edition of the *Cihai* is as follows:

> Verbiest (Ferdinandus Verbiest, 1623-1688), missionary of the Catholic Society of Jesus, Belgian, came to China at the beginning of the Qing dynasty. In the sixteenth year of Shunzhi (1659), supported by Portuguese colonialists, he came to China. At first he did missionary work in Shaanxi. The next year he went to Beijing and became an important assistant of Adam Schall afterwards. Later he was sent to prison as a result of the Schall case. After his release he lived in seclusion in Beijing. In the seventh year of Kangxi (1668) he was reinstated. Then he administered the affairs of the Imperial Bureau of Astronomy. He made astronomical instruments and supervised the manufacturing of cannons for the Qing government. From 1675 to 1678, when the envoy of the tsar of Russia (Nikolai Govrilovich Spathary Milescu) (1636-1708) came to China, he took advantage of being an interpreter of the Qing government to deliver a lot of confidential and important information about internal political, financial, and military affairs of the Qing government to Spathary. He wrote *An Introduction to Religion*, *the Perpetual Calendar of the Reign of Kangxi*, etc.[1]

The article on Verbiest in the *Cihai* edition of 1979, besides giving biographical notes about him, insists on the fact that Verbiest passed confidential information on the Qing government to the envoy from tsarist Russia. This has caused the rumor that Verbiest was a spy of tsarist Russia (even though the word "spy," as such, is not used in the text).

Our question is: Does this accusation conform to the historical facts? During the time period from May 5 until September 11, 1675 (the fourteenth year of Kangxi), the government of tsarist Russia dispatched a diplomatic delegation led by Spathary to Beijing in order to negotiate with the Qing government. In order to solve the language problem among them (Russian, Chinese, Manchu and Mongolian), they turned to translation into Latin. The Kangxi emperor requested Verbiest to be the interpreter.

It is not within the scope of this short paper to discuss further the role of Verbiest in the negotiations between China and Russia. We only say that, by accepting this task, Verbiest made a contribution to the promotion of mutual understanding between China and Russia. We also know that Spathary learned to appreciate Verbiest, as appears from his

1 Verbiest, *Cihai*, one-volume edition (Shanghai: Shanghai Dictionary Publishing House, 1979), p. 187.

Report on the Diplomatic Mission to the Qing Empire. It is also a fact that Verbiest was praised by the Kangxi emperor. More detailed studies on this passage of Verbiest's and Spathary's lives have been done by Egly Alexandre of France, Hao Zhenhua of the Chinese Academy of Social Sciences in Beijing, Edward Kajdanski of Poland and Vladimir Miasnikov of Russia. As appears from the paper of Hao Zhenhua, he would not agree with the text on Verbiest in the *Cihai* 1979 edition[2].

Has it been proved historically that Verbiest did give Spathary any important and confidential information, to which the *Cihai* refers? Did Verbiest do anything that, in fact, harmed China? If not, what then is there in Verbiest's whole life that would justify the rumor that describes him as a spy? Where does the "spy version" about Verbiest originate?

When we check earlier editions of the *Cihai* we notice that the accusation against Verbiest as expressed in the 1979 edition did not exist in previous editions. The first edition of the *Cihai* was printed and published in 1936. The text on Verbiest in that edition was as follows:

> Verbiest (Ferdinandus Verbiest, 1623-1688), also named Xun Qing or Dun Bo, priest of the Society of Jesus of Belgium, came to China, together with Adam Schall. He came to Beijing to do missionary work during the Shunzhi period of the Qing dynasty. He was proficient in mathematics, so that he was appointed as vice-director of the Imperial Bureau of Astronomy. Instructed by the emperor to make astronomical instruments, he succeeded in making six new instruments. He also made drawings, and wrote down the theory for the use of the instruments in sixteen volumes, entitled: *Report on the Newly Made Instruments for the Observatory*. In the middle of the Kangxi years, he was promoted to director of the Imperial Bureau of Astronomy, and became vice-minister of the Sacrificial Worship Court (*taichang si shaoqing*). Furthermore, he was instructed to supervise the manufacturing of cannons - and he succeeded well. After Adam Schall's *Sidereal Year Catalogue*, he calculated eclipses and wrote down *The Perpetual Calendar of the Reign of Kangxi* in thirty-two volumes. Afterwards he was given the rank of (right) minister (*you shilang*), and accompanied the emperor to Shengjing and there he measured the altitude. After his death the emperor conferred on him the title "Diligence and Intelligence" (*qinmin*)[3].

---

2 *Zhong-E guanxi wenti* (Problems on Chinese-Russian Relations) (Beijing, 1985).

3 Verbiest, *Cihai* (the first edition) (Shanghai: Zhonghua Press, 1936), 10:479.

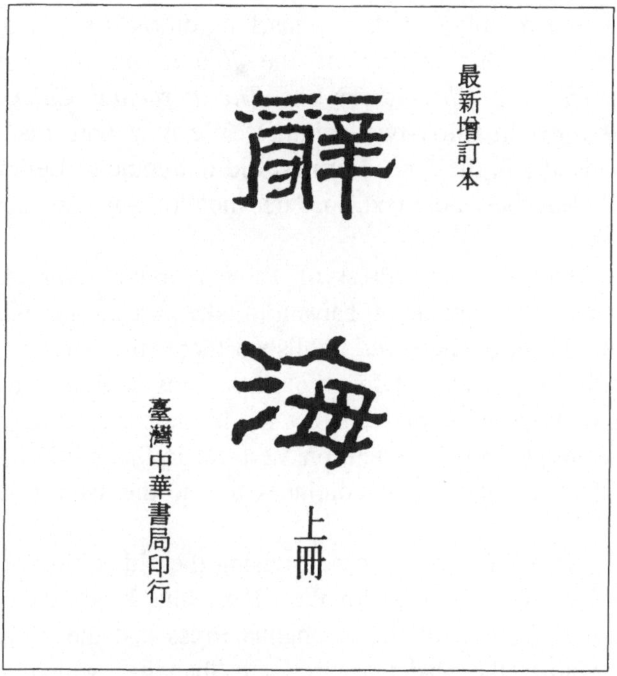
最新增訂本
辭海
上冊
臺灣中華書局印行

Ill. 64: Title page of *Cihai* (Taibei).

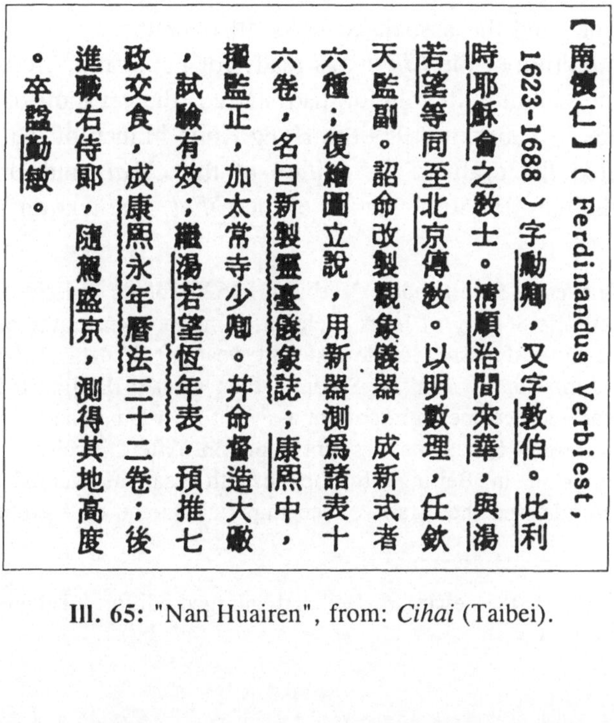
【南懷仁】（Ferdinandus Verbiest,
1623-1688）字勳卿，又字敦伯。比利
時耶穌會之教士。清順治間來華，與湯
若望等同至北京傳教。以明數理，任欽
天監副。詔命改製觀象儀器，成新式者
六種；復繪圖立說，用新器測爲諸表十
六卷，名曰新製靈臺儀象誌；康熙中，
擢監正，加太常寺少卿。并命督造大礮
，試驗有效；繼湯若望恆年表，預推七
政交食，成康熙永年曆法三十二卷；後
進職右侍郎，隨駕盛京，測得其地高度
。卒諡勤敏。

Ill. 65: "Nan Huairen", from: *Cihai* (Taibei).

In the first edition of the *Cihai*, the emphasis is put on the contribution of Verbiest to astronomy, such as remarking on the six new observatory instruments, and the writing of the *Report on the Newly Made Instruments for the Observatory* and *The Perpetual Calendar of the Reign of Kangxi* in thirty-two volumes. We may note that there is a historical mistake in the note. Verbiest did not come to Beijing together with Schall. But the whole text does not mention any "spy activities" of Verbiest.

In 1976, the Zhonghua Press of Taiwan invited over one hundred experts from in and outside of Taiwan to take part in the revision of the *Cihai*. The edition of the *Cihai* published there (the newest revised and enlarged edition) consists of four volumes. This edition sold more than one hundred thousand copies and set the highest record in the sales of dictionaries in Taiwan. The item on Verbiest in that edition (the newest revised and enlarged edition) continues to use the same text as in the first edition.[4]

In 1957, Mao Zedong proposed revising the old edition of the *Cihai*. Later, the former chief editor of the *Cihai*, Shu Xincheng founded the *Cihai* Editorial Section of the Zhonghua Press and the *Cihai* Editorial Board to organize the academic circles in Shanghai, which would carry out the revision of the *Cihai* according to different subjects. In 1960, when Shu Xincheng died, Chen Wangdao succeeded him as the chief editor. Cooperation range was extended to Beijing, the provinces in the east of China, and the associated units all over the rest of China. The draft of the edition of the *Cihai* was ready for publication in April 1965. But after the Cultural Revolution had started, the revision of the *Cihai* was stopped. In February 1979 the Hong Kong branch of the Zhonghua Press collated the draft of the edition of the *Cihai*, and changed the edition to the new 1965 publication of the *Cihai*. The text on Verbiest is as follows:

> Verbiest (Ferdinandus Verbiest, 1623-1688), missionary of the Catholic Society of Jesus, Belgian, came to China at the beginning of the Qing dynasty, in the sixteenth year of Shunzhi (1659). He first did missionary work in Shaanxi, then went to Beijing the following year and became an important assistant of Adam Schall. Later, due to the case of Schall, he was sent to prison. After his release, he lived in seclusion in Beijing. In the seventh year of Kangxi (1668) he administered the affairs of the Imperial Bureau of Astronomy, made

4 Verbiest, *Cihai* (the newest revised and enlarged edition) (Taibei: Zhonghua Press, 1976), 1:230.

astronomical instruments, and supervised the manufacturing of cannons for the Qing government. He wrote *An Introduction to Religion, The Perpetual Calendar of the Reign of Kangxi*, etc.[5]

> 南怀仁(Ferdinand Verbiest, 1623—1688) 清初来华的天主教耶稣会传教士。比利时人。顺治十六年(1659年)来华。初在陕西传教，次年到北京，成为*汤若望重要助手。后以汤若望案下狱，释放后幽居北京。康熙七年(1668年)管理钦天监监务，制造天文仪器，并为清政府监铸大炮。著有《教要序论》、《康熙永年历法》等。

Ill. 66: "Nan Huairen", from: *Cihai* (Hong Kong, 1965), 210.

> 【南怀仁】(Ferdinand Verbiest, 1623—1688) 清初来中国的天主教耶稣会传教士。比利时人。顺治十六年(1659年) 在葡萄牙殖民势力支持下来中国。初在陕西传教，次年到北京，成为*汤若望的重要助手。后以汤若望案下狱，释放后幽居北京。康熙七年 (1668 年) 复被起用，管理钦天监监务，制造天文仪器，并为清政府监铸大炮。1675—1678年沙俄派使臣斯巴法里-米列斯库 (Николай Гаврилович Спафарий-Милеску, 1636—1708) 来中国时，他以清政府译员身份为掩护，向斯巴法里-米列斯库递送有关清政府内部政治、财政、军事情况的机要情报多件。著作有《教要序论》、《康熙永年历法》等。

Ill. 67: "Nan Huairen", from: *Cihai* (Shanghai, 1977), 78.

5 Verbiest, *Cihai* (the new publication) (Hong Kong: Zhonghua Press, 1965), p. 210.

Even though the content is rather short, the appraisal of Verbiest in the new *Cihai* publication of 1965 is fair. In any case, it does not mention any "spy activities."

Would the item on Verbiest in the 1979 edition not be a product of the Cultural Revolution? In 1971, Zhou Enlai proposed to include the *Cihai* in the national publication plan. Afterwards the Shanghai People's Publishing House started to revise it, and published separated volumes according to subjects. The item on Verbiest was included in the *Volume on Religion*. During the Cultural Revolution, as a result of research on the history of the relationship between China and Russia, the references about this subject were translated and published in China. This included studies on *The Report on the Diplomatic Mission to the Qing Empire*, written by Spathary. That is how the story of Verbiest being a spy of tsarist Russia developed. In order to include the newest research of the Cultural Revolution, the story was included in the item on Verbiest in the *Volume on Religion*. Although the Gang of Four disappeared in October 1976, there was apparently no time to check the text on Verbiest in the volume on religion published in July 1977. This is regrettable, because this text has become the chief source of information for what has been written about Verbiest in China since then[6].

In December 1978 the *Cihai* editorial board was resumed and strengthened in China. With the death of Chen Wangdao, Xia Zhengnong became chief editor. After nine months of work, the *Cihai* was published in three volumes. The *Cihai* 1979 edition continues to use the text on Verbiest in the *Volume on Religion*. The abbreviated edition was based on the *Cihai* three-volume edition of 1979. Only some expressions and drawings have been slightly corrected and complemented. The item on Verbiest in this abbreviated edition was not corrected.

The *Cihai* is an authoritative dictionary which has been used for a very long time. The preface of the 1979 edition of the *Cihai* explains the following guidelines, which were followed in compiling it:

> To avoid quoting in translation those materials which have a strong time-bound character, or which are not yet confirmed, academic problems whose research has been concluded are introduced according to the final conclusion, while those without final conclusions, or which are still being discussed, are introduced objectively, recounted generally, or only introduced by giving the

---

6 *Cihai zongjiao fence* (Volume on Religion) (Shanghai: People's Publishing House, 1977), p. 78.

present different views, as much as possible, in order to avoid giving readers a one-sided or wrong view.

【南怀仁】(Ferdinand Verbiest, 1623—1688) 清初来中国的天主教传教士。号敦伯。比利时人。1641年入耶稣会。在葡萄牙殖民势力支持下，于顺治十六年(1659)抵中国。曾被派往陕西。顺治十七年(1660)奉召至北京，协助汤若望修历法。后以汤若望案下狱。释放后幽居北京。康熙七年（1668）复被起用，掌钦天监监务。十三年(1674)奉命制造天文仪器，加太常寺卿衔。十七年(1678)著成《康熙永年历法》三十二卷，加通政使司通政使衔。二十一年(1682)奉旨制成大炮数百尊后，晋工部侍郎。1675至1676年沙俄派使臣斯巴法里-米列斯库来中国时，借任清政府译员之便，曾向俄使透露有关清政府内部政治、财务、军事等多种情况。死于北京。著作还有《仪象志》十四卷、《赤道南北星图》、《教要序论》等。

Ill. 68: "Nan Huairen", from: *Zongjiao cidian* (Shanghai, 1981), 766.

Verbiest, Ferdinand
(1623—1688)
南怀仁，字敦白 韦比埃，费迪南

比利时耶稣会士。1656年他和耶稣会士柏应理等一起来华，船至地中海时为海盗所掳，赎出后继续航行，1658年7月始抵澳门。次年被派往西安传教。旋应顺治帝邀，于1660年到北京，为钦天监监正德国耶稣会士汤若望的重要助手，治天文历法。后因汤若望案下狱。获释后幽居北京。康熙帝亲政后复被起用，1669年受命为钦天监监正。康熙帝求知欲极强，相当钦佩西洋教士的科学素养，日与南氏研习几何、天文和历算。三藩之变发生时，且命其修旧炮，铸新炮。时清廷接待外国来使常以南氏充翻译。1675—1677年沙俄使臣尼果赖使华时，他为求得俄国应允开辟一条耶稣会士从陆路来华的交通路线，对俄使曲意逢迎，竟以清政府译员身份为掩护，把私绘的中国地图及清政府的政治、经济、军事机密情报多件送给尼果赖。1688年1月28日死于北京。南氏的汉文著作有三十九种，并编有一部满文字典。

Ill. 69: "Verbiest, Ferdinand", from: *Jindai lai Hua waiguo renming cidian* (Beijing, 1981), 490.

It is a pity that the text in the item on Verbiest in the edition of 1979 violated the above-mentioned compilation principle. It seems to me that indeed a lot more study needs to be done on the question of what Verbiest really gave to Spathary, but also on Spathary himself, and on the objectivity of his report, as well as on the whole range of contributions made by Verbiest to China as a country before it can be justified in describing Verbiest as a spy.

Also in present-day China the *Cihai* remains the most authoritative dictionary. It will inevitably and greatly influence the compilation of the other reference books in China. Some examples are:

In September 1981 the *Dictionary on Religion* was published by the Shanghai Dictionary Publishing House. In the item on Verbiest in this dictionary, the text for recounting the negotiations between China and Russia basically refers to the model of the *Cihai* 1979 edition. It says:

> From 1675 to 1676, when the envoy of tsarist Russia, Nikolai Spathary, came to China, Verbiest took advantage of being an interpreter of the Qing government to reveal a lot of information about internal political, financial, and military affairs of the Qing government to the envoy[7].

Meanwhile, the Publishing House of the Chinese Academy of Social Sciences published the *Dictionary of Names of Foreigners who came to China in Modern Times* in December 1981. It was compiled mainly by the Translation Section of the Modern History Research Institute of the Chinese Academy of Social Sciences. The part on Verbiest is written as follows:

> When the Qing government received foreign envoys, it often took Verbiest as an interpreter. From 1675 to 1677, when the envoy of tsarist Russia, Nikolai Spathary, came to China, in order to get permission from Russia to open a continental route for the priests of the Society of Jesus coming to China, Verbiest went out of his way to win the favor of the Russian envoy and, covered by the status of an interpreter of the Qing government, delivered the illegally-drawn Chinese map, as well as a lot of confidential information about the political, financial, and military affairs of the Qing government to the Russian envoy[8].

---

7 *Zongjiao cidian* (Dictionary on Religion) (Shanghai: Dictionary Publishing House, 1981), p. 766.

8 *Jindai lai Hua waiguo renming cidian* (Dictionary of names of foreigners who came to China in modern times) (Beijing: Publishing House of the Chinese Academy of Social Sciences, 1981), p. 490.

This text goes even further than the *Cihai* 1979 edition.

In 1983, the *Volume on Religion* of the *Cihai* was revised by the "revision-group of *Cihai*'s *Volume on Religion*" and a second edition came out. The item on Verbiest's life was more detailed[9]. It is worth mentioning that the phrasing of the fact of Verbiest passing confidential information to Spathary was changed. The first edition wrote that he did this "under the cover of his status of interpreter of the Qing government." This is now changed into "taking advantage of being an interpreter of the Qing government." Before it was said that Verbiest gave a lot of confidential and important information about internal political, financial and military affairs of the Qing government to Spathary; now it is written that Verbiest "disclosed the military and political situation under the Qing government."

The item on Verbiest in the volume on religion of *The Great Chinese Encyclopedia* published in 1988 calls for special attention. Besides giving a short history of Verbiest, it elaborates on Verbiest and the relations between China and Russia:

> In 1676 the government of the tsar dispatched the Nikolai Spathary diplomatic delegation to China. Verbiest wanted to obtain the support of Nikolai and the permission of the tsar to open a new road for West European missionaries to go to China via Siberia. On his own initiative he told Nikolai: "For the benefit of Christianity he would gladly offer his service and cooperation to the tsar." Nikolai then promised: "The tsar certainly will reward you." Consequently, Verbiest took advantage of being an interpreter of the Qing government, together with others in Beijing, to provide internal information about the political, financial, etc... situation of the Qing government to Nikolai. When Nikolai left Beijing, he also took along the *Kunyu quantu* (Map of the World in Chinese Characters) and a letter of Verbiest offering his services to the tsar[10].

The *Kunyu quantu* is a world map drawn by the Jesuits which already a long time before had been known publicly and was spread around far and wide. The item one-sidedly uses the words of Verbiest expressing his willingness to offer his services to the tsar ignoring that there are many more memorials which Verbiest presented to the Kangxi emperor and in which he expressed his gladness to offer his services. One can

---

9 *Cihai zongjiao fence* (Volume on Religion), 2nd ed. (Shanghai: Dictionary Publishing House, 1983), p. 111.

10 *Zhongguo dabaike quanshu. Zongjiao* (The Great Chinese Encyclopedia. Volume on Religion) (Beijing: Great Chinese Encyclopedia Publishing House, 1988), p. 286.

read this in the *Xichao ding'an* (Memoranda presented to the Kangxi emperor), which was published later.

At the end of the 1970s and the beginning of the 1980s, three revised editions of the *Cihai* appeared at the same time on the continent of China, Hong Kong and Taiwan. This is the so-called case of "the triplets of the *Cihai*." It is rather interesting that the items on Verbiest in the different editions reflect the versions of the 1930s (i.e., in the newest revised and enlarged edition published by the Zhonghua Press in Taiwan), the 1960s (i.e., in the new 1965 publication published by the Hong Kong branch of the Zhonghua Press) and the 1970s (i.e., in the 1979 edition published by the Shanghai Dictionary Publishing House) respectively.

In order to eliminate the influence of the "leftist" trend of thought, and to match the fast development of science, technology and culture, the Shanghai Dictionary Publishing House has since 1985 called upon more than a thousand experts and scholars of continental China, to revise the *Cihai* edition of 1979. The first draft of this revision, so we heard, is now ready. Recently the Shanghai Dictionary Publishing House, the Hong Kong branch of the Zhonghua Press, and the Yuanliu Publishing House of Taiwan signed a statement of intent in Hong Kong for jointly publishing a new edition of the *Cihai* using the non-simplified Chinese characters. It was planned to be published in Shanghai in September 1989, and could meet the readers in Hong Kong and Taiwan before July 1990. What will be said on Verbiest in this forthcoming edition of the *Cihai*? His fellow countrymen in Belgium, as well as many Chinese people and people around the world who admire him, wonder about this question[11]. Following is a summary of the ideas and observations on this matter which were expressed by experts, mentioned above, during the 1988 (September) Louvain International Conference on Verbiest.

Is it correct, and sufficiently based upon objective historical data, to call Verbiest a spy? Or would it not be closer to historical truth to see him simply as a man from the West, a missionary, and a man of science who wanted to contribute to cultural and scientific exchange between

11 I sent the text of this present lecture in 1988 to the scholars who prepared the new editon of the *Cihai* with the Shanghai Publishing House. Meanwhile, the 1990 revised edition has been published. The phrase referring to Verbiest "taking advantage of being an interpreter of the Qing government to deliver a lot of confidential and important information about internal political, financial, and military affairs of the Qing government to Spathary" has been left out of the new edition.

East and West? In order to achieve this goal better, he tried to open a road via Russia to shorten the distance between China and Europe. For that same purpose he was interested in establishing good relations with Russian friends. The visit of Spathary offered him an opportunity to do so.

What Verbiest did in order to achieve this goal did not imply any unusual activity. Exchanging information was the usual practice of diplomats and explorers of that time. Russian and Indian envoys during the same period did this quite regularly. Moreover, both the Russians and the Chinese knew very little at that time about the geographical areas north of China, and both sides were eager to exchange any useful information. It is historically certain that Verbiest received maps from the Russians. Whether or not he did give any maps about China to the Russians is not sure. It is affirmed by some historians, and denied by others. But even if he did, it would not be correct to interpret this fact in itself in terms of what today could be seen by some as "spying." It is the responsibility of the historian to avoid reading the historical facts of three hundred years ago through the concepts and ideas of our twentieth century. The author of spy or fiction stories obviously follows other norms.

Consequently, we suggest that in evaluating Verbiest's life and work, it might be closer to historical truth to avoid the use of any vocabulary that suggests his either being a spy or acting in a way that was harmful to China. There is no historical evidence to support that. On the contrary, there is a lot of historical evidence that proves that Verbiest was a good friend of China, a missionary and a man of science, a pioneer of cultural and scientific exchange between China and Europe. It would therefore be better to remain faithful to the evaluation which his personal friend, the Kangxi emperor, made about Verbiest by giving him the honorary title of *qinmin*, "Intelligent and Diligent."

East and West, in order to achieve this. [illegible] to [illegible] via Russia in search of [illegible] the same purpose he was interested in [illegible] Russian Church. The [illegible]

Why [illegible] [illegible] chemical activity [illegible] the same period [illegible] the Chinese knew very little at that time about [illegible] though it is historically [illegible] the Russians. [illegible] the responsibility [illegible] to avoid [illegible] through the concepts and ideas of [illegible] century. [illegible]

Consequently, we suggest that [illegible] [illegible] country [illegible] was a good [illegible] of China, a [illegible] cultural and scientific exchange between China [illegible]

Verbiest

Missionary

BETWEEN THE COURT AND THE CHURCH: FERDINAND VERBIEST IN THE CATHOLIC HISTORY OF CHINA

KU WEIYING

Professor Ku Weiying received his Ph.D. in history from the University of British Columbia, Vancouver, Canada. At present, he is professor at the Department of History, National Taiwan University.

I. Introduction

Scholars increasingly regard the late Ming and early Qing as an important era for Chinese history. Beginning from the late sixteenth century, several consequential phenomena emerged: the urbanization of the lower Yangzi, the commutation of labor services into money payments, the development of certain kinds of regional trade, the growth of mass literacy, the increase in the size of the gentry, and the commercialization of local managerial activities[1]. Moreover, the encounter between the West and China was also a major event in this crucial era. It contributed to the expansion of the Chinese world view, the increase of knowledge, and the change of research methodology. However, the extent of these influences remains to be studied.

The Jesuits represented the West in this encounter. The three most famous ones were Matteo Ricci (1552-1610), Johann Adam Schall von Bell (1592-1666), and Ferdinand Verbiest (1623-1688). Ricci has been studied in detail, while Verbiest - whose significance is also widely recognized - has not yet been studied systematically. This essay thus explores Verbiest's role in the Catholic experience in China.

1 See William T. Rowe, "Approaches to Modern Chinese Social History," in O. Zunz (ed.), *Reliving the Past. The Worlds of Social History* (Chapel Hill and London, 1985), pp. 240-241; Frederic Wakeman, Jr., "Introduction: The Evolution of Local Control in Late Imperial China," in F. Wakeman, Jr., and Carolyn Grant (eds.), *Conflict and Control in Late Imperial China* (Berkeley, 1975), p. 2.

This essay will cover several dimensions of Verbiest's activity in China. Following a brief introduction, his coming to the Manchu court, basic events in his life and his major work in the *qintianjian* (Bureau of Astronomy) are examined. Next, Verbiest's relationship to the Church and then his relationship to the Kangxi emperor. Comparing the situation faced by Ricci and Schall, I try to put Verbiest in an historical context. This is followed by a discussion of Verbiest's religious writings in Chinese. This shows the extent to which his knowledge about China influenced his missionary work. Verbiest's legacy is reviewed in the last section. What is outstanding is that, although Verbiest was caught under tremendous pressure from the Manchu court, the Church, other religious communities and the Jesuit Order itself, he still managed to make some significant contributions to the propagation of the Catholic faith in China.

II.
His Life and Time

Verbiest had several Chinese names, the most common one being Nan Huairen. The other two styles (*zi*) are Dun Bo and Xun Qing (Louis Pfister cited it as Xun Xiang. I think this was a simple misidentification on his part. It is much more common for a Chinese to have Qing, rather than Xiang, in his name)[2].

Born on September 29, 1623 at Pittem in what is now modern Belgium, Verbiest entered the Jesuit Order at the age of eighteen. He was educated in Belgium, Rome and Spain. In 1647 and again in 1655, he tried to become a missionary to South America, but did not succeed. His enthusiasm to be a missionary abroad moved his superior to send him to China. At this juncture, Martino Martini, returning from China, was recruiting confreres to join the China mission. As a result, led by Martini, Verbiest left for China in 1657. As was common in those days, they suffered pirate attacks and hardships, but arrived in China two years later[3].

Verbiest was first sent to Shaanxi province to work with Giovanni Francesco de Ferrariis after his arrival in 1659. He was summoned to

---

2 Fang Hao, *Zhongguo tianzhujiaoshi renwu zhuan* (Hong Kong, 1970), Nan Huairen, 2:174; and also Louis Pfister, *Notices biographiques et bibliographiques sur les jésuites de l'ancienne Mission de Chine, 1552-1773* (Shanghai, 1932), p. 338.

3 Jonathan Spence, *To Change China, Western Advisers in China, 1620-1960* (New York, 1969), p. 29.

Beijing on May 9, 1660 upon the recommendation of Schall to help him compile a calendar. As he traveled from Shaanxi to Beijing, he passed through three provinces and thirty-four cities. To his surprise, he found a remarkable uniformity in the architecture all over the *yamen* courts and palaces. This sense of uniformity and order strengthened Verbiest's belief that the selection of the date for doing something by the government was also part of common practice for the uniformity and order of the empire[4]. This belief seems to have later enabled Verbiest to defend Schall against the accusation of having become involved in superstitious activities.

In 1661, the young Shunzhi emperor died. His successor, the Kangxi emperor was only eight years old. Oboi and three other Manchu aristocrats were in charge in the beginning, but later Oboi became the most important regent. During the Oboi regency, the Manchu court reversed many policies initiated by the Shunzhi emperor. Perhaps in reaction to the previous pro-Chinese attitude and perhaps because of hostility toward Jesuit involvement in the exiled Southern Ming court, the Oboi regency took advantage of Yang Guangxian's accusations against Schall and persecuted the Catholics in the empire.

On the following counts Yang launched his charges against Schall: the selection of a wrong burial date for a prince, resulting in the untimely death of another prince; using heretical doctrine to confuse the Chinese people; banding together with other western missionaries to plot against the government; the remark on the cover of the calendar "Based on the New Western Method" (*Yi xiyang xinfa*) implying the replacing of the Chinese official calendar with the official western calendar (*zhengshuo*, this term carries very strong political implications in Chinese political culture, indicating the direction and object of loyalty); the inappropriateness of presenting the two hundred year calendar to the emperor (compared to the usual reference to the emperor as "ten thousand years" *wansui*, two hundred years indeed constituted a problem)[5]. Consequently, Schall, Verbiest and two other priests residing in Beijing, Ludovico Buglio (1606-1680) and Gabriel de Magalhaes (1609-1677), were put in jail for half a year. They were

---

4 John Witek, "The Missionary Leadership of Ferdinand Verbiest: Some Perspectives," in *Furen daxue, Nan Huairen shishi sanbai zhounian guoji xueshu taolunhui* (An International Scientific Conference organized by the Furen University in commemoration of the 300th anniversary of the death of Nan Huairen) (Taibei, 1987), p. 152.

5 *Qingshi gao*, Taibei, *juan* 272, p. 10021.

frequently interrogated during the first three months. Schall became sick and paralyzed and the entire defense was conducted by Verbiest, who was calm and devoted[6].

The case did not close until 1666. Schall was convicted and sentenced to death, but the death penalty was commuted. Five of his Chinese followers in the *qintianjian* were executed. Yang was appointed as vice-director of the *qintianjian*, but as he actually knew very little about astronomy, he declined. The Manchu dynasty insisted, however, and even appointed him as director. Yang could not but accept this job, and relied heavily upon the assistance of the Muslim astronomer Wu Mingxuan[7]. The calendar they compiled was inaccurate. This whole situation was resolved two years later, when the Kangxi emperor assumed power.

Discontented with the policies of the Oboi regency on the one hand, and energetic and intelligent on the other, the emperor wanted to find the true answer to this controversy. At the end of 1668, he asked all the members of the two parties to participate in a test supervised by the high officials. He said in his edict:

> Astronomy is most delicate, and calendar-making is an important affair for the country. Your two parties should not harbor previous animosities toward each other, nor be stubborn or self-righteous; neither should you challenge one another. You have to follow the right side after this test result comes out, and correct your mistakes. Be honest and whole-hearted in compiling the calendar and study astronomy in order to be error-free[8].

The test lasted three days and the winner was Verbiest. The emperor ordered the calendar compiled by Yang's party to be examined by Verbiest. Yang still resisted at this time by invoking political considerations:

> The calendar compiled by us followed the method inherited from the emperors Yao and Shun; the position Your Majesty ascended on was the one inherited from the emperors Yao and Shun ... The calendar issued by Your Majesty should use the one of Yao and Shun. Your Majesty follows the way of Yao and Shun in all other aspects. Why make an exception to calendar-making? Verbiest is a Catholic. No

---

6 Wei Te (Alfons Väth, S.J.), Yang Bingchen (transl.), *Tang Ruowang zhuan* (Johann Adam Schall von Bell, S.J.) (Shanghai, 1949), Ch. 13.

7 *Qingshi gao*, *juan* 272, Yang Guangxian, pp. 10022-10023.

8 Huang Bolu, *Zhengjiao fengbao* (Shanghai, 1904), p. 44.

> sagacious emperor who takes Yao and Shun as models would follow the Catholic way ...[9].

The emperor was very angry with this political rhetoric. In addition to reviling Yang, he held another test in early 1669. The final report, endorsed by twenty high court officials, stated: "Every experiment Verbiest conducted and the instruments he used were successful and correct. All the tests proved Wu Mingxuan's mistakes"[10]. Thus Yang was fired and punished. There were different opinions, however, as to how to reorganize the *qintianjian*.

The Board of Rites, of course, recommended that Verbiest be the director. But the emperor thought this promotion was too fast, and agreed only to bestow the vice-directorship on Verbiest[11]. The Board of Rites then suggested that the original vice-director, Hu Zhenyue, take the position of director. Apparently Hu belonged to the native Muslim party. Appointing him meant the restoration of the old method. Too sober to let this happen, the emperor rejected this suggestion. On the other hand, Verbiest time and again declined the new appointment for reasons which will be discussed later in this essay. The emperor finally agreed that Verbiest should take full charge of the *qintianjian* as director, though without official title and with the salary of a vice-director[12].

As soon as Verbiest began to work in the *qintianjian*, he set out to correct the previous mistakes made by the old party and to redress the miscarriage of justice done to Schall and his followers. He also got permission for the missionaries who were then exiled in Guangzhou to return to their respective mission posts. To encourage young Chinese to study astronomy as a career, Verbiest in 1670 successfully got the agreement of the Qing dynasty to give any student of astronomy a sub-ninth rank[13].

From the time Verbiest began working in the *qintianjian*, the emperor kept him very busy with various kinds of jobs. From 1669 to 1676 Verbiest was asked to make all astronomical instruments and build up an observatory. He also wrote a book titled *Yixiang zhi* (Records of

---

9 *Ibid.*, p. 46.

10 Nan Huairen, *Xichao ding'an*, in Wu Xiangxiang, *op. cit.*, p. 10.

11 *Ibid.*, p. 19; Huang Bolu, *op. cit.*, p. 52.

12 *Ibid.*

13 Nan Huairen, *Xichao ding'an*, in Wu Xiangxiang, *op. cit.*, pp. 25, 44, 59-67, 80; Huang Bolu, *op. cit.*, pp. 55-62.

Astronomical Phenomena). From 1676 to 1678 he again compiled the "Everlasting Calendar of Emperor Kangxi" (*Kangxi yongnian lifa*). During this entire period, Verbiest was from time to time summoned to the inner court to teach the emperor western astronomy and its relevant knowledge[14].

In the winter of 1673 the Rebellion of the Three Feudatories started. From 1674 to 1676, Verbiest was asked to make 120 cannons. In the latter stages of the rebellion he was ordered to manufacture more cannons and to train soldiers to use them. In 1681 the emperor participated in an exercise and examined the quality of the 440 cannons and soldiers in person. He was very happy with the results and had this to say: "In the past years you had made some cannons. They were used in Shaanxi, Huguang and Jiangxi, and proved to be powerful. The new cannons you made this time are even better." The emperor gave his own furcoat to Verbiest as a gift. Verbiest was awarded the title of Junior Vice-President of the Board of Works the following year[15].

In 1682, a year after the pacification of the Rebellion of the Three Feudatories, the emperor visited Manchuria and asked Verbiest to join his retinue. During this trip, Verbiest was able to do some geographical and astronomical investigations[16]. The following year the emperor traveled to Mongolia and again had Verbiest accompany him. In the late summer of 1686, Verbiest was asked to make a rocket (*chongtian pao*)[17]. He died on January 28, 1688, at the age of 65. He was given the posthumous title "diligent and sharp-witted" (*qinmin*), which no other missionary has ever enjoyed.

The regular workload had kept Verbiest very busy. On top of that, cannon-making was another important job. From time to time he was also asked to participate in irrigation projects and to solve various kinds of mechanical problems[18]. In short, from the time he was summoned to Beijing in 1660 until his death in 1688, Verbiest was continuously busy with all kinds of special assignments given him by the emperor. It is

---

14 Taibei Furen daxue shenxueyuan tushuguan, Nan Huairen, *Kangxi yongnian lifa*, 1678, pp. 4-5.

15 Huang Bolu, *op. cit.*, pp. 83, 86.

16 *Qingshi*, Taibei, 1961, *juan* 273, *xin* 3944.

17 Nan Huairen, *Xichao ding'an*, in Wu Xiangxiang, *Tianzhujiao dongchuan wenxian sanbian* (Taibei, 1966), p. 10.

18 Spence, *op. cit.*, p. 28; Nan Huairen, *Xichao ding'an*, in Wu Xiangxiang, *op. cit.*, pp. 137-141, 203-222; Nan Huairen, *Xichao ding'an*, 3, in Wu Xiangxiang, *op. cit.*, p. 11.

worth mentioning here that all these jobs involved science and technology.

III.
His Relationship with the Church

When dealing with Sino-western relationships, it is easy for us to assume the homogeneity of the West, but this often leads to misunderstanding. Upon close examination, in fact, many differences, contradictions and conflicts exist within the many aspects of western civilization. Take Christianity, for example, where the differences among the Churches are well-known. Even in the more homogeneous Roman Catholic Church there were many competing religious and hierarchical organizations. As a member of a very influential religious order, Verbiest was under tremendous pressure from many directions. To appreciate better the role he played in the Catholic Church of China, we have to know more about his relationship with his confreres and other Church organizations.

Founded by the devoted, effective and charismatic St. Ignatius of Loyola in 1540, the Society of Jesus played an important role in the Counter-Reformation. One of the features of this Society has been its emphasis on obedience and discipline. The problems faced by the Jesuits in China were complicated by different cultural backgrounds, difficult and slow Sino-western communication, and, of course, the Society of Jesus itself. The Jesuits in seventeenth-century China came from several different European countries, and had different intellectual backgrounds. When they approached China, the Jesuits faced quite a few alternatives and had to make some decisions which would become crucial in the Catholic history of China.

Ricci pioneered in many ways. It was he who approached the upper echelon of Chinese society, but was not without opposition even in his own Society. Working for the Manchu dynasty, and taking up a position in this government, was another problem for Schall. On top of this, he was a director of the Bureau of Astronomy, and to many of his confreres and missionaries, this only served to encourage superstitious activities among the Chinese. Verbiest also faced this kind of criticism. It is against this background that we can better appreciate Verbiest's insistence on not assuming any formal official post. Although he was in charge of the Bureau of Astronomy, Verbiest had never referred to himself as the director (*jianzheng*) of the *qintianjian*. I think Verbiest deliberately avoided using the formal title.

Another evidence of this was that the Manchu government, on formal occasions referred to him as "(the person) in charge of calendar-making" (*zhili lifa*) rather than as the "director." It is significant that in the portrait of Verbiest, he appears dressed in mandarin attire, but without any bird-like mark indicating official rank[19]. To be exact, it is not very accurate to refer to Verbiest as director of the Bureau of Astronomy. Even so, he was not free of criticism similar to that suffered by Schall. It was not until 1682, when the new leader of the Jesuit Order, Charles de Noyelle, exonerated him, that he was able to work in the Manchu court without any pressures from within the Order[20].

Perhaps because of Verbiest's influence in the Manchu government, and perhaps, also, because of his other abilities, he was appointed as vice-provincial from 1676 to 1680, an additional responsibility. He had to consider the welfare of the Jesuit community in China as his first priority. He also had to arbitrate all the differences among his confreres. In addition, he had to deal with the charges and criticism of other religious orders and Church organizations, resulting, mainly, from different theological viewpoints.

The Dominican and Franciscan opposition to the Jesuit approach to Chinese society is well-known. Although they sometimes enjoyed the privileges gained by their Jesuit counterparts in the Manchu court, and sought help from influential Jesuits in the court whenever they encountered troubles, they vehemently attacked the Jesuits' China policies in the Vatican. Another Church organization which shared many opinions on China with the Dominicans and Franciscans, and which had conflicts of interests with the Jesuits, was the Congregation for the Propagation of the Faith (*Chuanxin bu*). This organization was founded by Pope Gregory XV in 1622 with a view to integrating all Catholic movements around the world. In the late seventeenth century the Congregation sent a delegate to visit China and to inspect missionary activities. As the vice-provincial of the Jesuits in China, Verbiest now stood face to face with hostile and demanding people from Rome.

Insisting on the use of a Latin ritual, the Congregation was in conflict with Verbiest, who had strongly advocated the immediate formation of a native clergy for Chinese Catholicity to establish roots by conducting sacraments in Chinese instead of Latin. The predicament was aggravated by the demands from delegates from the Vatican and the Congregation, that all the Jesuits in China should take oaths of

---

19 Spence, *op. cit.*, p. 27.

20 Pfister, *op. cit.*, p. 343.

obedience to the pope. Verbiest was afraid that if the Manchu government heard this, it might close the missions. The next two passages summed up well the situation of the time:

> Verbiest, astronomer, mathematician and engineer, spent forty-four years in China, and earned the amity and admiration of the Chinese intellectuals. Verbiest had the misfortune to be the Jesuit Superior in China when François Pallu and Charles Maigrot arrived there in 1684, to give him instructions on how to convert the Chinese. His patience with the ignorant zeal of the Propaganda missionaries is very striking. But patience, said Balthasar Gracian (1601-1658), is the virtue of the Belgians[21].

The Portuguese government also challenged the missionary activities of the Jesuits in China. The Treaty of Tordesillas in 1494 established a meridian line 370 miles east of the Azores, through the arbitration of Pope Alexander VI, in order to solve the dispute between Spain and Portugal. It was decreed that all lands east of this line belonged to Portugal. In 1533 Pope Clement VII set up the diocese of Goa and declared that Portugal was the protector of the Catholic Church from the Cape of Good Hope to Japan. All the European missionaries going to the East had to have the permission of the king of Portugal in advance, and take Lisbon as their point of departure. However, in 1540 after the Society of Jesus was recognized by the pope, he entrusted the development of Catholicism in the East to the Jesuits. It is not surprising that in the long run the Jesuits would be in conflict with the interests of the Portuguese government[22].

To be sure, in the early seventeenth century the Society got a lot of help from the Portuguese before they entered China. The Jesuits often stayed in Macao and studied the Chinese language. For many of them, Macao was the port of entry to, as well as departure from China. Macao was attacked by the Dutch when Schall had just arrived there. He bravely joined the Portuguese army and helped defeat the invading Dutch soldiers[23]. However, conflict was inevitable owing to their very

---

21 See Malcom Hay, *Failure in the Far East. Why and How the Breach between the Western World and China First Began* (Philadelphia, 1957), p. 154.

22 Zhang Fengzhen, *Fuyin liuchuan Zhongguo shilüe*, *juan* 2, *shangpian* (Taibei, 1971), pp. 515-521; Luo Guang, *Li Madou zhuan* (Taibei, 1983), p. 28; Jiang Wenhan, *Ming Qing jian zai Huade tianzhujiao Yesu huishi* (Shanghai, 1987), pp. 4-8.

23 Väth, *op. cit.*, pp. 86-94.

different approaches to China, as well as Portuguese nationalism. The Portuguese approach did not respect traditional Chinese culture. All the Chinese converts had to take a Portuguese name[24]. The Portuguese members of the Jesuit Order played a difficult and sometimes suspicious role. As vice-provincial, Verbiest had to deal with all the attendant problems.

Furthermore, in the late seventeenth century, when Portugal was rapidly declining, Verbiest, when considering possible future confreres, suggested that French Jesuits be sent to China. In July 1687, the newly dispatched French Jesuits landed in Zhejiang province, bypassing Macao. Tomas Pereira (1645-1708), Verbiest's major assistant and a Portuguese confrere, strongly opposed this suggestion, and insisted that the 1494 Treaty be observed. Verbiest was very careful, and prudently avoided direct conflict with the Portuguese government[25].

All the aforementioned problems concerning the propagation of the Catholic Faith in China were mainly on the side of the Catholic Church. Verbiest had yet to confront the major problem of the other side - the government and emperor of China - which to many Jesuits was crucial for the success of the Chinese Catholic mission.

IV.
The Kangxi Emperor, the Manchu Dynasty and Verbiest

One positive element in the Ming-Qing dynastic change was the peaceful transition of astronomical research into the new era, and the continuing respect Schall had from the new ruling class. When the Manchus took over Beijing, many residents were forced to leave their homeland. With all the astronomical instruments Schall was working on, he was unable to move, and thus appealed to the most powerful man of the time - Dorgon. The latter immediately sensed the significance of Schall's work, and offered him help and protection. In this way, the Riccian approach could be continued.

On his deathbed Ricci dreamed that the Chinese emperor was baptized[26]. This indicated what was on his mind before he died. Ricci was not even able to see the emperor in person. Nevertheless, he did make some influential friends in the court, and finally, through their good office, his successors could get official positions to help compile

24 Jiang Wenhan, *op. cit.*, p. 10.

25 Witek, *op. cit.*, p. 157.

26 Luo Guang, *op. cit.*, pp. 224-225.

the calendar. With this legacy, Schall not only survived the Manchu takeover, but also had the opportunity to come very close to the power center.

Perhaps because of his work, and perhaps even more because of his moral example, Schall won the affection of the Shunzhi emperor. As is well-known, the emperor gave him many privileges and honors and treated him as a grandfather. It seems that under the circumstances, the conversion of the emperor was just a matter of time. But the emperor's affection proved to be short-lived. In his last years, his interests and affection switched to Buddhism and a Buddhist monk Mu Chenmin[27]. At the death of the emperor, Schall suffered persecution in the Yang Guangxian case, and Chinese Catholicism was on the brink of extinction. It is here that we can see the significance of Verbiest's contribution in the Catholic history of China.

In addition to the aforementioned description of the tasks the Kangxi emperor gave him to do, Verbiest's relationship with the emperor can better be seen in the detailed letters describing two journeys on which he escorted him to the north and northeast border. In these two letters it is evident that Verbiest was regarded by the emperor as an advisor of science and technology and nothing else. On one occasion, several Koreans caught a live sea-calf, which the emperor had never seen before. Verbiest was immediately summoned and asked to identify it[28]. Another incident took place during an informal open-air party at which several aristocrats and high officials were present:

> The night being very clear, the emperor would have me recite the names of all the chief stars which he embraced in his view of the hemisphere, as well in the Chinese as in the European language, and he himself displayed all the knowledge which he had acquired on the subject. After he had produced a small chart of the stars, which I had some years before prepared for him, he went on to tell by the stars the hour of the night, and took pleasure in thus displaying his science in the presence and hearing of the great men about him[29].

From the preceding passage, we can see the extent of the emperor's

---

[27] Chen Yuan'an, "Tang Ruowang yu Mu Chenmin", in *Chen Yuan'an shixue lunzhu xuan* (Taibei, 1982), p. 449.

[28] See "Translation of Father Ferdinand Verbiest's first Letter from the Dutch," in Pierre Joseph D'Orleans, S.J., *History of the Two Tartar Conquerors of China*, originally published in 1688 (London, The Hakluyt Society n. 17), p. 111.

[29] *Ibid.*, p. 116.

astronomical knowledge. It is indeed unprecedented in Chinese history. For the emperor, the ideal person to learn from was Verbiest.

To be sure, Verbiest did take some opportunities to introduce Catholicism to the emperor. At one time, Verbiest co-authored with another confrere a book entitled *Yulan xifang yao ji* (*An Outline of the Western World for the Imperial Reading*). In this small book, he commented on religion in the West:

> In the western countries, there is only one Catholic religion (*tianzhujiao*) and no other religion ... Since these countries embraced Catholicism, they have enjoyed great peace and prosperity for 1600 years. People are friendly and peaceful and social customs are good. Everyone has enough to live and happily does his own business[30].

It is understandable that in the seventeenth century the Catholic missionaries would do something like this to strengthen the Catholic cause in China. On several other occasions, Verbiest lectured to the emperor on the absurdities of superstitions and the importance of salvation. The emperor's response to these admonitions was only "I heard," or "I know this"[31], which showed no sign of excitement.

Three years before Verbiest's death, the emperor began to consider a successor to this resourceful advisor, and issued the following edict:

> Now Nan Huairen is old. I heard that there are some young men in Macao who know calendar-compiling and are as competent as Nan Huairen. You (the grand secretaries), together with the Board of Rites, should ask Nan Huairen about their names, and submit a memorial. Moreover, if there are any others who are specialized in medicine, be sure to include it also in the memorial[32].

According to Louis Le Comte, Verbiest expressed his last wish shortly before he died as being, that in all that he had been doing, he had had no other intention than "to bring it about that the greatest Prince of the Orient would be a protector of the most holy religion in the universe"[33]. This wish was consistent with the Jesuit effort since the late Ming period. However, there were no such words in Verbiest's last formal memorial. It was vaguely mentioned that "the basic motivation for my coming from afar, which Your Majesty is already familiar with, I dare

30 Nan Huairen, Li Leisi, *Yulan xifang yao ji*, 1669, p. 12.

31 Pfister, *op. cit.*, pp. 345-346.

32 Nan Huairen, *Xichao ding'an*, 3, in Wu Xiangxiang, *op. cit.*, p. 6.

33 Pfister, *op. cit.*, p. 380.

not repeat again and again"[34]. The difference here, I think, was the result of Verbiest's predicament. On the one hand, it was only natural for him to try to convert the emperor. But on the other hand, the authoritarian character of the Manchu despot kept Verbiest from preaching too much, so as not to displease him. It is possible that before he died, Verbiest expressed his last wish one way or the other. However, he could not dare to put it into his last formal memorial, which usually carried more weight than any other kind of expression.

In short, the relationship between the emperor and Verbiest was very different from that between the Shunzhi emperor and Schall. The Shunzhi emperor did not know anything about western sciences which were Schall's specialty and indispensable for government business. The benevolence shown to Schall was only the emperor's emotionalism (*ganxing*), whereas the Kangxi emperor built up his confidence in Verbiest through painstakingly laborious study of astrology and mathematics. His attitude toward Verbiest was very rational (*lixing*), calculated, and consistent. We could not expect that the Kangxi emperor would show dramatic affection towards Verbiest. But on the other hand, Verbiest would not be likely to suffer a sudden fall from the emperor's grace either. Throughout the emperor's life, his employment and trust of Verbiest remained unfaded. There continued to be religious persecutions after the time of Verbiest, but cases like Yang Guangxian would never happen again. It seems safe to say that Verbiest secured official posts in the Manchu court which survived in times of severe persecution until the nineteenth century.

V.
A Review of Verbiest's Religious Writings in Chinese

There are six books on Catholicism by Verbiest: *Jiaoyao xulun* (An Introduction to Fundamental Catholic Doctrine); *Shengti dayi* (An Answer to Skeptics about the Eucharist); *Daoxue jiazhuan* (Hagiologies); *Gaojie yuanyi* (The Original Meaning of Confession); *Shan e bao lüe shuo* (An Introduction to Retribution); *Wang tui jixiong zhi bian* (To Correct the Errors in Fortune-telling). All except the last are in the Jesuit library in Taibei. It is very interesting to know what these writings reveal about Verbiest's religious ideas.

---

34 Nan Huairen, *Xichao ding'an*, 3, in Wu Xiangxiang, *op. cit.*, p. 15. However, the last sentence did not appear in the same memorial recorded in Huang Bolu, *Zhengjiao fengbao*, p. 88.

The books available now in Taibei are not original editions. They were either copied or reprinted in the 1840s or later, when Catholicism was again legalized in China. This indicates that Verbiest's writings were still useful two hundred years after they were published. Moreover *Jiaoyao xulun* was so popular that it has 1677, 1799, 1848, 1867, 1875, 1893, 1905 and 1914 editions. It was also translated into local dialects, French, Manchu and Korean[35]. No wonder someone said: "This book is the best catechism of the time ... Fr. Verbiest's writings are eloquent and easy to understand. It is most suitable for a young missionary to study"[36].

The contents of the books warrant several observations. They show a deep understanding of the Chinese political system. In explaining arcane religious doctrines unfamiliar to the Chinese, Verbiest would always employ simple examples from Chinese politics. For instance, monotheism is a basic tenet in Christianity, but it was not very easy for Chinese of the time to accept. Verbiest used the emperor of China as an allegory to propagate the monotheistic idea:

> We should know that there is only one Lord, not two, who is the master of everything. Just like a lord in a family; if there were two lords in a family, it would be difficult to manage the family. In a nation, there is only one supreme ruler. If there were two, the nation would be in chaos. In the universe, if there were two lords, it would also be chaotic[37].

He also used examples such as tax-evasion and the civil-service examinations to make Catholicism easier for Chinese to embrace[38]. On the other hand, in the books Verbiest vehemently attacked Buddhism and Taoism, especially the *xingtian* god, the belief in the transmigration of the soul and retributions (*lunhui baoying shuo*), and correspondence from the uppermost (*taishang ganying pian*). Again, Verbiest used Chinese allegories to rebuke them[39].

It is evident that the audience Verbiest addressed was the *literati*

---

35 Fang Hao, *op. cit.*

36 Joseph Jennes,CICM, English trans. Albert Van Lierde, Chinese trans. Tian Yongzheng, *Tianzhujiao Zhongguo jiao li chuanshou shi* (Four Centuries of Catechetics in China. Historical Evolution of Apologetics and Catechetics in the Catholic Mission of China from the Sixteenth Century until 1940) (Taibei, 1976), pp. 85-86.

37 Nan Huairen, *Jiaoyao xulun*, pp. 4-5.

38 *Idem*, *Shan e bao lüe shuo*, pp. 2, 4, 6; *Gaojie yuanyi*, pp. 1, 3.

39 *Ibid.*, passim.

class in China. The examples were easy to comprehend for anyone who knew a little about Chinese politics. The gentry-officials would find the books even more suited because the examples were part of their daily life. In one book Verbiest even publicly addressed "all the princes, Beile, nine ministers and censors" (*zhuwang beile dachen jiuqing kedao*)[40]. Although one finds Verbiest familiar with Chinese politics, one cannot but notice that he seldom seeks the conversion of the Chinese upper class.

However, Verbiest, unlike Ricci, did not or could not frequently refer to Confucian classics to support his arguments. This was perhaps due to Verbiest's aptitude, and more likely, perhaps, to his continuous engagement in government business. His numerous assignments from the Kangxi emperor, left him little time to study the Chinese Classics had he even wished to do so. In dealing with Confucianism, Ricci embraced ancient Confucianism and rejected Neo-Confucianism, which he regarded as contaminated and corrupted by Buddhism and Taoism. Verbiest simply wrote from the perspective of common sense.

A last point to be made about Verbiest's writings is that according to the late Maurus Fang Hao, Verbiest's calligraphy was so good that it became a collectible item. However, it was all lost by the Mid-Qing[41]. This is questionable though, because the missionaries always had assistants in the court to help them with their Chinese[42]. As a person in charge of the *qintianjian* and respected by the Kangxi emperor, it is quite possible that Verbiest had competent assistants to beautify his Chinese. It is quite doubtful that Verbiest could write model calligraphy.

VI.
Verbiest's Contribution in Chinese Catholic History

In order to do justice to Verbiest, we have to evaluate his legacy against the background of the time. From the perspectives of the Manchu dynasty and the long traditional Chinese political culture, the conversion to Christianity of the Chinese people in general, and the emperors in particular, was very unlikely. The Manchu dynasty was, moreover, an alien dynasty. To establish its legitimacy, it had to use and manipulate

---

40 *Ibid.*, p. 1.

41 Fang Hao, "Nan Huairen zhi Hanzi shufa yu Hanwen chidu", *Fang Hao liu-shi ziding gao* (Taibei, 1969), p. 1589.

42 Jennes, *op. cit.*, p. 33.

all means to avoid antagonizing the Chinese. This is why the Manchu government had to be sinicized to the extent that the Chinese would identify with it. The cultural policy of the Qing was to reconfirm the orthodoxy of Zhu Xi's Neo-Confucianism and initiate several large projects on Chinese culture. Furthermore, imperial China had a long tradition of hostility toward non-Confucian ideas and thought[43]. Throughout the history of China since the Han, many emperors claimed to follow the way of the ancient sages, such as Yao, Shun, the Duke of Zhou and Confucius. As a foreign conqueror of China, the Manchu court saw that it was in its best interest to use Confucianism as the ideology of the empire. It is unimaginable to think that the Manchu dynasty would have adopted Roman Catholicism as the national religion, because this would have caused extreme negative responses from the Chinese people, for whom the difference between ideology and religion was not very easy to tell.

With this understanding in mind, we should not be surprised to know that, despite the affection shown to Schall by the Shunzhi emperor, the latter had this to say in the "Inscriptions on the Catholic Church by the Emperor" (*yuzhi Tianzhutang bei*):

> I made an inspection tour to the South Garden, accidentally stopping by the Church. I saw that the appearance of their God resembled the fellow-countryman (of the missionary), and the decorations inside the Church were those of a foreign country. I asked about the books on the table. They said: "They are the books on Catholicism." What I have followed is the way of emperors Yao and Shun, the Duke of Zhou and Confucius; what I have tried to observe is the principle of being essential, integrated and complying with the Mean (*jingyi zhi-zhong zhi li*). I did some casual reading in the Buddhist sutras and Taoist classics, but could not understand their meaning. In the case of western books and the Catholic religion, I have never read them, so how can I know their arguments?[44]

In spite of the emperor's trust in Verbiest, he expressed in his later years his suspicion of the western countries: "The overseas western countries, a hundred or a thousand years from now, will cause the Middle Kingdom a lot of troubles. This is my prediction"[45].

---

43 P. A. Cohen. *China and Christianity. The Missionary Movement and Chinese Anti-Foreignism, 1860-1870* (Cambridge, 1963), pp. 4-20.

44 Huang Bolu, *op. cit.*, p. 31.

45 *Qing shengzuren zong shilu* (Taibei, 1964), Kangxi 55th year, 10th month *renzi*, *juan* 270, p. 16.

Given all this unfavorable background[46], what was Verbiest's most important contribution to the Catholic history of China? I would argue that, in addition to the catechism he wrote in Chinese, his most important contribution was to secure a stronghold for Catholics in China for later development. It is true that there were many underground missionary activities, especially in Sichuan province. Catholics were not eliminated during the Qing with or without the Church in Beijing. But if the governor-general of Sichuan had been a persecutor of the Catholics, such as happened in Fujian in the 1700s, it would have hardly been possible for the Chinese Catholics to survive much longer without the help of the Beijing Church. In short, Verbiest consolidated the presence of the Church in the capital of China, which at least partly contributed to the survival of Chinese Catholics[47].

It is true that the Catholic Church in China was not officially legalized until 1692, when Verbiest had been dead for four years. Nevertheless, no one would deny that it was Verbiest who laid the solid foundation for this achievement. It was very unlikely that the emperor would have issued the "Edict of Toleration" had Verbiest not worked so hard to win his confidence. In fact, in late spring of 1687, shortly before his death, there was, through Verbiest's effort, a presage of the legalization. The emperor formally authorized the request by Verbiest that the clauses in the prohibition against heretical religions in some areas, equating the Catholics with White Lotus followers, be stricken[48]. One point worth noting here is that Verbiest presented this request together with a memorial about the cannons[49]. Here we could see how Verbiest maximized his influence in the Manchu court.

The duration of the legalization of the Chinese Catholics was very short (from 1692 to 1707). What was more lasting was the official position Verbiest earned for the later missionaries. Although Verbiest did not assume the post of director in *qintianjian* officially, some of his

46 In addition to the aforementioned, there are some other incompatibilities between Confucianism and Christianity in language, logic and mode of thinking. However, the "incompatibility" between a religion and a culture is disputable.

47 I revised my argument here on Verbiest's contribution to the Catholics in China thanks to the opinions of Profs. John Wills and Edward Malatesta. Verbiest could be regarded as crucial for the survival of the "High Church," meaning visible Church institutions, the grand imposing Church buildings, etc.

48 Nan Huairen, *Xichao ding'an*, 3, in Wu Xiangxiang, *op. cit.*, p. 6.

49 *Ibid.*

successors did assume the office. In the biography of Verbiest in the *Qingshi gao* (A Draft History of the Qing Dynasty), there is a passage indicating Verbiest's legacy:

> (Since the death of Verbiest) the Bureau of Astronomy employed westerners. Some distinguished themselves as the director or vice-director. One succeeded another without interruption ... During the Daoguang reign (1821-1850), westerners such as Gao Gongchen[50] returned home or died of illness. At that time, the Chinese officials in the Bureau were already familiar with the new method. There was no need to employ westerners any more. Emperor Daoguang issued the edict stopping the appointment of westerners to the Bureau. When Shengzu (the Kangxi emperor) employed Nan Huairen, he allowed him to practise Catholicism and follow the customs of his own country; the Catholic religion, however, was still banned in local areas. At that time more than thirty churches existed in all the provinces. During the Yongzheng reign (1723-1735), the ban was strictly enforced, and all the churches were destroyed except the one in the capital for the westerners in the Bureau to live in. All who entered the countryside to preach Catholicism were persecuted. Shortly after the westerners left the Bureau, the ban on sea-going was lifted. The propagation of the Catholic faith was legalized in the treaties, and new churches, together with the old ones, were everywhere[51].

This passage shows that before 1837 the missionaries possessed a legal status in the Manchu court one way or the other, thanks to Verbiest's efforts. Also because the missionaries were legally staying in Beijing, the presence of the Catholic Church was necessitated. Despite the severe persecutions of the Yongzheng and Qianlong emperors, a visible Catholic Church was kept open in the capital of China. This is very significant. Kenneth Scott Latourette made the following statement: "Had missionaries after 1835 gradually ceased coming to China, instead of increasing in numbers, the Church would probably have passed out of existence within a few generations, leaving no permanent work behind"[52].

---

50 This is a mistake. Gao Gongchen should be Li Gongchen (Joseph Nunes Ribeiro). The last missionary to serve in the *qintianjian* was a Vincentian by the name of Bi Xueyuan (Gaetan Pires-Pereira). See Octave Ferreux, C.M., Wu Zongwen, transl., *Qianshihui zai Hua chuanjiao shi* (Taibei, 1977), pp. 144-147.

51 *Qingshi gao*, *juan* 272, p. 10025.

52 K. S. Latourette, *A History of Christian Missions in China* (London, 1929), p. 196.

The Nestorianism of the Tang and the Christianity of the Yuan once prospered but later disappeared. It is beyond the scope of this essay to discuss their demise. Nevertheless, we could imagine that if the emperors of China had allowed them to keep practising their faith in the capital, they might have survived and even developed further when the time was ripe. With this in mind, we can perhaps better appreciate the contribution of Verbiest to the Catholics in China. Some may argue that Schall was more crucial because it was he who assumed the directorship of *qintianjian* in the first place. However, as discussed earlier, Schall's position was quite fragile, and his relationship with the Shunzhi emperor was precarious. It was Verbiest who won the confidence of the Kangxi emperor because of his superior knowledge of science and technology, and thus secured an unchallenged foothold for his successors until 1837[53]. Five years later, Roman Catholicism was legalized in the treaty systems. Since then, the Catholic Church in China has had tribulations, but it is there to stay.

---

53 Pires-Pereira took a sick-leave and died in late 1838. See Ferreux, *op. cit.*, p. 147.

THE INFLUENCE OF FERDINAND VERBIEST ON THE POLICY OF THE KANGXI EMPEROR TOWARDS CHRISTIANITY

LIN JINSHUI

Lin Jinshui currently teaches Church history and Christianity in China. He has been doing research work on the history of Chinese-Belgian relations at the invitation of the Ferdinand Verbiest Foundation (1986-1988). He was a visiting scholar of the Fairbank Center for East Asian Research, Harvard University (1990-1991). He published several articles on the history of East-West cultural exchange.

Most Jesuits in China were both missionaries and scientific experts. Today, when we study their achievements, the most often described aspect of their lives is their scientific work. Ferdinand Verbiest (Nan Huairen) is a typical example of this. As a famous missionary, his biography was included in the main encyclopedias of the world, together with his predecessors, Matteo Ricci (Li Madou) and Johann Adam Schall von Bell (Tang Ruowang). In these very limited biographies, Verbiest emerges not as a missionary, but as an astronomer. Even the *New Catholic Encyclopedia* characterized him as an astronomer. This corresponds with the views of Chinese historians, who consider the study of the Jesuits an interesting subject. This is because they agree on their importance as scientists and their role in Chinese astronomy, mathematics, geography, mechanics, and so on. It is necessary to survey and sum up the historical role and lessons of early missionaries during the Ming and Qing dynasties for a better understanding of today's Open Door Policy in China. After all, Verbiest came to China as a missionary and during his twenty-eight years of activity in China, he had an important impact on the policy of the Kangxi emperor towards Christianity. The situation of Christianity in the Kangxi reign, both contemporary lay and ecclesiastical historians, invariably consider as the golden age of the Jesuits in China[1]. This was the most flourishing period for propagating the holy religion, and everyone agrees that this situation was due to Verbiest's efforts. Philippe Couplet has said that Verbiest was very appreciated by the emperor while he was in charge of astronomy and the calendar. The outstanding Christian female leader (Madame Xu) thought that Christianity's flourishing or decline depended on him[2]. Lei Sitian (Louis Gabriel Delaplace) said: "The persecution was gradually weakened, and holy churches of various provinces slowly returned to activity, because Verbiest, and other fathers, were favorites of the Kangxi emperor, after his taking over the reins of government upon coming of age"[3]. The famous French sinologist Jacques Gernet also said it was Verbiest who strengthened the position of the missionaries in the empire[4]. Chinese scholars thought that the safety and danger of the mission in China were in the hands of Verbiest. He was a firm rock in midstream, stemming a raging tide going backward[5]. These conclusions show clearly the results of Verbiest's contribution to the China mission.

But how did Verbiest achieve this influence over the emperor's policy towards Christianity? There are few articles which deal with this subject. This essay tries to survey the situation, and is based on both western and Chinese material.

Verbiest's first name was transliterated as Fudinantesi and his surname as Fubisituo in the Chinese official records *Qingshi gao*, the draft history of the Qing dynasty[6]. His Chinese name was Nan Huairen. He

---

1 "Verbiest, Ferdinand", *New Catholic Encyclopedia*, 14, 609; Guo Tingyi, *Jindai Zhongguo shi gang* (An Outline of Modern Chinese History) (Hong Kong, 1986), p. 28.

2 Bai Yingli (Philippe Couplet), *Xu tai furen zhuan lüe* (Brief Biography of Madame Xu), translated from *Historia nobilis feminae Candidae Hiu Christianae Sinensis*, 1688, by Xu Caibai (Shanghai, 1882), p. 12.

3 Lei Sitian (Louis Gabriel Delaplace), *Shengjiao tongkao* (General Studies on the History of Christianity), Jingdu jiushi tang, 1874, 2:95a.

4 Jacques Gernet, *A History of Chinese Civilization* (Cambridge, 1982), p. 517.

5 Lü Shiqiang, *Nan Huairen de keji gongxian yu xuandao fengfan* (Ferdinand Verbiest's Technological Development and His Ideal Method of Evangelization), *International Conference in Honor of Ferdinand Verbiest. Commemoration of the 300th Anniversary of his Death (1688-1988)* (Taibei, 1987), p. 75.

6 *Nan Huairen zhuan* (Biography of Ferdinand Verbiest), *Qingshi gao* (Draft History of Qing Dynasty), Vol. 272 (Beijing, 1977), pp. 10024-10025.

was born on October 23, 1623, in Pittem in West Flanders, near Bruges. His father Josse Verbiest was a bailiff of Pittem and Coolscamp[7]. The bailiff was the representative of the noblemen, and was a well-known person in Flemish villages. He was the local magistrate who administered the law, governed the cities and regulated labor and economic life[8]. Although his official rank was not high, the position was rather important.

Verbiest's mother was Ann Van Hecke. There were six brothers and sisters in his family. Verbiest was the fourth child. He studied at Jesuit schools in Kortrijk and Bruges, and then went to Louvain to study philosophy at the College of the Lily. He admired the Society of Jesus very much, and entered the order after the examination in Mechlin in 1641[9]. He taught Greek and Latin in Brussels in 1648. Two years later he was sent to Rome and later to Seville to study theology.

In the beginning of the seventeenth century, Belgium was ruled by the Spanish. Belgium, in the center of Europe, was the crossroads of many European countries, so that many characteristics of European culture appeared there. Although Belgian culture is related to French and German culture, it is neither French nor German, but has its own national character. Some scholars regard Belgian culture as Flemish culture[10]. Bruges, where Verbiest once studied, was one of the centers of Flemish culture. Louvain, a university city, was the academic and educational center of the Low Countries and the home of many famous scientists and social scientists. It was in Louvain that Verbiest got his knowledge of astronomy and mathematics.

Living in such a good educational and cultural environment laid the foundation for Verbiest's later becoming famous in the history of communication between China and Europe. His abilities were already blossoming before he left for China. When he studied in Rome and Seville, he was regarded as "a brilliant intellectual, enjoying the confidence of his superiors." He could have looked forward to a distinguished life in the highest circles of Catholic Europe[11].

---

7 C. Carton, *Notice biographique sur le Père Ferdinand Verbiest, missionnaire à la Chine* (Bruges, 1839), p. 64.

8 Marc Schreider, *Belgium* (Macdonald, 1944), p.39.

9 Carton, *op. cit.*, p. 64.

10 Emile Cammaerts, *Belgium - From the Roman Invasion to the Present Day* (London, 1921), p. 21.

11 Jonathan Spence, *To Change China: Western Advisers in China 1620-1960* (Boston, 1969), p. 23.

At first in 1647 Verbiest had the approval of religious superiors to go to Mexico, but the Spanish authorities did not allow him to embark. Upon completing his theological studies in 1655 in Seville, he hoped to reach the West Indies but again the Spanish government rejected his request. Since two Belgian confreres, Couplet (1623-1693) and François de Rougemont (1624-1676) were already in Lisbon waiting for passage to India and China, Verbiest applied to go to China. The Jesuit General, Goswin Nickel, wrote: "I approve of your excellent disposition. The desires of Your Reverence will, I hope, be granted. You will happily enter into China for the good of a great number of souls"[12].

Verbiest arrived in China in 1659 (the sixteenth year of Shunzhi). At first he went to Shaanxi from Macao to spread Christianity in Xi'an. Soon, through Schall's recommendation, he was summoned to the capital by the Shunzhi emperor[13]. He left Xi'an on May 9, and arrived in Beijing on June 9[14].

The China mission was in great danger at the time he arrived in Beijing. Gabriel de Magalhaes, the superior of the residence in Beijing, was imprisoned, having been accused of bribery. It was feared that other fathers would also be arrested[15]. Facing this situation, Schall, who had once been a very well-known priest, felt quite helpless. This *tong wei jiaoshi* (the religious teacher who comprehends the mysterious) was declining from the peak of favor of the Shunzhi emperor. Furthermore, after the latter's death, the Kangxi emperor ascended the throne as a young ruler, but power was vested in four regents, namely Soni, Suksaha, Ebilun and Oboi. The greatest threat to the China mission was the rise of the anti-Christian campaign. Its representative was Yang Guangxian, who attached himself to Oboi and invented false stories, accusing many people[16]. From 1660, he wrote the *Pixie lun* (Essays on Refuting Heresy), "to rebuke Catholicism, which ranted and seduced the

---

12 H. Josson and L. Willaert (ed.), *Correspondance de Ferdinand Verbiest* (Brussels, 1938), p. 5 as translated in Yves Raguin, "Ferdinand Verbiest. One of the Giants of the Apostolate in China", *International Conference in Honor of Ferdinand Verbiest*, p. 33.

13 Xu Risheng (Tomas Pereira) and An Duo (Antoine Thomas), *Nan xiansheng xingshu* (The Life of Ferdinand Verbiest), Bibliothèque Nationale, Paris (BNP), mss. chinois 3033, pp. 1-2.

14 *Correspondance*, p. 38.

15 *Correspondance*, pp. 103-104.

16 Wang Xianqian, *Donghua lu* (Records of Donghua), 1899, "Kangxi jiu", p. 35a.

people"[17]. He also claimed that "heretical books and words are contrary to reason and truth", and "a decent person must not convert to heterodoxy, for one who converts to heresy must not be a decent person"[18]. On September 15, 1664 (KH 3:7:26) he sent a memorial to the emperor that he should indicate "that the heretics dispersed their partisans both throughout the capital and in the province", and "that they spread heretical books to mislead people under heaven."

On September 24 (KH 3:8:5), a secret edict ordered the Board of Rites to join the Board of Civil Office to try Schall. This was acted on the following day[19]. Later, Verbiest and the others were also condemned because of their connection with Schall. On April 15, 1665 (KH 4:3:1), the Board of Rites and the Board of Civil Office pronounced the death sentence for Schall, and flogging with a stick for the others. The next day, while the subjects were going to their punishments, there was an earthquake in the capital at *wuke* (11 a.m. to 1 p.m.)[20]. "It happened five times and continued for several days. Meanwhile, a change in the position of the stars, *xingbian*, occurred twice"[21]. Warned by these signals, the officials begged to redress all mishandled cases. Verbiest and the other fathers were dismissed, and tentatively allowed to stay in the capital. Missionaries who were arrested in various provinces were banished to Guangzhou. Schall was released by an edict of the queen mother[22].

On April 29 (KH 4:3:15), Yang Guangxian compiled *Zhai miu (shi) lun* (Quotation of Fallacies) to indicate ten errors in Schall's calendar, as well as *Xuanze yi* (Commentary on Selection) to point out that Schall had wrongly applied the *Hong Fan wuxing* (the Five Elements of Hong Fan) for the selection of burial dates for Prince Rong Qin[23]. These were sent to the emperor as memorials. In this way, he caused the "calendar conflict" with the Jesuits. This conflict started with Yang striking first,

---

17 *Yang Guangxian zhuan* (Biography of Yang Guangxian), *Qingshi gao*, Vol. 272, pp. 10022-10024.

18 Yang Guangxian, *Budeyi* (Feeling Constrained), pp. 31-33.

19 *Ibid.*, pp. 4-7.

20 *Donghua lu*, "Kangxi wu", p. 16b.

21 Li Leisi (Ludovicus Buglio), *Budeyi bian* (Refuting "Feeling Constrained"), reprinted in 1847, preface, p. 1.

22 Huang Bolu, *Zhengjiao fengbao* (Public Praise of the True Religion) (Shanghai, 1894), p. 44.

23 *Donghua lu*, "Kangxi wu", p. 16b.

but ended with Verbiest's victory by means of a surprise move. This, finally, averted the threatening collapse of the China mission, and the first recognition of Catholicism by the imperial government was won.

The "calendar conflict" had, in fact, already begun in the third year of Kangxi (1664) with the estimation of the exact time of the eclipse of the sun. Although Verbiest's calculation proved to be correct, it could not change the prejudice against the "heretical" viewpoint of the Jesuits. However, their triumph only commuted the punishment. They were allowed to stay on together under house arrest in Beijing[24]. This was, so to speak, the first round of the "calendar conflict."

Yang Guangxian's memorial to the emperor accusing Schall of ten errors was actually the second round. The prince and ministers in the state council investigated his charges one by one. Yang Guangxian and Schall each defended the accuracy of their statements. However, because the method of calendar-making was very complicated, nobody could distinguish who was right and who was wrong[25]. Only because the four regents were in power and tended to protect Yang[26], did the princes and ministers finally declare Schall guilty. They dismissed him and the other officials of the Board of Astronomy and punished them. The new western method was stopped, and the old method of *Da tong* was again applied[27].

Yang temporarily got the upper hand and was promoted from vice-director to director of the Board of Astronomy. But after the Kangxi emperor took control from the regents on August 25, 1667, he gradually discovered errors in the old method of *Da tong*, which did not always correspond with the movements of the celestial bodies. So, on December 26, 1668 (KH 7:11:23) he issued an edict to Yang Guangxian, Verbiest, and the others that "Astronomy is very complicated, and the calendar relates to important affairs of state. You should not have your long-cherished hatred influence your own ideas, which you consider to be right and, consequently, the opponents', wrong, and so continue to dispute with each other. To follow what is right and to remedy what is wrong: this requires honesty of you"[28].

---

24 Spence, *To Change China*, p. 24.

25 *Donghua lu*, "Kangxi wu", p. 16b.

26 *Yang Guangxian zhuan*, pp. 10022-10024.

27 *Huangchao wenxian tongkao* (Imperial Encyclopedia of the Qing Dynasty), 256 "Xiang wei kao", p. 1b.

28 *Xichao ding'an* (Memorials by Ferdinand Verbiest and others), ARSI, *Jap.*-

On the same day he also instructed the princes and ministers to calculate the length of the shadow that would be thrown by a given object at a given time of day, together with Bu E, the president of the Board of Rites and Hao Weina. The result of this test proved the accuracy of Verbiest's calculation for three days in succession (i.e. the 24th, 25th and 26th day). Verbiest was victorious on this important occasion. After hearing this news, the emperor again ordered Verbiest to check the calendar of *qizheng* (the seven rulers of time) as well as *minli* (the calendar used by the people), and point out errors. Upon checking it, Verbiest found that it was very different from his own calendar. Although he did not like to discuss the faults of others[29], he did jump at the chance to use his knowledge of astronomy to fight the accusations of Yang and Wu while being tested by the emperor. On January 27, 1669 (KH 7:12:26), Verbiest accused the vice-director of the Board of Astronomy, Wu Mingxuan, of mistakes: that the intercalary twelfth month of the eighth year of Kangxi should actually be the first month of the ninth year of Kangxi, and that Wu had included two vernal equinoxes (*chunfen*), and two autumnal equinoxes (*qiufen*) plus various other errors, in the calendar of *qizheng* and *minli* of the eighth year of Kangxi[30]. On February 26, 1669 (KH 8:1:26), the Grand Secretary and other officials were appointed to test the predictions of *lichun* (the beginning of spring, first solar term) and *yushui* (rain water, second solar term), to test the phases of *Tai Yin* (the moon), *Huoxing* (Mars), and *Muxing* (Jupiter). The detailed data of their test agreed with Verbiest's calculations, whereas the statements of Wu Mingxuan disagreed. "So we (the princes and ministers) order Verbiest to calculate all dates of the calendar for the ninth year of Kangxi"[31]. Because the ministers had to retract their previous deliberations, the emperor was a bit angry. He questioned them by edict: "Earlier Yang Guangxian accused Schall, and the princes and ministers determined on what points Yang Guangxian was right, and on what points Schall was wrong. Formerly you decided to abolish the western method. Now you decide to restore it. What reasons do you have?[32] From this edict, we can also see that the emperor was very conscientious in the study of science, and

---

Sin., II, 67.

29 *Ibid.*

30 *Donghua lu*, "Kangxi ba" and "Kangxi jiu", pp. 30b-31a.

31 *Ibid.*

32 *Yang Guangxian zhuan*, pp. 10022-10024.

very prudent in policy making.

On March 8, 1669 (KH 8:2:7), the princes and ministers of the state council sent twenty officials to the observatory to test them again according to the above edict. The result once again proved the accuracy of Verbiest's calculations and the errors in the statements of Wu Mingxuan[33]. In addition, Verbiest's method for calculating one day with ninety-six quarters (*jiushiliu ke*) also agreed with astronomical facts. So the calendar according to the method of *jiushiliu ke* was practised from the ninth year of Kangxi.

So far facts had proved the correctness of the western method. In April 1669 (KH 8:3), it was again formally applied by the Qing government[34]. From then on, Verbiest gained the trust and the favor of the Kangxi emperor. On April 17 (KH 8:3:17), he was appointed vice-director of the Board of Astronomy[35], to administer the calendar. Yang Guangxian was almost sentenced to death, but the emperor sympathized with his old age and spared his life. He was banished to his home town but died on the way[36]. Originally Wu Mingxuan was dismissed and heavily punished. But again, due to the emperor's sympathy, he was finally allowed to retain his position. The emperor, however, warned him that in the future he should remedy his previous faults, collaborate earnestly and sincerely with Verbiest, and obey the truth in setting the calendar. If he should continue to confuse right and wrong again, while knowing very well someone else's talent, but envying him, he would be severely punished[37]. This inevitably raised Verbiest's position, while also serving as a warning for other officials who were jealous of the missionaries.

The significance of the victory in the "calendar conflict" with the missionaries exceeded that of the calendar itself. Taking advantage of his achievement, Verbiest began to practise his missionary function, making his first step to influence the emperor's policy towards Christianity. This was to reinstate Schall's honor and build up the reputation of Christianity. He expounded views of Catholicism which originally the emperor had not considered heretical. Verbiest wrote him a memorial concerning Catholicism:

---

33 *Donghua lu*, "Kangxi ba" and "Kangxi jiu", pp. 30b-31a.

34 *Huangchao wenxian tongkao*, 256, "Xiang wei kao", p. 2a.

35 *Donghua lu*, "Kangxi ba" and "Kangxi jiu", pp. 30b-31a.

36 *Yang Guangxian zhuan*, pp. 10022-10024.

37 *Zhengjiao fengbao*, p. 50.

> It is just like what the *Shijing* (The Book of Poetry) described: Great is God, beholding this lower world in Majesty! He is the master of everything. In the Wanli reign of the former Ming dynasty, Catholic missionaries compiled books to expound their religious ideas. To adore God and love people was regarded as Catholicism's chief aim. Nothing is important except a few things, such as, cultivating one's human nature, being loyal, filial, honest and chaste. Because of these upright principles, many famous scholars and officials came to revere it. Emperor Shi Zuzhang (i.e. the Shunzhi emperor) himself visited our church several times, granted silver to build it and bestowed a stone inscription as well as a horizontally-inscribed board entitled *tong wei jiajing*, and conferred on Schall the honorable title of *tong wei jiaoshi*. If Catholicism is heretical, why did the former emperor not forbid it?[38]

By imperial order, the princes and ministers deliberated on Verbiest's memorial again and again. Finally, they not only restored Schall's title, but also affirmed for the first time that Catholicism was not heretical. They said: "The villain Yang Guangxian made up a story to accuse Catholicism of being heretical. Now, it has been denied by our deliberations. We think that those who believe in Catholicism have, so far, demonstrated no trace of evil deeds. We therefore allow them to follow their Catholic practices as before." Similarly the *Donghua lu* described this: "Their worship of Catholicism follows the custom of their own country. It demonstrates no trace of evil deeds"[39].

Unlike his princes and ministers, the emperor was not as tolerant at that time. His edict stated: "Only Verbiest and the others (the Europeans in the capital) may practise Catholicism as before. However, we fear that in Zhili province, and the other provinces, they may again set up churches and convert our people to their religion. We therefore order that this practice be strictly prohibited. We approve the other points as you have deliberated"[40]. *Huangchao wenxian tongkao* recorded this as: "They are allowed to practise their religion, but it is prohibited to set up churches and establish their religion in Zhili and the other provinces"[41]. *Qingshi gao* stated: "While Sheng Zu (the emperor's temple name) was employing Verbiest, he allowed him to practise Catholicism, according

---

38 *Ibid.*, p. 55.

39 *Ibid.*, p. 58; *Donghua lu*, "Kangxi jiu", p. 35a.

40 *Ibid.*

41 *Huangchao wenxian tongkao*, Vol. 292, "Yidaliya zhuan" (Account of Italy), p. 45a.

to the custom in his country. However, he prohibited the setting up of churches to convert people in various provinces"[42]. This is a well-known edict that forbade Christianity in the eighth year of Kangxi, despite the different statements made in the above-mentioned official documents. As to its basic content, this 1669 edict did not allow the Chinese people to believe in Catholicism, nor any missionary other than Verbiest, to propagate Christianity. This edict, so to speak, was the basic policy of the Qing government before the 1680s. But the statement of the emperor - "concerning the other points, we approve as you have deliberated" - shows that this edict differed from the policy of forbidding Christianity, because it was not based on the opinion that Catholicism was a heresy. On account of this one point, his policy of forbidding Christianity was not decidedly strict from the beginning. It demonstrated much more flexibility and adaptability for some things and for certain persons. For example, this edict allowed Verbiest "to practise Catholicism as before", so that Verbiest enjoyed much freedom to spread Christianity. More apparently, it was demonstrated in the emperor's attitude towards twenty-five missionaries, including Antonio a Santa Maria Caballero (Li Andang) and the others who had been banished to Guangzhou. The princes and ministers deliberated about sending them to the capital, but the emperor did not agree with this. The edict declared: "Twenty-five persons, including Maria Caballero and the others, should not be summoned back to the capital"[43]. Due to the request of Verbiest, however, the emperor finally not only agreed with the Board of Rites' deliberation, which let Christian Herdtrich (En Lige) and Claudio Filippo Grimaldi (Min Mingwo), who were familiar with the calendar, come to the capital, but the other nineteen persons, including Jean Valat (Wang Ruwang), who were not familiar with the calendar, could return to their original churches; he also allowed Francisco Varo (Wan Jiguo) to return to Fujian, together with Antonio de Gouvea (He Dahua). This was not agreed to by the Board of Rites. Because Varo did not belong to those who had suffered an unjust accusation by Yang Guangxian, he was arrested in Funing. There were no reasons for him to ask to redress a mishandled case. But at that time the Kangxi emperor was more lenient than the Board of Rites; his edict stated: "Since Antonio de Gouvea had prepared to bring Francisco Varo

42 *Nan Huairen zhuan*, pp. 10024-10025.

43 *Donghua lu*, "Kangxi jiu", p. 35a.

to Fujian to settle, we approve this"[44].

To allow the missionaries to return to their own local churches meant tacitly approving their freedom to spread Christianity. In other words, since missionaries returned to their own churches, who could keep them from doing otherwise[45]? Their only purpose was to go to various places and spread the gospel, but, it was impossible to do so in public. In addition, the Qing government would not punish those Chinese who converted to Catholicism, because the princes and ministers had said that it was wrong that, due to their conversion to Catholicism and writing prefaces for Catholic books, Xu Zuanzeng and the others were dismissed[46]. It therefore also provided an opportunity for missionaries. The emperor was well informed about this situation, "The emperor knows what is going on, and is not ignorant that in Beijing alone you annually convert two thousand Christians. But he pretends to know nothing of it, and it would not be safe to attempt to make him change his line of conduct"[47]. Due to the emperor's tacit approval and his favor towards the missionaries, officials of various provinces were able to take a *laissez-faire* attitude, and let the missionaries spread the gospel freely, and convert the Chinese people. "They turned a blind eye to it all"[48]. This treatment, which was strictly prohibited in principle, may be regarded as a tacit approval of the unwritten law of the Qing government towards Christianity during the early Kangxi reign, or may be regarded as "an uncertain policy"[49], as explained by Zhang Weihua. Both in principle and in fact, it demonstrated a trace of Verbiest's influence over the Kangxi emperor and the princes and ministers.

The calendar conflict not only changed the fate of the China mission, but also stimulated the emperor's desire and interest to study western science.

Sheng Zu claimed that while the calendar conflict was still going on, he did not yet study western science. So he could not affirm which

---

44 *Xichao ding'an*, pp. 143-145.

45 Zhang Weihua, *Ming Qing zhi ji Zhongxi guanxi jian shi* (A Concise History of the Relations between China and the West) (Jinan, 1987), p. 136.

46 *Xichao ding'an*, p. 115.

47 P.- J. D'Orleans, *History of the Two Tartar Conquerors of China* (London, 1854), p. 95.

48 Fang Hao, *Fang Hao liushi ziding gao* (The Collected Works of Maurus Fang Hao Revised and Edited by the Author on His Sixtieth Birthday), 3 vols. (Taibei, 1969), I:147.

49 Zhang Weihua, p. 136.

points were right or wrong. Therefore, he put all his energy into studying. Finally, he mastered the knowledge of astronomy and the calendar and understood them thoroughly[50].

It was from the Jesuits that the Kanxgi emperor learned western science which provided an opportunity for the Jesuits to spread the gospel to him. Strategically, the target of the Jesuits in China was to conquer the emperor in spirit and make him convert to Catholicism. They knew very well that "The emperor's attitude towards the problem of religion was decisive"; following his example, lower civil and military officials would also favor Christianity"[51]. In his last memorial to the emperor, Verbiest described his task clearly:

> Your Majesty,
> I will die happily, because I have spent almost my whole life in Your Majesty's service. But I humbly beg Your Majesty to remember after my death, that I had no other goal in all I did other than to provide a protector, you being the greatest King of the Orient, for the holiest Religion on earth[52].

Indeed, it was necessary to import knowledge of the Christian faith to the emperor if missionaries wanted to convert him. There were several ways and means for Verbiest to do this, such as:

1. Taking advantage of the opportunity to teach western science

As the emperor's teacher, Verbiest instructed him during two years in the methods of using the main equipment for astronomy and mathematics. He also taught geometry, as well as the most interesting and elementary subjects of astronomy. During his lessons, Verbiest talked with the emperor about all kinds of topics, and imbued him with an elementary knowledge of Catholicism. Due to his curiosity about western science, the emperor's interest was stimulated to study Catholicism[53]. Verbiest was confident that the emperor, being

---

50 *Nan Huairen zhuan*, pp. 10024-10025.

51 Bai Jin (Joachim Bouvet), *Kangxi di zhuan* (Biography of the Emperor Kangxi), translated from *Histoire de l'Empereur de la Chine* by Ma Xuxiang, *Qingshi ziliao* (Sources of Qing History), n° I (Beijing, 1980), pp. 244-245.

52 *Correspondance*, p. 551.

53 Bai Jin, *Kangxi di zhuan* (Biography of the Emperor Kangxi), translated by Feng Zuomin, included in *Qing Kang Qian liangdi yu Tianzhujiao chuanjiao shi* (The Kangxi and Qianlong Emperors and the History of Evangelization) (Taibei, 1966), pp. 94, 99.

impressed by astronomy and delighted by mechanics, would swing to the faith behind the science. He said: "Thanks to Divine Providence, we hope that this (imperial) favor will grow, and will finally bring our joy to a peak. But we must still be patient, awaiting the suitable moment, and moving little by little along the way which will lead to the difficult and important goal"[54]. He made a very vivid suggestion: "A star, a long time ago, guided the Three Kings to adore the True God. In the same way, the science of the stars might also be a guide for the princes of the Orient to worship their Lord"[55].

2. Indirectly propagating Christianity by using science

In 1671 (KH 10), the Board of Works wanted to set up a memorial archway of stone for Xiaoling (The tomb of the Shunzhi emperor). It needed six pillars and other large blocks of granite. Some of them weighed more than a hundred thousand *jin* (about one-third of a ton). It was feared that it would damage the structure of the Marco Polo bridge. But by imperial edict, Verbiest transported them by using western pulleys. Nothing at all was damaged, and the cost of strengthening the bridge was saved[56]. When the emperor heard the good news, he was just returning from a hunt during which he had killed two stags. In gratitude, he presented one to Verbiest and the other to the Chinese officials in charge. Obviously, what had been done was only an application of practical physics. Verbiest, however, exaggerated it enormously, and mystified it, as if its success was to be attributed to faith in Christianity. To quote Verbiest's own words: "When all was ready, the signal was given to start, and under the patronage of Saint Gregory the Thaumaturge, of Saint Symphonorosia the Martyr, and of the seven sons, the stone was set in motion by using pulleys. Then, with bands playing and flags waving, the crossing was made"[57].

From 1674 to 1676, Verbiest manufactured one hundred and twenty cannons. After finishing each cannon, he clothed himself very lavishly in surplice and stole, sprinkled the cannon with holy water and offered prayers for the success of the imperial arms, while the dignitaries of the

---

[54] Spence, *To Change China*, p. 28.

[55] Henri Bosmans, "Ferdinand Verbiest, directeur de l'Observatoire de Pékin", *Revue des Questions Scientifiques*, 71 (1912), p. 386.

[56] *Xichao ding'an*, 1982, pp. 139-140.

[57] Arnold H. Rowbotham, *Missionary and Mandarin - The Jesuits at the Court of China* (Berkeley, 1942), p. 99.

empire looked on. On each cannon he engraved the name of a Christian saint, according to the fashion of the time in Europe[58]. This role of the Jesuits as cannon makers was bitterly criticized by other Catholic orders. Pope Innocent XI, however, supported it wholeheartedly. The Pope knew about this when the news was disseminated in Europe. He highly praised it and considered that it was done in an appropriate way. He issued an edict encouraging and rewarding Verbiest[59]. In his letter to Verbiest, on December 3, 1681, he wrote: "The most delightful thing of all was to read in the same letter how wisely and opportunely you have applied the use of profane sciences to the salvation of the Chinese people and to the increase and utility of the Christian faith"[60]. Using profane sciences to propagate Christianity was a practical target of the Jesuits in China from Matteo Ricci on. "Because of the practices of more than a century, people understood that to propagate Christianity and introduce it in China, the dissemination of science was one of the chief ways which they had to follow", Joachim Bouvet stated[61].

3. Opportunity of contacting the emperor while traveling

Accompanying the emperor, Verbiest traveled in Tartary twice. The first time was to eastern Tartary in 1682. He explained:

> During their long expedition, I got the opportunity to spread the Gospel among the nobles and high officials. I let them understand our task, as well as the purpose and character of the European spirit. Not limited to the circle of noble and high officials - and in order to dispel their loneliness - the imperial people also asked me about the knowledge of the sky, stars, meteorological phenomena, as well as sailing. When I answered them, I seized the chance to introduce our faith and the creed of Catholicism in addition[62].

More specifically focusing on the emperor, Verbiest commented,

> Looking at me riding one of the emperor's horses, people find that the emperor is discussing the faith of Christianity with me as if he is

---

58 *Zhengjiao fengbao*, p. 79.

59 *Ibid*.

60 *Correspondance*, p. 369.

61 Bai Jin, *Kangxi di zhuan*, Ma Xuxiang's ed., p.249.

62 Nan Huairen, *Dadan lüxing ji* (Account of a Journey of the Emperor into Eastern Tartary), translated by Xue Hong from the Japanese version, *Qing-dai xiren jianwen lu* (Westerners' Accounts of the Qing Dynasty) (Beijing, 1985), pp. 80-81.

> giving a lecture on the rostrum, and it seems as if I am giving a speech to an assembly[63].

The second trip was to western Tartary in the summer of 1683. Verbiest discussed the Catholic faith with the emperor. The fruit of this journey was once again to the advantage of Christianity[64]. Here it is necessary to point out that Verbiest was very prudent in spreading the gospel in the presence of the emperor. He knew well that if he went a bit too far, the consequences would be too ghastly to contemplate. He claimed: "The will of the emperor was in all cases a limit to us. To oppose it, or to make the slightest demonstration of doing so, would be at once to imperil our whole mission"[65].

In order to impress upon the emperor how advantageous Catholicism would be for the Qing dynasty, Verbiest exaggerated its social role in European history:

> European countries have had peace and security for one thousand six hundred years, since they converted to Christianity. People live together peacefully, folk customs are respected. Each family is supplied, each person has enough. None of them contend or scramble. Everyone is very happy with his work[66].

In addition, the subject which Verbiest most often discussed with the emperor was the worship of God, which was the key to accepting the Christian faith.

> God is the great master who created heaven and earth as well as all things, the great father-mother of all people in the world. There is a master of heaven and earth, just as there is a sovereign for each country[67].

The Kangxi emperor, however, might not have understood what Verbiest and the other fathers told him.

> ... His majesty asked the clergymen a heathen question about the name, the worship, and the power of God, and when these

63 *Ibid.*

64 D'Orleans, *History*, pp. 130-131.

65 *Ibid.*

66 Nan Huairen and others, *Yulan xifang yao ji* (A Description of the Most Important Kingdoms in Europe for the Emperor's Reading), ARSI, *Jap.-Sin. II*, 158, pp. 6b-13a.

67 *Ibid.*

> were answered this prince concluded with very derisive and slanderous words[68].

The emperor also asked Verbiest why God had not forgiven his son without making him die, but though Verbiest tried to answer, he did not understand him. The story of the flood described in the Bible was also hard for the emperor to understand. He said: "Though in China there had been a flood at about the same time as Noah, those on the plains drowned, while those who escaped to the mountains were saved"[69].

In order to make the Chinese people understand and accept the faith easily, Verbiest wrote several Chinese books on Catholicism. Among them, the most important is *Jiaoyao xulun* (Summaries of the Christian Religion), written in 1669. "This book is not too long, but its main principles are rather complete. People read about the origin of Catholicism and easily understand, after they read it"[70]. Therefore this book was printed again and again. It has been translated into the Shanghai dialect, Manchu, Korean and French. I have not found direct sources indicating whether or not the emperor read this book, but I am sure it is quite possible that he did. According to Bouvet's account,

> Missionaries in China have written many books on Catholicism. I think His Majesty has read most of them. In particular, he paid most attention to Matteo Ricci's book entitled *The True Meaning of God*. He always had it in his hand and read it for a period of six months[71].

Due to this, and the close relation between Verbiest and the emperor, my supposition is not difficult to accept. If so, Verbiest had much influence on the emperor concerning the doctrine of Christianity.

Although the emperor was not converted to Catholicism after Verbiest's attempts to explain the Gospel, his attitude towards Christianity did change. This is reflected in the favor and respect he showed Christianity. Indeed, as a protector of Confucianism, it was impossible for the emperor to adopt the Christian faith. But in order to show his kindness to the foreigners as a ruler of the Orient, he might favor and respect Christianity by his attitude. This explains, perhaps,

---

68 John E. Wills, Jr., "Some Dutch Sources on the Jesuit China Mission, 1662-1687", *Archivum Historicum Societatis Jesu*, 54 (1985), 89.

69 Jonathan Spence, *Emperor of China, Self-Portrait of K'ang-hsi* (New York, 1974), p. 84.

70 Nan Huairen, *Jiaoyao chuyan* (Shanghai dialect version of Verbiest's *Jiaoyao xulun*) (Shanghai, 1903), preface, p. 1a.

71 Bai Jin, *Kangxi di zhuan*, Feng Zuomin ed., p. 99.

many of his seemingly contradictory deeds and words.

In the first place, the emperor did not believe in God, but he revered Him. In 1671, the emperor went to the Jesuit residence to extend his regards to the fathers. He presented them a horizontally-inscribed board bearing two characters, *jing tian* (Revere Heaven), written by himself. The fathers hung it in the center of the church. The emperor explained that he wrote "Revere Heaven", because he revered the Lord[72]. All the missionaries in China then placed this board in front of their churches, as a tacit approval given by the emperor to the true religion[73]. In the beginning of the winter of 1686, two Russian ambassadors begged the emperor to approve their going to the church to pray. The emperor not only agreed with this, but also ordered the senior vice-president of the Board of Rites Sun Guo, the junior vice-president of the *lifan yuan* (a board in charge of affairs relating to Mongols), La Bake, and the senior secretary of the Board of Rites, Ba Hai, to prostrate themselves before the picture of Christ. On February 4, 1689 (KH 28:1:15), when the emperor visited Jinan, Shandong province, the eunuchs Zhao and Wu brought twenty *liang* of silver to the church. First, they prostrated themselves before the picture of Christ, and then announced the imperial edict: "Wansui Ye (His Majesty) ordered us to prostrate ourselves before the picture"[74]. In 1714, the emperor made a written instruction on Joseph Suares's (Su Lin) memorial: "For each word of 'God' in this memorial the characters are not placed at the head of the following line. I order you to correct this"[75]. This was also a recognition of God as the highest.

The emperor did not necessarily understand the creed of Christianity, but he revealed his favorable impression by praising it. Bouvet said: "The emperor really highly commended our doctrine in the extreme. People have heard that he admires Christianity and its progress in China more than ever." The emperor not only revealed his favorable impression of Christianity at each opportunity, but also imbued the

---

72 *Qinming chuanjiao yue shu* (Collection of the Imperial Edicts for Evangelization), Xujiahui, p. 5.

73 D'Orleans, *History*, p. 94.

74 *Xichao ding'an*, BNP., mss. chinois 1331, pp. 10, 29.

75 *Zhengjiao fengbao*, p. 127. In Chinese official documents, all names and terms related to the emperor and the empire, such as "his majesty," "holy dynasty," "holy rulers," have always been put at the head of the following line, wherever they appeared, in order to show the highest and the most dignified position of the emperors.

ministers, the nobles, and particularly the crown prince, with respect for it. He especially tried to influence the crown prince with his favorable attitude towards the missionaries. One day, he showed the crown prince all the elegant equipment of astronomy and mathematics, and guided him on a visit of the Beijing Observatory. He said to the crown prince: "We have to thank Father Verbiest. Because of him, we have such elegant equipment." Then, he told how Verbiest and his colleagues gave their assistance to the previous emperor[76].

The emperor strictly prohibited Chinese people from converting to Christianity, but he was very pleased when subjects of a feudal ruler converted to Catholicism. According to one report, on March 10, 1688, "the Muscovite Prelate lets it be known that he wishes to become Roman Catholic with his people; this he does so that he may be held in the same honor and respect by the emperor as are the Jesuit Fathers." When this was made known to the emperor four days later, he "showed his satisfaction"[77].

Although the Kangxi emperor ruled China according to Confucian rites, he not only approved missionaries following their own religious rites and customs, but most of the time showed them every consideration. When Magalhaes died on May 6, 1677 (KH 16:4:5), the emperor granted two hundred *liang* of silver and ten *pi* of satin, to show his mourning for this subject from far away. After Magalhaes was put in his coffin, he ordered three eunuchs (including Gong Sa) to go to the Jesuit residence to submit his edict. They asked Verbiest and Ludovico Buglio (Tian Leisi): "When is the burial of Magalhaes? What rites will you practise according to your Catholic custom?" According to Catholic funeral rites, Verbiest wrote a program, which the eunuchs submitted to the emperor. After reading it, he was very pleased. He asked if the grant of silver was enough to cover the expenses of such a funeral. On the day of the burial, the emperor ordered the same eunuchs: "Go to the cemetery and carefully observe all the rites practised by the Catholics"[78]. In their report to the emperor, the eunuchs carefully described all their acts, such as the kneeling position of the praying Catholics. The emperor also knew well the habits of the Jesuits of keeping fast days, and would often ask whether the day was a fast day or not. If

---

76 Bai Jin, *Kangxi di zhuan*, Ma Xuxiang ed., pp. 243, 247, 249.

77 Wills, *op.cit.*, p. 289.

78 *Zhengjiao fengbao*, pp. 14-15.

so, he would send them dishes in compliance with their wishes[79].

The emperor expressed more trust and praise about the Jesuit deeds of charity. Xu Zuanzeng, the son of Madame Xu, indicated that "There were hospitals for unwanted children in the capital and the provinces at that time"[80]. In his biography of Madame Xu, Couplet adds, "They adopted many unwanted children, looked for wet nurses, and a charity cemetery was set up"[81]. In the province of Yunnan, from 1675 to 1696, "they adopted 5,480 unwanted children found on the road"[82].

On April 7,

> the Fathers likewise gave the said Prince a list of the foundlings dead and buried from 1686 until April 1687, consisting of 2,321 children from both inside and outside the city who were buried, 373 who were over half eaten by the dogs, and 191 who are still alive. But of these, few are raised to maturity, since most of them die young from poison given them by their own parents. This list pleased the emperor wondrously well, and he ordered the Fathers to carry on with this charitable work[83].

In 1704, Shandong suffered famine, and innumerable victims were thus begging in the capital. The emperor said to the missionaries who lived in the three churches of Beijing:

> You came from remote countries. It is your original duty to love people. Now I give you this government money and order you to go to Shandong to provide relief to the victims. May they genuinely enjoy the imperial kindness[84].

The emperor's trust in, and friendship for, the missionaries, including Verbiest, is well known in the history of communication between East and West. His favors towards Verbiest exceeded all others. Among all the missionaries of China, this very seldom happened. The official positions at the Qing court to which Verbiest was appointed by the emperor were in order of priority: vice-director of the Imperial Board of Astronomy (*qintianjian jianfu*) in charge of the calendar, in 1669;

---

79 D'Orleans, *History*, pp. 129-130.

80 Xu Zuanzeng (son of Madame *Xu*), *Yuyingtang quan shan wen* (Descriptions of Advising People on Philanthropy for Foundling Hospitals), quoted in Fang Hao, *Zhongguo tianzhujiaoshi renwu zhuan* (Biographies of Catholics in Chinese History) (Hong Kong, 1970), 2:76.

81 Bo Yingli, p. 12.

82 Xu Zuanzeng, quoted in Fang Hao, *Zhongguo tianzhujiaoshi*, 2:76

83 Wills, *op. cit.*, p. 289.

84 *Qinming chuanjiao yue shu*, p. 9.

president of the Supreme Court of the Imperial Sacrifices (*taichang shiqing*) in 1674; commissioner of the Office of Transmission (*tongzheng shisi tongzheng shi*) in 1678; senior vice-president of the Board of Works (*gongbu youshilang*) in 1682. The highest grade of this official rank was the second grade[85]. Verbiest was the only missionary who was granted the posthumous title of *qinmin*. This was one of the few cases in Qing history, which broke *shi fa* (the rules of conferring a posthumous title), drawn up by the Qing government.

As a rule, if a posthumous title is to be conferred upon officials of the first rank or above, it has to be determined by imperial decree. For the second rank or below, there are no posthumous titles. In case one is conferred after all, it has to be by special imperial decree. For hard work in office, for dying in battle, for literary achievement or for military exploits - in all these cases exceptions can be made[86].

In the history of the Qing dynasty, in 1752, Han Tan was granted the posthumous title of *wenyi* because of his contribution to *zhiyi* (stereotyped writing); and in 1765, Wang Shizhen was granted a posthumous title as *wenjian* because of his poetic contributions. Liang Zhangju thus considered that both of them may be regarded as having received an outstanding honor, if one investigates history from ancient times until now[87]. Here, I have to add that Verbiest, likewise, may be

---

85 Verbiest never was a director of the Imperial Board of Astronomy. The position of the director was still open before Verbiest died. Then the Board of Rites was supposed to promote Hu Zhenyue as a director. But the emperor considered that since it was Verbiest who administered the calendar, it was not necessary to appoint a new director (see *Xichao ding'an*, BNP, mss. chinois 1331, p. 20). However, why did some Chinese official documents regard Verbiest as a director? It may be a mistake. After Verbiest finished the manufacturing of the celestial globe, the emperor ordered the Board of Civil Office to deliberate preferentially Verbiest's promotion. So the Board gave a proposal to promote Verbiest as a director of the Imperial Board of Astronomy, in addition to the title of vice-president of the Supreme Court of the Imperial Sacrifices (see *Xichao ding'an*, BNP, mss. chinois 1330, pp. 51-52). But the imperial edict said that it was very worthy to praise Verbiest for working hard in making astronomical instruments. It ordered that Verbiest should still administer the calendar, in addition to having the title of president of the said Supreme Court. Obviously, the emperor did not appoint Verbiest as a director of the Board. So, the mistake was caused by the fact that some people have taken the proposal of the Board of Civil Office for a real fact. Then it circulated and was added to erroneous reports.

86 Liang Zhangju, *Langji congtan* (Accounts of Wandering) (Fuzhou, 1983), 4:45.

87 *Ibid.*

regarded as such. Besides, there were also few in the history of the Qing dynasty who, like Verbiest, administered the calendar at the Board of Astronomy while being a senior vice-president of the Board of Works. Wang Shizhen's *Chibei outan* (Accounts at the House by the North Side of the Pond) said:

> It could hardly happen that the officials of the Imperial Board of Astronomy added a position of vice-president of other boards in this dynasty during the Kangxi reign. Verbiest administered the calendar at the Imperial Board of Astronomy, added a position of vice-president of the Board of Works, and became vice-president of the Board of Rites after his death[88].

The relations between the emperor and Verbiest exceeded normal external relations between an emperor and remote foreigners as well as the feudal order for the sovereign and his subjects. The special treatment of the Kangxi emperor towards Verbiest could not be expected to be accepted by many high officials and nobles. He regarded Verbiest as his amiable and trusted friend.

For traveling, the other officials had to arrange a lot of money, but Verbiest did not have to worry at all, since all his costs were paid by the emperor. He took more than ten horses which had been ridden by the emperor, and stayed in the imperial uncle's tent. When they met trouble on the journey, the emperor always took care of Verbiest. For example, one day, in order to cross a river before dark, he ordered Verbiest to take a boat with him, the crown prince, and the other people of the imperial lineage. The same thing happened the following day. One of the imperial uncles wanted to cross the river early. So he looked for a reason and said: "Verbiest lives in my tent and eats with me at the same table. So I would like to cross over the river." The emperor answered: "Granddad, don't hurry up to cross the river, Verbiest may stay in my tent and eat with me"[89]. In addition, he frequently sent Verbiest and the other fathers dishes from his own table to their tent. He sometimes even ordered them to be entertained in his own tent. Concerning the benevolence and attention of the emperor, Verbiest said: "He displayed it to no one else, not even to the princes or his own kinsmen, and this both in word and deed in the presence of the whole army"[90].

---

88 Quoted in *Fang Hao liushi ziding gao*, 2:1590.

89 Nan Huairen, *Dadan lüxing ji*, pp. 71-80.

90 D'Orleans, *History*, p. 129.

In 1681, Verbiest received an edict to make eight bronze cannons, each one weighing a thousand *jin*. They were not finished in 1685. The emperor noticed that he did not look well. He worried about him deeply and gave him two jars of pills, ordering him to get better. When Verbiest was again ill in 1687, the emperor repeatedly sent him to be looked after by the imperial doctor. As a subject coming from far away, "Verbiest enjoyed such extreme graces from the emperor. He was honored during his lifetime, as well as mourned after his death. From ancient times till now, in all treatment for remote subjects, there were no such boundless hospitalities," commented Tomas Pereira and Antoine Thomas[91].

Towards the other missionaries, the emperor was likewise favorable. Some missionaries served at the Beijing court by recommendation of Verbiest, such as Grimaldi, Herdtrich, Pereira, Thomas, and so on. "The emperor ordered the westerners Grimaldi, Pereira and Thomas to administer the calendar, and appointed them as *jia la zhang*"[92]. After the death of Verbiest, Grimaldi replaced him. Herdtrich originally served at the court and then went to Jiangzhou, in Shanxi province. After his death, the emperor granted a personally inscribed horizontal board with the words *hai yu zhi xiu* (An excellent man from a remote place)[93]. Pereira received even more favors from the emperor. The greatest honor was that the emperor presented him with an ivory fan before the throne, on which there were pictures of clocks, trees, buildings and a poem written by the emperor himself: "Cycling day and night, the clock is better than the clepsydra; it indicates the correct time sentimentally and tactfully. Their royal mind never changes, no matter if it is sunny or cloudy"[94].

Thomas was the second Belgian missionary whom the Kangxi emperor favored. When he was ill, the emperor sent a doctor, Sun Huibai, to check his pulse and the eunuch, Zhao Chang, to take care of him and asked if Thomas's sickness was the same as his sickness in traveling. He also issued an edict: "I order Wang Yuanzuo to cure Thomas's disease. Please be careful." Later the emperor again asked: "How is Thomas's disease these days? Is he better?" The eunuch

---

91 Xu Risheng and An Duo, pp. 3-4.

92 *Qinming chuanjiao yue shu*, p. 6. *Jia la zhang* may be translated as executive secretaries.

93 *Zhengjiao fengbao*, p. 81.

94 *Xichao ding'an*, ARSI, *Jap.-Sin.* 1167, 2a, unclear page.

answered: "Due to the kindness of your Majesty in sending the imperial doctor to assist in healing him, he now feels better"[95]. Because Verbiest asked for approval from the emperor, five French Jesuits - Jean de Fontaney, Jean-François Gerbillon, Joachim Bouvet, Claude Visdelou and Louis Le Comte - were summoned to the capital. "His Majesty interviewed them at the Hall of Qianqing Palace, entertained them with refreshments, granted each one fifty *liang* of silver, and ordered Bouvet and Gerbillon to stay on in the capital and serve at his court"[96]. Later, all of them received the emperor's bounties too. He issued an edict:

> Since the westerners Verbiest, Magalhaes, Pereira and Buglio have served at my court, all of them tried to do their best for public affairs. They had no faults. But most of our Chinese people did not believe this. I knew them well; they were trustworthy and honest. I have investigated their deeds carefully for many years. They did not do anything that was unlawful. Why should we point out their faults?[97]

The emperor's trust in, and favor to, the missionaries provided a tacit umbrella for them. "Since missionaries were employed respectfully, their religion could not be prohibited"[98]. Those at court could guess the intentions of the emperor so that Catholics openly practised their religion and "anti-Christians did not dare to persecute them on account of Verbiest's position at court and his being favored very much by the Kangxi emperor. There was no loophole to exploit"[99]. An anti-Christian official in the provinces learned this relevant point fairly quickly, according to Wu Dezhi.

During the Kangxi reign, the magistrate of Huanggang district hated Catholicism very much, and it pleased him to destroy the church and expel the missionaries and to punish heavily those officials who converted to Catholicism. Less than ten days later, his superior gave an order criticizing him for (meddling) too much[100].

---

95 *Xichao ding'an*, BNP, mss. chinois 1331, pp. 4, 6.

96 *Qinming chuanjiao yue shu*, p. 6.

97 *Ibid.*, p. 29.

98 Zhang Weihua, p. 136.

99 Xiao Ruose, *Tianzhujiao chuanxing Zhongguo kao* (Studies on Introducing Catholicism into China), 3rd ed. (Xianxian, 1931), pp. 313, 318.

100 Wu Dezhi, *Tianzhujiao shushi* (A Written Account of Catholicism), quoted in Liang Zhangju, p. 66.

In commenting on this event, Wu claimed that this was "the result of bribing." In fact, however, it was because the magistrate's superior "guessed the intentions of the emperor," and feared that the persecution would cause trouble. The emperor's friendship for the missionaries was not influenced by his prohibition of Christianity. In an interesting comment Verbiest observed that the emperor "hitherto treated us with such benignity and familiarity and honor, even more so than the lamas themselves"[101].

Before Verbiest's death, the emperor's policy of the prohibition of Christianity was not concealed, but it changed gradually under his influence. If we say that the imperial prohibition edict in 1669 mainly reflected the emperor's own state of mind and intentions, then we will find that this attitude and policy towards Christianity was influenced increasingly by Verbiest's ideas. In particular, it was most striking at the end of the eighties of the seventeenth century. Then the emperor's perspective of right and wrong concerning missionary affairs depended almost completely on Verbiest's viewpoint. Almost any request of Verbiest was granted by the emperor, no matter what it was.

One imperial edict for the Jesuit Simon Rodrigues (Li Shouqian) was a typical case which illustrated the gradual change in the emperor's policy. In 1679, at the recommendation of Verbiest, Rodrigues was summoned to the Beijing court to help Verbiest in administering the calendar. The following day the emperor received him at the court with refreshments. He recognized his worth, and granted him an imperial endorsement: "To spread the gospel under the imperial edict, and allow him to go to various provinces to propagate (it)"[102]. Obviously, this was against his own prohibition policy. But it demonstrates that the prohibition policy was not only flexible, but basically beginning to change. The only problem was that it did not apply to all missionaries, but only to one individual person.

In order to promote the rescinding of the prohibition policy, Verbiest did his utmost before the emperor. Although the imperial government had come to the conclusion that "Catholicism demonstrates no trace of evil-doing," it could not prevent the publication of opinions which held Catholicism as a heresy. On May 3, 1687 (KH 26:3:22), Verbiest asked for the deletion of a passage from a book printed by a mandarin, in which the law of Christ was said to be false and rebellious:

---

101 D'Orleans, *History*, p. 127.

102 *Zhengjiao fengbao*, pp. 11-18.

> When one remembers some of the good acts for the conservation of the empire, and more specifically, that he (Verbiest), in the time that several provinces rebelled, taught the imperial craftsmen the art of casting big cannons, by which the rebels were defeated and subjected, then one can draw therefrom an efficient argument to refute the calumnies of Yang Guangxian (a former accuser of the divine law) in which he said that the Europeans planned an invasion of the empire, and that they were previously expelled from Japan for the same reason.
> Furthermore, since it is clear from these facts that they worked, not for the ruin of the empire, but for its conservation and strength, he (Verbiest) asks - since the accusation is refuted by his acts of many years - that it be allowed by an imperial edict that anyone may accept and profess the divine law, and that every insult of certain accusers be destroyed - accusers who, even in printed books, contend that Christianity is false and rebellious, so that even men right before the emperor, as domestic helpers, insult them as propagators of a false and rebellious law[103].

Only a part of a Chinese document corresponding to this memorial has thus far been located. It is an outline of one undated memorial that Verbiest, as administrator of the calendar and senior vice-president of the Board of Works, presented: "I beg Your Majesty to allow Catholicism to be practised freely, and not to keep people from converting to Catholicism, in order to break off slander and nonsense"[104]. This was sent soon after the emperor prohibited all kinds of heresies. On March 28, 1687 (KH 26:2:16), the emperor said : "*Seng dao*[105] (heresy) is always contrary to the rites and laws (of the empire). In particular, heretics are so foolish as to mislead people. It is quite worthwhile to sneer if a foolish man meets necromancers and hears their made-up words, for he believes that what they say is true, and worships them as gods. So all must be prohibited strictly"[106].

"According to Jesuit manuscripts and printed sources, in 1686 the emperor had issued an edict against heterodox sects, and in some provinces this was interpreted to include Roman Catholicism"[107]. As

---

103 *Correspondance*, pp. 535-537.

104 *Xichao ding'an*, BNP, mss. chinois 1330, unclear page.

105 *Seng* - Buddhist, *Dao* - Taoist. Here, *seng dao* refers to heresy in general, not to Buddhism and Taoism in particular.

106 *Kangxi qiju zhu* (The Diary of the Kangxi emperor) (Beijing, 1984), 2:1595.

107 Wills, *op.cit.*, p. 283.

promised to Verbiest, and in order to dispel this misunderstanding, he issued an edict on May 24 (KH 26:4:14) declaring clearly that Catholicism was excluded from heresies. It said: "At present, some regional officials, in proclaiming the prohibition, have called the teaching of the Lord of Heaven (Roman Catholicism) rebellious, the same as the White Lotus teaching; these characters should be deleted." The Board of Rites correspondingly passed on the imperial document to the governors of Shandong, Henan, and others, as well as to all persons concerned[108]. As to spreading the gospel freely, the emperor still insisted on only allowing westerners to practise Catholicism as before. On May 15 (KH 26:4:5), Verbiest once again had made a petition "with tears in his eyes, about the persecution of the Christians, which now is so strongly pursued by various Lipous (officials of the Board of Ceremonies [*libu*]), that neither he nor they (the Christians) could endure it or resist any longer; in relation to which His Majesty promises this prelate satisfaction." Half a month later, on June 4 (KH 26:4:25), "the Emperor summoned the fathers to the Great Assembly and ordered them to point out which mandarins so bitterly opposed and persecuted them in their religion, whereupon Father Verbiest only gave a melancholy account, complaining in general terms, but named no one in particular ..." On June 23 (KH 26:5:14), the emperor again summoned the fathers to the court and "consoled them for the persecution that had taken place, and promised them further sustenance." The emperor sought to satisfy the request of Verbiest since just two days after Verbiest's memorial of May 15, he learned that some Muscovite parties were close to the Tartar frontiers, and the ambassador's troops were increased to 5,000. Alarmed by this news he had a discussion with Verbiest on May 19 about practices of war and siegecraft in which he received good advice. In addition, Verbiest made an extraordinary water-fountain in the court, before the emperor's throne, "in which (fountain) there is (a mechanism like) an organ, so that when it moves and one uses the keys many kinds of beasts and figures are made to appear in the water"[109]. This pleased the emperor so much that he ordered that the image of Verbiest be engraved on a copper plate and placed on it as an "eternal memorial."

From the above facts, it is not difficult to understand why the emperor might respond to Verbiest's demand. On August 3 (KH 26:6:26), he once again

108 *Xichao ding'an*, ARSI, *Jap.-Sin.* II, 67, unclear page.

109 Wills, *op. cit.*, pp. 290-292.

> had Fathers Verbiest and Pereira summoned to the Great Assembly, in the presence of all the Lipous, and read out the privilege that they shall henceforth enjoy in his Empire, consisting of, first, that Father Verbiest, even if he absents himself from the Assembly still shall continue to hold his mandarin's post, with the honor and benefits pertaining to it, until his death, [and second], that the Catholic religion may be practised publicly and without hindrance all over the Empire, at the pleasure of the fathers, and priests for this purpose under Father Verbiest's seal shall be granted free passage...

On August 5, this privilege was "published and posted by the said Great Assembly, to which it pertains, and the fathers have hung up in their church an official text of it, signed in the emperor's own hand, which gives them great esteem and respect among this nation"[110]. Without doubt, it was the most important change in the emperor's policy towards Christianity before Verbiest's death. However, he only granted the privilege of free passage by Verbiest's seal. It would no longer be valid after Verbiest's death a half year later.

Several years after Verbiest's death the emperor formally rescinded the prohibition policy. In the ninth month of the thirtieth year of Kangxi (October 21 - November 19, 1691), "a persecution took place in Hangzhou, the province of Zhejiang, in which the governor of the said province ordered the local magistrate to demolish the church, to destroy books of scripture, to denounce Christianity as heresy, and to banish the Catholics from the territory (of China)"[111]. Because of this, on February 2, 1692 (KH 30:1216), Pereira and Thomas memorialized the emperor. They said:

> Verbiest received the favor of the emperor in administering the calendar, so he said what he knew and said this without reserve. He (and the other fathers) wrote all kinds of western books, including those on the sources of the calendar, on calculation, on the origin of *lülü* (a series of standard bamboo pitch-pipes used in ancient music) and on the investigation of things (i.e. sciences), for more than twenty years at the court. They were not finished until now. Since Your Majesty has known each of them clearly, it cannot be blas-

110 *Ibid.*, p. 292. Prof. Wills questioned the validity of these materials because he has found no other source that described any substantial change in August. He suspected that Flettinger mistook a Jesuit recital of what had been proposed to the emperor for an account of what they had received (p. 283). But from what these materials describe in detail, it does not seem to be a proposal. It still remains to be identified. In anticipation, this paper uses it tentatively.

111 *Xichao ding'an*, BNP, mss. chinois 1330, unclear page.

phemy. We cannot accept that (our religion) is regarded as heresy.

On February 4, 1692 (KH 30:12:18), the emperor ordered the Board of Rites to deliberate[112]. On March 7, 1692 (KH 31:1:20), the Board passed the previous edict to the governor of Zhejiang that the prohibition had called the teaching of the Lord of Heaven rebellious, the same as the White Lotus teaching; these characters should be deleted. Besides, they stressed that the church of Hangzhou should be preserved as before. However, only the westerners were allowed to practise Christianity[113]. On March 10 (KH 31:1:23), the imperial edict noted, "I approve what you have deliberated." Apparently, this tolerance towards the missionaries was very limited and it was not much beyond the prohibition of Christianity as given in the edict of 1669. But, this limited tolerance had been hard to gain. The French Jesuit, de Fontaney, recorded this in detail. He said the emperor promised to put down this persecution quietly. He ordered the governor to recall his instruction and keep Father Prospero Intorcetta (Yin Duoze) and the other Christians safe. The emperor suggested that the Jesuits send a memorial, so that this persecution might be solved by the court, and the solution extended to the other provinces[114]. So the emperor sent a draft of a memorial to the Jesuits secretly. Meanwhile, he told Pereira and Thomas, who were in charge of astronomical affairs, to give a lecture in public. The emperor feigned ignorance on purpose, as if he knew nothing at all. When he received this memorial, along with the others, he ordered the Board of Rites to deliberate on it according to the normal procedure, and to report the result of its deliberations to him. Although he had given a hint to respect the westerners, the officials of the Board of Rites thought the emperor still hated the missionaries, because of the edicts against Catholicism he had issued in the early years of his reign. So the officials considered that the emperor might not allow Christianity to spread within the empire. Surely, two reports of the Board of Rites did not satisfy the emperor. Therefore, he found that no solutions from various Boards could suit him. He decided to approve the reconsideration of the Board of Rites, i.e. "We allow Father Intorcetta to stay in Hangzhou. The European people may still worship their God and

112 *Xichao ding'an*, ARSI, *Jap.-Sin.* 1168, unclear page.

113 *Xichao ding'an*, BNP, mss. chinois 1330, unclear page.

114 Hong Ruo (Jean de Fontaney), "Faguo Beijing chuanjiao tuan de chuangshi" (The Founding of the French Mission in Beijing), translated by Geng Sheng from *Lettres édifiantes et curieuses*, *Qingshi ziliao*, n° 6 (Beijing, 1985), pp. 163-164.

spread Catholicism at their Church, but it is strictly forbidden to convert Chinese people"[115].

The validity of the past edicts was confirmed once again. As soon as the Jesuits heard this decision, "they were shocked as if a bolt from the blue hit them; they were simply dumbfounded." The emperor felt shocked. He tried to comfort the fathers by every possible means, and suggested they send someone to various provinces, with the purpose of showing all the people about his respect for missionaries and his favor towards Catholicism[116]. However, all of these means could not reduce the Jesuits' anxiety; it seemed to grow with each passing day. The emperor summoned Prince Songgotu to the palace and asked him for ideas on how to comfort the fathers. Songgotu remembered their great contributions to the emperor again and again, and that they were still working in the imperial service, so the Qing court could only repay them by allowing them to spread the gospel in public within the empire. The emperor considered it sound and reasonable. But he anxiously asked the prince again: "If all Boards insist on their original idea not to recognize the western religion, what means shall we take to comfort them?" Songgotu answered: "Your Majesty, you should show them you are a master of one country. If you issue an edict, I shall take an uncompromising stand to negotiate with the officials. Then no one can act in defiance of your edict"[117]. With the support of Songgotu, the emperor issued an edict on March 17, 1692 (KH 31:1:30) to rescind his policy of prohibiting Christianity:

> The westerners administer our calendar-making. In previous military campaigns they have worked hard to manufacture military equipment and cannons. Furthermore, Catholicism has no trace of evil-doing. Those who attend the churches to burn incense should be allowed to do as before. Your previous deliberations are ordered to be cancelled. All of you, deliberate with the Manchu officials of the Board of Rites and the Manchu Grand Secretary, and then report to me[118].

Two days later, on March 19, the emperor issued a new edict:

> Earlier, the Board (of Rites) decided that the various Catholic churches should be preserved. However, we allow only the

115 *Ibid.*

116 *Ibid.*, p. 165.

117 *Ibid.*, p. 166.

118 *Xichao ding'an*, BNP, mss. chinois 1330, unclear page.

> westerners (not the Chinese) to practise Christianity. This has already been approved. At the present time the westerners are managing the administration of our calendar-making; in previous military campaigns they worked hard to manufacture military equipment and cannons. Lately they have also offered services by becoming part of our mission to Russia. They have committed no crime. It seems unfair if we denounce Christianity as a heresy and forbid it, since they are innocent. The grand secretariat should join with the ministers in the Board of Rites to deliberate upon this problem further, and then report to us!

In this way, because of these two imperial edicts, the Board of Rites could never dare to resist the emperor's order again. They immediately took an active response. They made more explanations and further detailed supplements for the imperial policy of not prohibiting Christianity. On the following day (March 20), a group of ministers jointly memorialized:

> The westerners admired our sacred culture when they sailed to China from ten thousand *li* away. At the present time, they administer our calendar-making. In previous military campaigns they worked hard to manufacture military equipment and cannons. When they were sent to Russia with our mission, they donated their services and completed their assignment. They have offered many services to this nation. The westerners who live in the provinces have not committed crimes or behaved badly. They do not commit the heresy of misleading people or causing trouble. Since we permit our people to burn incense and worship in the temples of Lamas, Buddhists and Taoists, and the westerners do not actually violate the laws (of China), it therefore seems unfair that their religion should be prohibited. We should order that the Catholic churches in all the provinces be preserved as before. We should allow those who attend the churches to burn incense and to worship as usual. It is not necessary to prohibit Catholicism. With Your Majesty's approval, your servants will proclaim this order to (the magistrates of) all provinces throughout the empire.

On March 22, they received the imperial endorsement: "I approve your deliberation (of March 20)." So, the above series of documents of March 17, 19, 20 and 22, formed the contents of the Edict of Toleration[119]. From these documents, we see that the imperial policy of tolerating Christianity was based on three matters in which Jesuits were involved : 1) administering calendar-making; 2) making military equipment and cannons during military campaigns; and 3) becoming part of

119 *Ibid.*

the Qing mission to Russia. No doubt, the first one was mainly directed to Verbiest. For the second one, it is well-known that Verbiest contributed to the making of cannons. For example in their biography of Verbiest, Pereira and Thomas said: "The emperor praised you that the cannons which you manufactured in the previous year had been effective. The cannons made in this year shoot even more accurately than before"[120].

Western cannons played a role in unifying Taiwan and ending the Albazin campagin. A Chinese military record notes:

> We received the imperial edict in the eighteenth year (1679) which said we should use western cannons (which were quite effective and light, and easy to carry) to attack barracks of pirates. In order to aid in the attack and defense, we ordered all the twenty cannons in Huguang to be sent to the military front in Fujian by officials[121].

In the Albazin campaign it is known that

> Duke Peng Chun commanded troops to march. They reached the walls of the city of Albazin (Yakesa) on the twenty-second day of the fifth month (June 23). On the twenty-third day, they divided into a land force and a navy force, two army routes. On the night of the twenty-fourth day, they moved the *shenwei jiangjun* (cannons made by Verbiest), and the other military equipment to the front. At dawn of the twenty-fifth day, they marched together to attack. There was panic inside the city. The head of the Luocha (Russian) garrison, Elikeshe (Aleksei Tolbusin) and the others were forced to surrender. They went to the front of the Qing army and kowtowed, begging to surrender[122].

The emperor himself decided to dispatch troops against the Russians. He believed that the Russian army could not match the Chinese strong soldiers and sturdy horses, as well as strong and sharp weapons. It is thus clear that the emperor had already estimated the power of the Verbiest cannons while he devised the strategy of the campaign.

Verbiest did not participate in the Qing mission to Russia, but Pereira and Gerbillon donated their services. Specifically, Pereira took part in the negotiations of the Treaty of Nerchinsk, as was arranged by Verbiest. Joseph Sebes has written:

> The high point of Pereira's career was, undoubtedly, the role he

---

120 Xu Risheng and An Duo, p. 2.

121 *Huangchao wenxian tong kao*, vol. 194, "Bing kao shi liu", p. 22

122 *Donghua lu*, "Kangxi ershiwu", pp. 44b-45a.

> played, at the Nerchinsk negotiations, in the conclusion of the first Russo-Chinese peace treaty. His participation in these negotiations, however, is most unexpected and most surprising. He was sent on this trip, as seen in one of the foregoing chapters, on the recommendation of Father Ferdinand Verbiest[123].

This shows that Verbiest also had an influence on the conclusion of the Treaty of Nerchinsk.

Verbiest was related to three major affairs: 1) putting down the Rebellion of the Three Feudatories; 2) unifying Taiwan; and 3) fighting against the aggression of Russia. These were the most important events of the early decades of the Kangxi period. He gave his assistance to most of the needs of the Qing government either directly or indirectly, and made an important contribution. Consequently, the emperor and some subjects made a concession to Christianity in public. We may surely conclude that the imperial policy towards Christian toleration in 1692 was mainly the result of Verbiest's influence during his lifetime. His role was exactly as one historian has claimed in stating that Verbiest "had helped to pave the way for the Treaty of Nerchinsk between China and Russia in 1689 and the edict of 1692 granting religious toleration to Chinese Christians"[124].

Roman Catholicism developed and flourished for a time after 1692. The Chinese Rites controversy beginning in 1705 forced the Kangxi emperor to try and control Christianity. Those missionaries who were willing to receive identity cards might be allowed to stay in China; those who refused them would be banished to Macao[125]. In 1717, because Catholicism set up churches and gathered masses of people inside and outside the city of Guangzhou, where foreign ships entered, it was feared that they lured Christians to create disturbances, so Christianity was prohibited again according to the regulation of 1669[126]. In the subsequent Yongzheng and Qianlong periods, prohibiting Catholicism became more and more strict. In this way, the

---

123 Joseph Sebes, *The Jesuits and the Sino-Russian Treaty of Nerchinsk (1689)* (Rome, 1961), p. 139.

124 Donald W. Treadgold, *The West in Russia and China*, 2 vols. (Cambridge, 1973), 2:20.

125 Forty-six missionaries received the identity cards issued by the Qing Government; thirteen missionaries did not receive them and were expelled from China. Gao Zhentian, "Kangxi yu xiyang chuanjiaoshi" (Kangxi and the western missionaries), *Lishi dang'an*, 1 (1986), 90.

126 *Huangchao wenxian tongkao*, vol. 298, "Si yi kao liu", p. 45b.

golden age of the Jesuits in China, made possible by Verbiest - the forty years between the calendar conflict and the Chinese Rites controversy - came to an end.

BRIEF INTERSECTION

CHANGING CONTEXTS AND PROSPECTS OF THE CHINESE-CHRISTIAN ENCOUNTER FROM MATTEO RICCI TO FERDINAND VERBIEST

JOHN E. WILLS, JR.

John E. Wills, Jr. earned his Ph.D. at Harvard and is Professor of History at the University of Southern California and Director of the U.S.C.- U.C.L.A. Joint East Asian Studies Center. He is author and editor of several books on Chinese history and early Sino-Western relations.

Ferdinand Verbiest in the last years of his life stood near the last of three false summits of hope in the history of seventeenth-century Roman Catholic missions to China. The first was the fruit of Matteo Ricci's work, ripening in the days of Xu Guangqi's greatest influence and of the first successes of Johann Adam Schall von Bell in Beijing; the second of Schall's influence at the Shunzhi court and the power of the Tong family and other friends of the missionaries; the third of the growing toleration of Christianity and prestige of the court Jesuits in the 1660s, culminating in the famous Edict of Toleration. Why was it that these great hopes led to such great disappointments? One might argue, as Jacques Gernet and John Young[1] have, that the missionary endeavor was doomed from the beginning by the unbridgeable basic differences between Chinese and European Christian values and world views. One might argue, as do proponents of the Riccian project, that missionary optimism was justified until the heavy-handed enforcement of a too-narrow European Catholic orthodoxy and orthopraxis antagonized the Kangxi emperor and severely limited the appeal of Christianity to the scholar-official elite. One might argue, with the Holy See, the Dominicans, the Franciscans, the fathers of the *Missions Étrangères*,

---

1 Jacques Gernet, *China and the Christian Impact* (New York, 1985); John D. Young, *Confucianism and Christianity: The First Encounter* (Hong Kong, 1985).

and the doctors of the Sorbonne, that the new restrictions were necessary to prevent the mongrelization of Catholicism in China, and were not incompatible with the emergence of a fervent popular Catholicism in China that even attracted some converts from the local elite. I happen to think that all these arguments are partly correct, but that none of them pays sufficient attention to changes in the broader contexts of Chinese and European culture and politics, to the ways in which broad trends made possible a brief intersection and then contributed to an overdetermined parting of the ways. The trouble with raising these questions is that they open up around our already challenging but possibly manageable subject quite impossible vistas of things one ought to know and books one ought to read. But I really do not see any other adequate way of explaining this brief intersection.

At a number of levels, the years from 1590 to 1620 were an unusually propitious time for Chinese and European intellectuals to get acquainted. Ricci and his distinguished colleagues in the pioneering years - Giulio Aleni, Lazzaro Cattaneo, *et al.* - found in China many things, good and bad, that they already experienced in late sixteenth-century Europe: rapid urbanization and commercialization; an increasing flood of printed books; vigorous intellectual interchange often focused in large public meetings at academies; respect for highly formal literary pursuits, for classical learning, and for the arts of memory; a questioning of received traditions of learning combined with a veneration for the most ancient wisdom, which later traditions were thought to have obscured; receptiveness to inventive syncretisms seeking to reunify an increasingly diverse and compartmentalized intellectual life; hedonism and luxury and moral condemnations of them[2]. After 1590 and especially after 1600, an unusually large proportion of the high bureaucratic elite was out of office, as a result of the Wanli emperor's refusal to fill vacancies, dismissals resulting from factional strife, and voluntary retirements or failures to return after mourning leaves as principled officials avoided the factional morass of the court. These scholar-officials at leisure had plenty of time to get involved and argue about the future of their state and civilization, to read and write, and to listen to the Jesuits and explain the Chinese

---

2 Some of these points and many others about these first encounters are developed in a fascinating way in Jonathan D. Spence, *The Memory Palace of Matteo Ricci* (New York, 1984). See also David E. Mungello, *Curious Land: Jesuit Accommodation and the Origins of Sinology* (Stuttgart, 1985), Ch. 1.

traditions to them. The growth of the Jesuits' understanding of China was remarkably rapid, but I know of no evidence that they had much historical perspective on Ming elite culture or appreciated just how unusual the sense of crisis and openness to new departures of these decades was. Let me simply highlight a few relevant features of a very complex scene.

One useful way of starting to think about the high culture of this period is to begin with the ambiguous legacy of the great Wang Yangming (Wang Shouren, 1479-1529)[3]. One can find passages in Wang's writings and his disciples' records of his conversations where he seems to focus on issues of the nature of being and of human knowledge congenial to the western philosophical tradition. But the central thrust of his thought was a new stage in working through the implications of the Confucian tradition, in asking not "How can I know what is real?" or "How can I be saved?" but "How can I be good?", that is, "How can I realize my limitless potential for selfless sociability in this corrupt world?" Wang was reacting against the approaches to moral growth developed by "Neo-Confucian" scholars of the Song (960-1279), which prescribed a long process of study and of disciplined self-cultivation before an individual could hope to be good and do good amid the stresses of public and private life. Wang attacked this recipe of first knowledge and then action or application on two fronts. First, he argued that the only kind of knowledge that matters is moral knowledge which is fundamentally innate in us, so that in their moral fundamentals "All the ten thousand affairs are within me". Second, he argued that the only real knowledge is that which issues immediately in morally appropiate action; to know and yet not to do what is right is not to know at all. The first of these points could be extended in the direction of a subjectivism that denied the authority of classical texts and received traditions when they conflicted with what one finds in one's own mind; there are sayings of Wang's that would support such an extension. Nor did such a view give the Classics any authority to exclude ideas drawn from Buddhism and Daoism. Wang himself had been deeply involved with these teachings in his youth, and continued to mix some Chan Buddhist attitudes and methods with his deeply Confucian fundamentals. The

3 Reliable starting points for studying Wang Yangming include L. Carrington Goodrich and Chaoying Fang (eds.), *Dictionary of Ming Biography* (New York, 1976), pp. 1408-1416; Wang Yangming, *Instructions for Practical Living and Other Neo-Confucian Writings*, tr. Wing-tsit Chan (New York, 1963); Tu Wei-ming, *Neo-Confucian Thought in Action: Wang Yang-ming's Youth (1472-1509)* (Berkeley and Los Angeles, 1976).

second teaching, that knowledge and action should be one, encouraged a bias toward practical studies and practical, especially governmental, activity, and a suspicion of abstract specialization.

Both sides of the Wang Yangming heritage contributed to the challenges and opportunities the Jesuits faced. The first side of Wang's teachings contributed to a continued speculative temper, as for example in the endless discussions of human nature in the academies noted by the Jesuits, and to some conspicuous currents of moral subjectivism and relativism. Buddhist and Daoist teachings also enjoyed independent revivals, fueled by urbanization and the growing book trade and by the multiform quest for personal authenticity. There were many "syncretic" attempts to reconcile all this diversity, perhaps most notably the "Three Teachings" movement of Lin Zhao'en in Fujian[4]; the Jesuits knew about these movements and loathed them, but surely it was the openness to new ideas and new mixtures of ideas in the general elite culture that made it possible for Chinese scholars to be interested in the missionaries and to take seriously what they had to say. In content, as contrasted with form, the Jesuits drew from and ranged themselves with the growing reaction against Confucian-Buddhist-Daoist syncretism and abstract speculation, with the condemnations of the heritage of Wang Yangming as subjectivist, morally relativist, irresponsible, "empty" (*xu*), not "solid" (*shi*). This search for solidity, paradoxically, had been stimulated in part by Wang Yangming's moral seriousness and focus on knowledge that leads directly to action. Wang's own insights had been very strongly shaped by the risks and crises of his own political career; for at least one prominent late Ming intellectual, moments of risks and crises were of vital importance in the moral life, and all too often lacking in the comfortable life of the late Ming elite[5].

Much remains to be done to show how the Jesuit approach was stimulated by specific contacts of the Jesuit pioneers with Chinese intellectuals. We know that Ricci was acquainted with the circle of Jiao Hong and Chen Di, where he may have encountered Jiao's questioning of Neo-Confucian metaphysics and Chen's pioneering work in the demolition of the Song interpretation of the Classics[6]. He knew of Li

---

4 Judith A. Berling, *The Syncretic Religion of Lin Chao-en* (New York, 1980).

5 Joanna F. Handlin, *Action in Late Ming Thought: The Reorientation of Lü K'un and Other Scholar-Officials* (Berkeley, Los Angeles and London, 1983), pp.168-171.

6 Edward T. Ch'ien, *Chiao Hung and the Restructuring of Neo-Confucianism*

Zhi, and saw him, as many Chinese did, as the prime example of Buddhist-influenced subjectivism and irresponsibility rather than as an isolated and eccentric critic of the Neo-Confucian tradition and proponent of the idea that moral "solidity" would never be possible until moralists adopted a positive attitude toward private, even "selfish" desires. (Ricci would have deplored that idea for his own reasons, of course). In the basic intellectual strategy developed by Ricci and followed by most seventeenth-century Jesuits, Christianity would be presented as offering an alternative to Buddhist teachings on ultimate reality, life after death, and so on. It would, in a phrase repeated in several forms by missionary and convert writers, "supplement Confucianism and replace Buddhism" (*buru yifo*). The Christian approach would begin by leading Chinese intellectuals as far as possible through arguments based on natural reason before showing them how the insights gained in that way were completed by Christian revelation. They would show them that the ancient Chinese had known much of what was accessible to natural reason, had in fact worshipped the true God, before their culture was corrupted by Buddhism[7].

The pioneers' approach to China was also shaped by their European backgrounds. It is well-known that the Jesuits' focus on evangelization of the elite and dreams of converting the emperor were products of the strategies and experiences of the Society in Europe. Their sympathy for the proponents of moral revival in China also owed a good deal to the participation of the Society in the post-Tridentine "Catholic Reformation" of Church organization and ethics. The appeal to natural reason made excellent sense for products of an intellectual environment in which reason took late Scholastic forms, and the ideas of Galileo Galilei, René Descartes, Baruch Spinoza, John Locke - accounts of reason or products of reason very hard to reconcile with Catholic faith - had not yet emerged. The appeal to natural reason also resonated with a widespread conviction that many pre-Christian sages had, by natural reason and perhaps by divine inspiration, some knowledge of the true God and even of the Messiah to come. Christian Platonism and Renaissance Hermeticism occasionally tipped over into heresy, but were accounts of human meaning and divine revelation that could be kept

---

in the Late Ming (New York, 1986); Goodrich and Fang, *op. cit.*, pp. 180-184.

7 Matteo Ricci, *The True Meaning of the Lord of Heaven (Tien-chu shih-i)*, tr. Douglas Lancashire and Peter Hu Kuo-chen; ed. Edward J. Malatesta (St. Louis, 1984); Mungello, *op. cit.*, Ch. 2.

compatible with Catholic belief, not the radical challenges to the fundamentals of that belief that would be presented by the new philosophies of the seventeenth century. And if one believed even a small part of what Renaissance scholars wrote about Hermes Trismegistus and the others, it was no great intellectual leap to add Confucius to the list of the *prisci theologi*[8].

Between 1590 and 1620, several kinds of cultural and political change contributed to a widespread sense of cultural crisis which enhanced the openness of some intellectuals to the Jesuits' message. Around 1590 none of the accepted modes of cultural and moral "solidity" seemed to be working very well. Li Zhi and He Xinyin were two conspicuous representatives of a widespread speculative and non-conformist temper that had nothing to offer the world of political and moral engagement[9]. The court was caught in the institutional-cultural gridlock so vividly described by Ray Huang in his *1587: A Year of No Significance*[10]. The "emptiness" of much elite culture was widely acknowledged, and a keen need was felt for guidance toward "solid" programs of moral action and institutional reform. Chinese scholars were impressed from the beginning by the solidity of the Jesuit knowledge of natural phenomena. More important, I think, was the impression made by their moral and spiritual solidity: the selflessness with which they left their homes and families and braved incredibly long sea voyages to bring their teachings to distant peoples; the rigor with which they kept the fasts and other ritual and moral requirements of their faith and insisted that their converts do so. All this was reinforced, of course, by the immense moral and spiritual solidity of the Gospel story of Jesus' self-sacrifice and by the stories of the saints. It also is possible that Christianity's picture of the immortal soul at risk of eternal damnation appealed to the late Ming longing for moral and existential risk noted above, but I do not know of any text that would offer direct support for this conjecture. It sometimes has seemed to me that the missionaries made a terrible mistake in intellectual strategy when they

---

8 Daniel P. Walker, *The Ancient Theology: Studies in Christian Platonism from the Fifteenth to the Eighteenth Century* (Ithaca, 1972); Frances A. Yates, *Giordano Bruno and the Hermetic Tradition* (London, 1964).

9 W.T. de Bary, "Individualism and Humanitarianism in Late Ming Thought", in de Bary (ed.), *Self and Society in Ming Thought* (New York, 1970), pp. 145-148; Ronald G. Dimberg, *The Sage and Society: The Life and Thought of Ho Hsin-yin* (Honolulu, 1974).

10 New Haven, 1981.

began with abstract Scholastic reasoning about the nature of God and man (which many Chinese intellectuals found incomprehensible) rather than with the Gospel and the stories of the saints. Of course they did both, but the number of tracts devoted to abstract Scholastic reasoning in the Louis Pfister bibliographies is striking. Nicolas Standaert has shown us, however, that for at least one intellectual of this generation the abstract reasoning about substance reinforced a conviction that Christianity offered a profound approach to moral substance and solidity and an alternative to Buddhist "emptiness"[11].

For some intellectuals the sense that the received tradition no longer provided adequate guidance toward moral solidity, that the traditional roles no longer worked, continued down to the era of the Qing conquest; Fang Yizhi's "Seven Solutions" (*Qi jie*, 1637)[12] is a brilliant example of this sort of frustration and alienation. A few found what they were looking for in Roman Catholicism. Many others, however, discovered neglected resources of intellectual solidity in their own tradition and new opportunities to demonstrate their moral solidity in changing political circumstances. There were several strands of this revival and reshaping of Confucian thought and action. They were not without tensions and contradictions, but frequently more than one of them can be observed in an individual or a group of close associates. First, scholar-officials now pursued their political causes at court - pressuring the emperor not to change his heir apparent, urging him to fill bureaucratic vacancies, opposing eunuch power - not only by individual protest and exploitation of bureaucratic position and connection in the capital but also by organizing societies (*she*) of like-minded scholars in various provincial cities and linking them in networks of organization and communication[13]. Second, members of the local elite and scholar-officials out of office became intensely interested in improving conditions and combatting government abuses in their own areas, collected documents and wrote analyses of them, and gave renewed focus to "statecraft" (*jing shi*) as a genre of scholarly writing. Third, revival of a Confucian moral seriousness of religious intensity,

---

11 Nicolas Standaert, *Yang Tingyun, Confucian and Christian in Late Ming China* (Leiden, 1988), pp. 152-161.

12 Williard J. Peterson, *Bitter Gourd: Fang I-chih and the Impetus for Intellectual Change* (New Haven, 1979).

13 William S. Atwell, "From Education to Politics: The *Fu She*", in W.T. de Bary *et al.* (eds.), *The Unfolding of Neo-Confucianism* (New York, 1975), pp. 333-365.

frequently drawing on the teachings of the Song masters and of Wang Yangming, was fostered through meetings at academies, especially the famous Donglin Academy, and by the intensely individual quests and writings of men like Gao Panlong[14]. Fourth, the new critical scholarship on the Classics continued to develop. Fifth, the ontological dimensions of the quest for "solidity" was resolved for many not by the Jesuits' Neo-Scholastic formulae but by the emergence of a monism of *qi* (material forces), denying any equivalent or superior ontological status of *li* (organizing principle), insisting that *li* are simply relatively constant patterns of change and development of *qi*, frequently supported by renewed attention to the appendices of the *Yijing*[15]. Only the last of these five trends was in serious conflict with Catholic belief, and Xu Guangqi and other converts were notable participants in several of the others. The influence of Xu Guangqi, the astronomical successes of Schall, the deepening impact of the missionary presence in Shanghai and a few other localities, the establishment of centers of evangelization and conversion in places as distant as Xi'an and Chengdu, all were sources of missionary optimism in the 1630s. But it seems to me that in the wider cultural context sketched here the moment of greatest opportunity had passed. The conditions that had given the Jesuits their opening around 1600 - the desperate sense that the received tradition had lost its capacity for moral solidity, the openness to syncretic and other non-traditional ideas - were waning.

It is now clear from the work of Nathan Sivin that Chinese specialists in mathematical astronomy recognized the superiority of methods introduced by the Jesuits and abandoned their old algebraic approach for the European geometrical one[16]. It was of course this astronomical knowledge that gave the Jesuits the secure position at the court that led eventually to their being employed in Verbiest's time and thereafter in other useful and ornamental ways and was so helpful in securing tacit and eventually explicit toleration of the missions and converts in the provinces. But this change in astronomy did not have anything like the impact that the new science of Nicolaus Copernicus, Galileo, and Isaac Newton had in Europe. Despite all we have learned

---

[14] Rodney L. Taylor, *The Cultivation of Sagehood as a Religious Goal in Neo-Confucianism: A Study of Selected Writings of Kao P'an-lung (1562-1626)* (Missoula, Montana, 1978).

[15] Ch'ien, *op. cit.*, Ch. 5, 6.

[16] Nathan Sivin, "Copernicus in China", *Studia Copernicana*, 6 (1973), 63-122.

recently about the scientific abilities and cosmological interests of Ming-Qing scholars[17], I still think it fair to say that the mathematical astronomy was more compartmentalized and peripheral in late imperial elite culture, where the central issues were political, moral, and historical, than in early modern European culture with its obsessions with knowledge and reality.

The wars against the Manchus gave the Jesuits new roles as founders of cannons and (abortively) recruiters of Portuguese gunners for the Ming. The wars of the Ming-Qing transition[18] caused at least transitory disruption in many mission areas, but new and promising relations soon emerged with the Manchu-Chinese elite of the new dynasty. In the broader political and cultural contexts I am trying to keep in focus here, the transition probably reduced prospects of evangelization of the elite. The scholar-officials of Jiangnan, where some of the most positive responses to Christianity had been found in late Ming, found their lives and fortunes badly disrupted for several decades and never entirely regained the kind of political domination they had had in late Ming. Many of them died in suicidal resistance or defiance of the conquerors. The selflessness and solidity of their own loyalty certainly confirmed the possibility of living and dying by traditional values, but left some aghast at the uselessness of these sacrifices[19], so that traditional morality and the search for practical ways of expressing it were both reinforced. Classical scholarship, statecraft scholarship, and *qi*-monist ontology all found some of their greatest monuments in the works of great loyalist scholars like Gu Yanwu and Wang Fuzhi. The texts of the Classics became both objects of detailed research and the final arbiters of all ethical and philosophical questions. Song Neo-Confucianism survived as a standard of state orthodoxy and a guide to some forms of self-cultivation, but there was little interest in its cosmological ideas, many of which were seen as non-classical[20]. Scholars who had criticized late

---

17 See especially John B. Henderson, *The Development and Decline of Chinese Cosmology* (New York, 1984).

18 For starting points in surveying this period see Jonathan D. Spence and John E. Wills, Jr. (eds.), *From Ming to Ch'ing: Conquest, Region, and Continuity in Seventeenth-Century China* (New Haven, 1974); Lynn A. Struve, *The Southern Ming, 1644-1662* (New Haven, 1984); Frederic Wakeman, *The Great Enterprise: The Manchu Reconstruction of Imperial Order in Seventeenth-Century China* (Berkeley and Los Angeles), 1985.

19 Weiming Tu, "Yen Yuan: From Inner Experience to Lived Concreteness", in de Bary, *Unfolding*, pp. 511-541, at 521-522.

20 Henderson, *op. cit.*, Chapters 6-9; Wing-Tsit Chan, "The *Hsing-li ching-i*

Ming abuses of government and proposed reforms found that some of these reforms now were put into effect by a government that was alien, ruthless, but remarkably effective. This was a world of chastened political realism, hard-headed search for evidence, and more literal allegiance to the Classics that left little room for late Ming abstractions and syncretisms or for the Catholic message that had found its openings in the complexity and openness of the late Ming.

The relations of the Jesuits with the Qing court is a relatively well-known story, with its promising beginning, years of persecutions in the 1660s, favor under Kangxi but then rapid cooling, but with astronomy and a few other functions remaining in Jesuit hands when Christianity was prohibited in the Qing empire. Less well-studied is the broader picture of the substantial number of converts among the Manchu and Chinese Banner elites. Western scholarship is just beginning to take seriously the Manchuness of the Qing ruling house and the peculiar complexities of the bicultural Chinese Banner group. The famous Tong family had a number of converts and sympathizers; the Portuguese Jesuit who spent Holy Week of 1670 in Tong Guoqi's mansion in Nanjing was very much impressed by the pious atmosphere[21]. The tremendous amount of material on Catholicism among these elites in Beijing to be found in the volumes of the *Lettres édifiantes* awaits systematic study by someone who understands the Chinese and Manchu cultures and societies. It probably was easier for these people to consider the Christian message than it was for classically-educated Chinese scholar-officials, since the Manchu and Liaodong Chinese cultures already included Manchu beliefs and ceremonies and a great deal of interest in Lamaism as well as Chinese elements, so there was little resistance to adding another disparate element to the mix. And, of course, they could hardly endorse Chinese anti-foreignism without calling into question their own position in the empire.

The story of Schall and the Shunzhi emperor needs to be read again in the light of our growing understanding of this period. A contest for religious influence between the lamas and the fathers is one important dimension. Another is the immensely important role of the emperor's mother, the Empress Dowager Xiaozhuang. In the late 1660s the Kangxi emperor clearly used his favor toward the Jesuits to emphasize the

---

and the Ch'eng-Chu School of the Seventeenth Century", in de Bary, *Unfolding*, pp. 543-579.

21 John E. Wills, Jr., *Embassies and Illusions: Dutch and Portuguese Envoys to Kang-hsi, 1666-1687* (Cambridge, 1984), pp. 232, 234-235.

change from the Oboi regency to his personal rule[22]. The fall of the Mingzhu group in 1688 and the political changes that followed may also have contributed to the high favor in which the fathers stood in the 1690s.

It was of course the incredible mismanagement of the controversies over the ceremonies permitted to the Chinese converts and the terms used for God that turned the emperor from cordiality to hostility. John Witek has analyzed some of this with his usual care[23]. Again trying to keep a broader context in view, it seems to me that we see in these decisions as we do not see in the Church of c. 1600 an organization preoccupied with the control of deviation, sensing threats of unprecedented depth from new ideas abroad in the elite culture. This seems especially clear in the struggles against Quietism carried on both in Rome and in Paris; the doctrinal deviations involved seem to an outsider rather small, but the threats to the standing of the Church as necessary intermediary between devout believer and God seemed very real. And some of those who were exploring new modes of accommodation between Christianity and the Chinese tradition, like Joachim Bouvet and Jean-François Foucquet, or were responding in Europe to Jesuit reports of China, like Pierre Daniel Huet, were much more closely involved than Ricci had been with a hermetic-numerological tradition increasingly at odds with Catholic orthodoxy. Even a less rigidly defensive and control-oriented Church, it seems to me, would have found some of the ideas of the Figurists unacceptable. In Paris the whole attack on the Jesuit position also was linked to the great influence both of Jacques Bossuet and of Françoise d'Aubigné, Marquise de Maintenon, and came to involve the perennial theological problem of the salvation of the heathen as well as the issues of missionary policy in China[24].

Popular and local elite conversion to Catholicism did not stop after 1700. The Jesuits in Jiangnan, the Dominicans in Fujian, and the fathers of the *Missions Étrangères* in Sichuan were among the notable success stories. But in a China much better governed than it had been under the late Ming, with an elite considerably chastened and turned toward

---

22 *Ibid.*, Ch. 3.

23 John W. Witek, *Controversial Ideas in China and in Europe: A Biography of Jean-François Foucquet, S.J. (1665-1741)* (Rome, 1982), Ch. 2.

24 Witek, *op. cit.*; Walker, *op. cit.*, Ch. 6; Virgile Pinot, *La Chine et la formation de l'esprit philosophique en France (1640-1740)* (Paris, 1932), pp. 92-109; Mungello, *op. cit.*, pp. 292-353.

textual scholarship and practical statecraft, there were no more Xu Guangqis and Yang Tingyuns, no more members of the high scholar-official elite making a desperate leap into a new foreign religion in the hope of thereby saving some of the core values of the Confucian tradition. Nor could any missionary now embrace the latest trends in European intellectual life, reach out to admire China and adjust his evangelical efforts to its expectations, and remain within the discipline and organization of the Church.

Gottfried Wilhelm Leibniz was the greatest European intellectual to take China seriously in any century[25]. He was very favorable to the Riccian project and even to some Figurist ideas. But he was not Catholic, not a missionary, and his uncompleted works and failed projects, including the reunion of Christendom, suggest that by his time the European intellectual world was too fragmented, its political and ecclesiastical worlds too defensive and preoccupied with control, the Chinese and European high cultures too far apart on their separate trajectories after their brief intersection, for such wide and noble visions to have any hope of realization.

---

25 David E. Mungello, *Leibniz and Confucianism: The Search for Accord* (Honolulu, 1977); Gottfried W. Leibniz, *Discourse on the Natural Theology of the Chinese*, tr. Henry Rosemont, Jr. and Daniel J. Cook (Honolulu, 1977).

THE INVESTIGATIONS OF THINGS AND THE FATHOMING OF PRINCIPLES (*GEWU QIONGLI*) IN THE SEVENTEENTH-CENTURY CONTACT BETWEEN JESUITS AND CHINESE SCHOLARS

NICOLAS STANDAERT, S.J.

N. Standaert, S.J., holds a Ph.D. degree in Chinese Studies, Leiden State University (1984). He is the author of "Yang Tingyun, Confucian and Christian in Late Ming China" (Leiden, E.J. Brill, 1988).

Abbreviations

B.: Henri Bernard, "Les adaptations chinoises d'ouvrages européens: Bibliographie chronologique depuis la venue des Portugais à Canton jusqu'à la mission française de Pékin", *Monumenta Serica*, 10 (1945), 1-57, 309-388.

C.: M. Courant, *Catalogue des livres chinois, coréens, japonais, etc.* (Paris, 1912). (= reference number in the *Bibliothèque Nationale de Paris*).

FVP: Ferdinand Verbiest Project, Catholic University of Leuven (Belgium).

Xu: Xu Zongze, *Ming Qing jian Yesu huishi yizhu tiyao* (Summary of the Works Translated by the Jesuits in Late Ming and Early Qing) (Taibei, 1957).

21: Wu Xiangxiang, ed., *Zhongguo shixue congshu xubian* (Sequel to the Collection on Chinese History); n° 21: *Tianzhujiao dongchuan wenxian sanbian* (Third Collection of Documents on the Spread of Catholicism to the East), 6 vols. (Taibei,1972); *Zhongguo shixue congshu* (Collection on Chinese History).

23: *Tianxue chuhan* (First Encyclopedia of Heavenly Studies), Li Zhizao, ed., 6 vols. (Taibei, 1965).

24: *Tianzhujiao dongchuan wenxian* (Documents on the Spread of Catholicism to the East) (Taibei, 1965).

A generally accepted theory in the analysis of contact between cultures states that when a foreign element is introduced into a culture, and this element is relatively easily taken up by this culture, it is most often due to the fact that this culture already presented an internal movement which enabled it to accept the new element[1]. This theory was confirmed by the cultural exchange between China and the West in seventeenth-

1 P.E. Josselin de Jong, *Contact der Continenten* (Leiden, 1978), pp. 69 ff.

century China. According to several scholars, the western science introduced by the Jesuits enjoyed a certain acceptance because of the rediscovery of "concrete studies" or "real learning" (*shixue*) in early seventeenth-century China[2]. This movement of stress on practicality was characteristic of the Donglin thinkers, and got further development in early Qing philosophers like Fang Yizhi (1611-1671), Wang Fuzhi (1619-1692), Yan Yuan (1635-1704), Liu Xianting (1648-1695) and Li Gong (1659-1733).

I.
The Investigation of Things

One of the main issues in the intellectual elaboration of this "real learning" was the interpretation of the term *gewu*, "the investigation of things"[3] in one of the famous sentences of the *Daxue* (Great Learning):

> The ancients who wished clearly to exemplify illustrious virtue throughout the world would first set up good government in their states. Wishing to govern their states well, they would first regulate their families. Wishing to regulate their families, they would first cultivate their persons. Wishing to cultivate their persons, they would first rectify their minds. Wishing to rectify their minds, they would first seek sincerity in their thoughts. Wishing for sincerity in their thoughts, they would first extend their knowledge. The extension of

---

2 Cf. M. Übelhör, "Geistesströmungen der späten Ming-Zeit, die das Wirken der Jesuiten in China begünstigten", *Saeculum*, 23 (1972), 172-185; Jacques Gernet, *Chine et Christianisme: action et réaction* (Paris, 1982), pp. 36-37, 81 ff; W.J. Peterson, "Fang I-chih: Western Learning and the Investigation of Things", in Wm. Th. de Bary, ed., *The Unfolding of Neo-Confucianism* (New York, 1970), pp. 369-411.

3 On this theme see: Chen Li-fu, "The Investigation of Things", in *The Confucian Way: A New and Systematic Study of the Four Books* (London, 1987), Ch. 2; D.C. Lau, "A Note on Ke Wu", *Bulletin of the School of Oriental and African Studies*, 30 (1967), 353-357; Shimada Kenji, *Die Neo-Konfuzianische Philosophie* (transl. M. Übelhör) (Hamburg, 1979), pp. 108-109, 135-137; Zhang Liwen, *Song Ming lixue yanjiu* (Study of Song Ming Neo-Confucianism) (Beijing, 1985), pp. 60 ff, 326 ff, 416 ff, 566 ff, 646-647. During the writing of this article we were happy to learn that a similar theme was treated by Zhang Yongtang at the Verbiest Conference in Taibei in 1987; because of sufficient differences in approach, both articles can be seen as complementary. Zhang Yongtang, "Mingmo Qingchu xixuepai dui «gewu qiongli» guannian de xin jieshu - jian lun Nan Huairen zhi qiongli-xue" (International Conference in Honor of Ferdinand Verbiest: Commemoration of the 300th Anniversary of his Death [1688-1988]) (Taibei, 1987), pp. 83-91.

their knowledge lay in the investigation of things.[4]

In the Song dynasty, Zhu Xi crystallized the thought of the Cheng brothers and reinforced their shift from the "investigation of things" to "fathoming principles" (*qiongli*). He commented that *gewu* means "to reach to the utmost the principles of activities and things" (*qiong zhi shiwu zhi li*). This interpretation, which became orthodox, implies that principles in things and activities are particularized aspects of the one Principle (*li*), which is the basis for moral action. The goal in knowing Principle is to know what to do. This moralistic interpretation of *gewu* remained one of the central concerns in the subsequent development of the *Cheng Zhu* School. In the sixteenth century it developed into a greater interiorization. According to Wang Yangming, there was nothing to be investigated in the things of the world. The Principle for moral action had to be found within the mind-and-heart (*xin*) and not outside. The "investigation" meant "rectification" of the mind-and-heart, especially through the acts in which man's mind-and-heart is engaged. By locating Principle in our mind-and-heart, and thus making what is outside irrelevant for the purposes of moral self-cultivation, Wang Yangming laid the basis for the flowering of intuitionism in the sixteenth century.

The seventeenth century reestablished the intellectual importance of the "things in the world." Fang Yizhi[5] is a good illustration. He advanced an inclusive definition of *wu*, embracing the cosmos, physical objects, human events, and mind-and-heart. Together with other contemporaries, he criticized those who neglected to pay attention to the physical world. In stressing "things," he was not negating the goal of fathoming "principles." He merely contended that such knowledge had to be rooted in a detailed understanding of *wu*. His attention was on things and their principles: material investigations such as music, the calendar, astronomy, medicine and the like, as well as the "comprehending seminal forces" (*tongzhi*), which reveal the "extended principles of things" (*wu zhi zhili*). Neither things nor their principles can be neglected. Under Fang's view, our investigations must be directed at "things," and, at least in part, those "things" must be external and not merely the products of our own mental processes.

This rediscovery of the importance of the physical world,

---

4 Wm. Th. de Bary, ed., *Sources of Chinese Tradition* (New York, 1963), 1:115.

5 Cf. Peterson, *op. cit.*, pp. 379 ff.

characteristic of many seventeenth-century philosophers, is viewed as having facilitated the introduction of western science, which the Jesuits brought together with Christianity. In this paper we would like to look at the contact between the Jesuits and the Chinese from the perspective of the "investigation of things in order to fathom principles." To what extent were the Jesuits and the Chinese who were in relationship with them involved in reflection on this issue?

II.
Finding God in All Things

Before analyzing this question through the Chinese writings of both Jesuits and Chinese, one could ask a preliminary question. If the above theory of cultural receptivity is right, were there also internal characteristics of the Jesuits (the foreign element) that encouraged this reception? In other words, was their preoccupation with the physical world merely a strategy to attract interest, or were they themselves also interested in "things?" This question brings us to the core of Ignatian spirituality. The dynamic of the Jesuits' view of the world can be summarized by the famous sentence of St. Ignatius (1491-1556): "to seek and find God in all things" (*buscar y hallar a Dios en todas las cosas*)[6]. In founding the Society of Jesus, Ignatius conceived, for the first time in the history of religious life, a purely apostolic order. The new style of the Society, which evoked much opposition, broke with a religious life that was based on monasticism and its distinctive signs: cloister, habit, and prayer in choir. For Ignatius, God could not only be experienced in personal or community prayer, but could be found in "all things"[7]. What is presented in his *Spiritual Exercises* comes out of his own religious experience and mystical view. These *Spiritual Exercises* are a pedagogy to help people (religious as well as lay) to an experience of God. Things are considered as both internal and external: the elements, plants, animals, men, consciousness, intelligence, in sum, all creatures[8]. One can notice a reappraisal of the external world and all human activities; they do not oppose, but rather assist, the goal of finding God.

---

6 On this theme see R.M. Saldarriaga, *La Dinámica de la Integración Espiritual (buscar y hallar a Dios en todas las cosas)* (Rome, 1980), 493 pp; J. Stierli, *Chercher Dieu en toutes choses* (Paris, 1985), 185 pp.

7 Cf. Stierli, *op. cit.*, pp. 97-98.

8 Cf. *Spiritual Exercises*, transl. T. Corbishley (Wheathampstead, 1963), n° 235.

All of creation is good and leads to God.

This spiritual intuition can also be found in the *Constitutions of the Society of Jesus*. The *Constitutions* pay great attention to the formation of Jesuits. For the ministry of transmitting the experience of God, young Jesuits prepare themselves in studies. The end of such studies is to aid others to attain the knowledge and the love of God and the salvation of their souls. The study of theology is most suitable to this end. But theology requires a long preparation of humanistic studies (including grammar and rhetoric), languages (classical languages and the languages of the cultures with which they would interact), logic, the arts and natural sciences as well as moral philosophy and metaphysics[9]. As in medieval universities, the arts and other branches learned from natural reason are conceived of as both worthy of serious study in themselves, and also as handmaids to theology[10]. Moreover, studies can reinforce a spiritual experience: God can also be found in studies[11]. Therefore, young Jesuits "should often be exhorted to seek God our Lord in all things, stripping off from themselves the love of creatures to the extent that this is possible, in order to turn their love upon the Creator of them, by loving Him in all creatures and all of them in Him, in conformity with His holy and divine will"[12].

As can be observed, there apparently existed an analogy between the Jesuits' interest in "all things" and the Chinese revaluation of "things." For both, the "investigation of things" was also connected with another goal: "to find God," in the case of the Jesuits, and "to fathom principles" (to know moral conduct), in the case of the Chinese. The adjustment of these two aspects has often been a point of tension for Jesuits as well as for Chinese. What happened when the "investigation of things in order to fathom the Principle" met with "in all things seek and find God?" As mentioned above, we will try to analyze this meeting from the Chinese sources, as much as possible using those of Ferdinand Verbiest's time (early Qing). References to the Ming period will be necessary to understand the change in ways of interpretation. We will distinguish the Jesuit from the Chinese interpretation.

---

9 Cf. *The Constitutions of the Society of Jesus*, transl. G. Ganss (St. Louis, 1970), pp. 351 (IV, 5,1), 446-452 (IV, 12,1-4).

10 *Ibid.*, p. 214, n° 4.

11 Cf. "Trouver Dieu dans l'étude", in *Notes ignatiennes*, n° 6 (Braine-le-Comte, n.d.), 6 pp..

12 Cf. *The Constitutions of the Society of Jesus*, p. 288 (III, 1, 26).

III.
Jesuit Interpretation

The Jesuit interpretation of the investigation of things was narrowly linked with their distinctive training and, more specifically, with Aristotelian philosophy. Generally speaking, one can observe the following succession in their works: through the investigation of things, one can fathom principles in order to find God (*gewu - qiongli - zhitian*). Investigation of things concerns natural sciences as mathematics, the calendar, geography, irrigation, mechanics, and the like. Although the Jesuits introduced new elements concerning these sciences into Chinese culture, their conception and practice of these sciences corresponded very much to that of the Chinese. But to find a personal God, who created all things, and to glorify Him by the proper use of things was surely a new element of another kind. We will attend especially to the Jesuits' interpretation of the notion of "principle" (*li*).

The fathoming of principles in order to find the underlying Principle of existence was no longer the ultimate goal for the Jesuits, as it was for the Chinese. *Li* was indeed rejected as being the origin of "creation." Here one has to return to the unfortunate misunderstanding of Neo-Confucianism by Matteo Ricci, which had a great influence on the Chinese writings of the Jesuits in the seventeenth century. In his *Tianzhu shiyi* (The True Meaning of the Lord of Heaven), Ricci rejected the Supreme Ultimate (*taiji*) and *li* as having produced heaven and earth on the basis of the Aristotelian distinction between "substance" (*zilizhe*) and "accident" (*yilaizhe*). These terms are explained as follows:

> There are two categories of things: substance and accident. Things which do not depend on other things for their existence, such as heaven and earth, ghosts and spirits, men, birds and beasts, vegetation, metals, stones, the four elements, and the like, are all classed as substance. Things which cannot stand on their own and which can only be established subject to other things as, for example, the Five Basic Virtues, the Five Colors, the Five Notes, the Five Flavors, the Seven Emotions, and the like, are all classed as dependent. Let us take a white horse as an illustration. Here are two things: whiteness and horse. Horse is substance. Whiteness is an accident, since even without its whiteness the horse could continue to exist. If there were no horse, however, the whiteness of the horse would be totally incapable of continued existence. We therefore say that it is an accident. When we compare these two categories, we find that substance has prior existence and is of value, whereas accident is

comparatively secondary and of little consequence[13].

This distinction of accident and substance, here presented by Ricci with typically Chinese illustrations, was an important help in scholastic argumentation. We find it in similar terms in Johann Adam Schall's theological treatise *Zhujiao yuanqi* (The Origin and Progress of Christianity, 1643) for the explanation of the Trinity[14], and in Verbiest's *Shengti dayi* (Answer to Doubts on the Eucharist, 1677) concerning the transubstantiation of the host[15]. Both works refer to the distinction as being defined by the "specialists of investigating things" (*gewujia*). Ricci used the theory in order to reject the Supreme Ultimate and the Principle:

> When we come to the Supreme Ultimate, we find that it is only explained in terms of principle. It cannot, therefore, be the source of heaven, earth, and all things, because principle also falls into the category of accidents. Since it is not substance, how can it establish other things? When men of letters and learned men in China discuss principle, they only speak of it in two ways: they either say that principle resides in the minds of men, or else they say it is to be found in things. They only say that things are real when their mode of being harmonizes with the principles in men's minds. When the human mind is able to study, penetrate and completely understand the principles inherent in things, this is called "the investigation of things." It is clear, on the basis of these two ways of speaking about principle, that principle is dependent, and cannot be the source of things. Principles, whether in the human mind or in things, are all subsequent to those things; and how can that which is subsequent be the source of that which existed prior to it?[16]

This text, which is fully based on western philosophy, greatly influenced the Jesuit interpretation of *li*. In western texts on this topic, one can find much more elaborate interpretations, which also provoked intellectual discussion in Europe, e.g., Nicolas de Malebranche (1638-1715), Gottfried Wilhelm Leibniz (1646-1716)[17]. In Chinese texts, how-

---

13 Ricci, *Tianzhu shiyi*, 23 I, p. 406, transl. D. Lancashire & P. Hu Kuo-chen, *The True Meaning of the Lord of Heaven*, Variétés Sinologiques N.S. 72 (Taibei/Paris, 1985), p. 109.

14 Schall, *Zhujiao yuanqi*, j. 1, p. 31b (C. 6937).

15 Verbiest, *Shengti dayi*, p. 3a (C. 7284).

16 Ricci, *op. cit.*, 23 I, p. 407, transl., *op. cit.*, p. 111.

17 Cf. N. Longobardo, "Traité sur quelques points de la religion des Chinois", in G.W. Leibniz, *Opera omnia* (Geneva, 1768), 4:89-145; cf. "L'idée de

ever, the limitation of *li* to accident by Ricci was taken over by several authors. It was, for instance, quoted by Ludovico Buglio in his defence of Christianity against the attacks of Yang Guangxian, *Budeyi bian* (I Cannot Do Otherwise, Refuted, 1665)[18], as well as in his *Zhujiao yaozhi* (Essential Meaning of Christianity, 1668)[19]. Verbiest's catechism *Jiaoyao xulun* (Introduction to the Essentials of Christianity, 1670), being less apologetical, does not contain a discussion of *li*.

Worth noting in Ricci's quotation is the presence of the two Chinese interpretations of *gewu*: the intuitionalist interpretation of the Wang Yangming School (*li* resides in the mind-and-heart), and the rediscovery of the presence of *li* in things. These two versions will be accepted through reinterpretation by the Jesuits. For them *li* will correspond to *ratio* (reason)[20]: a tool of the intelligent nature of man by which he can fathom and investigate things and events[21]. It is through this consciousness, by which he differs from all other living beings, that man can reason. *Li*, understood as principle, can also be found inside things. In the preface to his translation of Clavius's commentary on *Euclid's Elementorum libri XV*, *Jihe yuanben* (1607), Ricci links it with the "extension of knowledge" (*zhizhi*) as explained in the *Daxue*:

> When (Confucian scholars) study, they consider the extension of their knowledge as very important, and this extension of knowledge necessarily has to start from a clear understanding of the principles in things. Since the principles in things are minute and hidden, and man's ability is obstinate and obscure, and since it does not rely on what is already clear and can hardly find out what is unclear, how can my knowledge come to extension? Although our distant western countries are very small, the method of investigating things and fathoming principles as practised in the schools is regarded by all countries as perfection. Therefore, the books which analyze and inve-

---

Dieu chez Malebranche et l'idée de *li* chez Tchou Hi", and "Leibniz et la pensée chinoise", in Etiemble, *L'Europe chinoise* (Paris, 1988), chapters 23-25, 29.

18 Buglio, *Budeyi bian*, 24, pp. 303 ff; three explicit references to the *Tianzhu shiyi*.

19 Buglio, *Zhujiao yaozhi*, p. 2a (C. 6917); clearly based upon the *Tianzhu shiyi*.

20 See also the translation of the *Daxue* in *Confucius Sinarum Philosophus* (Paris, 1687), p. 4. "Ad extremeum hac ratione perficere, seu ad summum apicem perducere vim-intellectivam, consistit in penetrando, sive exhauriendo res omnes, seu rerum omnium rationes."

21 Cf. Buglio, *Zhujiao yaozhi*, p. 2b; see also Verbiest *infra* note 51.

stigate the principles of things are extremely numerous and rich[22].

Both aspects were integrated into the concept of western philosophy[23], and expressions like "the study of the principles" (*lixue*) will be used to translate the word *philosophy*. We find all these meanings condensed in the explanation of philosophical studies in Giulio Aleni's *Xixue fan* (A General Account of western Studies, 1623):

> After finishing literary studies (*wenxue*), one continues in philosophy (*lixue*). Philosophy is the great learning of right conduct (*yi*) and of *ratio* (*li*). It is because of his right conduct and his *ratio* that man is above the myriad things, and that he is the intelligent one (*ling*) among the myriad things. Through investigation of things and fathoming principles, one attains the fullness of humanity and comes close to Heaven. But the principles of things are hidden inside the things, like gold in gravel and jade in unpolished gem. One has to sift and to split them with the study of philosophy (*feilusuofeiya*)[24].

Philosophy is thus considered as the investigation of things and the fathoming of principles by human reason. In his work, Aleni gives an ample description of Jesuit formation in philosophy. Philosophical studies generally lasted for three or four years and were based on Aristotelian philosophy. The program was divided as follows:

1st year: Logic or the "way of clear discussion" (*luorijia*; *mingbian zhi dao*).
2nd year: Physics or the "way of scrutinizing the principles of nature" (*feixijia*; *cha xingli zhi dao*).
3rd year: Metaphysics or "scrutinizing the principles above nature" (*meidafeixijia*; *cha xingyishang zhi li*).
4th year: Further studies in mathematics (geometry, arithmetic, music, astronomy) or the "study of how much" (*mademadijia*; *jihe zhi xue*); and studies in ethics or the "study of cultivation, regulation, government and peace" (*edijia*; *xiuqi zhiping zhi xue*)[25].

---

22 Ricci, *Jihe yuanben*, 23 IV, p. 1929.

23 See also F. Noël's translation of the *Daxue* in *Sinensis Imperii libri classici sex* (Prague, 1711), p. 11 "... modus denique assequendi perfectam Boni & Mali notitiam, consistit in perscrutandis rerum naturis ac rationibus, sive in philosophiae studio."

24 Aleni, *Xixue fan*, 23 I, p. 30, 9-31, 5.

25 *Ibid.*, 23 I, pp. 31-41; compare this description with the *Ratio atque Institutio Studiorum Societatis Iesu* (1599), which determined Jesuit formation during the whole seventeenth century in L. Lukacs, ed., *Monumenta Paedagogica Societatis Iesu* (M.H.S.I. 129) (Rome, 1986), 5:355 ff.

It is interesting to see how Aleni presented western studies. He is fully faithful to the western program, but he uses a vocabulary which was taken from Chinese tradition. The reference to the "investigation of things in order to fathom principles" is clearly inspired by the *Daxue*. This is especially confirmed by the vocabulary used for "ethics." For his translation of Aristotle's *Ethica*, Alfonso Vagnoni also used the expressions *xixue xiushen*, *xixue qijia*, *xixue zhiping*[26]. It refers to the sentence of the *Daxue* following the one quoted at the beginning of this article:

> ... The extension of knowledge lay in the investigation of things. For only when things are investigated is knowledge extended; only when knowledge is extended are thoughts sincere; only when thoughts are sincere are minds rectified; only when minds are rectified are persons cultivated (*xiushen*); only when persons are cultivated are our families regulated (*qijia*); only when families are regulated are states governed (*zhiguo*); only when states are well governed is there peace in the world (*ping tianxia*).

It is also important to notice the place of the sciences in Jesuit education. One only started studying the sciences after the study of logic, physics and, most often, metaphysics. Here intervenes the importance of "nature" (*xing*). Aleni explains it in his preface to his treatise on nature *Xingxue cushu* (Brief Explanation of Nature, 1624):

> St. Augustine said: "Those who want to investigate things have to attend to major subjects: the doctrine on human nature and the doctrine on the Creator. Human nature leads man to know himself, and the Creator leads man to know his origin and end. The first is called physics (*xingxue*), the second is called metaphysics (*chaoxing zhi xue*). The first enables us to perceive true happiness, the second to obtain it in reality[27].

This explanation corresponds to Jesuit spirituality. Before fully using things, things have to be perceived as being given by the Creator. As Manuel Dias says in his preface to his *Tianwen lüe* (Short Questions about Heaven, 1615): "It is by recognizing and serving the Creator that studies are concrete (*shixue*) and not void (*xuxue*)"[28]. This theme was especially treated in the third year of philosophy with Aristotle's *De*

---

[26] On these works see also: H. Verhaeren, "Un livre inédit du P. Vagnoni, S.J. (1605-1640)", *Bulletin catholique de Pékin*, 22 (1935), 36-40.

[27] Xu, p. 213.

[28] Dias, *Tianwen lüe*, 23 V, p. 2630.

anima and *Metaphysica*. The *Metaphysica* treated the question of God from a philosophical point of view, and not yet on a theological or scriptural basis. *De anima* was also connected with moral philosophy. Its moral influence on the whole world can be perceived from Francisco Sambiasi's preface to the Chinese version of this work (*Lingyanli shuo*, 1624). After having explained the importance of knowing oneself by acknowledging the possibility and the beauty of one's *anima* (*yanima*), he says:

> That is why the scholars who investigate things and fathom principles, display the excellence (of *anima*); and in esteeming it, they regulate families, govern countries and bring peace to the world. All those who want to be teachers or leaders of others certainly have to practise this study of *anima*.[29]

From this explanation of western studies appears clearly the importance Jesuits paid to the investigation of things and the fathoming of principles, and how they felt it corresponded to Chinese tradition. They also agreed with Chinese tradition in the moral influence of these investigations. As noticed before, the difference was situated in the role of theology (*douluriya*) or "heavenly studies" (*tianxue*), for which Aleni preferred the term "study of the way" (*daoxue*)[30]. Therefore, theology teaches people the beginning and the end of all existence, and without it, other studies are without basis[31]. Here again, the vocabulary used by Aleni will have a great influence on the contact between Jesuits and Chinese concerning their studies. Verbiest, Buglio and Gabriel de Magalhaes similarly used the terms *gewu qiongli zhe xue* or *qionglixue* for "philosophy" in their presentation of western studies in their *Yulan xifang yao ji* (Summary Record of the West, for the Emperor's Inspection, 1669)[32]. One also finds these terms in the Chinese biography of de Magalhaes by Verbiest and Buglio, *An xiansheng xingshu* (1677)[33], and in the Chinese biography of Verbiest by Tomas Pereira

---

29 Sambiasi, *Lingyanli shuo*, 23 II, p. 1137; he defined "philosophy" as *gewu qiongli zhe xue*, p. 1127.

30 Aleni, *Xixue fan*, 23 I, p. 49.

31 *Ibid.*, p. 50.

32 Verbiest, Buglio, Magalhaes, *Yulan xifang yao ji*, p. 6a, 13a (*Jap.-Sin.* II 158: photocopy FVP). These passages were copied from Aleni's *Xifang dawen* (Questions and Answers about the West); 1637 cf. J. Mish, "Aleni's *Hsi-fang Ta-wen*", *Monumenta Serica*, 23 (1964), 11, 26 (*j. shang* p. 12a; *j. xia* p. 14b).

33 Verbiest, Buglio, *An xiansheng xingshu*, p. 1a, 4; p. 2a, 2. (C. 1024).

and Antoine Thomas, *Nan xiansheng xingshu* (1688)[34]. In the preface to his *Budeyi bian*, Buglio calls "masters in fathoming principles and the investigation of things" as witnesses[35]. Related to this vocabulary is also the expression *gezhi* "to investigate (things) in order to extend (one's knowledge)." It is included in several titles of Jesuits' writings on natural sciences: Vagnoni's *Kongji gezhi* (1633)[36], Schall's *Kunyu gezhi* (1640)[37], and Verbiest's *Kunyu gezhi lüe shuo* (1674)[38]. Another very good example is a list of Chinese writings by the Jesuits published in Beijing around 1684[39]. The 212 works are divided into two sections: first, the religious works (*Tianzhu shengjiao shu*, *Libri sacri*, 123 works); then, the astronomical and philosophical works (*Lifa gewu qiongli shu*, 89 works with some biographies as well). This list of publications was made to prove the truth and reality of Christianity. It shows the Jesuit interpretation of the investigation of things: on the one hand, the importance they paid to the investigation in itself (seven books have the character *ce* in their title)[40], on the other hand, its connection with the sacred.

Verbiest fits into this picture (the previous list was established near the end of his life). Three of his religious writings are included in the first list, and nine of his scientific works in the philosophical list[41]. Verbiest was indeed a man of "practicality." He put much stress on practical investigation and verification (*ceyan*). He was primarily a man of action who spent most of his time in making instruments, calculating

---

34 Pereira, Thomas, *Nan xiansheng xingshu*, p. 1a, 4 (C. 1032); see also in *Shengjiao xinzheng*, 21 I, p. 361.

35 Buglio, *Budeyi bian*, 24, p. 226.

36 B. 227 (21 II, p. 843 ff); probably translation of *Libros meteorum* of Coimbra.

37 B. 298; Mss. presented to the emperor, on the exploitation of mines.

38 Although the title is very similar to Schall's work, it is a work on geography (C. 4922).

39 C. 7046 vii, viii; published in H. Bernard, "Les adaptations chinoises d'ouvrages européens: deuxième partie (1689-1799)", *Monumenta Serica*, 19 (1960), 351-359; see also H. Cordier, *L'imprimerie sino-européenne en Chine* (Paris, 1901), p. ix; Raccolta Generale-Oriente, n° 13 : photocopy FVP.

40 *Ceshilüe*, *Dace*, *Xiyang ce rili*, *Celiang fayi*, *Ceyan ji lüe*, *Cetian jishuo*, *Celiang quanyi*.

41 *Shengti dayi*, *Gaojie yuanyi*, *Jiaoyao xulun*, *Kangxi yongnian lifa*, *Jianping guizong xingtu*, *Yixiang zhi*, *Yixiang tu*, *Ceyan ji lüe*, *Xichao ding'an*, *Lifa budeyi bian*, *Kunyu tushuo*, *Chidao nanbei xingtu*.

eclipses and establishing calendars. He did not give all that much explanation of his action in his Chinese works, but something can be perceived from his controversy with Yang Guangxian. In the rejection of Yang Guangxian's *Budeyi* (I Cannot Do Otherwise), Buglio had answered the religious arguments, and Verbiest was engaged in the scientific response. This again confirms Verbiest's preoccupation with science. The "investigation and verification" (*ceyan*) is at the core of his argumentation:

> Investigating things and verifying them is the correct method. But investigation without verification is not a method. "To exhaust the human in order to correspond to heaven," these are my humble words. But "to force heaven to correspond to man," these are the words of (Yang) Guangxian. ... He considers the old method (*jiufa*) as good and the new method (*xinfa*) as perverse. ... According to his words, he only knows the principles of astronomy (*lili*) and does not know the method of astronomy (*lifa*). But the method comes forth from the principles, and the principles are proved by the method. This is a fixed theory from ancient times. Now, (Yang) Guangxian only wants to know principles, and not methods. It seems that his mind and heart fear that the method of astronomy can be investigated and verified through practice[42].

"To exhaust the human in order to correspond to heaven" can be called characteristic for Verbiest's life. This sentence also illustrates the difficult adjustment between "human" and "heaven." His religious preoccupation can, of course, be viewed from his religious writings. But about his way of combining these two elements, much can be learned from one of his major projects: the compilation of a collection of western philosophy. In full accord with his predecessors, Verbiest called his work *Qionglixue* (Study of Fathoming Principles)[43]. It is difficult to know how much translation work was done by Verbiest himself. He mainly collected the Chinese translations of the commentaries on Aristotle by the Jesuit University of Coimbra: *Commentarii Collegii Conimbricensis*[44]. Among these translations were the above-mentioned

---

42 Verbiest, *Budeyi bian*, 24, p. 35, p. 337.

43 B. 533, Xu 190-191; T.T. Dunin-Szpot, *Collectanorum pro Historia Sinica* (*Jap.-Sin.* 105 i), IV, ff. 38v-39*r*; the microfilm of FVP contains 7 *j*: 5 *j* of the first edition of Furtado's and Li Zhizao's *Mingli tan*; 2 *j* of the *Litui zhi zonglun* which is probably a continuation of the translation of Aristotle's *Dialecticam.* All are presented under Verbiest's name (*jishu*). On this work see also "Nan Huairen Qionglixue" in *Yishibao*, 19/11/1939.

44 On the translation of the works of Aristotle: cfr. H. Verhaeren, "Aristote en

works *De anima* (Aleni, Sambiasi), as well as the translations by Francisco Furtado of the *In universam dialecticam* (*Mingli tan*, 1636) and some other philosophical works by Vagnoni and Buglio[45]. The whole work is said to have counted sixty *juan*.

The memorial Verbiest sent to the emperor, with the request to have the work officially published, shows much of his intention (dated October 16, 1683, Kangxi 22/8/26)[46]. He explains to the emperor how he has worked for many years in his service in the Bureau of Astronomy. Astronomy does not only consist of figures, but there is also a principle (*li*) that establishes this method. Figures alone, without principles, are like a man with a material body (*xingti*) but without a spiritual nature (*lingxing*). According to Verbiest, much damage had been caused to Chinese astronomy by the fact that one only knew figures and no principles, and this was due to the fact that one did not know the method of deducing principles (*tuili zhi fa*). Philosophy is necessary for astronomy:

> From old times, those who studied astronomy had first to be experienced in the complete study of philosophy (*qiongli*). Indeed, astronomy is one branch among the philosophical studies. Without philosophy, there is no real study of astronomy: it would be like a tree without roots, from where could it produce branches?[47]

As can be observed here, astronomy was only one section of mathematical studies. Other sections were music, arithmetic and geometry. They were all classified under the *artes contemplativas* or *speculativas* (*theoreticas*) in the classification of the different sciences (*artes*) in the *Commentarii Collegii Conimbricenses*[48]. As mentioned before, in Jesuit formation mathematical studies were preceded by philosophical studies, and this was also the curriculum followed by Verbiest himself. As Verbiest further explains in his memorial, philo-

Chine", *Bulletin catholique de Pékin*, 22 (1935), 417-429.

45 According to B. 533 and Dunin-Szpot, *op. cit.*, *ff.* 38*v*.: Furtado's *Mingli tan*, Aleni's *Xingxue cushu*, Sambiasi's *Lingyanli shuo*, Vagnoni's *Feilu dahui* (1636), Buglio's *Wanwu yuanshi* (1677); maybe also Vagnoni's translation of the *Ethica* (*xixue xiushen, qijia, zhiping*) (1631), of the *Libros meteorum* (*Kongji gezhi*) (1633) and Furtado's translation of the *De coelo et mundo* (*Huan you quan*) (1628).

46 See Xu, pp. 191-192; *Xichao ding'an*, Rome, ARSI, *Jap.-Sin.* 67, 196a-199b.

47 Xu, p. 192; *Xichao ding'an*, 197a-b.

48 *Commentarii Collegii Conimbricenses*, Prima Pars (Lyons, 1610), 9-10; Verbiest, *Qionglixue (Mingli tan)*, p. 7b (*Mingyi*).

sophy is the root and basis of all studies. It is of prior necessity for military science, engineering, medicine, music, mathematics and other studies. It is also an important value in bringing peace to the world:

> In all discussions of events and things, if one does not take the *ratio* (*li*) as guide, one will often not be able to know how to discern where the truth or falsity of the principle is located. As a result, one will have contrary explanations and will not be able to come to a union. Therefore, one needs one sure method of deducing principles. This method of deducing principles can really serve men's mind and heart, and can become a help to the whole world - it can become a method to bring peace to the whole world (*ping tianxia*)[49].

One can notice the final reference to the expression of the *Daxue*. For Verbiest, philosophy applied to very concrete studies as astronomy also had an important moral value. Verbiest finally hopes that the emperor will support his work. This support will be comparable to the influence of Confucius and Mencius. The following quotation shows the centrality of reason for the Jesuits, and their hope for spreading this notion in China.

> The learnings of Confucius and Mencius have never been obliterated, and it is the same with the study of deducing principles. Surely, the *ratio* is a basic element of human nature (*xing*) and has always been engraved in the mind-and-heart of mankind. Now, if His Highness will spread the fame of philosophy (*lixue*), He will similarly engrave it in human beings' mind-and-heart as an eternal consolidation. The reason why human nature will never be destroyed is situated in this task. As such, the honor of His Highness will be similar to the brightness of Confucius and Mencius in heaven and on earth[50].

Verbiest's argumentation clearly shows how he fits in with the Jesuits' interpretation of the investigation of things and the fathoming of principles. The astronomical work, when preceded by the philosophical understanding of principles, could bring peace to the world, and was oriented towards the transmission of the knowledge of a personal God. For this end, the investigation of things through astronomical work had to be taken seriously. Verbiest's main missionary work was located in such work. He was not so much interested in Chinese philosophy. His concept of *li* and his stress on the importance of reason were, therefore, also typically western.

---

49 Xu, p. 192; *Xichao ding'an*, 198b; see also Dunin-Szpot, *op. cit.*, *ff.* 38*v*, where the book is described as *ars ratiocinandi*.

50 Xu, p. 192; *Xichao ding'an*, 198b-199a.

IV.
Chinese Interpretation

We will now focus on the Chinese reaction. When we speak of "Chinese," we can only speak of intellectuals who expressed their ideas in books. By the large number of publications by Jesuits as well as by Chinese, the intellectual flourishing of this mission was at the center of all studies. It should, however, be observed that most Jesuits spent their time with non-erudite and poor people, and this tendency developed as the seventeenth century progressed[51]. The Chinese reaction was both positive and negative, although a complete negative reaction was rather exceptional in that century. In fact, people mixed positive and negative responses in their interpretation of the Jesuits' studies and way of action. Generally speaking, the following evolution can be observed. In the early seventeenth century, a certain number of Chinese scholars accepted the Jesuit manner of the investigation of things, by which they fathomed principles in order to know God. Jesuits and Chinese scholars cooperated very closely in translating and editing scientific and religious works. By the end of the seventeenth century and the time of Verbiest, the situation had changed. Jesuits and scholars cooperated much less. This can be proved by the very small number of prefaces written by Chinese scholars to Jesuit writings in that period. Their interest in finding God, and even in fathoming principles (in the western sense), had diminished - but contact on the level of investigation of things continued. We will now elaborate this evolution in chronological order.

Xu Guangqi (1562-1633) is one of the most famous scholars who accepted the Jesuits' trilogy: investigation of things - fathoming principles - knowing God. He had a very clear idea of this succession of values and of the unity of Jesuit studies. In his preface to Ricci's *Jihe yuanben*, he explains these studies as follows:

> Generally speaking, (Ricci) has three kinds of studies: self-cultivation and serving Heaven (*xiushen shitian*), the investigation of things and the fathoming of principles, which is the least of the three.

---

[51] This is also confirmed by an interesting traveling journal of a lay visitor in Beijing in 1691, Gemelli Careri, *Voyage autour du monde* (Paris, 1727), 4:197: "D'ailleurs (les missionnaires) ne tirent pas grand secours des Chrétiens chinois qui sont pauvres; car les riches et les Mandarins n'embrassent point le Christianisme, pour ne pas renoncer à ce grand nombre de femmes."

Mathematics is one branch in the principles of things[52].

In Xu Guangqi's preface to the *Jianping yishuo* (Treatise on the Planisphere, 1611), Ricci's study is described as "knowing heaven, serving heaven, fathoming principles and exhausting one's nature (*zhitian shitian qiongli jinxing zhi xue*)"[53]. The expression *jinxing* probably refers to Mencius VIII A, 1: "By exhausting one's mind-and-heart, one will know one's nature and so be able to know heaven." Xu Guangqi, even if he is known for his scientific work, gave a positive response to all these aspects. In the next quotation, taken from his preface to Sabatino de Ursis's work on irrigation methods *Taixi shuifa*, 1612, he explains the use of the whole of western teachings. We find, also, one of the typical formulations of Xu Guangqi on the benefit of Christianity:

> ... I have often said that their (religious and moral) teachings can help Confucianism and change Buddhism (*buru yifo*). Moreover, they have a study of investigating things and fathoming principles, by which principles of all things and events in and outside the world can be completely answered and be explained even to the smallest detail. If one reflects further about it, after some years, one will moreover see that their sayings are necessary and cannot be changed[54].

What can be observed in these quotations is the use of a typically Chinese vocabulary. It is, of course, through people like Xu Guangqi that the Jesuits got to know the importance of Chinese notions such as "serving and knowing heaven," "investigation of things and fathoming principles ..." The important presence of these notions in the texts of Chinese scholars shows how much it corresponded to their own interest. This is also very clear in the writings of Li Zhizao (1565-1630). His stress on practicality (*shixue*)[55] shows how the introduction of western science fitted well in the movement for real learning in late Ming China. Well known is his memorial to the emperor, wherein he asks the emperor to translate the western books. He proves how the Jesuits really engage in concrete studies (*shixue*) with their books on the calendar, mathematics, geography, medicine, the investigation of things and fathoming principles, music, and the like. The translation of their

52 Ricci, *Jihe yuanben*, 23 IV, p. 1927.

53 S. de Ursis, *Jianping yishuo*, 23 V, p. 2723.

54 Xu, p. 308.

55 For example, the preface to Ricci's *Hungai tongxian tushuo* (1607), Xu, pp. 263-264; preface to Ricci's *Tong wen suan zhi* (1613), 23 V, p. 2783.

books, which contain many new aspects, would be of great use[56].

Li Zhizao also had an inclusive interpretation of western religion and science. This can be seen in several of his prefaces[57], but appears most clearly in his collection of twenty western writings, *Tianxue chuhan*, 1629. Writings on studies in the West, the Nestorian stèle, friendship, God, self-cultivation, geography, water conservation, mathematics, astronomy, and the like, all belong to the "heavenly studies" (*tianxue*). "To know and to serve Heaven" (*zhitian shitian*) is presented as the essence of heavenly studies and is not regarded as opposed to the meaning of the Six Classics[58]. Besides his main interest in the western sciences, Li Zhizao opened himself to the idea of a Creator:

> In order to study, one has to know Heaven; and that is to know the mystery of creation, to know that creation has a master, and to know the Creator's mercy[59].

The importance of Heaven was most often expressed in Confucian vocabulary, but this sentence illustrates how new ideas could also be introduced into Chinese culture.

One can also look at the interpretation of the idea of *li* by these Chinese scholars. We have seen the negative interpretation of Ricci, which corresponded to his refusal of Song Neo-Confucianism. Chinese scholars who converted to Christianity such as Xu Guangqi, Li Zhizao and Yang Tingyun (1562-1627), were well-versed in Neo-Confucianism. According to Neo-Confucianism, the origin of all things was explained by the interaction of *li* (principle) and *qi* (matter-energy). If these scholars, because of the acceptance of a Creator, rejected the primary role of *li*, they still reserved an important place for *li*. This is very clear in the writings of Yang Tingyun, who was mainly interested in philosophical questions[60]. It can also be observed in Li Zhizao's compilation of the above-mentioned *Tianxue chuhan*. He divided the twenty works into two sections, each of ten works: *li* (principle) and *qi* (instrument). This is one pair of basic concepts of the categorical system of traditional Chinese philosophy. It has its equivalences in pairs like: *li* (principle) -

---

56 Li Zhizao, *Qingyi xiyang lifa dengshu shu*, Xu, 253-256.

57 See note 56; see also preface to Furtado, *Huan you quan*, in Xu, 198-200; preface to Ricci, *Huan rong jiao yi* (1614), in Xu, 275-276.

58 *Tianxue chuhan*, 23 I, 1, 5-6.

59 Preface to Furtado, *Huan you quan*, in Xu, p. 198.

60 Cf. N. Standaert, *Yang Tingyun, Confucian and Christian in Late Ming China* (Leiden, 1988), pp. 112-114.

qi (matter-energy); *dao* (way) - *qi* (instrument); *tian* (heaven) - *ren* (man); *xin* (mind-and-heart) - *wu* (matter, thing); *qiongli* (fathoming principles) - *gewu* (investigation of things)[61]. These are contradictory but complementary categories; one cannot exist without the other. It is remarkable to see how Li Zhizao, in his "heavenly studies", saw the whole of western learning as both divided and united in these two aspects.

It is more difficult to perceive the response to the western interpretation of *li* as *ratio*. *Li* surely also had the meaning of "reason" in Chinese thought, but not in the rational sense of western philosophy. The Chinese interpretation of *li* as "principle" seems to have prevailed. Li Tianjing in his preface to the translation of Coimbra's *In Universam dialecticam* (*Mingli tan*, 1636), when referring to the intelligent capacity of man, attributes it in a Chinese way to the human mind-and-heart (*renxin zhi ling*). He stresses the importance of deduction and discussion (*tuilun*) instead of dialectics (*tuili*) for the knowledge of truth and reality and as a help for one's intellect. *Li* remains basically "principle."

> The *Mingli tan* essentially wants people to understand the principles of truth and reality and to be a help for one's capacity of understanding. Therefore, deduction and discussion are a ladder. Reading its meaning, it seems mysterious; but when one appropriates it, its principles are all correct. It is really fundamental and basic for the investigation of things and the fathoming of principles ... In the past, people often placed principles inside material things, like gold in gravel. The gold seeker has to scour and to rinse it to avoid having it covered by earth. The one who seeks principles needs to have a method of deduction and discussion, to avoid having it covered by mistakes and errors. The *Mingli tan* is a method of deduction and discussion[62].

One can notice that the illustrations are similar to those used by Aleni in explaining western philosophy in his *Xixue fan*. According to Li Tianjing, even the method corresponds to the Confucian study of fathoming principles and exhausting one's nature.

For Chinese scholars, the investigation of things was also directly

---

61 Cf. Zhang Liwen, *op. cit.*, p. 685; Tang Yijie, "Questions Concerning the Categorical System of Traditional Chinese Philosophy", *Social Sciences in China* (1982/4), 190-211.

62 Preface to Furtado, *Mingli tan*, in Xu, pp. 194-195 (Li Tianjing was a disciple of Xu Guangqi and his successor at the Bureau of Astronomy); compare also with the preface of Li Cibin (Li Zhizao's son) to the same work, Xu, pp. 194-195.

related to the "extension of knowledge" (*zhizhi*) as mentioned in the *Daxue*. Qu Shibo discusses in his preface to Aleni's *Xingxue cushu* the Aristotelian division of three kinds of souls as presented by this work (vegetative soul of plants, *shenghun*; the sentient soul of animals, *juehun*; the rational soul of the human being, *linghun*). He explains how the extension of knowledge finds its origin in the investigation of things.

> For Confucians, in order to extend one's knowledge, one first has to investigate things, and things have a sentient soul. One's intelligence (*ling*) can really master it. But, to recognize what intelligence is, one should first know what sensation is (*jue*)[63].

If the Chinese appreciated or accepted the knowledge of Heaven and the exhausting of one's nature as proposed by the Jesuits, they especially appreciated that the Jesuits' fathoming of principles and extension of knowledge passed through the investigation of very concrete things such as mathematics, the calendar, medicine, and the like[64]. For those who accepted Christianity, the natural sciences were not merely a pole of attraction which could be avoided, but had a real significance. It contrasted sharply with the many philosophical groups inspired by Buddhism and the Wang Yangming School, which spent their time in empty discussions and quibbling on principles without any investigation of things. The pedagogical manner of the Jesuits helped some Chinese to rediscover their own tradition, as can be seen in Xu Xuchen's preface to the *Xixue fan*.

> The *Xixue fan* suddenly made me realize that the investigation of things by our Confucians was originally not at all overflowingly aimless, and that the extension of knowledge had not to be something completely abstract, but that the investigation (of things) for the extension (of knowledge) really led to government and peace, and that the government and peace necessarily started with the investigation (of things) for the extension (of knowledge). ... According to the *Xixue fan*, study starts with the investigation of things for the extension of knowledge; this also corresponds to the teaching of the old sages[65].

---

63 Preface to *Xingxue cushu*, in Xu, p. 211.

64 See also: Bi Gongchen's preface to Terrentius's work on the human body *Taixi renshen shuogai*, Xu, pp. 303-304; Wang Zheng's (1571-1644) preface to Terrentius's work on western mechanical instruments *Qiqi tushuo*, in Xu, p. 298.

65 Aleni, *Xixue fan*, 23 I, pp. 21, 24.

This preface also contains a reference to the moral influence of the investigation of things: governing the countries and bringing peace to the world. This is the last aspect accepted by those who give a positive response. A person like Han Lin (1601-1644 or 1648) was mainly interested in ethics. He described western moral studies in the same traditional vocabulary as was adopted by Aleni: self-cultivation, regulation of the family, governing the country and bringing peace to the world. According to Han Lin, the studies of western scholars are similar to those of Chinese scholars[66].

It is important to notice that most of these prefaces refer to the Chinese tradition. Western investigation of things, fathoming principles, knowing Heaven and morality can be accepted because they correspond to the ancient sages, the Confucian tradition. This does not mean that no differences existed, but the import of new elements had to be justified with regard to the native culture. The correspondence between East and West was stressed by several scholars by quoting the famous sentence of Lu Jiuyuan (1139-1192): the sages in the Eastern and in the Western Seas have the same mind and the same principle (*tongxin tongli*)[67]. So far the positive reactions. All these quotations date from late Ming times. As noticed before, at that time there existed a relatively extensive cooperation between the Jesuits and the Chinese, and a considerable number of these scholars also converted to Christianity. There were also a number of those scholars who cooperated with the Jesuits without converting to Christianity. They were interested in the Jesuit manner of "investigating things," but not so much in their "fathoming of principles and finding Heaven." This group will become the most important one in early Qing period.

Chinese who completely rejected in an active way western religion and science are exceptional. The place of the investigation of things is not at the core of the conflict with Yang Guangxian. The discussion was mainly focused on the use of the "new method" of astronomy, as opposed to the "old method." In fact, this conflict reflected the difficulty

---

66 Preface to Vagnoni, *Tong you jiaoyu* (1628), in Xu, p. 217; see also *Shengjiao xinzheng* (1647), 21 I, p. 280.

67 Cf. Yang Tingyun, in Standaert, *op. cit.*, pp. 200-201; Li Zhizao, preface to *Tong wen suan zhi*, 23 V, p. 2782; preface to *Tianzhu shiyi*, 23 I, p. xx; Hu Shi'an, preface to *Chaoxing xueyao* (1650-1680), in Xu, p. 189; Mi Jia, preface to *Xifang dawen*, in Xu, p. 300; Bi Gongchen, preface to *Taixi renshen shuo*, in Xu, p. 304; Li Jiugong, preface and introduction to *Lixiu yijian*, 21 I, p. 435, 5; p. 442, 8; anonymous preface to *Daoxue jiazhuan*, p. 1a, 6; also in Buglio, preface to *Budeyi bian*, 24, p. 331.

of accepting a considerable amount of new elements in a culture. Most scholars continued to accept western science, but rejected western philosophy and religion. Fang Yizhi is an important representative of this group. Fang argued that "comprehending seminal forces" (*tongzhi*) and understanding the "extended principles of things" (*wu zhi zhili*) had to be founded on "material investigation" (*zhice*). The two approaches are concomitant. Neither can be neglected, although "material investigation" has the primacy by being based more immediately on "things." With his contemporaries, Fang was critical of those who sought to extend their knowledge without reference to "things" in the sense of objective phenomena. At the opposite extreme are, according to him, the Jesuits, who passed beyond the "material investigations" into other realms, particularly religious. "The knowledge from the Far West which entered (China) in the Wanli period is detailed in 'material investigations' but deficient in speaking about 'comprehending seminal forces.' This being so, in the estimation of truly knowledgeable scholars, their 'material investigations' are still not adequate"[68]. Similar opinions were expressed by the famous philosopher Huang Zongxi (1610-1695) and the philologist Zhang Erqi (1617-1678)[69].

This is also the context Verbiest lived in. Chinese who wrote prefaces to his writings are very limited. Xu Erjue, Xu Guangqi's grandson, wrote a preface to his *Kunyu gezhi lüe shuo* in which he praises the Jesuits for their investigation of things and their fathoming of principles by geographical research[70]. In his preface to the *Xifang yao ji*, Zhang Chao (circa 1697) expresses his admiration for the scientific work of the Jesuits but his reservation towards their religious ideas[71]. This shows that the theme was still actual. Verbiest himself hoped to transmit to the Chinese the western concomitance of investigating things and fathoming principles. We have seen the importance he paid to his major compilation *Qionglixue* by asking the emperor to have it published. The answer from the court, however, was negative. The emperor said that "the style of the book was perverse, erroneous and illogical." Grand Secretary Mingzhu (1635-1708) and others reported: "Saying that man's knowledge and memory belong to

---

68 Cf. Peterson, *op. cit.* pp. 382, 399; quotation from Fang Yizhi, *Wuli xiaozhi* (Notes on the Principles of Things, 1660), p. 1b.

69 Cf. Gernet, *op. cit.*, pp. 56-57.

70 Verbiest, *Kunyu gezhi lüe shuo*, p. 1a, 4 (C. 4922).

71 Gernet, *op. cit.*, pp. 57-58; see *Zhaodai congshu*, *jia*, *juan* 27.

his brains is in complete contradiction with the reality of the principle"[72]. In Chinese philosophy, as we observed in the preface of Li Tianjing, knowledge was attributed to one's mind-and-heart (*xin*). The emperor finally ordered the sending back of the original set of books to Verbiest, without giving a reason for it. This refusal clearly shows the discrepancy between the western and Chinese methods for the fathoming of principles. Western philosophy was judged as incompatible with Chinese philosophy and too different from the ancient wisdom. The disappointment and frustration of Verbiest must have been very deep, especially because of the value he had attached to this voluminous work[73].

This tendency, however, will continue to prevail. The clearest example is surely the *Siku quanshu zongmu* (The Table of Contents of the Complete Library of the Four Treasuries)[74]. This work contains the description of the books which have been included in the imperial library, and also a commentary on books which have not been included. Thirty-six western writings received a description, of which twenty have been included in the Library: the included books are only scientific works. The books of the *qi*-section of Li Zhizao's division of the *Tianxue chuhan* were incorporated, but the books of the *li*-section have been refused, except for Aleni's geographical work *Zhifang waiji* (1623). Although the *Zhifang waiji* is considered as exaggerating and ostentatious, it is kept for its extraordinary and unusual aspects. Likewise, Verbiest's explanation of the world map *Kunyu tushuo* is included in the Library as curiosa, whereas two others of his description of the West *Kunyu waiji* and *Xifang yao ji* only received a commentary because they are taken as exaggerated and boastful. No other works of Verbiest were taken into consideration. Much evidence of the rejection of Verbiest's geographical writings in his time could not be found. These works surely contained a lot of extraordinary data. But the reaction of the compilers of the *Siku quanshu zongmu* illustrates well what already existed in Verbiest's time. The best description concerning this attitude with regard to the investigation of things and fathoming principles can be found in the commentary on Aleni's *Xixue fan*. After

---

72 *Kangxi qiju zhu (Beijing, 1984), 2: 1104 (Kangxi 22/11/14 = December 31, 1683).*

73 Cf. Dunin-Szpot, *op. cit.*, f. 39r, *frustratus*.

74 Cf. Zhong Mingdan (N. Standaert), "*Siku quanshu zongmu* duiyu *xixue* de pingjia" (The Evaluation of Western Learning by the *Siku quanshu zongmu*), *Zhongwai guanxi shi xuehui tongxun*, 4 (1983), 4-11.

the short description of the several branches of western studies, as described in Aleni's work, the commentary says:

> In their education, there is an order; generally speaking, from literature to the study of principle, and the latter is considered as fundamental. The branch of literature corresponds to the minor studies in China and the branch of philosophy to the higher studies. The branches of medicine, law and education (in fact, canon law) are their professions. The branch of the Way is, according to their method, the extreme of exhausting one's nature and coming to the ultimate goal of one's destiny (*jinxing zhiming*). In using their strength, the investigation of things and fathoming principles is the basis; the understanding of the essence (*ti*), which leads to praxis (*yong*), is merit. This pattern of proceedings is similar to that followed in the Confucian studies. However, the things they investigate are secondary aspects of tools and figures, and the principles they fathom are even more irrelevant, prodigious and intractable. That's why theirs is a heterodox study (*yixue*)[75].

This was written about a hundred years after Verbiest. Not only was western religion rejected; even the attitude towards science had become critical.

V.
Conclusion

The purpose of this article was to analyze the presence of the notions of "investigation of things, fathoming principles and finding God" in the Chinese writings of Jesuits and Chinese, and to see if these notions could be a pattern for the analysis of the development in the seventeenth-century contact between the Jesuits and the Chinese. We have focused our attention on these terms and, by their large presence in these writings (more than thirty-five references), it can be concluded that they are at the core of the encounter between China and the West at that time. The vocabulary used is very important. The Jesuits, through their contacts with Chinese scholars, discovered in Chinese tradition the proper expression of their way of acting, and could therefore use the terms *gewu, qiongli, zhitian, shitian*, and the like. The use of this vocabulary, however, does not mean a completely identical interpretation. The new sciences brought by the Jesuits surely corresponded to what the Chinese expected. Their Aristotelian interpretation of *li* was rather different, and their notion of Heaven concerned a personal God

---

75 *Siku quanshu zongmu* (Beijing, 1981), p. 1080 *xia*.

(*Tianzhu*). Inversely, through their contacts with Jesuits, Chinese scholars found in Jesuit thought and action, in their writings and explanation of their formation, and in their practical scientific work, the confirmation of their own interest.

We have also noticed the difficult adjustment between "investigation of things" and "fathoming principles." They have been considered as a unity: "fathoming principles" had to pass through the "investigation of things" and the "investigation of things" had to lead to "fathoming principles." The separation of these two aspects in our present world does not correspond to the seventeenth century. We have seen how the Chinese stressed the importance of practicality (*shixue*) in order to avoid empty discussions. Likewise, the missionary work of the Jesuits, who tried to find God in all things, can be found not only in their religious action, but in their scientific work as well. This does not mean that individual persons could not be more interested in one aspect or another. But all of them sought in some way the combination of both.

This analogy also allows us to analyze the evolution of the contacts between Jesuits and Chinese scholars in the seventeenth century. In the scholarly activity of the Jesuits of that century there seems to have existed a definite pattern in three steps: from the investigation of things, to the fathoming of principles, to the transmission of the experience of God. Ricci and Aleni are examples of Jesuits whose publications contain in a balanced way all three aspects. When the Jesuits were entrusted with demanding tasks in the Bureau of Astronomy and at the court, greater specialization was necessary. Schall and Verbiest spent most of their time in scientific work, while Buglio wrote more religious works. Verbiest, of course, also compiled the enormous *Qionglixue*. Their contacts with Chinese scholars went through a different evolution. In late Ming China, an interesting group of scholars joined the Jesuits in their pattern: they were interested in science, philosophy, morality, and the knowledge of Heaven. But as the seventeenth century progressed, Chinese scholars who continued to be interested in the westerners' investigation of things were no longer interested in their way of fathoming principles, although they continued to fathom principles in a Chinese way. This discrepancy shows how difficult the dialogue was that existed in Verbiest's time, and why his work on philosophy was refused for publication by the emperor. The religious dialogue with scholars became even more limited. This tendency seems to have continued, to develop further during the eighteenth century and, because science had developed on its own, even the importance of the introduction of western science had diminished considerably.

The pattern of *gewu - qiongli - zhitian* can also be applied to present-day contacts with China. The theory of cultural exchange explained in the beginning of this article is confirmed. Because of its internal development and search for economic progress, China has opened itself to the West and accepted western science and technology. The entire contact between China and the West is based on this level of *gewu* (investigation of things). China's "fathoming of principles" is very limited: even creative Marxist thought hardly exists. The West still has some philosophy, but is only interested in exporting science and technology - the "knowing of God" is not the concern of most contemporary westerners who enter into relationship with China. This contact on the level of *gewu*, however, can be an occasion for China and the West to rediscover the rich *qiongli* level of each tradition. This rediscovery is necessarily linked with dialogue and can, therefore, be an incentive for mutual enrichment.

THE RELIGIOUS WRITINGS OF FATHER FERDINAND VERBIEST

JOSEPH HSING-SAN SHIH, S.J.

Joseph Hsing-san Shih, S.J. is a native of Ningbo, Zhejiang province, China. He entered the Society of Jesus in 1944. Besides degrees in philosophical and theological studies, he holds a master's degree in East Asian studies and a doctorate in missiology. Since 1965 he has been a professor of mission catechetics and Chinese religion and philosophy at the Pontifical Gregorian University, Rome, Italy. Since 1978 he has also been a redactor in the Chinese section of Vatican Radio. His publications include a number of articles on Chinese religion, chiefly in "Studia Missionalia."

The year 1940 was the fourth centenary of the foundation of the Society of Jesus. To commemorate the event, Father Xu Zongze, then in charge of the Zikawei Library, published a book entitled *Ming Qing jian Yesu huishi yizhu tiyao* (An Epitome of Jesuit translations and writings from the late Ming and early Qing Period)[1]. The book is divided into ten chapters. The first chapter serves as an introduction; the last is something of an appendix, containing three lists of pertinent books preserved in the Zikawei Library, the National Library of Paris, and the Vatican Library. The ninth chapter of the book, entitled "Brief Biographies of Translators and Authors," owes its materials to Louis Pfister's *Notices biographiques et bibliographiques sur les Jésuites de l'ancienne Mission de Chine*[2]. The substance of the book is, however, to be found in the other seven chapters.

These seven chapters are divided according to the seven categories of works reviewed. These are *shengshu lei* (catechetical and liturgical works), *zhenjiao bianhu lei* (apologetical works), *shen zhexue lei* (theological and philosophical works), *jiaoshi lei* (church histories), *li-*

1 Shanghai, 1949.

2 Two volumes, Shanghai, 1932-1934.

suan lei (astronomical and mathematical works), *kexue lei* (other scientific works), and *geyan lei* (ethical works). Of these seven categories, two refer to natural sciences and technology while the other five treat religion, namely Catholicism. Following the method adopted by the famous *Siku quanshu zongmu tiyao* (The General Catalogue of the Four Sections of the Entire Imperial Library)[3], Xu wrote separate introductions to the seven categories of works and reviewed and annotated each work listed under the individual categories. The annotations consist, in his own words, in indicating the name of the translator or author, giving the title and place of publication, and reproducing the preface if the work in question has one. If there is no preface, the table of contents is reproduced instead.

Ferdinand Verbiest dealt mostly with scientific and technological matters. Pfister mentions only seven of his religious works. Xu does the same. However, of the seven works, only five were found in the Zikawei Library. They are also the only ones annotated in Xu's book. To borrow Pfister's translations, these are *Jiaoyao xulun*[4] (Abrégé de la doctrine chrétienne), *Gaojie yuanyi*[5] (Du sacrement de pénitence), *Shengti dayi*[6] (Réponses à des questions sur la Sainte Eucharistie), *Shan e bao lüe shuo*[7] (Petit traité de la rétribution du bien et du mal) and *Daoxue jiazhuan*[8] (Traditions de famille sur la doctrine).

Working from these two lists, I did some research of my own in the Archives of the Jesuit headquarters in Rome. Thanks to the excellent annotated catalogue compiled by Father Albert Chan, the task was easy. I soon found not only the two works lacking in the Zikawei Library, namely *Wang tui jixiong zhi bian*[9] (Discussion sur les vains calculs des présages bons ou mauvais) and *Wang zhan bian*[10] (Discussion sur la fausseté des sorts), but also a third work, one not even mentioned in Josson-Willaert's more extensive bibliography[11]. The title of this book

---

3 *An Annotated Bibliography of Selected Chinese Reference Works* (compiled by Ssu-yu Teng and Knight Biggerstaff), Peiping, 1936, p. 23, n° 1.

4 Louis Pfister, *Notices*, p. 352, n° 1.

5 *Ibid.*, p. 353, n° 2.

6 *Ibid.*, p. 353, n° 3.

7 *Ibid.*, p. 353, n° 4.

8 *Ibid.*, p. 353, n° 7.

9 *Ibid.*, p. 353, n° 5.

10 *Ibid.*, p. 353, n° 6.

11 H. Josson and L. Willaert (eds.), *Correspondance de Ferdinand Verbiest*

is *Tianzhujiao sangli wenda* (Questions and Answers on the Funeral Rites of the Catholic Church)[12]. In his annotated catalogue, Chan makes the following comment on the subject:

> Neither the catalogue of the Zikawei Library nor those of the Vatican or Courant give the title of this book, which contains five questions, together with their respective replies:
> 1. the meaning of the prayer;
> 2. prayers at the funeral are for the souls in purgatory;
> 3. although the penance in purgatory is determined by God, human efforts can be of help to lessen the penance;
> 4. the meaning of Catholic funeral ceremonies;
> 5. the reasons why the Catholic Church forbids burning paper money for the deceased.

Folio 6a has a comment in red ink written by the Kangxi emperor: *yan zheng li shun* (the words are well chosen and the ideas convincing); and folio 7a, *shuode shuangkuai touli* (straightforward and thorough); and another comment: *youli youli dan kong dezui yu ren* (right, right, but it may offend someone)[13].

Tianzhujiao sangli wenda sheds much light on how the Jesuits handled practically the problem of "the Chinese rites." Unfortunately, this work has been generally neglected. However important as it is, the present study does not deal with this book. My knowledge and interest are limited rather to another book, *Jiaoyao xulun*, which Pfister translates as "Abrégé de la doctrine chrétienne." I would prefer to paraphrase it as "An Orderly Exposition of the Essentials of the Christian Teaching"[14]. On the wide diffusion and, consequently, also the importance of this book, Chan has this learned annotation:

> Pfister, p. 352 n°1 quotes from Van Hee that this book was incorporated in the *Siku quanshu* for its simple, pure and logical style.

(Brussels, 1938), pp. 569-585.

12 ARSI, *Jap.-Sin., I*, 38/1. Supplementary note of N. Standaert: H. Bernard, "Les adaptations chinoises d'ouvrages européens: deuxième partie 1689-1799," *Monumenta Serica*, 19 [1960], 476) mentions a fifth work: *Tianzhujiao sangli wenda* (Beijing between 1672 and 1686; revised by Grimaldi and Pereira). As reference he gives the Chinese Beitang Library n° 1939.

13 Annotated catalogue, p. 37.

14 N. Standaert supports this interpretation, stating that Verbiest called this book *Ordo Catechizandi*, "because the catechists often did not follow an order" (Letter of Verbiest to Philippe Marini, August 24, 1678).

The *General Catalogue*, however, has no trace of this book. Van Hee may have been mistaken. A. Wylie, *Notes*, p. 177: A version of this book was afterwards printed in Manchu, which was denounced in an imperial edict of 1805. Pfister, p. 353 n° 1 quotes from Wylie, *Notes*, p. 142: "Il a été aussi traduit et imprimé en coréen vers 1864." Our edition of Wylie *Notes* does not give this text.
This text was reprinted in Zikawei (ed. 4a, 1914); another edition was printed in Xianxian (1932); in Hong Kong (1905); in Yanzhoufu, Shandong (ed. 3a, 1931); and in Beiping (1931).
Renwu zhuan II, 178, gives a number of early editions of this book: 1677, 1799, 1848, 1867. It was also translated in the Shanghai dialect under the title *Fangyan jiaoyao xulun*. There was a French translation[15].

The main text of the book which I have examined at the Jesuit Roman Archives consists of sixty-seven folios. The first twenty folios deal with the three fundamental truths regarding the existence of God, the immortality of the human soul, and retribution in the afterlife. They correspond substantially to the contents of Matteo Ricci's famous catechism, or *Tianzhu shiyi* (On the Existence of the Lord of Heaven). Of the remaining forty-seven folios, thirty are dedicated to the explanation of the Apostles' Creed, four to the Lord's Prayer, one each to the Hail Mary and the prayer of the Sign of the Holy Cross, and, finally, six folios are used to state the rules for receiving the sacrament of Baptism. The body of the book is, therefore, dedicated to the exposition of the Apostles' Creed; and no explicit explanation of the seven sacraments is to be found in it. In the two folios of preface preceding the main text, the author explains the order he followed in writing the book[16].

What was the nature of the *Jiaoyao xulun*? How did it relate to the catechetical tradition of the Jesuit missionaries in China? What influence did it have on the formation of Chinese Christianity? These are, indeed, the questions that interest us in the present study.

To begin with, we recall a declaration of Ricci regarding the first publication of religious books by the Jesuits in China. He stated:

> At that time permission had already arrived from the inquisitors for us to print the necessary books for our faithful; what Father Matteo Ricci first did was to publish a new version of the prayers and other things necessary for the new Christians, which he did with the help of

---

15 Annotated catalogue, p. 699.

16 ARSI, *Jap.-Sin. II*, 47; Annotated catalogue, p. 699.

> the other priests. In addition to these prayers, other things of the Christian teaching were included with brief annotations, especially on the seven sacraments[17].

In this, and in subsequent passages, Ricci mentioned three books: two were already written and published; the third was still in preparation. Ricci called the two books already published: the *Dottrina Christiana* and the *Catechismo*. The former, he said, was reserved to the catechumens and Christians, while the latter was addressed to the general reading public[18].

Ricci's Catechism, or *Tianzhu shiyi*, is well known to sinologists[19]; but what he called *Dottrina Christiana* may puzzle many of them. Actually, one of the first copies of this booklet is to be found in the Archives of the Congregation for the Evangelization of Peoples. It bears the title *Tianzhu jiaoyao* (The Essentials of the Teaching of the Lord of Heaven)[20]. The Jesuit Roman Archives has a similar copy with the title *Shengjing yue lu*[21]. "From the title of this book", notes Chan pointedly, "one may easily make the mistake of thinking that it is a synopsis of the Scriptures. In reality, it is a small prayer book, rightly translated by Courant as *Petit recueil de prières*"[22]. However, it must be noted that this is a particular collection of prayers, giving the Our Father, the Hail Mary, the Ten Commandments, the Creed, the Sign of the Cross, the seven works of corporal mercy, the seven works of spiritual mercy, the eight beatitudes, the seven capital sins, the seven remedies against the seven capital sins, the three theological virtues, the five senses, the three faculties, and the seven sacraments. It was meant to be a manual for the instruction of the catechumens and neophytes[23].

Both copies - the *Shengjing yue lu* and the *Tianzhu jiaoyao* - have the same editorial postscript. Chan translates it as follows:

---

17 Pasquale M. d'Elia, *Fonti Ricciane* (Rome, 1949), 2: n° 708. Translation borrowed from Albert Chan, Annotated catalogue, p. 172.

18 *Ibid.*, n° 709.

19 An excellent English translation of the book is *The True Meaning of the Lord of Heaven (Tianzhu shiyi)*, translated with introduction and notes by Douglas Lancashire and Peter Hu Kuo-chen, edited by Edward J. Malatesta, St. Louis, 1985.

20 D'Elia, *op. cit.*, 2:289, n° 2.

21 ARSI, *Jap.-Sin. I*, 126.

22 Annotated catalogue, p. 173.

23 D'Elia, *op. cit.*, n° 708.

> The above is a translation of the essential Christian teaching. After repeated correction, it finally agrees with the original. We can now print it with assurance, and the translation is different from that of the first edition. Some of the terms are rendered by transliteration and, to make them clear, we have added some annotations to explain the meaning of each of them[24].

Hence it is clear that what Ricci called *Dottrina Christiana* was a catechism in the modern sense of the word, while what he called *Catechismo* was, in fact, an apologetical treatise. As for the third book, Ricci described it as "a more copious explanation of the new *Dottrina Christiana*"[25]. According to Pasquale M. d'Elia, Ricci intended to write such an explanation in 1605 and, in 1610, Diego Pantoja had something of the kind ready for print[26]. We have found no trace of the "explanation" which Ricci intended to write in 1605, but Pantoja did publish two volumes of his *Pangzi yi quan* which, as Pfister puts it, contains the exposition of the Apostles' Creed, with two small treatises on the angels and the soul[27]. He started, thus, a series of works which would constitute "a copious explanation of the new *Dottrina Christiana*" to which Verbiest's *Jiaoyao xulun* belonged. In fact, as its title suggests, the *Jiaoyao xulun* was a sort of "copious explanation of the new *Dottrina Christiana*." This answers our first question concerning the nature of the book.

But there had been other books of the same kind. Two of them merit particular attention. They are the *Tianzhu shengjiao qimeng* (First Instructions on the Holy Religion of the Lord of Heaven), by Luo Ruowang (Joao da Rocha)[28] and the *Tianzhu jiaoyao zhilüe* (Simple Explanation of the Essentials of the Christian Teaching), by Wang Fengsu (Alfonso Vagnoni)[29]. On the former, Chan makes this comment:

> The translation is based on the famous *Cartilha* of Marco Jorge (?-1571), a Portuguese Jesuit who wrote this little book, published in 1566, for the instruction of the young and rural people. The book was widely used throughout Portugal and the Portuguese colonies and

---

24 Annotated catalogue, p. 173.

25 D'Elia, *op. cit.*, n° 709.

26 *Ibid.*, 2:291, n° 1.

27 Pfister, *op. cit.*, p. 72, n° 5.

28 *Ibid.*, p. 69, n° 1.

29 *Ibid.*, p. 91, n° 1.

> won the title of "The Golden Book." The translation was adapted for the use of the Chinese, especially for the beginners; hence the name *qi meng*, instruction of the young. The book is in the form of dialogue, in colloquial style, and clearly explained. The Chinese text was most probably revised by Yang Tingyun[30].

Concerning the other work, namely, the *Tianzhu jiaoyao zhilüe*, d'Elia gives the following description:

> It explains word for word:
> 1. the Our Father,
> 2. the Hail Mary,
> 3. the Ten Commandments,
> 4. the Apostles' Creed,
> 5. the Sign of the Holy Cross,
> 6. the seven sacraments of the Church,
> 7. the fourteen works of corporal and spiritual mercy,
> 8. the eight beatitudes,
> 9. the seven capital sins,
> 10. the three theological virtues,
> 11. the four moral virtues,
> 12. the five senses,
> 13. the three faculties[31].

Verbiest's *Jiaoyao xulun* was published for the first time in 1670, in the ninth year of the reign of the Kangxi emperor, at a time when the Church in China was still under another persecution but would regain renewed confidence and vigor. Only a few years earlier, Verbiest had edited, in collaboration with Father Ludovico Buglio, a new prayer book entitled *Shengjiao rike* (Daily Exercise of the Holy Religion)[32]. It was a revised edition of the collection of the texts of prayers written by various Jesuit missionaries. Paul Brunner, in his doctoral thesis *L'Euchologe de la Mission de Chine*, calls it "l'édition *ne varietur*", and states:

> En effet, encore aujourd'hui les prières se récitent dans le texte qui fut fixé à cette époque. Après cette date le Manuel nouveau gagnera encore en extension par l'agrégation d'éléments nouveaux, mais on ne touchera plus au texte des anciennes prières qui passe pour un petit chef d'oeuvre de style, et fait à juste titre la fierté des Catholiques chinois[33].

---

30 Annotated catalogue, p. 102.

31 D'Elia, *op.cit.*, 2:291, n° 1.

32 ARSI, *Jap.-Sin. I*, 172.

33 Paul Brunner, *L'Euchologe de la mission de Chine: Editio princeps 1628 et développement jusqu'à nos jours* (Münster, 1964).

A close examination of the texts reveals a striking similarity between the *Jiaoyao xulun* and the *Shengjiao rike*. A case in point was shown by Nicolas Standaert, who identified an almost word-for-word reproduction of an integral passage taken from Vagnoni's *Tianzhu jiaoyao zhilüe* in Verbiest's *Jiaoyao xulun*. The text dealt with interpreting the expression *erguo lin'ge*, or "your kingdom come", found in the Lord's Prayer. Vagnoni saw in it a threefold meaning and consequently stated it as follows:

> *Erguo lin'ge* or "your kingdom come." This expression has three meanings:
> 1. We wish and hope that the Lord of Heaven bestow on us his holy protection, and he be present in our spirit, considering it the capital of his kingdom where he holds his dominion and directs every activity.
> 2. "Your kingdom" means paradise. We hope that, after our death, we may reach this heavenly kingdom and be blessed with seeing the Lord of Heaven.
> 3. God is the Lord of Heaven, Earth, and all men. However, tempted by the evil devils, many people do not believe and obey him. We wish and hope that God exercise his authority to subdue them, so that peace may be restored on this earth as it is in his heavenly kingdom[34].

Verbiest reproduced the same passage with very few modifications in his *Jiaoyao xulun*. Thus he stated:

> *Erguo lin'ge* or "your kingdom come." This expression allows three different interpretations.
> 1. We wish and hope that the Lord of Heaven protect us, be present in our soul, as in the Court in the Imperial City. Hence he governs our soul and directs all our actions, so that we may commit no errors.
> 2. The so-called "your kingdom" means heavenly paradise. We hope to reach there after our death, to meet the Lord of Heaven and to enjoy forever the heavenly bliss.
> 3. God is the Lord of Heaven, Earth, and all nations. But among the nations, there are many people who have followed the temptations of the devils. They do not believe and obey God. To fulfil my filial devotion, I hope ardently that God would demonstrate his great authority to subdue those evil multitudes to such peaceful conditions under which nobody dares to venture any rebellion.

---

34 ARSI., *Jap.-Sin. I*, 61.

> We wish that your will be done on earth as it is in heaven[35].

This, and many similar examples, confirm our conviction that Verbiest's *Jiaoyao xulun* was, indeed, comparable to his *Shengjiao rike*. It, too, was a kind of "l'édition *ne varietur.*" Its publication showed the maturity of the Church in China in the field of religious instruction.

Now that we have determined the nature of the *Jiaoyao xulun* and established its place in the catechetical tradition of the Jesuit missionaries in China, we may proceed to assess its influence on the formation of Christianity in China, both regarding the method of its teaching and the content of its belief.

Regarding the method, we note that, like Ricci, both Pantoja and da Rocha wrote their catechisms in question and answer form. But while in Pantoja's *Pangzi yi quan* it is the student who asks the master, in the *Tianzhu shengjiao qimeng* of da Rocha it is the teacher who questions the pupil. Thus we read, in the former, the following conversation:

> **Student**: "I have no presumption to look into all the mysteries contained in your holy books. I would be happy enough if you could only indicate some essential points so that I might know how to conduct my life."
> **Master**: "Every word in the holy books is greatly beneficial to both the body and soul. As for the essential points you have mentioned, there are four: 1) the Apostles' Creed, 2) the Our Father, 3) the Ten Commandments of God, and 4) the seven sacraments. Once you have grasped these four points, you are near to your aim[36].

Da Rocha introduces the same procedure in a slightly different way:

> **Teacher**: "Christian, now that you have reached the age of understanding, tell me what the things are that you need to know."
> **Pupil**: "I need to know three things."
> **Teacher**: "What are these three things?"
> **Pupil**: "They are how to pray, how to believe and how to behave properly."
> **Teacher**: "What is to pray properly?"
> **Pupil**: "It is to pray as taught in the Lord's Prayer."
> **Teacher**: "What is to believe properly?"
> **Pupil**: "It is to believe the twelve articles of the Apostles' Creed."
> **Teacher**: "What is to behave properly?"
> **Pupil**: "It is to keep the Ten Commandments and to avoid the seven capital sins[37].

---

35 *Op. cit.*, ff. 53-54.

36 *Op. cit.*, *I*, ff. 1-2.

37 *Op. cit.*, ff. 1-2.

Verbiest, on the other hand, proceeded differently. His *Jiaoyao xulun* began with twenty folios which explain the three fundamental truths regarding the existence of God, the immortality of the human soul, and the retribution in the afterlife. Neither Pantoja, nor da Rocha, nor Vagnoni dealt with similar matters in their writings mentioned above. A further difference regards the place of the explanation of the prayer of the Sign of the Cross, which reads: "With the sign of the holy Cross God, our Lord, save us from our enemies, in the name of the Father, and the Son, and of the Holy Spirit"[38].

Not unlike his two confreres, Verbiest saw in this prayer a synopsis of the Christian mysteries, as it not only indicates the mystery of the Holy Trinity, but also expresses with pathos the condition of man's earthly existence. However, he disposed of the matter differently. In his preface, he had insisted on orderly exposition. The order he had proposed to himself followed a twofold trail. On the one hand it was, as with Ricci, to proceed from the truths knowable through the proper use of the human intellect to those that could be known only through divine revelation. On the other hand, Verbiest shared the conviction of many missionaries that the Chinese Christians were generally more given to the exercise of piety than to the understanding of the faith. Therefore he further insisted that the teaching of the Christian mysteries should be preceded and prepared for by the practice of the Christian virtues[39]. So he did not use the prayer of the Sign of the Cross as an introduction to the exposition of the Apostles' Creed, but put it near the end of his book, considering it rather as the culmination of a gradual and orderly exposition of the Christian doctrine and the immediate preparation for receiving the sacrament of Baptism[40].

With regard to the content of its belief, let us bear in mind that the preaching of the Jesuit missionaries in the sixteenth and seventeenth centuries laid particular emphasis on the Passion of Jesus and the primacy of the pope. Both of these features are prominent also in Verbiest's *Jiaoyao xulun*. The theme of the Passion is frequent already in his *Shengjiao rike*. Apart from the well-known *Rules for the Recitation of the Rosary*, written by Gaspard Ferreira[41], and the

---

38 Brunner, *op. cit.*, p. 277.

39 This point is very well explained by John W. Witek in his contribution "Presenting Christian Doctrine to the Chinese. The *Jiaoyao xulun* of Ferdinand Verbiest."

40 *Jiaoyao xulun*, ff. 59-60.

41 Brunner, *op. cit.*, pp. 38-43.

beautiful literary piece, *The Beginning and the End of the Passion of Jesus*, by Pantoja[42], the prayers to be said during Mass are introduced by the following explanation:

> At the beginning, when the priest enters to say Mass, all kneel down together and, having humbly made the Sign of the Cross, they watch attentively the priest saying Mass. This is to watch Our Lord Jesus suffering his Passion, dying on the cross, and giving his Blood in oblation to God the Father for the ransom of the sins of men. If the priest says Mass every day, it is to offer to God the Holy Body of Jesus, so that he forgives the sins of men and grants men abundant graces. It is also a memorial which reminds men of the great benefits received from Jesus[43].

The *Shengjiao rike* further included the prayer to the five wounds, very popular among the Chinese Catholics, perhaps because it is more akin to the Confucian moralizing spirit. Here is the text of this prayer:

1. I adore you, wound of the right hand; grant me the virtue of strength to despise the happiness of this world, so that I may never succumb to pride. Lord save me, body and soul.
2. I adore you, wound of the left hand; grant me the virtue of patience to overcome the adversity of this world, so that I may never lose hope. Lord, save my parents, my friends and my benefactors.
3. I adore you, wound of the right foot; grant me the virtue of diligence, to do every kind of good work, so that I may merit heaven. Lord, liberate the souls in purgatory.
4. I adore you, wound of the left foot; grant me the virtue of fear of God to avoid all sins, so that I may not fall into hell. Lord, save sinners.
5. I adore you, wound of the chest; grant me the virtue of charity to love God in heaven and all men on earth. Lord, save my enemies and those who do harm to me[44].

Realizing the prominent place the Passion of Jesus had in the devotion of Chinese Christians, Verbiest capitalized on it also in expounding the Christian faith, not only to explain the union of human and divine natures in the one person of Jesus, but also to confirm the truth of three distinct persons in the one and undivided nature of God. Thus, in his *Jiaoyao xulun*, before entering into the explanation of the cause for the

---

42 *Ibid.*, p. 36.

43 Brunner, *op. cit.*, p. 278.

44 *Ibid.*, pp. 190-191.

Passion, he dwelt upon the two following preliminary considerations:

> 1. When we speak of the Passion suffered by God, we do not mean the nature of God's Holy Trinity, but we mean the human nature of the incarnate Second Person of God. On the other hand, owing to the incarnation of the Lord of Heaven, the divine nature and the human nature (of the Second Person of God) are so intimately united that, though they differ from each other in the reality of their natures, they share, nevertheless, the same person and the same name. Hence, although it was only the human nature which suffered the Passion and was crucified and died on the Cross, we may speak of the Passion, crucifixion and death of God. It is like a man who is wounded. The wound affects only his body, not his soul; but since the body and the soul are united in one man and one person, we do say that the man is wounded.
>
> 2. It is to be noted that Jesus was not bound to suffer the Passion; he suffered it voluntarily. Had Jesus forbidden others to do him harm, no matter how numerous were the men on earth and the devils in hell, none of them would have been able to do so. In fact, even before the incarnation, the Passion of Jesus had been foretold by many holy prophets; their prophecies are all clearly recorded in their books. After the incarnation, during his public ministry, Jesus often quoted these prophecies and declared publicly that he would suffer, be humiliated, scourged, crucified, and would die on the cross, but that, after three days, he would rise again. More than five hundred men were sent by the enemies to arrest him. On hearing but one word of his, they all fell down as if dead. If Jesus had not told them to get up, they would never have stood up and laid hands on him. So, if Jesus allowed himself to be arrested, scourged, and put to death, it was because he was willing to be treated thus by his enemies, in order to accomplish the work of redemption and to render honor to the infinite dignity of God, offended by the sins of mankind. As to redemption, the truth is that a single gesture on the part of Jesus had an infinite merit, enough to redeem all mankind. It was, therefore, by no means necessary for Jesus to suffer so much pain and even the pain of death. If he chose voluntarily to do so, it was only to show how he loved God. He loved God so much that he considered it fitting to suffer any pain, any death, to honor God whose infinite dignity was offended by the sins of mankind[45].

Besides the emphasis on the Passion of Our Lord Jesus, the Christians in China have always given much importance to the role and the person of the pope. Though not a catechism in the proper sense of the word, Ricci's *Tianzhu shiyi* speaks already of the pope in its last chapter,

45 *Jiaoyao xulun*, ff. 33-35.

which describes the customs of Christian Europe[46]. Verbiest had one advantage over Ricci. By the time he wrote the *Jiaoyao xulun*, Christianity was better known in China. So he did not need, like Ricci, to introduce the theme of the papacy through the description of the good customs of western countries. He could speak of the pope simply as the head of the Catholic Church. Following the general practice of his time, he transliterated the Church as *egelexiya* and interpreted it in Chinese as *jiaohui*, meaning a religious community. Thus, in explaining the ninth article of the Apostles' Creed, "I believe in the holy and catholic Church and in the communion of saints", he began his commentary with these words:

> What we call *egelexiya* may be said in Chinese *jiaohui*, for *egelexiya* is a *hui*, a community. All those who follow the religion of the Lord of Heaven share the same truths. Though they are the citizens of the various countries of the world, it is as if they were gathered in one and the same place. The head of this community is the Lord Jesus. He has founded the community. Before his ascension into heaven he chose from among the twelve disciples one named Peter, whom he ordered to replace him as the head of the community. At that time the various states in the great West were united in one country. With territories as vast as several thousands of miles from south to north, the several hundreds of states were all governed under one great dynasty. Rome, which is located in the far westerly Italy, was the capital of that great dynasty. Saint Peter, who was then the head of the Catholic Church, chose, therefore, Rome for his residence and the residence of his successors. From there they have spread the teaching of God to the whole world. The head of the Church is what we call the vicar of God. He is the king who is charged to spread the religion of the Lord of Heaven.
>
> During his earthly life the Lord Jesus instructed Saint Peter, telling him that he and those who would succeed him as the head of the Church, would govern the whole Church and command the whole world. In the things regarding religion, their orders are equivalent to His. Hence, the Christians of the whole world must listen to them as to Himself. If anyone fails to do so, he commits the grave sin of disobeying God's order.
>
> Since the pope takes the place of God, he enjoys the highest dignity but also bears the heaviest responsibility. Only a person endowed with eminent virtues, solid learning and great ability, is equal to such a task. All the popes have vowed to keep celibacy. They have no children to whom to hand down their office; nor are they allowed to

46 *The True Meaning of the Lord of Heaven*, pp. 408 sq.

> choose their own successor. They are assisted in their governing of the Church by dozens of virtuous and learned persons. After their death, these same persons are to gather themselves in one place. They are first to pray and then to elect from among them one to become the pope. Once the new pope has taken his office, the kings and the peoples of all the Christian countries honor him as God and love him as father. So he is generally hailed as the Holy Father[47].

Verbiest's *Jiaoyao xulun* was not equal to Ricci's *Tianzhu shiyi*. It made little appeal to the Chinese reading public in general. But it had a very considerable and lasting influence on the Chinese Christian community. Up to now the two tenets it stressed, namely the Passion of Our Lord Jesus and the primacy of the pope, have characterized and are still characterizing the teaching of the Catholic Church in China.

The same book, by the way, made a notable contribution to the formation and development of the Chinese ecclesiastical and theological terminology as well. To appreciate this achievement, one should remember that in the new *Dottrina Christiana* which Ricci published in 1605, most of the ecclesiastical and theological terms were not properly translated into Chinese, but simply transliterated into Chinese characters[48]. The transliterations naturally reminded the missionaries of something they had known in their native languages, but they conveyed no meaning to the Chinese readers. To remedy this nonsense, Ricci said that "he added some annotations to explain the meaning of each of" those words. But in the annotations he added, he necessarily needed to have recourse to the native Chinese words. Thus, with no pretension on his part, he actually coined some basic Chinese ecclesiastical and theological terms. Pantoja, da Rocha and Vagnoni followed suit; each of them made considerable contributions in this field. Verbiest inherited their accumulated treasures and put them to good use. In his *Jiaoyao xulun* he adopted a method similar to the one generally used by the Neo-Confucianist masters in commenting on the Confucian classics. He, too, took care to declare at once the meaning of the individual words used in the text of the *Dottrina Christiana* and to develop the dogmatic message they were meant to convey. Thus he paved the way for replacing gradually the earlier transliterations of western jargon with the more readily recognizable native Chinese expressions, and initiated a process known today as "indigenization" of the Christian faith, or simply as "inculturation." Here are a few examples:

---

47 *Jiaoyao xulun*, ff. 43-45.

48 D'Elia, *op. cit.*, 2:289, n° 2.

1. This is what we read in the *Dottrina Christiana*:

> **Text**: "Twelve *yabosiduoluo xingbolu*."
> **Annotation**: "*Yabosiduoluo* means one who is sent. When Jesus was in this world he chose twelve principal disciples (*zong tu*) and sent them to various places to preach his doctrine. So they are called *yabosiduoluo*."
> **Again**: "*Xingbolu* means common declaration. Before the principal disciples left each other, they made a common declaration and had it put down on paper. This regards the twelve essential points concerning the realities of the Lord of Heaven. All who follow the holy religion must believe them wholeheartedly"[49].

The following passages are taken from Pantoja's *Pangzi yi quan*:

> **Question**: "To believe is the beginning of all virtues, and *xingbolu* teaches us what to believe. Now, please tell me what *xingbolu* is, and what the meaning of the word *yabosiduoluo* is."
> **Answer**: "*Yabosiduoluo* means ones who are sent. When Jesus was in this world, he chose among his many disciples twelve men whom he sent to various places to preach his doctrine. So Jesus called them *yabosiduoluo*. And *xingbolu* means common declaration. Before their departure to the various places, *yabosiduoluo* selected the twelve most important points in the teaching of the Lord of Heaven and put them down in one canon. All those who embrace sincerely the religion must believe these twelve points. *Xingbolu* consists of one chapter which is divided into twelve articles; each article corresponds to one of those twelve points"[50].

In Verbiest's *Jiaoyao xulun* the words *xingbolu* and *yabosiduoluo* disappear; they are replaced simply by the Chinese words *xingjing* and *zong tu*. The former means literally the canon of belief; and the latter, the principal disciples. These are the words now generally used in the Catholic Church in China. And here is how Verbiest explained the Apostles' Creed:

> The canon of belief (*xingjing*) was handed down by the twelve principal disciples (*zong tu*). They once followed Jesus. They saw most of what Jesus did and heard personally what he said. They recorded faithfully what they had seen and heard. So the canon of belief speaks of the events that have happened. We who live in a later age cannot personally see what he did or hear what he said, but we must necessarily rely on the canon of belief. The Lord of Heaven is truthful. He is infallible and never deceives anybody. We must all

---

49 *Tianzhu jiaoyao*, f. 4.

50 *Pangzi yi quan*, ff. 5-6.

> believe Him constantly. To believe Him is to gain merits; not to believe Him is to commit sins[51].

2. Another pertinent example is the exposition of the second article of the Apostles' Creed. In the *Dottrina Christiana* we read:

> **Text**: "I believe in the one only *feilüe Yesu qilisidu*, our Lord."
> **Annotation**: "*Feilüe* means son; it is the name of the Second Person (*wei*) of the Lord of Heaven."
> **Again**: "*Qilisidu* means one who is anointed. In former times, kings and priests used to be anointed. So Jesus is called *qilisidu*[52].

Verbiest explains the same article more extensively, by stating:

> **Text**: "I believe in the one only *feilüe Yesu Jilisidu*, our Lord."
> **Explanation**: "It must be remembered that there is but one Lord, the Lord of Heaven, though the one Lord has three persons (*wei*). The First Person is called *badele*, which means Father; the Second Person is *feilüe*, meaning Son; and the Third Person is *sibiliduo sanduo*, meaning Holy Spirit. The three persons are not three Lords. They are only the one only Lord, the one only substance (*ti*). The human mind cannot grasp completely such a truth. In fact, the nature of the Lord of Heaven is infinite. Anything directly related to Him surpasses infinitely our human intellect. The intellect of men is limited. It cannot comprehend the infinite nature of the Lord of Heaven. Otherwise He would not have been the Lord of Heaven"[53].

In short, as in revising the *Shengjiao rike*, so too in composing his *Jiaoyao xulun*, Verbiest claimed no originality. He intended only to present the collective teaching of the Catholic missionaries in China and to lend to it his personal authority. In doing so, he contributed to forming and stabilizing the religious tradition of the Church in China, with regard to both the contents of her teaching and the expressions of her faith. The Church in China owes much to this energetic and amiable Belgian missionary.

---

51 *Jiaoyao xulun*, ff. 20-21.

52 *Tianzhu jiaoyao*, ff. 5-6.

53 *Jiaoyao xulun*, ff. 25-26. It is very important to note that Father (*fu*), Son (*zi*), Holy Spirit (*shengshen*), as well as the one nature (*ti*) and the three persons (*wei*), are all expressions which are still used in the Catholic Church in China to this day.

PRESENTING CHRISTIAN DOCTRINE TO THE CHINESE: REFLECTIONS ON THE *JIAOYAO XULUN* OF FERDINAND VERBIEST

JOHN W. WITEK, S.J.

John W. Witek, S.J. is an associate professor of East Asian history at Georgetown University. Among his publications are a biography of Jean-François Foucquet, several essays for the Chantilly conferences, and a number of biographies for the Jesuit encyclopedic dictionary.

From Pittem, Belgium to Beijing, China, the traveler faces a long road over the Eurasian landmass and even longer if one goes by sea. Survey coverage of the early life of Ferdinand Verbiest, his entrance into the Society of Jesus in September 1641 and his training as a Jesuit in Belgium, Italy and Spain are covered elsewhere in this volume. Perhaps the most striking event in his life in Spain was the rebuff he received from the government when he requested permission to work on the missions in South America. The Jesuit General suggested that he should instead begin to consider going to the East. For the history of China that negative decision in Spain eventually was to have a positive effect halfway around the globe in Beijing.

In assisting Johann Adam Schall von Bell (1592-1666) during a turbulent era that Christian missionaries, and more so, western astronomy, faced, Verbiest overcame serious obstacles and eventually the Qing court accepted western scientific methods in astronomy and calendar calculations. After the astronomy controversy was settled and the missionaries, once exiled in Guangzhou, were allowed to return to their mission stations in 1671, Verbiest kept making further efforts to change the legal status of Christianity from that of a heterodox sect to an orthodox one. He never lived to achieve that goal which finally was realized in the 1692 Edict of Toleration.

The dimensions in the life of Verbiest are so multifaceted that attempts to keep all of them in perspective may seem difficult. Perhaps

Verbiest himself offers a clue in his study of the stars through telescopes. In taking accurate measurements he would focus on several minute data before entering them onto a larger picture - an accurate calendar for China. The aim of this essay is somewhat similar as it concentrates on but one of Verbiest's Chinese works. As Henri Bosmans wrote "with some apprehension," to cite the opening phrase of his major study on Verbiest, so too is there some concern that this essay may be unable to discuss thoroughly this catechism of Verbiest in the limits of the conference proceedings[1]. Emphasizing that Verbiest was the director of the Imperial Observatory in Beijing is, according to Bosmans, to overlook the religious dimension of his presence in China. After a presentation of a short summary of the catechism as found in reference works about Verbiest, the focus will shift to a synopsis of Verbiest's own preface and a discussion of the major topics of the catechism. Analyzing the importance of the catechism in Verbiest's days and its later impact through various editions and translations can then contribute to an assessment of the larger historical context of Verbiest's role in China.

An original copy of the 1670 edition of the *Jiaoyao xulun* in the Vatican Library is the basis for this study. Although the title does not differ in the other copies in this library, the three in the *Bibliothèque Nationale* in Paris have the variant title, *Tianzhu jiaoyao xulun*[2]. Translations of the title have also differed. In his *Astronomia Europaea* Verbiest refers to it as *Ordo proponendi mysteria fidei* (Order of Proposing the Mysteries of the Faith) and this title is handwritten on the outside cover of the copy in the Vatican Library. In the fifth edition of 1935, however, the title is translated as *Rudimenta religionis ordinate exposita* (The Fundamentals of Religion Explained in Proper Order)[3]. An argument could be made that neither of these is satisfactory and that a more literal translation "An Introduction (*xulun*) of Doctrinal Essentials (*jiaoyao*)" might be preferred. Instead of pursuing these

---

1 Henri Bosmans, *Ferdinand Verbiest, directeur de l'Observatoire de Pékin* (Louvain, 1912), p. 3. This is a separately printed extract from the article that appeared in *Revue des questions scientifiques*, 71 (1912), 195-273, 375-494.

2 *Jiaoyao xulun*, R.G. Oriente III, 222 (5), Biblioteca Apostolica Vaticana which is the basic text used in this essay; its abbreviation will be *JYXL*. For the copies in the *Bibliothèque Nationale*, see Chinois 6976-I, 6976-II and 6976-III.

3 *Jiaoyao xulun* (Shanghai, 1935) is in the Library of Congress, with the catalogue number J 791.05 V 58.

translational avenues, the more generic term *catechism* will be used in referring to this Chinese work.

Even a cursory glance at the table of contents leads to an understanding of the complexity of this work. There are sixty-two numbered sections (*pian*), not chapters, whose shorter titles appear above and outside the commonly found borders of the Chinese text. At times two or more titles are listed under one section, but the numbering sequence is inconsistent since there are no sections 11, 25, 35, 46, 52, 53, 54, 56, 57, 58, 60, 61, actually numbered in the table of contents. In fact the numbering of sections is only in the table of contents, not in the text itself[4].

In his preface Verbiest states that the study of sacred doctrine compared to other studies should have order. If the speaker discusses the topics in strict order, he can focus on the essentials and will not deviate into many tangents. The listener in turn more easily comprehending the basic message can keep it in his mind. Verbiest indicates that a question that a guest often asks is "What is Catholicism?" Those who are already Catholics are well versed in obeying doctrinal precepts but are not necessarily equally versed in explaining doctrinal principles. The example Verbiest offers is that of a person building a house, but, without any plans, has only a pile of bricks, stone and timbers. With no aim at finishing the structure, such a person cannot make the guest happy who comes to that house. The mind can be content with the important aspects of the sacred doctrine once it understands the order involved, just as an accumulation of timbers and stone must be subject to a sequence of above and below, right and left, front and back in order to construct a house. Stressing that his language in this catechism will lack embellishment and fine rhetoric, Verbiest notes that his focus is on truth which ultimately is beautiful in itself. He adds that his work will concentrate on the basic truths of Catholic doctrine without presenting details. If someone wants such details, he recommends that such a reader carefully scrutinize those Chinese books already in print and widely circulated[5].

---

4 *JYXL*, *mulu*, pp. 1-5.

5 *Ibid.* Verbiest closes with the usual dating system, *Kangxi jiunian, zhengyue, xiahuan*, the ninth year of Kangxi, first month, last decade of the month, that is, February 9-19, 1670. The preface is also found in Xu Zongze, *Ming Qing jian Yesu huishi yizhu tiyao* (Taibei, 1958), pp. 162-163.

The opening sentence in the text illustrates the clear methodology that Verbiest consistently uses. "Whoever wishes to become a member of the Catholic Church must know who God is." He explains that God who made everything - heaven, earth, spirits, man and all things - is unique, without beginning and without end. After some further clarification of God's relationship to things, Verbiest adds that in ancient China Pan Gu, Buddha, Laojun (that is Laozi) and others existed after heaven and earth and all were born of a mother and a father. All Chinese of his day knew the difference that Verbiest was highlighting[6].

"God created heaven, earth, man and all things. What does this mean?" The question Verbiest asks he puts in colloquial Chinese, *shi shenma yisi*, a phrase still recognizable today. God shows his love to man by his creation not only of man himself but of heaven, earth, and all things, since man is to use all of them. These are God's blessings to man and God does not derive any benefit from them since he is self-sufficient[7].

Man's ancestors, created by God, after the creation of the rest of the world, were Adam and Eve. God created both of them, not because of his need but because of his goodness. Originally both of these ancestors were good. Yet each person's nature and each person are not the same but reflect in different ways divine omnipotence, goodness, and omniscience. Verbiest also adds that, if God does not possess these attributes, then there is no explanation how and why things come to be, how and why the sun and the moon are so regular in their appearance nor how and why anyone could enjoy the use of beautiful things. But one should know that the maker of all things has only one nature, not many. Just as each family can only have one head, else chaos will ensue, so too can a person understand God's relationship towards all men by examining the regularity of heavenly phenomena and the seasons of the year[8].

Presenting three types of soul (vegetative, animal and rational) Verbiest stresses that the human soul, united to the body, lives beyond the body's life span. Using traditional Chinese terminology as *wuguan* (five regulations of bodily actions), *siti* (four parts of the human frame: head, trunk, arms and legs) and *baizhi* (one hundred limbs of the body), he notes that all of them cannot move once death occurs. The soul is

---

6 *JYXL*, pp. 1a-1b.

7 *Ibid.*, p. 2b.

8 *Ibid.*, pp. 3b-5b.

totally different yet had been dependent on the body. With these premises as a foundation Verbiest is able to underscore how Catholic doctrine on the soul differs from Buddhist teaching of paths or conditions of existence, called *jati* in Sanskrit[9]. A human being can only beget another human being, not a horse nor an ox. Those born during this year came into existence in that period and are not dependent on those of last year. The soul has two possible places of rest, heaven or hell. While heaven is our home, it is not the same as the Buddhist erroneous exposition of the four heavens, *sitian*. By not entering into details on this Buddhist belief, Verbiest assumed that his readers readily understood the contrast. Actually the four heavens refer indirectly to their guardians (*sitian wang*), who, in later Buddhist tradition were said to rule the legions of supernatural beings, guarding the slopes of Mount Meru (paradise) and thus they were worshipped as protective gods of Buddhist temples[10]. But to describe heaven itself, Verbiest stresses, is impossible for according to sacred scripture "Eye has not seen nor ear heard nor has the mind thought of the happiness that God has prepared for those who are good." Man desires to ascend to heaven with its joy, but first he must take the heavenly road (*tiantang daolu*) by following the Ten Commandments. After briefly describing how God gave Moses the commandments on two stone tablets, Verbiest is then set to enter into a fairly lengthy discussion of each commandment[11].

Since the First Commandment emphasizes monotheism, Verbiest shows that the Chinese believer cannot simultaneously worship Buddha, bodhisattvas, Laojun, *Tai yang* and *Tai yin* or others. Such monotheistic belief expected of the Chinese Catholic creates for Verbiest an occasion to explain in some detail not so much theoretically the three virtues of faith, hope and charity, but how a person possessing such virtues will act since all three have God as their central focus. Veneration of angels and saints (both men and women, *shengren* and *shengnü*) differs from worship of Buddha, bodhisattvas and others, because originally those who worshipped the latter could not recognize the true God. Although everyday one should always be good internally, on the Lord's Day and on certain days in honor of Our Lady and the saints, there is an

---

9 *Ibid.*, pp. 6b-7b. For data on *wuguan*, *siti* and so forth, see W.F. Mayers, *Chinese Reader's Manual* (Shanghai, 1874; reprint, Taibei, 1964), pp. 310, 316, 327.

10 *Ibid.*, p. 310.

11 *JYXL*, pp. 9b-10b. The quotation from Sacred Scripture refers to St. Paul's account, 1 Cor. 2:9.

obligation to go to church to hear Mass (*Misa*) and to practise what is taught at that time. Although Verbiest makes comparisons of Chinese tradition to Christianity in presenting the First Commandment, he does not express any allusions to the teachings of Confucian filial piety in the section on the Fourth Commandment. Instead he describes the so-called parents (*fumu*) under three aspects as 1) those who gave me birth, 2) those who govern me and 3) those who counsel me[12].

From the Commandments the focus moves to the Apostles' Creed which allows Verbiest to enter into a more elaborate theological discourse. Monotheism in the First Commandment now takes on a new dimension in the Trinity, the identity of Jesus Christ as God and man and other related topics.

For instance, to explain the three Persons in one God (*Tianzhu sanwei yiti*), he notes that all three acted together in the creation of heaven and earth. There is no difference of time among them so that one cannot state that God the Father was originally the first or existed prior to the other two persons. In creating heaven, earth, angels, etc., there was a priority that God established. The fall of man occurred when Lucifer tempted Eve, then Adam. Man, in contrast to all else created by God, has an immortal soul and a relationship to the triune God. In describing the Trinity Verbiest uses both transliterations and translations, *badele* (*padre*), that can be translated as *fu* (father), *feilüe* (*filio*) translated as *zi* (son) and the third person, *sibiliduo sanduo* (*Spiritu Sancto*) translated as *shengshen* (Holy Spirit)[13]. Jesus, the Son of God, is also the son of the Mother of God, for at a point in time he was born as a man. The Holy Spirit comes from both the Father and the Son because of the love existing between the two. Man cannot probe the depths of this relationship but he is to accept this on faith. *Yesu* (Jesus) is both the *ming* and the *hao* of the one who came down from heaven and became man. His name can be translated as "savior of the world" (*jiushi*). In calling him *jilisidu* (Christ) Verbiest indicates that a translation as the "Anointed One" (*fuyou*) is proper. By reason of his origin Christ is the Lord of all people (*tianxia wanguo ren zhi gongzhu ye*). The terminology Verbiest employs, *tianxia wanguo ren*, has overtones with the *wanwu tianxia* that was so familiar in the Chinese lexicon of his day[14].

---

12 *JYXL*, pp. 10-14; for his discussion of the other Commandments, see *ibid.*, pp. 14b-19b.

13 *Ibid.*, pp. 20-25.

14 *Ibid.*, pp. 25b-28a.

The relationship of Christ to sinners Verbiest explains in presenting the reasons why Christ came down from heaven. His motives were to establish personally in the body a demonstration of all goodness and virtue, to set in motion the preaching of the sacred teaching to all nations and to take upon himself grievous afflictions (*kunan*) and thus to save and atone for the sins of mankind. Verbiest dwells at some length on this last point in order to concentrate on the suffering and then the crucifixion and death of Christ[15]. He was quite aware that the life and death of Christ baffled many Chinese who questioned his predecessors on the mission about such strange events.

After the resurrection Christ interacted with his apostles, but before ascending to heaven, he ordered them to proclaim his message to all the people. According to the Bible (*Shengjing*) at the Last Judgment there will be a separation of good persons on the right hand of God and of evil persons on the left. After a detailed description of the differences between these two types of persons, Verbiest clarifies the unity of the Church, *egelexiya (ecclesia)*. Yet he adds that this term can be translated as *jiaohui* in that all countries will come together along one road following the true principles (*zhengli*). This was the mission that Christ gave to his apostles[16].

To understand the Church, Verbiest describes that the capital of this Church is in the Great West, in Rome, Italy where Saint Peter was the first pope. This status he received from Christ so that the pope takes the place of God and his commands are those of God. There are three reasons why such commands are to be followed. First, the original head of this religion is Jesus Christ who manifests the virtues he wants us to achieve. Secondly, Catholicism is a matter of faith commanding what should be and a matter of true principle commanding what should be done. History records, he adds, that in the last 1,600 years there are tens of thousands of Christians. Thirdly, the saints, both men and women, who are in heaven, are saints of the Church and its exemplary models[17]. Each of these reasons is elaborately discussed in the catechism but cannot be presented in more detail at this point.

---

15 *Ibid.*, pp. 31b-33b. For a discussion of the impact of the preaching of the coming of Christ and the crucifixion and resurrection, especially on Yang Guangxian and his *Budeyi* (I Cannot Do Otherwise), see George Dunne, *Generation of Giants* (Notre Dame, 1962), pp. 275-277.

16 *JYXL*, pp. 34b-42b.

17 *Ibid.*, pp. 43-44b.

The belief in the forgiveness of sins is dependent upon man's having faith internally in his heart. He will receive forgiveness by baptism in case of the convert or by penance in case of a person who is already a member of the Church. As Verbiest notes, "knowledge of the principle of the sacred doctrine is indispensable, for it is able to regulate the interior and the exterior of a sincere person"[18]. The exterior of man is his eye since it can see, but the interior is man's soul which the eye of the body cannot see. At the time when one enters into this religion by conversion, pure water cleanses man's forehead which has an interior effect although man can only see the external action. The internal effect is the ridding the soul of its sins so that simultaneously God's grace might dwell in that person's soul. God forgives that person's sin, for while Christ was in the world, he took on all the sins of men whether ancient or modern, of the past or the future. To clarify for the Chinese reader what the resurrection of the dead meant, as the second last item in the Apostles' Creed, Verbiest describes the separation of the soul from the body at death. But this leads him to highlight that since the body had a role in man's actions when the two were conjoined on earth, it will one day possess eternal bliss or eternal damnation. It is in God's plan that the body of a good person be glorified, that is, that God will grant it four characteristic qualities or as Verbiest states, "four kinds of great favors" (*sizhong da'en*). These are impassibility or the inability to suffer again (*wu shangsun zhi en*), clarity (*guangming zhi en*), agility (*shensu zhi en*) and subtility (*shentou zhi en*). He returns to these four characteristics in his discussion of the last point of the Creed, the belief in everlasting life. But he uses a variant form for all characteristics (*si da'en*) and for impassibility (*bu shoushang*), that is, not receiving injuries. For an evil person, however, there is only bitterness and suffering. After the "Amen" Verbiest comments that past and present events are thus related to the Creed[19].

In the discussion of the *Our Father*, Verbiest covers some of the themes already presented but this time in more depth. These include God as Father, man's relationship to him, and man's requesting God to forgive man his sins which is the counterpart to the same theme in the Creed. In the section on the *Ave Maria*, he explains that Ave (*Awu*) is a

---

18 *Ibid.*, pp. 45-46.

19 *Ibid.*, pp. 46b-50b. On the four characteristic qualities of the glorified body which Verbiest lists in the order usually found in theological treatises, see C.J. Corcoran, "Glorified Body", *New Catholic Encyclopedia*, VI, pp. 512-513.

western sound that can be translated as a morning greeting or congratulations. This was the action of the angel in greeting Mary, for the name of Mary, both as *ming* and as *hao* can be translated as star of the sea (*haixing*). But the sea is not to be understood in the terms of the four directions, but as the Dipper (*beidouxing*) guides a person along a road. The Dipper, Verbiest continues, is one name of the star of the sea. Most of the rest of Verbiest's explanation of the *Ave Maria* repeats his views found in the Apostles' Creed. But one cannot overlook a question of terminology that is basic for understanding the faith, that is, how to express grace. Using *elajiya* transliterated from *gratia*, Verbiest translates it into *shengju*[20]. The latter term *ju* means "to favor" and its nominative form means "kindness." In adding *sheng* (holy) Verbiest was once more creating a Chinese Christian neologism.

The last two sections of the catechism are somewhat unevenly divided partly because of the subject matter and partly because Verbiest wants to avoid repetition. There is a short section on the sign of the Cross. This offers the author an opportunity to repeat some of the data already found in the discussion on the Creed. He then explains how to make the gesture correctly and what the sign means in itself, namely the belief in a triune God and the passion and death of Christ on earth[21].

Since the catechism was intended for a Chinese trying to learn about Catholicism, Verbiest ends his work with a section on the sacrament of Baptism, which is the means of formally entering into the Church. His term for baptism, *shengshui*, today means "holy water" whereas currently *shengxi shengshi* is used for baptism. In describing the act of baptism Verbiest stressed that it differs fundamentally from the belief in Buddha, since the Catholic accepts the teaching on blessings in life after death. In administering the sacrament olive oil is used as the base for the holy oil *shengyou*. Just as Christ's name indicates the Anointed One so too had kings in ancient days been anointed and so too is the convert to be anointed and thus joined to the Lord Christ[22]. After explaining the various parts of the liturgical act of baptism, Verbiest ends his account by indicating the need for the convert to have godparents (godfather for the male, godmother for the female) and to choose a saint's name, e.g., Peter or Paul for a male, Anna or Lucia for a female. With that, he ends the formulas for Baptism, one a transliteration from Latin, the other a

---

20 *JYXL*, pp. 50b-58b.

21 *Ibid.*, pp. 58b-61b.

22 *Ibid.*, pp. 61b-63.

translation with the use of *fu*, *zi* and *shengshen* for the Father, the Son and the Holy Spirit[23].

To assess at least to some degree the impact of the *Jiaoyao xulun*, a short description of the various editions is in order. Besides additional copies in the Vatican Library, there are three copies in the *Bibliothèque Nationale*, Paris. All are from the early Qing period but they are not identical even though bound together. The first two copies have only the preface by Verbiest, the table of contents separately paginated and then the text running to sixty-seven pages. The third copy is singular in that it has a preface dated Shunzhi *dinghai* that is, 1647, by Han Lin, a convert baptized by Giulio Aleni (1582-1649) and also a statement of approval by Li Leisi, the Chinese name of Ludovico Buglio (1606-1682) dated Kangxi 16, that is, 1677. Both of these precede Verbiest's own preface of 1670 and the text[24]. Although the 1647 preface begins by asking about the origins of Catholicism, it does not seem to be a preface relevant to Verbiest's catechism. In fact it was incorrectly bound, for Verbiest had not come to China until 1659, that is, a dozen years after this preface was written. Buglio's approval dated 1677 has led several bibliographers to list this as the first edition[25]. Since the 1670 Vatican Library edition does not have a statement by any Jesuit book censor, it can be argued that the inclusion of the Buglio approval in 1677 constitutes the second edition. Reprints of this catechism may have ap-

---

23 *Ibid.*, pp. 63b-67.

24 *Tianzhu jiaoyao xulun*, Chinois 6976-III, pp. 1-2b, *Bibliothèque Nationale*, Paris. Han Lin was a native of Jiangzhou, Shanxi province who received his *xiucai* degree in 1617 and his *juren* degree in 1621. Probably in that same year he was baptized in Jiangzhou and had a Christian name, Thomas. For further details on his apologetical works, see F. Margiotti, *Il cattolicismo nello Shansi dalle origini al 1738* (Rome, 1958), pp. 311-319; for a biography, see A. Hummel (ed.), *Eminent Chinese of the Ch'ing Period* (Washington, 1943-1944; reprint, Taibei, 1967), pp. 274-275.

25 See Henri Cordier, *L'imprimerie sino-européenne en Chine. Bibliographie des ouvrages publiés en Chine par les Européens au XVIIe et XVIIIe siècles* (Paris, 1901; reprint San Francisco, 1979), p. 59; A. Wylie, *Notes on Chinese Literature* (Shanghai, 1867; reprint Taibei, 1972), p. 177. The 1647 preface of Han Lin is that of his own *Shengjiao xinzheng* (Proofs of the Christian Religion) which, with its preface of that year, was printed in Beijing in 1668 and reprinted in 1679. See Hummel, *op. cit.*, p. 274. Han Lin's work is reprinted in *Tianzhujiao dongchuan wenxian sanbian* in *Zhongguo shixue congshu xubian*, n° 21 (Taibei, 1972), 1:267-293. This is from the copy in the Vatican Library with its handwritten, but inaccurate, annotation on the title page, "Tom. 10. *Christianae Legis testimonia, P. Ferd. Verbiest.*"

peared during the later decades of the Kangxi era although none seem to be extant.

The next edition known to bibliographers is that of 1799. After the Opium War, an edition was printed near Shanghai in 1848. This was followed by further editions in 1867, 1875, 1903 and 1914, all printed at Tushanwan (T'ou-se-wei) near Shanghai. Other editions appeared in Hong Kong in 1905, at Changde in Hunan province in 1922, at Xian-xian in Hejianfu in 1923, and at the Beitang (North Church) in Beijing, undated, at Chongqing also undated, at Yanzhoufu, undated and at Fuzhou, undated. The *Jiaoyao xulun* was rewritten into literary Chinese by Matthew Chen, S.J. and entitled *Jiaoyao chuyan*, printed at Tushanwan in 1886 and reissued in 1903. A copy in the Shanghai dialect entitled *Fangyan jiaoyao xulun* appeared at Tushanwan in 1888. An edition entitled *Fangyan jiaoyao xulun fa* published at Tushanwan in 1910 contains the Chinese text with a romanized transcription and a French translation. In 1864 a Korean version under the title *Kyoyo Syerou* appeared in Seoul. A Manchu translation was published in Beijing perhaps at the end of the eighteenth century[26].

Comments on Verbiest's catechism have appeared in works by Jesuits. Louis Pfister states that the Manchu version was condemned by an imperial edict of 1805. This is based on a remark of Alexander Wylie in his bibliographical researches. The text of the edict, however, under the Jiaqing emperor, does not indicate whether a Manchu or a Chinese version was read that led to the condemnation[27]. The claim of Louis Van Hee, followed by Bosmans, that Verbiest's work was reviewed in the *Siku quanshu* (Complete Library in the Four Branches of Literature) has no foundation[28]. Research in various indexes of this

---

26 Data for this paragraph is drawn from Cordier, *op. cit.*, p. 59 and R. Streit et al. (eds.), *Bibliotheca Missionum*, 30 vols. (Münster, 1916-1974), 5:809. See also L. Van Hee, *Ferdinand Verbiest, écrivain chinois* (Bruges, 1913), 55 n. 3. This is a separately printed extract from the *Mélanges de la société d'émulation de Bruges*, VII.

27 Louis Pfister, *Notices biographiques et bibliographiques sur les Jésuites de l'ancienne Mission de Chine*, 2 vols. (Shanghai, 1932-1934), pp. 352-353; Wylie, *op. cit.*, p. 177; Fu Lo-shu, *A Documentary Chronicle of Sino-Western Relations, 1644-1820* (Tucson, 1966), pp. 355, 596. The 1805 condemnation was part of a campaign of censorship of Catholic publications by the imperial court at that time. Verbiest's catechism was one of the works singled out in the edict.

28 Van Hee, *op. cit.*, p. 55; Henri Bosmans, *Les écrits chinois de Verbiest* (Louvain, 1913), p. 25. This is a separately printed extract from the *Revue des questions scientifiques*, 74 (1913), p. 292.

famous collection in sinology and in the text itself has shown that it is not included. Apparently Van Hee confused this catechism with that of Matteo Ricci.

To place the catechism in a historical context one cannot assert that Verbiest's work is unique in the genre of the catechetical tradition in the late Ming and early Qing period. The *Tianzhu shiyi* (True Meaning of the Lord of Heaven) by Ricci (1552-1610) and the *Wanwu zhenyuan* (True Origin of All Things) by Aleni (1582-1649) were typical of a natural theology approach to the Chinese, that is, one which did not attempt to explain in detail all the supernatural truths of Catholicism[29]. But even a glance at a list of Chinese catechisms and similar writings before Verbiest's work in 1670 indicates that this study was an integral part of that tradition in China. Michele Ruggieri (1572-1607) wrote a compendium on the Holy Law and on the Incarnation and the Passion; Lazzaro Cattaneo (1560-1640), on contrition and also on the soul's relationship to God; Niccolò Longobardo (1565-1655), on the soul, on death, on the litany of the saints, and on the Rosary; Diego de Pantoja (1571-1618), on the origin of man, on the Apostles' Creed, on God and his attributes, on the Passion of Christ, on the good and bad angels; Francisco Sambiasi (1582-1649), on the three kinds of soul (vegetative, animal and rational) and on the divine attributes. Others such as Iñacio da Costa (1603-1666) and Giacomo Rho (1592-1638) wrote on topics found in Verbiest's catechism. In fact, Schall presented to the Shunzhi emperor a memorial that described illustrations of Christ[30]. Yet this brief listing shows that Verbiest may have had at his disposal almost a shelf of books printed by his confreres before 1670 and in them had a vocabulary of Chinese Catholic terminology that had gradually evolved. The *Jiaoyao xulun*, written just shortly after the Yang Guangxian affair had been settled, reflected these past publications. It was intended to be a summary of Catholic doctrine taught in China until then, but put into Verbiest's own style.

---

29 Dunne, *op. cit.*, p. 96; *The True Meaning of the Lord of Heaven (T'ien-chu shih-i)* (St. Louis, 1985), p. 43; on Aleni's *Wanwu zhenyuan* (True Origin of All Things), see Wylie, *op. cit.*, p. 175; J.W. Witek, *Controversial Ideas in China and in Europe. A Biography of Jean-François Foucquet, 1665-1741* (Rome, 1982), p. 116.

30 See the list of Chinese works by Jesuits compiled by Philippe Couplet and published in Henri Bernard-Maitre, "Les adaptations chinoises d'ouvrages européens. Bibliographie chronologique depuis la venue des Portugais à Canton jusqu' à la Mission Française de Pékin", *Monumenta Serica*, 10 (1945), 47-56.

The earliest historical evaluation of the impact of the *Jiaoyao xulun* comes from Verbiest himself. In answer to a request from Giovanni Domenico Gabiani (1623-1694) who wanted views of the fathers in Beijing concerning the *Tratados* of Domingo Navarette, O.P., Verbiest wrote a long report entitled *Responsa ad varia dubia P. Navarrete*[31]. In Scholastic philosophical format Navarette proposed a doubt (*dubium*) about the three Chinese language catechisms published by the missionaries of the three religious orders in China. He stated that they claimed that the Buddha was the lawgiver of an idolatrous sect and was thus condemned by name. In the time of persecution, he added, when the civil magistrates of the Board of Rites (*libu*) interrogated the missionaries about this statement, they replied that it should be understood not absolutely, but only conditionally, that is, if Buddha had not known God and had not carried out his precepts. Navarette continued by asking whether this was a good and proper reply, since the Catholic books do not speak conditionally, but absolutely. The faithful and others thought that the missionaries gave such tergiversations out of fear. Navarette's response to this doubt was that if the lawgiver of this sect was notorious and it could equally be shown that he did not know God, then one should respond absolutely and say that Buddha was condemned.

Against this doubt and response of Navarette, Verbiest argued that Buddha, as the author of this sect, not even his sacrificing clergy knew sufficiently. His sect was introduced into China from India after the coming of Christ. Some of his followers referred to Buddha as the son of a king in India and that he established an eremitical way of life in the mountains. Very probably many of the things attributed to him by the monks and the Chinese he himself did not intend. The gentiles often said that Buddha was the minister and friend of God and as such he could be revered just as the Catholics revered beyond the emperor licitly and usefully those others who prevail by reason of favor and authority in the imperial presence. To Verbiest this meant that neither for the missionaries nor for the gentiles was their sufficient knowledge about what kind of leader Buddha was. Indeed among the Buddhists there were not lacking those who said that in using the term *Fo* (Buddha) they wanted to signify that divinity (*numen*) which the Chinese Christians called *Coeli Dominus* (*Tianzhu*, Lord of Heaven). It followed, Verbiest contended, that the missionaries proceeded prudently when they spoke conditionally about Buddha. This was especially true when stating

---

31 F. Verbiest, *Correspondance de Ferdinand Verbiest* (Brussels, 1938), pp. 279-342.

absolutely that Buddha was in hell, since it was evident that a danger of a harsher persecution existed. In fact, Verbiest points out, the past persecution against the Beijing missionaries was much longer than all the others, since the Buddhist monks and other adversaries were close to the Empress Dowager (*reginam aviam*). A most devout follower of Buddhism, she took up its cause with her nephew, the emperor, when she heard that the Buddhist leaders accused the Christians of saying that Buddha was in hell[32]. At this point Verbiest interjected the role of his catechism in this scenario.

> Our very own catechism, which Navarette has mentioned here, was called into question because of this, and our adversary Yang Guangxian in that book he wrote against the Christian religion cited this very catechism, and from this one chapter he aroused severe hatred against us[33].

This refers to the remarks of Yang in his *Budeyi* (I Cannot Do Otherwise), which attacked western astronomy and Christianity itself. Surveying the astronomy controversy is outside the scope of this essay, but Verbiest's assessment of later events is also noteworthy. After the fathers were liberated because of the earthquake in the capital, some court officials and princes who supported the missionaries with a singular affection, admonished them to abstain from speaking in a similar way about Buddhism if they wanted to continue their efforts in propagating Christianity. These friends openly stated that the Jesuits had strong adversaries in the palace and insinuated that the most powerful was the Empress Dowager.

It is curious that the connection Verbiest draws between his catechism and Yang Guangxian's attacks on Christianity and western astronomy have escaped notice in the recent works on Confucianism and Christianity by Jacques Gernet and John D. Young[34].

---

32 *Ibid.*, pp. 299-300.

33 The Latin text: Ipsemet ille catechismus noster, de quo Navarrete hic mentionem fecit, ob hanc causam in iudicium vocatus est, et adversarius noster Yang Koang-Sien appellatus in libro illo quem contra religionem christianam conscripsit hunc ipsum catechismum citavit, et ex uno hoc capite grave in nostros odium excitavit. *Ibid.*, p. 301. From this statement it appears that Verbiest's catechism, or at least parts of it, were known to Yang. For data on Verbiest's reference to Yang's views, see the latter's *Budeyi* in *Tianzhujiao dongchuan wenxian xubian* in Wu Xiangxiang (ed.), *Zhongguo shixue congshu* n° 40 (Taibei, 1966), 3:1091-1092.

34 Jacques Gernet, *Chine et Christianisme. Action et réaction* (Paris, 1982); its English version, *China and the Christian Impact* (Cambridge, 1985); John

In studying Chinese publications in the late Ming and early Qing the sinologist is confronted with the task of trying to determine the total number of copies actually printed. Christian publications not only should be included but perhaps even more so since the status of Christianity had not yet been secured until 1671 when the missionaries were allowed to return to their mission stations in the interior of China. During that countrywide persecution it is not improbable that Christian books had been confiscated or destroyed by government officials. Verbiest's catechism may have fulfilled a need as he probably perceived it.

*

Several conclusions then can be drawn from studying Verbiest's *Jiaoyao xulun*. Not only was there a connection between it and the astronomy controversy, but later the Jesuits were advised to be less assertive against Buddhism. Even Verbiest was still undergoing a learning process in adapting himself to the Qing court. The effect of his catechism on the development of Christianity in the early Qing period may be more difficult to assess since more reading of Chinese and western sources is needed, even though there seem to be no more editions after the 1677 and then the 1799 ones appeared. In the nineteenth and twentieth centuries, however, more editions, at times for different audiences as in Shanghai or the French or Korean translations would argue towards a greater receptivity of Verbiest's work. The fifth edition in Shanghai in 1935 is in the style of a booklet known in China at that time, as the copy now in the Library of Congress reflects[35]. Undoubtedly much effort would be required to locate copies of the various editions listed by bibliographers. Even if that were never accomplished, it is to the credit of Verbiest that besides his work on astronomy and calendar-making, he became known to Chinese, Manchu and Korean

D. Young, *Confucianism and Christianity* (Hong Kong, 1983).

35 *Jiaoyao xulun* (Shanghai, 1935), bound in dark blue covers with the title on the outside cover: *Rudimenta religionis ordinate exposita, 5° editio*, and the author's name appear on the verso of the title page. There is no indication how many copies were printed. The highlighting of certain topics by larger calligraphy in the Vatican Library text is discontinued in this later version. This does not affect its argumentation but is an interesting aspect in the history of the printing of Christian books from the early Qing period until this century.

readers for his explanation of the basic doctrines of Christianity. It was fitting that the Kangxi emperor ordered that his eulogy proclaimed to the court be inscribed in Manchu and in Chinese on the reverse side of Verbiest's tombstone still extant today.

JING TIAN - THE KANGXI EMPEROR'S GIFT TO FERDINAND VERBIEST IN THE RITES CONTROVERSY

CLAUDIA VON COLLANI

Dr. Claudia von Collani studied Catholic theology in Würzburg (1970-1976) and missiology (Prof. Dr. Bernward Willeke), and completed a doctoral thesis about Joachim Bouvet, S.J., and his figurism in China. She is now preparing her habilitationsschrift about the "Brevis Relatio in the Chinese Rites Controversy."

I.
The Kangxi Emperor's Visit to the Jesuit Church

There are several reports about the visit of the Kangxi emperor to the Jesuit church in 1675. During his visit the emperor offered a gift of great consequence to Ferdinand Verbiest and the whole Catholic Church in China. The following report is based on an undated letter of Gabriel de Magalhaes, S.J. (1610-1677), an eyewitness to the event[1].

The Kangxi emperor was only twenty-one years old when, on July 12, 1675, he went on horseback to the church and residence of the

---

1 The basic report is in: *Notizie circa l'uso delle Tabelle colle parole cinese King tien coelum colito*, Bibliothèque Mazarine, Paris, *Ms. 1667*, ff. 84r-85v. It was written by Gabriel de Magalhaes (1610-1677), in China since 1640 and in Beijing since 1648, founder of the Dongtang in Beijing and author of *Nouvelle relation de la Chine* (Paris, 1688). Another report, the *Brevis narratio adventus Imperatoris Chinarum et Tartarii ad domum et ecclesiam PP. Societatis Jesu*, Hauptstaatsarchiv, Munich, was not at my disposal (see Bernward H. Willeke, "Missionshandschriften im Hauptstaatsarchiv", *Euntes Docete* 21 (1968), 336). See also Philippe Couplet, "Relazione dello stato e qualita delle missioni della Cina ..." *Sinica Franciscana (SF)* 5:458, n° 31; Caspar Castner, *Relatio de statu sinicae ecclesiae et progressu S. Fidei in Imperio Sinarum ab anno 1692 ... ad annum saecularem 17OO*, ARSI, *Jap.-Sin. 125, ff.* 201-202; *Confucius Sinarum philosophus ...* (Paris, 1687), cxii; David E. Mungello, *Curious Land: Jesuit Accommodation and the Origins of Sinology*, *Studia Leibnitiana Supplementa* 25 (Stuttgart, 1985), p. 266.

Jesuits in Beijing[2]. He was accompanied by his elder brother, Fuquan (1653-1703), his brother-in-law, his younger brother and, finally, by an uncle of the empress and great friend of the Jesuits, a man called "kolao" (*gelao*) and "esteemed adviser of the emperor" by Magalhaes. Quite probably this was Songgotu, later to be China's representative at the Russian-Chinese negotiations leading to the well-known Treaty of Nerchinsk in 1689[3]. Besides these members of the imperial family, many high-ranking persons and courtiers also were present.

In front of the church all the members of the party dismounted and passed through the gate, where the Jesuits were lined up to receive and welcome the emperor who, with his attendants, respectfully entered the church through the main door. The church was shown to the young emperor who proved to have some detailed knowledge of Christianity. From the church, the "garden of the spiritual enjoyments" as it was called, they passed to the "house-garden," which was laid out by Claudio Filippo Grimaldi (1638-1712). Afterwards, the emperor proceeded to the residence and was shown into the room of the superior, with its large number of mathematical instruments. The emperor was highly interested in them, for he had studied mathematics with Verbiest. The whole party finally entered the reception room, where the emperor said to his brother and attendants: "Look at these Fathers who came from four different European nations to our Chinese empire on the ships of these two Fathers" (meaning Magalhaes and Tomas Pereira[4], and thus on Portuguese ships)[5]. Then the emperor, together with Verbiest,

---

2 According to Kilian Stumpf (1655-1720), the date was in 1672. See his *Tractatus Patris Kiliani Stumpf Pekino submissus anno 1701*, ARSI, *FG 724*, n° 5. This residence was the Xitang, later called Nantang (Southern church). The "Belgian" church Dongtang, founded in 1655, had been destroyed in 1664 and was rebuilt only after 1692; Sebald Reil, *Kilian Stumpf 1655-1720. Ein Würzburger Jesuit am Kaiserhof zu Peking* (Münster, 1978), p. 75.

3 Sosan, Songgotu or So-e-tu, was later in charge of the Treaty of Nerchinsk (1689) between China and Russia, together with Jean-François Gerbillon (1654-1707) and Tomas Pereira, S.J.(1645-1708). He also supported the Jesuits in their efforts to obtain the Edict of Toleration in 1692. Later he fell into disgrace with the emperor who considered him responsible for the bad behavior of his heir. Arthur W. Hummel, *Eminent Chinese of the Ch'ing Period* (Washington, 1943, repr. Taipei, 1970), p. 663; *SF*, 5:460, 464.

4 Pereira (1645-1708) was an important person in the imperial court and in the Jesuit Order in China.

5 I.e., on Portuguese ships. Magalhaes, as both a Portuguese and a descendant of the family of the famous navigator Magellan, was pleased to cite such a

Grimaldi, Magalhaes and Ludovico Buglio (1606-1682) entered the latter's room[6]. There he took a brush and quickly wrote down two characters, *jing tian*, each of them one yard high. They were written very well, and everyone praised them highly. The emperor approved them with his imperial seal, looked at them for some time, and then said that riding and eating had made his arms and hands tremble a little bit. Finally he and his attendants returned to the imperial palace.

The next day the emperor had the same characters written, this time two-and-a-half yards high and approved with the imperial seal, as well as two other characters, *yu bi* (written with the imperial brush), and the Chinese characters for the year 1675. Additionally, he remarked that he had already decided to present these characters to the Jesuits four years ago.

The Jesuits knelt down, as was the custom, and bowed their heads to express their gratitude for such a great and unique present. They assumed that because of the persecution of Christians, the emperor wanted to show his personal goodwill and to assure their safety. The two characters were fixed in the best place of the Jesuit church in Beijing. Copies were also printed and distributed to all Jesuit churches in China, in order to protect Christianity against its enemies[7].

In the following years the story of the emperor's gift to Verbiest and the Jesuits was reported many times with slight modifications. In most of these reports Verbiest's part was emphasized much more than in the one of Magalhaes who was not Verbiest's friend[8], and so diminished the latter's role. In a letter written in 1697 by Joachim Bouvet S.J. (1656-1730), for instance we find a better description that the gift was for Verbiest, but also that he had even requested it:

> Ce fut pour l'autoriser dans son Empire lorsqu'elle y estoit deffendue que l'Empereur se sentant un jour pressé par le P. Verbiest qui lui demandoit quelque grace en faveur du Christianisme, ce prince écrivit de sa main ces deux lettres chinoises *king tien* (*caelum colito*) tirées de leurs anciens livres qu'ils reverent comme leurs livres sacrez, et

---

remark made by the emperor. Besides, China was considered as part of the Portuguese *padroado*, where only Portuguese ships were allowed to call.

6 Georg Bernhard Bülffinger, *Specimen doctrinae veterum Sinarum moralis et politicae* ... (Frankfurt, 1724), p. 321.

7 Bibliothèque Mazarine, Paris, *Ms. 1667*, ff. 84r-85v; see also Castner: *Relatio de statu* ..., ARSI, *Jap.-Sin. 125*, f. 201; Stumpf, *Tractatus* ..., ARSI, *FG 724*, n° 5, f. 135.

8 Louis Pfister, *Notices biographiques et bibliographiques sur les Jésuites de l'ancienne Mission de Chine* (Shanghai, 1932), p. 343.

> qu'il les donna à ce Pere, avec permission de s'en servir comme d'un temoignage par lequel il reconnoissoit que la religion chrétienne s'accordoit avec la loy de la Chine et consequemment qu'elle ne devoit pas y estre deffendue.
> Et c'est depuis ce temps là que les missionnaires de Péking et à leur exemple ceux des provinces, aiant fait graver ces deux lettres sur de grands cartouches en forme de tableaux, commencerent à les exposer dans leurs églises pour autoriser par le temoignage de l'Empereur même qui est, comme le souverain pontife et le chef de la religion, celle dont les chrétiens faisoient l'exercice, et empecher par là le peuple et les magistrats de les molester dans leurs fonctions que le souverain parassoit approuver par là[9].

There is another fact indicating that the imperial gift was thought to honor Verbiest in particular. For the next twelve years (i.e., from 1675 to 1688, when Verbiest died) the two characters *jing tian* were regarded as a "protective shield" for the Jesuit churches and residences, and even for those of other religious orders[10]. Note that the emperor's visit took place in 1675, long before the famous Edict of Toleration was announced in 1692. At that time, Christianity was only tolerated because of the Jesuits' merits at the imperial court. But the situation was rather uncertain, and persecution could start at any time. That is why the emperor's gift was distributed throughout China, and served as a provisional protection for the missionaries and the churches of the Jesuits, the Franciscans and the Augustinians; even one Dominican used it, namely, Luo Wenzao (Fray Gregory Lopez, 1616-1691), who later became the first Chinese bishop. None of them seemed to fear that these two characters could be directed against the Christian faith or honor the "material heaven"[11].

---

9 Joachim Bouvet, Letter dated August 30-October 15, 1697, Paris, Archivum Societatis Jesu, Paris (ASJP), Fonds Vivier, p. 52, published in Claudia von Collani, "Ein Brief des Chinamissionars P. Joachim Bouvet S.J. zum Mandat des Apostolischen Vikars von Fu-kien, Charles Maigrot MEP", in *Neue Zeitschrift für Missionswissenschaft (NZM)* 43 (1987), 203.

10 See Nicolas Standaert, *Yang Tingyun, Confucian and Christian in Late Ming China*, *Sinica Leidensia* 19 (Leiden, 1988), p. 103; *Breve Ristretto delle notizie gia dedotte circa l'uso delle Tabelle colle parole Cinesi King tien caelum colito ...*, Biblioteca Nazionale Centrale Vittorio Emmanuele (BNC VE), Rome *41. 3B. 6*, n° 6, IV-XI; *Bibliotheca Missionum* (BM), 30 vols. (Freiburg, 1929), 5:954, n° 2786.

11 Letter of the Jesuits of Beijing, November 12, 1701, Archivio Segreto Vaticano (ASV), *Albani 248*, f. 59.; *Analecta argumentorum pro sententia Missionariorum Societatis Jesu super ritibus Sinensibus,* BNC VE, *Fondo Gesuitico* (FG) *1257/27*, n° 58-62.

Jing tian or *coelum colito*, as it was translated officially, was not at all invented by the emperor, rather he had taken the phrase from the old Chinese Classics[12]. For example, there is the following passage in the *Shijing*: "Revere the anger of Heaven and presume not to make sport or idle. Revere the changing mood of Heaven and presume not to drive about at your pleasure"[13]. In the *Shujing* there is another passage: "Porro jussit Hi et Hovo, reverenter obsequentes immensi coeli (legibus)..."[14], and many other similar phrases[15]. Certainly these passages may have various interpretations, e.g. Giovanni Francisco Nicolai a Leonissa, O.F.M. (1656-1737), believed that their meaning was just a demand for attention to the observation of astronomical phenomena[16].

But *jing tian* was not only nice calligraphy from the emperor's hand, used only in Christian churches, for the Daoists, Buddhists, Muslims, and Jews in China, also used the same two characters in their temples. The synagogue in Kaifengfu, for example, had a tablet with the two characters *jing tian* at the first entrance[17]. All these tablets show clearly the Chinese conception of religion and state. As long as one respects the emperor and the state, one is allowed to venerate whatever God one

---

12 Collani, "Ein Brief ...", p. 203.

13 James Legge, *The Chinese Classics*, 5 vols. (Hongkong, 1985), *The She King*, 4:503; see the Latin translation in Stumpf's *Tractatus* ... , 22: "Qui reveretur coeli iram, non audet voluptatibus indulgere. Qui reveretur coeli, seu ejus decretorum mutabilitatem, non audet habenas cupiditatibus laxare".

14 Séraphin Couvreur, *Chou King. Les Annales de la Chine* (Paris, 1950), p. 3; see also *SF,* 6:243; 7:243.

15 Stumpf, *Tractatus* ..., 22: "Reverere Coeli intelligentiam, eius decretum non est facile observare. Ne dicas altissimus sursum altius sedet, sed deorsum aeque ac sursum fertur, quotidie intuens, quae hic circa me sunt"; Legge, *Classics*, 4:598; see also a quotation from the *Shujing*: "Nunc vero rex Cheou non colit supremum coelum." And the explication: "Nemo est tam rudis ac perditus qui nesciat colendum esse coelum: at nunc rex Tcheou non colit coelum. Igitur consequens est ut ejus peccata in dies magis ac magis gravia sunt." Joachim Bouvet, *Coelestis doctrinae vera notitia*, Biblioteca Capitolare Fabroniana, Pistoia, *Ms. 53*; see Claudia von Collani, "Tianxue Benyi - Joachim Bouvet's Forschungen zum Monotheismus in China" and David Mungello, "Unearthing the Manuscripts of Bouvet's *Gujin* after nearly Three Centuries", *China Mission Studies Bulletin*, 10 (1988), 9-33; 34-61.

16 Joannes Francisco Nicolai (a Leonissa), *Responsum ad S. Officium*, July 1, 1699, in *SF,* 6:243.

17 Joseph Dehergne and Donald D. Leslie, *Juifs de Chine à travers la Correspondance inédite des Jésuites du XVIIIème siècle*, Bibliotheca Instituti Historici S.J., 41 (Rome, 1980), pl. VII, pp. 54, 57, 62.

wishes[18]. In this sense, we should understand the phrase by Han Tan, (1637-1704), president of the Hanlinyuan, in his preface to Bouvet's *Tianxue benyi*: "When they (i.e. the old Chinese philosophers) say that one has to revere Heaven and one has to revere the king, then both are the same"[19]. Han Tan wanted to confirm that the Christians and the missionaries were loyal subjects of the Kangxi emperor, and that their teachings were in accordance with the old Chinese books. The role of *jing tian* in the Rites Controversy should be understood in this way.

II.
Jing tian and the Rites Controversy

The quiet gladness and exuberant expectations following the Edict of Toleration of 1692 gradually vanished after the year 1693. On March 26, 1693 the apostolic vicar of Fujian, Charles Maigrot, M.E.P. (1652-1730), announced to his vicariate his well-known mandate containing seven prohibitions for the missionaries and Christians, concerning the rites in reverence of the ancestors and Confucius, and also the rendering of God's name in Chinese[20]. The second point of this mandate dealt with the characters *jing tian*. Together with *Tian* and *Shangdi* as names for the Christian God, Maigrot also forbade the exposition of tablets with the characters *jing tian* in any church: "We severely prohibit that the tablet with the two characters *coelum colito* (*king tien*) be fixed in any church." The tablets had to be removed within two months, as Maigrot thought them to be some sort of idolatry. He was not really certain about it, but even the suspicion of a danger (*periculi suspicio*) to the pure doctrine was sufficient for him. According to his experience, those missionaries not using these tablets were as successful as the others[21]. Later Maigrot tried to explain his attitude: "... the Chinese have some sort of religion, or better, some vestige of religion. The material heaven is the most important object of their religion, their highest idol." According to Maigrot, it was not at all clear that the

---

18 Jacques Gernet, *Christus kam bis nach China* (München, 1984), p. 161; for further information I thank Dr. Nicolas Standaert.

19 Collani, "Tianxue Benyi ...", p. 22.

20 *BM*, 5:925, n° 2686.

21 *Bullarium Romanum Clemens XI* (Graz, 1735), 1:129: "Districte prohibemus, ne in ulla Ecclesia Tabella quaedam, cui hae duae litterae Coelum Colito King Tien inscriptae sunt, appendatur, & ubicunque appensa fuerit, intra duos menses aufferri mandamus."

emperor had thought of the Christian religion when writing the two characters; rather, he had aimed to have the physical heaven worshipped in Christian churches. Moreover, Maigrot believed that the emperor felt responsible only for the Jesuits at his court. Furthermore, the emperor's thoughts about the two characters were irrelevant; only the general view of the Chinese people counted. To Maigrot it seemed paradoxical that Confucius and the ancient Chinese would already have worshipped God[22].

Maigrot's aim was to carry out his mandate not only in Fujian, but throughout China. He therefore sent Nicolas Charmot (1645-1714), one of his confreres, to Rome in 1693 to hand over the text of his mandate, together with some explanations (dated November 10, 1693) to Pope Innocent XII. Charmot reached Rome in early 1697[23].

In his covering letter, Maigrot proposed that the non-Jesuit missionaries in China were of the opinion that the emperor did not have in mind the Christian religion when writing the two characters, but that he was considering the Chinese religion, of which he was the head. His aim was to have the material heaven worshipped. Most of the superiors in China had not issued any special regulations concerning these tablets except the superior of the Dominicans, who had forbidden them under pain of excommunication (ca. 1686/1687, i.e. before Luo Wenzao died)[24].

Early in 1697 Charmot put forward his concern to Pope Innocent XII, who handed it over to the Holy Office and established a Particular Congregation (*Congregatio Particularis*) to investigate the affair[25]. With that step, Maigrot's mandate started anew the Chinese Rites Controversy, followed by a veritable avalanche of papers both for and against the rites. Charmot's reports offered to the Particular Congregation contained nothing new. They forbade the expressions *Tian* and *Shangdi* and simultaneously outlawed the exposition of the two characters *jing tian*[26], arguing that since *tian* meant the material heaven,

---

22 Charles Maigrot, *De Synica Religione Dissertatio II*, cap. VI, sect. 23, Biblioteca Casanatense, *Ms. 2453*, *f.* 357v.

23 *BM*, 5:927-928, n° 2687-2688; Collani, "Ein Brief ...", p. 192.

24 *Historia cultus Sinensium seu varia scripta de cultibus Sinarum* ... (Cologne, 1700), p. 352.

25 Ludwig von Pastor, *Geschichte der Päpste* (Freiburg, 1930), 14:1136; Collani, "Ein Brief ...", p. 192.

26 *Historia cultus Sinensium* ..., p. 352; for the titles of Charmot's papers, see *BM*, 5:941-942, n° 2740-2745.

jing tian, consequently, meant to worship the material heaven[27].

The Jesuits, who only gradually came to know about Charmot's accusations, needed some time to be able to refute Maigrot's and Charmot's articles[28]. The Jesuits first pointed out that nobody in China could misunderstand the tablet *jing tian*. Furthermore, the Edict of Toleration clearly proved that the emperor had not at all intended to have the material heaven worshipped, but the God of Creation, who was frequently addressed allegorically by the emperor himself[29]. Finally, they stated that it was well-known throughout China that all churches (about 200) were devoted to God and not to the material heaven[30].

In 1699 the Jesuits had compiled a great deal of material concerning the theological questions raised in Maigrot's mandate[31]. With respect to *jing tian* the following two manuscripts are of particular interest: *Notizie circa l'uso delle Tabelle colle parole Cinesi King tien coelum colito*[32] and *Breve Ristretto delle notizie gia dedotte circa l'uso delle tabelle colle parole Cinesi King tien coelum colito, presentato alla Sacra Congregatione del S. Officio in settembre 1699*[33].

The *Notizie* are very detailed, aiming to refute Charmot's charge that

---

27 See Nicolas Charmot's paper dated March 19, 1697 (see *BM*, 5;941, n° 2741): "Etenim Secta Literaria, quae apud Sinas latissime patet, & cujus caput est Imperator, coelo materiali cultum divinum defert. Itaque cum coelum colere praecipiunt aut coelum Imperatorem vocant, non continens pro contento accipiunt, id est coelum pro Deo in coelo existente tanquam in sua sede, ut Patres Societatis Jesu interpretantur, sed pro coelo ipso materiali, ut dictum est, quod ex multis colligitur. Nam I. Sinae, praesertim ex Secta Literaria, quae ubique dominatur in China, profitentur se discipulos Confucii, qui materiale coelum coluit. II. Imperator ipse Literariae Sectae Princeps & doctor palam Coeli materialis cultorem se exhibet." *Historia cultus Sinensium* ... p. 444.

28 Collani, "Ein Brief ...", p. 194.

29 *Historia cultus Sinensium* ..., pp. 614-625.

30 *De Ritibus Sinensium* ... (Augsburg-Dillingen, 1701), pp. 142-160.

31 For the name of God in Chinese, see *Ristretto delle notizie circa l'uso delle voce Cinesi Xam-ti ... e delle voce Tien* and *Notizie intorno all'uso delle voci cinesi Thien e Xam-Ti* ..., *BM,* 5:954, n° 2785, both of these works are in BNC VE.

32 *BM*, 5:954, n° 2786. Nicolas Charmot wrote against those four essays concerning the name of God and *jing tian* in his *Secundae Vindiciae* ..., dated November 18, 1699 (*BM,* 5:955, n° 2791), in *Continuatio Historiae cultus Sinensium* ... (Cologne, 1700), pp. 65-146.

33 A printed copy of the *Breve Ristretto* ... is in BNC VE *41* 3B, 6, n° 6, whereas one copy of the *Notizie circa l'uso* ... is in the Bibliothèque Mazarine, Paris, *Ms. 1667*.

the emperor of China was an atheist whose purpose in announcing the *jing tian* tablets was to propagate the worship of the material heaven[34]. To prove that the contrary was true, a letter from Bouvet is added. Dated December 1, 1697, it testifies that the Kangxi emperor was certainly no atheist[35]. The *Ristretto* is essentially shorter and much more precise. Maigrot is accused of wrongly casting suspicion on the emperor's supposed idolatry and superstition. It is true that *tian* can mean the material heaven (i.e., the visible heaven), but there is also the meaning of the spiritual heaven, i.e., the Creator of heaven and earth. Finally, the teaching of Origenes (185-254) is cited, which states that each nation should call God by the name designated to its highest and most esteemed being (*Ens*)[36]. The *Ristretto* also contains quotations from a letter of Verbiest concerning the story of the emperor's gift:

> ... after he had taken the brush, he (the emperor) himself wrote in the presence of the Fathers of Beijing the Chinese characters (*jing tian*), to be fixed as a tablet in our churches. The next day he had it printed and, furnished with the imperial seal, sent it to us ... This tablet was not only fixed in our churches, but our example was followed and continued even to this day by missionaries of other orders scattered over different Chinese provinces..."[37].

However, in the view of the Holy Office, all these long and detailed papers written by the Jesuits did not prove the untenability of Maigrot's charge against the expression *Tian*. Hence the Jesuits conceived the idea of asking Chinese authorities about China's own customs and terminology. The Jesuit General, Tirso Gonzalez, encouraged the Jesuits in China to get as many reports as possible from learned Chinese in order to be able to refute each of the points of Maigrot's mandate[38]. The Jesuits at the Beijing court even dared to ask the emperor for his opinion, for he was the first representative of Confucianism. On November 30, 1700 they presented a petition written in Manchu, containing and summarizing each of the points in dispute. The emperor approved the correctness of the petition by a short declaration. Both

34 Bibliothèque Mazarine, *Ms. 1667*, f. 87v.

35 Joachim Bouvet, Letter dated December 1, 1697, ARSI, *Jap.-Sin. 166*, ff. 259-260 and Bibliothèque Mazarine, Ms. 1667, ff. 98r-99v (copy).

36 *Breve Ristretto ...*, p. VI.

37 *Ibid.*, p. IX-XI.

38 Joseph Dehergne, "L'exposé des Jésuites de Pékin sur les cultes des ancêtres présenté à l'empereur Kangxi en novembre 1700", in *Actes du IIe Colloque international de Sinologie, Chantilly 1977* (Paris, 1980), p. 186.

documents are known as the *Declaratio rituum*[39], which was published later in Europe[40].

The imperial declaration runs as follows, according to a translation prepared by François Noël, S.J. (1651-1729):

> That, what you have written is written very well, and agrees completely with the Great Learning (*dadao*). To revere heaven (*jing tian*), to serve rulers and parents, to revere the teachers and superiors is common usage and custom all over the world. Everything contained in this paper is completely true and there is nothing to be changed or to be corrected[41].

The petition, together with the emperor's reply, was published in China[42]. Together with a Latin translation, several copies of the *Declaratio rituum* were sent immediately to Rome on different ships[43]. In 1701 the *Declaratio rituum*, together with its story and the testimonies of ten prominent Chinese and Manchus, was printed and is known as the *Brevis Relatio eorum quae spectant ad Declarationem Imperatoris Kam Hi circa caeli, Cumfucii, et Avorum cultum, datam anno 1700*[44]. The two Jesuits, Noël and Caspar Castner (1665-1709), took this booklet to Rome and presented it to the Particular Congregation and to Pope Clement XI[45].

---

39 *BM,* 7:44, n° 2158.

40 Paul Pelliot, "La Brevis Relatio", *T'oung Pao*, 23 (1924), p. 363; John W. Witek, *Controversial Ideas in China and Europe. A Biography of Jean-François Foucquet (1665-1741),* Bibliotheca Instituti Historici Societatis Jesu, 43 (Rome, 1982), p. 113; Pastor, 15:296.

41 "Ea quae scripsisti, optime scripta sunt, magnaeque viae (seu rectae & communi omnium Hominum doctrinae) penitus cohaerent. Coelum colere, Principibus ac Parentibus inservire, Magistros ac Superios revereri, ista est orbi universo communis consuetudo & sensus. Omnia, quae hoc scripto continentur, verissime sunt nec est quod mutari aut emendari debeat." François Noël, *Philosophia Sinica tribus tractatibus, primo cognitionem primi entis, secundo ceremonias erga defunctos, tertio ethicam, juxta Sinarum mentem* ... (Prague, 1711), tract. I, 174; see *Brevis Relatio eorum quae spectant ad declarationem Imperatoris Kam Hi circa Caeli, Cumfucii, et Avorum Cultum* ... (Peking, 1701), 13r-14v; Antonio Sisto Rosso, *Apostolic Legations to China in the Eighteenth Century* (South Pasadena, 1948), pp. 143, 145.

42 *Brevis Relatio* ..., p. 15.

43 Pastor, 15:296; J. Brucker, "Chinois (rites)", *Dictionnaire de Théologie catholique (DTC)* (Paris, 1930), 2:2375.

44 Pelliot, pp. 355-372.

45 Pastor, 15:299.

At the same time the Jesuits in China did not cease compiling more and more evidence to support their interpretation of *jing tian*. As a special feature, they considered the fact that Maigrot's episcopal ordination[46] took place in the Jesuit church of Jiading below the tablet of *jing tian* on March 14, 1700. Although there were many opponents of the rites present (among them Artus de Lionne, MEP, 1655-1713, and Basilio da Glemona, O.F.M., 1648-1704, not to mention Maigrot himself) none of them entered a protest[47].

In the meantime the Holy Office discussed Charmot's charges in Rome. The first debate about *Tian* and *Shangdi* and about the characters *jing tian* took place on May 13, 1698[48]. The discussion continued during August and September[49]. The following event which happened during the meeting of the Particular Congregation on April 2, 1703 (after almost five years of discussion) may serve to characterize the meeting. Cardinal G. Marescotti, one of the four members of the Particular Congregation, admitted confusion and asked why the expression *coelum colito* was sometimes used, and why at other times *coelum colo* or *coelum cole*, or even *coelum colite* was used. The explanation finally agreed upon was very simple: in China there was no distinction between the one and the other[50].

The results of the many meetings of the Particular Congregation are given in the so-called *Quaesitum* and *Responsum*, both of which basically follow Maigrot's mandate. It was vaguely assumed in the *Quaesitum* that *tian* was then regarded as the material heaven by the Chinese literati[51]. The statement of the *Responsum* was much more precise: it was forbidden to fix the tablets in question in any Christian church. In any case, this practice had to be stopped, as Pope Leo the Great (440-461) had already condemned it in *Sermone 7 De Nativitate Domini* (i.e., worshipping heaven together with God's creatures the sun

---

46 *BM*, 5:927.

47 Dehergne, p. 204 n° 32; *Analecta argumentorum* ..., n° 82; see also the letter of Alessandro Ciceri, dated November 27, 1701, ASV, *Fondo Albani 248*, f. 65v.

48 Caspar Castner, *Historica Relatio controversiae de ritibus aliquot Sinicis*, Munich, B. Stat., *Codd. Ms. lat. tome II, I, 8689*, f. 22.

49 *Ibid.*, f. 23.

50 *Ibid.*, f. 43.

51 *Bullarium Romanum Clemens XI*, 1:131.

and the moon)[52]. The *Responsum* became the basis for the papal decree, *Cum Deus optimus*, announced by Pope Clement XI on November 20, 1704, more or less confirming Maigrot's mandate[53].

In China the course of the affair was not known for a long time. At Pope Clement's request two members of the Jesuit Order were to be sent to Rome. The Jesuits had sent Noël and Castner (1655-1709), together with the *Brevis Relatio* and many other documents, to Rome. They reached Rome in December 1702[54], and, pressed for time, they tried to bring the requested material in order and to write a juridically correct petition to be presented to Pope Clement XI. The first part, *Memoriale circa veritatem* ..., dated March 27, 1703, was presented on January 8, 1704[55]. The second booklet, entitled *Memoriale et summarium novissimorum testimoniorum Sinicorum* ..., is dated August 27, 1704[56]. To defend the tablets with the inscription *jing tian*, they stated that the emperor's intention could be clearly understood from the imperial commentary *Rijiang*, which was published some years before in the name of the emperor. The following quotation was intended to prove this statement: "Whenever heaven (*tian*) is worshipped, then not the material or visible heaven is to be understood, but the Creator and Lord of Heaven, Earth and all things." That is what Chinese people believed, according to Noël and Castner.

They next quoted Charmot himself, who had written in 1699 that, in the event that the emperor would declare clearly and in public that *jing tian* was not used by him in an atheistic or superstitious sense, but that it meant the true God, then Charmot would believe in the emperor's words (*quod tamen nequaquam facit*)[57]. The very same opinion is expressed in the *Apologia Dominicanorum* ... (Cologne, 1699, cap. 6, p. 99, Italian ed.):

> The Jesuits should be allowed to keep their tablets in case the emperor would publish throughout China a declaration saying that the

---

52 *Ibid.*, p. 136; Migne, *Patrologia Latina*, 54, col. 219.

53 *BM*, 7:85-86, n° 2314.

54 Pastor, 15:299.

55 *BM*, 7:73, n° 2272; Pastor, 15:299; one copy of the *Memoriale* in Rome, ARSI, *FG 729*.

56 *BM*, 7:80-81, n° 2296; Munich, B. Stat., *Jes. 174°*.

57 *Secundae Vindiciae Scriptorum Nicolai Charmot* ..., dated November 18, 1699, printed in *Continuatio Historiae cultus Sinensium* ... (Cologne, 1700), p. 126; *Notizie circa l'uso delle tabelle colle parole* ..., n° 127.

> characters *jing tian* stood for nothing else than what the Christians understand when saying "adore the God of Heaven."

It is precisely this demand that was fulfilled by the *Declaratio rituum* of 1700, which was published in the *Brevis Relatio*[58].

In their second *Memoriale*, Noël and Castner presented the testimonials of twenty Chinese scholars. The summary states:

> The fact that the emperor and those who believe in the true Chinese teaching do not worship the material heaven is attested by the Chinese pagans as well as by Chinese scholars ... That is why the emperor, the sovereign of the people, convinced of our Lord's unlimited charity and bearing it in his mind, had written down those two characters *jing tian*, in order to teach all human beings - the authorities as well as the people - that Heaven should be worshipped and not that blue sky above us, not that physical *qi*, or *li*, or *taiji* ...

This is attested by twenty heathen scholars and five Christian literati:

> Therefore *jing tian*, worshipping heaven, *jing shangdi*, worshipping the Supreme Lord, *jing laotianye*, worshipping the Old Lord of Heaven, are synonyms. And this is the general and commonly used explanation, as far as we know[59].

In the *Summarium* (no. 7) it is said that the emperor had used the characters *jing tian* in the same sense as Confucius: "The superior man reveres the mandate and will of heaven; as the foolish man does not know the mandate and will of heaven, he does not revere it." And finally, it is said in the *Shujing* "revere the dignity of heaven"[60].

Besides the printed petitions presented to the Holy Office, there were some further handwritten documents. The Jesuits had, for instance, interviewed some of the Jews living in China about the meaning of *jing tian*. An eighty-eight year old Jew of Kaifengfu had told Giampaolo Gozani, S.J. (1619-1732) that *jing tian* was understood by the Jews as

---

58 François Noël and Caspar Castner, *Memoriale* ... 27; *La veritá e l'innocenza de' missionari della Compagnia di Giesú nella Cina* ... (n.p., 1710), p. 8; *Apologia* refers to that of Noël Alexandre, *Apologia de Padri Dominicani Missionarii della Cina* ... (Cologne, 1699).

59 Noël and Castner, *Memoriale et summarium novissimorum testimoniorum Sinicorum* ... 7; German translation in Georg Pray, *Geschichte der Streitigkeiten über die chinesischen Gebräuche, worin ihr Ursprung, Fortgang und Ende in drey Büchern dargestellt wird*, 3 vols. (Augsburg, 1791-1792), 2:75.

60 Noël and Castner, *Memoriale* ..., p. 48.

meaning to adore Heaven, Jesus, and the Cross[61].

In the meantime Carlo Tomasso Cardinal Maillard de Tournon (1668-1710), papal legate *a latere*, was sent to China in order to settle the differences among the missionaries and dispose them to accept and observe the forthcoming decree concerning the Chinese Rites[62]. It must be noted that Tournon had already left Rome in 1702, two years before the Holy Office came to a decision concerning the Chinese Rites. Although he could not know the text of the decree, he acted as if he did when he arrived in China in April 1705[63].

He did not tolerate the tablets with *jing tian*, and forbade the Jesuits to have the tablets displayed in their new church (Beitang) in Beijing[64]. When the emperor heard about Tournon's attitude towards the rites, he had two Manchu courtiers sent to him with the following message:

> People under your Christian law say *jing tianzhu*, worship the Lord of Heaven ...; the Chinese people say *jing tian*, worship Heaven. Although the expressions are different, their meaning is one and the same[65].

In spite of this clear message Tournon did not change his attitude, but quoted Maigrot as an authority to support his prohibitions[66]. Maigrot was ordered to come to Beijing, where he arrived on June 29, 1706[67].

The emperor proceeded to arrange what would nowadays be called a summit meeting. This memorable meeting took place at the imperial summer residence in Jehol on August 2, 1706, with the emperor and Maigrot as participants. This encounter had devastating consequences for the Christian mission in China. The very promising development of the mission came to a halt. This decisive meeting was taken down word for word by Kilian Stumpf, S.J. (1655-1720), who included his protocol

---

61 Letter by Giampaolo Gozani, dated August 25, 1705, in Dehergne and Leslie, *Juifs de Chine*, p. 54.

62 Rosso, p. 149.

63 Letter of Alvaro de Benavente, OESA (1646-1709), dated December 7, 1707, in *Supplément des mémoires pour Rome sur l'état de la religion chrétienne dans la Chine* (1709), 7; Pray, 2:395; Pastor, 15:298.

64 Pray, 2:329; *Mémoires pour Rome sur l'état de la religion chrétienne dans la Chine* (1709), 1:11.

65 Noël, *Philosophia Sinica* ..., tract. I, 177.

66 Pray, 2:395.

67 Rosso, p. 165; *SF,* 5:508, n° 12; *SF,* 6:603, n° 27.

in his famous *Acta Pekinensia*[68].

Even before the meeting started, however, Tournon, who did not himself participate in the meeting, had laid down some restrictions on Maigrot. He forbade him to debate the controversial rites or any other religious topic with the emperor[69]. Hence Maigrot failed to act as an authority to prove the incompatibility of Christianity and Confucianism. The emperor showed himself extraordinarily patient, but nevertheless drove Maigrot into a corner. What follows is a word for word passage dealing with *jing tian*.

> **Emperor** : Why do you not mention anything about that which you once disapproved and condemned as being ill-natured? Namely, about the characters *jing tian*, about the reverence for Confucius, and about the tablets for the ancestors ... What are your objections to the characters *jing tian*?
> **Maigrot** : *Tian* does not mean the Lord of Heaven.
> **Emperor** : I am very surprised at you. Did I not already state that *tian* is a much better expression for the Lord of Heaven than *tianzhu* or *tian di wanwu zhi zhu*? For *tian* universally means the Lord of Heaven, and *tian di wanwu zhi zhu* is the Lord of Heaven, Earth and the ten-thousand things. Tell me, why do the people call me *wansui*, (ten thousand years) ?
> **Maigrot** : This is to indicate that they wish Your Majesty innumerable years of life.
> **Emperor** : Well, learn from that: the true meaning of Chinese words does not always coincide with their literal meaning[70].

The very same day the emperor publicly announced Maigrot's ignorance and incompetence:

> Those who have adopted your law use the characters *jing tian*, revere Confucius ...; but because of your ignorance, you pretend that this is

---

68 Kilian Stumpf, *Acta Pekinensia* ..., Archivum S. Congregationis de Propaganda Fide (APF), Rome, *Inform. liber 162. Pro Miss. Sin.* Vol. 7; Stumpf, *Compendium Actorum Pekinensium* ... 1705 and 1706, Beijing, Nov. 1706, ARSI, *Jap.-Sin. 169, ff.* 295v-317r; German translation, Pray, 2:308-434.

69 *Mémoires de la Congrégation de la Mission*, 9 vols. (Paris, 1868), 4:23, 365; Stumpf, *Acta Pekinensia* ..., APF, *Inform. liber 166, Pro Miss. Sin.* Vol. 11, *f.* 150.

70 Pray, 2:391, *Acta Pekinensia* ... liber 162, *f.* 491v; see C.G. de Murr, "Collocutio Imperatoris Sinarum cum Reverendissimo Carlo Maigrot, Episcopo Cononensi ...", *Journal für Kunstgeschichte* (Nürnberg 1778), 6:180-190; Jonathan Spence, *Ich, Kaiser von China* (Frankfurt, 1985), p. 133.

> worthless. We, the emperor, have explained to you that *jing tian* has the same meaning as *jing tianzhu*, according to your law: honor the Lord of Heaven ... If the 5000 Christians (in Fujian, Maigrot's vicariate) believe in you, and think that they should not revere Confucius anymore, etc. ... then they are certainly not Chinese, and are not my subjects anymore...

The emperor handed this note over to Maigrot on August 11, 1706, giving him another chance to come round[71].

Next the emperor had Maigrot's two teachers in Chinese summoned[72]. The two scholars, Li Yifen (Leontius) and Jiang Weibiao, (Xaverius), had been *xianggong* at Maigrot's church[73]. The interrogation about their Chinese lessons and Maigrot's knowledge took place on December 15, 1706[74]. Jiang made the following statement: "I have been with Maigrot for two years and have explained Chinese books to him. But he stubbornly stuck to a different European interpretation"[75]. Jiang had clearly declared that it was wrong not to practice the rites. In spite of his words, Maigrot had removed the imperial tablets with the characters *jing tian*, and by this had caused a great tumult among the Christians[76].

Assured that Maigrot's mandate and Tournon's legation were political affairs directed against his sovereignty, the Kangxi emperor announced on December 17, 1706, i.e., two days later, a decree by which not only Maigrot and some of his followers were deported, but which also ordered that, in the future, every missionary in China should have a residence permit (*piao*)[77]. To get a *piao*, the missionaries had to promise to follow Ricci's practice, i.e., to tolerate the Chinese rites. Those who did not make this promise would have to leave China at once[78]. Tournon reacted against this imperial decree by his Decree of

---

71 Noël, *Philosophia Sinica ...*, tract. I, p. 178; *Supplément des mémoires pour Rome ...* (1709), pp. 50-52.

72 Pray, 2:399.

73 *SF,* 8:505, n° 14; *SF,* 5:525, n° 35. *Xianggong* is an honorable title for a youth; in Catholic terminology, it means a catechist or seminarist.

74 *SF,* 5:525.

75 Maigrot had used the interpretation of Francesco Varo, O.P , *L'état présent de l'église de la Chine* (1710), p. 127.

76 *Acta Illmi Maigrot Episcopi Cononensi in Aula Pekinensi Imperatoris Sinarum*, BNC VE *FG 1246/1*, *f.* 17r.

77 Text in *SF,* 5:504, n° 2; *Acta Illmi Maigrot ...*, *f.* 18r.

78 Rosso, pp. 171-172.

Nanjing (February 7, 1707), in which the Papal Decree *Cum Deus optimus* (which in the meantime had arrived in China) was even intensified[79].

During the ensuing years, the Jesuits in China did not give up. While they worked diligently to have things changed, everything was in vain. On March 19, 1715 the Chinese Rites were forbidden by the Apostolic Constitution *Ex illa die*, bringing to an end the successful Jesuit mission in China[80]. Two years before *Ex illa die* was announced in China (November 1716)[81], the Kangxi emperor had tried again to contact the pope, and to this end had written a letter to Clement XI in November 1714. In it the Chinese viewpoint was stated very clearly:

> With regard to the expression *jing tian*, worship heaven, it is not that we regard heaven (*Tian*) as the Lord of Heaven (*Tianzhu*), but rather when we raise our eyes we see heaven (*Tian*), but are unable to see the Lord of Heaven (*Tianzhu*). Since among the innumerable things, both great and on high, created by the Lord of Heaven (*Tianzhu*), there is none equal to heaven (*Tian*), when we gaze at heaven we are only moved by meditation and reverence[82].

Some eighteen years later the Yongzheng emperor (1723-1735) proved the Chinese tolerance towards Christianity, on the one hand, and the firm attitude towards foreign interference, on the other, when he told the Jesuits at court on the occasion of an audience on March 18, 1733: "I have up to now not forbidden anyone to follow your law (i.e. Christianity), or refused to give my permission. You worship *Tianzhu*, and who does not revere Him? We worship Heaven (*jing tian*), and that is just the same"[83].

---

79 *SF*, 6:610, n° 10.

80 "Idcirco tabellas cum inscriptione Sinica King Tien, caelum colito, in Ecclesiis Christianorum appendi, seu jam appensas in posternum inibi retineri permittendum non esse", *Bullarium Romanum Clemens XI,* 2:50; *BM*, 7:194, 196, 205; n° 2807, 2812, 2849.

81 Reil, pp. 170-172.

82 Rosso, p. 301; *Mémoires de la Congrégation de la Mission* 5:216-220 (French 203 ff.). On p. 200 in Latin: "Quod attinet ad literas 'coelum colito' illa reverentia non est pro caelo, sed pro Domino coeli; elevamus ad coelum oculos, quia non possumus videre Dominum coeli; inter infinitas res quas Deus condidit, suprema et omnium maxima est coelum; idcirco illud aspiciendo Deum cogitamus, et excitatur in nobis eum colendi affectus." For the circumstances, see *ibid.*, pp. 205-207.

83 Letter of Antoine Gaubil in common with others, April 14, 1733, Antoine Gaubil, *Correspondance de Pékin, 1722-1759* (Geneva, 1970), p. 351; see

Let me come to an end by quoting Voltaire's *Dictionnaire philosophique*, in which the tragic Chinese Rites Controversy is described in the following, rather summary fashion:

> Le plus singulier exemple de cet abus des mots, de ces équivoques volontaires, de ces malentendus qui ont causé tant de querelles, est le *King-tien* de la Chine. Des missionnaires d'Europe disputent entre eux violemment sur la signification de ce mot. La cour de Rome envoie un Français nommé Maigrot, qu'elle fait evêque imaginaire d'une province de la Chine, pour juger de ce différend. Ce Maigrot ne sait pas un mot chinois; l'empereur daigne lui faire dire ce qu'il entend par *King-tien*; Maigrot ne veut pas l'en croire, et fait condamner à Rome l'empereur de la Chine[84].

---

also M. Huc, *Le Christianisme en Chine, en Tartarie et en Thibet*, 4 vols. (Paris, 1857-1858), 4:52.

84 Henri Cordier, *Bibliotheca Sinica*, 5 vols. (Paris, 1904-1924; reprint, Taipei, 1966), 2:869.

FERDINAND VERBIEST AND MARTINO MARTINI

GIORGIO MELIS (1925-1990)

Giorgio Melis resided for many years in the Far East. From 1972 to 1976 he worked as expert-cultural attaché in the Italian Embassy in Beijing. From 1980 on he was professor of history and institution of African-Asian countries at the University of Lecce and at the University of Naples. He is the author of numerous books and critical essays on contemporary China and modern Southeast Asia. Professor Melis died on January 5, 1990.

Ferdinand Verbiest (1623-1688) and Martino Martini (1614-1661) were both missionaries engaged in scientific work in China during the seventeenth century. Verbiest was younger than Martini, and the two did not operate at the same time: Martini died in 1661, eighteen years after his first arrival in China in 1643, while Verbiest first arrived in China in 1659 and began his work as Johann Adam Schall's aide in 1660. Nevertheless, they belong roughly to the same period of the (European Jesuits') China enterprise, and one finds a number of significant meeting points between the two men. Recalling some details about the connections between the two men will cast further light on the milieu they operated in, and on their personalities and work.

Verbiest traveled from Genoa to Macao in 1657-1658 with a group of Jesuits headed by Martini who was returning to China after achieving a number of significant scientific and religious results in Europe. He traveled again with Martini in 1659 from Macao to Nanjing, heading for his first China mission station at Xi'an. The next year, 1660, he was transferred to Beijing as helper to Schall, an assignment which Martini had unsuccessfully tried to get for himself ten years earlier. Even Martini's brief adventure with cannons while serving the Ming pretender Longwu in the late 1640s, can be viewed against the successful casting of cannons by Verbiest for the Imperial Army thirty-five years later in the mid-1670s.

The major work of Martini, his *Geographical Description of China*, as he called it, was published in Europe in the mid-1650s. In the mid-1670s, Verbiest - by then in the Beijing Mathematical Bureau - published in Chinese some explanatory notes on the *World Map* he edited, from which one can collect some data to compare with Martini's major geographical work on China.

Before discussing some of the significant details of the personal and scientific intercourse between the two men, I shall recall some data about Martini. Those about Verbiest are better known, and have already been mentioned quite often during this conference.

Martini was born in 1614 at Trento. He was an Italian both by family and by cultural formation, although at times he is also considered a German on account of Trento's peculiar political and geographical position: it belonged to the Habsburg domains, and was a gateway from northern Italy and central Europe across the Alps. He became a Jesuit in Rome in 1632, and reached Hangzhou in 1643, at the eve of China's takeover by the Manchus. His missionary work in Zhejiang province was temporarily interrupted by the advancing Manchu armies, but he was quick to realize that the new masters were there to stay, and adapted to the changed political situation. He was thus able to resume his Church activities, to further pursue his Chinese studies, and to travel as far as Beijing[1].

At the beginning of 1651, Martini set out for Europe as a representative ("Procurator") of the Jesuits working in China at their Order's headquarters in Rome. He traveled via Manila and Batavia, and reached Amsterdam in August 1653 by way of Norway and Hamburg - quite an unusual itinerary during those times of the Portuguese *Padroado*'s supremacy[2].

Martini carried along on his trip about fifty Chinese books, and on them he based the works he was composing in Latin for the European readers, (namely, a detailed geographical account of the Chinese empire, and a lengthy historical summary of the "Chinese nation," from prehistory up to the beginning of the Christian era) in order to be able

---

1 B. Bolognani, *L'Europa scopre il volto della Cina. P. Martino Martini* (Trento, 1978), pp. 20, 29, 68-69, 78-79; F. Ghetta, "M. Martini in his Town", in Giorgio Melis (ed.), *Atti del Convegno Internazionale 1981*, Bilingual Italian-English Edition (Trento, 1983), pp. 357-362; Giorgio Melis, "M. Martini Travels in China", in Melis, *op. cit.*, pp. 421-432.

2 Joseph Sebes, "M. Martini's Role in the Controversy of Chinese Rites", in Melis, *op. cit.*, pp. 477-479; Henri Bernard, "Les Sources mongoles et chinoises de l'Atlas Martini (1655)", *Monumenta Serica*, 12 (1947), 137, note 42, p. 138, note 48.

to acquaint them with China directly from Chinese sources. After leaving Manila and passing through the Celebes, he was detained in Batavia for about eighteen months - up to February 1653 - by the Dutch, who, nevertheless, allowed him ample freedom for work. During the journey he also prepared a first-hand account of the recent change of dynasty in China due to the Manchu conquest, so that upon arriving in northern Europe, he had the two relevant geographical and historical manuscripts almost ready for publication[3].

Before proceeding to Rome, he published his *De Bello Tartarico Historia* (Antwerp, 1654) on China's conquest by the Manchus, and the *Novus Atlas Sinensis* (Amsterdam, 1654); he also arranged for the printing of his *Sinicae Historiae Decas Prima*, published in Munich in 1659 when the author was already back in China. In Rome he was instrumental in securing from the Church authorities a decree by *Propaganda Fide*, dated March 1656, favorable to the prevailing Jesuit stand in the Chinese Rites Controversy[4].

Such a professional publisher and printer as Johannes Blaeu decided at once to include Martini's *Atlas* in his famous geographical series. The Dutch scholar Jacobus Golius[5] wanted, by all means, to meet him, and testified to the fame which in those days surrounded Martini's manifold and rare knowledge of things Chinese. Actually it is really amazing how Martini could write such important works on China, drawing directly from Chinese books, after only a very short period spent in China

---

3 Bolognani, *op. cit.*, pp. 95-99; Bernard, "Les Sources mongoles", *op. cit.*, p. 139.

4 *De Bello Tartarico Historia, in qua, quo pacto Tartari hac nostra aetate Sinicum Imperium invaserint, ac fere totum occuparint, narratur; eorumque mores breviter describuntur.* Auctore R.P. Martino Martinio, Tridentino, Moreti, Antverpiae, 1654; *Novus Atlas Sinensis*, a Martino Martinio Societatis Jesu descriptus (Blaeu, Amstelaedami, 1655); Martini Martinii, Tridentini, e Societate Jesu, *Sinicae Historiae Decas Prima, Res a gentis origine ad Christum natum in Extrema Asia, sive Magno Sinarum Imperio, gestas, complexa* (Blaeu, Amstelaedami, 1659).

5 Jacobus Golius (1596-1667) wrote, in his "De Regno Catayo Additamentum" appended to the *Atlas* ("eximium Opus," he calls it) of Martini (to him, "intelligentissimus Martinius"): "Cum suo initio proximae aestatis fama increbuisset, multiplici scientiae virum, eaque rara et inaudita, quam inter Sinenses plurium annorum consuetudine acquisivisset, navibus nostris ex India Orientali delatum in hasce venisse oras; equidem vidende ejus et compellandi cupido me haud mediocris incessit ... Perhumaniter exceptus, ex paucis dictis et responsis facile perspexi, de insignibus viri dotibus minime vanam aut fallacem fuisse famam"; also J.J.L. Duyvendak, "Early Chinese Studies in Holland", *T'oung Pao*, XXXII (1936), pp. 298-303.

(1643-1651), and in spite of the lack of leisure during those eight years, due not only to the prevailing war climate, but also to his struggle with the Chinese language, and having to attend to church activities as well.

I.
The Travels of Martini and Verbiest to and in China

The young Verbiest obviously could not anticipate his own scholarly achievements in China twenty years later when, in January 1656, at the age of thirty-three, he boarded a Dutch boat in Genoa, together with nine other Jesuits, among whom was Martini, their group leader. The ship should have taken them to Portugal, where they would proceed to India and the Far East. Due to a number of incidents, Verbiest, Martini, and some of the other Jesuits with them, only reached Macao in July 1658. Thus Verbiest, for whom China was apparently a second choice as a missionary field after failing to get to South America, spent more than two years as a companion to Martini, an experienced and learned veteran who stood high as the author of ponderous best-sellers on China, and also as a capable negotiator in his dealings with the Holy Office in Rome[6].

After a few days of sailing, the ship boarded on January 11, 1656 in Genoa by the Jesuits, was confronted on February 1 by a French pirate vessel. A major fight ensued and the ship was forced to follow the pirates, who were more than happy to lay hands on a group of priests, and foresaw a higher price before giving back the ship and releasing the passengers. The pirate captain, towing the seized Dutch boat, approached the coast near Antibes, and ordered Martini and Giacinto De Magistris (a Jesuit assigned to India) to travel to Genoa in order to look for merchants who would pay the ransom.

While the Genoa merchants and Martini were having talks at Cannes with the pirate captain on the terms for rescuing both goods and prisoners, the Jesuits were allowed to land - provided they did not leave the town of Nice. Verbiest and Albert D'Orville, another Belgian Jesuit assigned to China, were at first left in the pirate boat as hostages, but two days later the pirate captain allowed the two to go ashore as well. But, as he was not getting all the money he wanted, he ordered the Jesuits to return to the ship. At this point, Martini decided on February 9, that the missionaries should hide, while he himself with D'Orville,

---

6 Sebes, "M. Martini's Role", *op. cit.*, pp. 486-489.

and De Magistris with Verbiest, made for Genoa by separate roads, occasionally concealing their flight and identity with half lies.

On those three weeks in the hands of the pirates, Verbiest wrote a detailed report[7] to the Jesuit superiors in Rome, which at times partially varies with the account of Martini. Verbiest abounds in praise and admiration for Martini's courage and ability; he also gives some hints on the complex personality of Martini, with his *savoir faire* and imagination, his daring and generosity, his flares of anger and his propensity to view it all as a thrilling experience for the missionary recruits.

The Jesuits and the other passengers lost their personal belongings and their money to the pirate crew; only the goods stowed in the hold were returned to the owners. Because of the stress of those hectic days, and also because of lack of money, the Martini-Verbiest group had to postpone their crossing to Asia to the following year[8].

We find aboard the ship *Bom Jesus de Vidigueira*, which sailed from Lisbon on April 4, 1657, Martini and Verbiest, together with fifteen other Jesuits assigned to Asia. On January 24, 1658 only fifteen of them landed at Goa, and on July 17, only Martini, Verbiest, Prospero Intorcetta, André Ferrao, and D'Orville arrived in Macao via Macassar, the only five survivors out of the Martini-Verbiest group of seventeen who had left Lisbon for the East sixteen months earlier: two died before reaching Goa, six stayed behind in Goa (and only two of these survived more than two years), three died in and about Celebes, where one more was left behind and only later joined the China mission. To these five should be added two more who also left Lisbon in April 1657 - Joao de Figueiredo, who remained at Goa, and Franz Xaver Schäffelmayer, who reached Macao later. The loss of ten out of seventeen in a single missionary expedition to China not plagued by a shipwreck, was, even for those times, a heavy toll[9].

---

7 Bolognani, *op. cit.*, pp. 116-123.

8 Letter from Martini to G. Nickel, general of the Society of Jesus, February 19, 1656, ARSI, *Jap.-Sin.* 162, ff. 182v; H. Josson and L. Willaert, *Correspondance de Ferdinand Verbiest de la Compagnie de Jésus (1623-1688)* (Brussels, 1938), pp. 6-38. Verbiest describes Martini as tall and robust, with an imposing flowing beard, and inspiring courage amid the destruction of the corsair attack.

9 Alfonso Aires, a Spaniard, and Giovanni Gregorio Parisi, an Italian from Calabria, died aboard the ship on the way to Goa. Four Portuguese (Joao Couceiro, Joao de Figueiredo, Antonio Preto, Amador Fernandes), one German (Franz Xaver Schäffelmayer) and one Frenchman (Jean Brandi)

The newly arrived missionaries[10] had to wait a few months before entering China. In Macao they found a climate of tension between the Jesuits in the Portuguese enclave and the Chinese authorities in Guangzhou. It all started when a Portuguese Jesuit in Macao, not belonging to the China mission, arrested inside a church a Chinese sailor who had misbehaved towards the ladies. The priest claimed he could act that way on the basis of special powers granted by the Chinese court, but the Chinese populace did not quite accept his point. The local mandarin was quick to seize the opportunity to intervene, and was able to get the backing of the Guangzhou authorities. There followed a number of moves and countermoves, until the Chinese official was able to imprison two Jesuits of the Chinese mission, and would release them only against some proper money arrangement. He also forbade under heavy penalties the local people to offer any passage to European missionaries intending to enter China from Macao[11].

Luckily before the issue could be settled, the resourceful Jesuit from Trento received from a Manchu general, with whom he made friends eight years earlier, a message welcoming him back to China, and providing him and his companions with a safe-conduct to travel inside the country. Martini's party of nine Jesuits left Macao by boat on March 5, 1659. They were able to avoid the danger of bandit gangs, and

---

stayed behind at Goa; they were all in bad shape and only two of them (Figueiredo and Schäffelmayer) survived long enough to reach the China mission. Thus, on January 30, 1658, only nine Jesuits set off for Macao. One of them (Giovanni Maria Guicciardi, an Italian) died shortly before the others landed at Macassar on the island of Celebes on the following May 21; another (François Clément, a Frenchman), died a few days after landing there; a third one (Jakob Dimer, a German) was left behind on the island and soon passed away; a fourth one (Christian Wolfgang Herdtrich, a German) was also left behind on Celebes because he was very sick when the other five left for Macao on July 17, 1658, and was able to travel later to Macao after recovering. Martini, Verbiest, Intorcetta and Ferrao had also been seriously sick during the long journey.

10 Shortly afterwards, four more Jesuits arrived in Macao from Europe: Bernhard Diestel (a German) with Johann Grueber (an Austrian), and Philippe Couplet with François de Rougemont (both Belgian). "Appulimus itaque Macaum 17 Julij ... Auxit hanc animi laetitiam non parum adventus P. Bernardi Diestel, et P. Joannis Grueber, qui paucis post adventum nostrum diebus, Anglica navi vecti, hunc portum feliciter tenuere. Paulo post etiam appulere P. Philippus Couplet et P. Franciscus Rougemont, Belgae ... Adjunxere ... in hoc Collegio Missionis Sinensis subjecta omnino 12 ..." D'Orville's letter to the Jesuit general, October 30, 1658, ARSI, *Jap. Sin. 134*, ff. 16-60.

11 Letter of D'Orville to the general, see note 10.

Martini, together with Intorcetta and François de Rougemont, arrived in Hangzhou on June 11, after meeting their Manchu protector at Songjiang. Verbiest, together with D'Orville and Giovanni Francisco de Ferrariis, left the group earlier in Nanjing, and headed for Shaanxi province, while the other three were assigned to various missionary stations[12].

Martini and Verbiest did not meet any more, as the former died two years later at Hangzhou. After a short stay at Xi'an, Verbiest was called to Beijing by the same Schall who ten years before strongly opposed Martini's self-candidacy for the post[13]. By 1660 the situation of Schall at court was quite different; he was badly in need of a helper, and the choice of Verbiest proved to be excellent. History is not based on "if's," so one will never know how Martini would have fared, in case he had been chosen for that work.

Instead, it seems to be proved that Martini did not really succeed with cannons[14] when trying to cast during his brief stay in the service of Longwu, the Ming pretender in Zhejiang-Fujian in the late 1640s. On the contrary, Schall had been quite successful in the casting of cannons

---

12 "Hang cheu (Hangzhou) tenui ii Junii 1659, antiquam nimirum meam residentiam ... Scripsi hoc eodem anno Macao ad Paternitatem Vestram de itinere iam parato ad interiorem Sinam, deque novem Sociis mecum profecturi ... quinta Martii Quangtung versus vela dedimus, non tamen sine aliqua formidine a latronibus, ... et quamvis ob turbam inter Macaum et Regulos Quangtumenses aliqui dubitassent de nostra profectione, tamen fretus Deo primum, deinde diplomate amplissimo, quod ad me Macaum miserat summus Tartarorum praefectus, olim amicus meus singularis quamvis nondum Christianus, omnia superavimus. Hic enim praefectus ubi cognovit me Macaum pervenisse, ... diploma illud ultro ex Sinis transmisit, imo ubi Nan king (Nanjing) provinciam (in qua ipse praefecturam agebat) me ingressum cognovit, milites ac naves submisit qua exciperent, et deducerent ad Urbem Sung kiang (Songjiang) ubi pro tempore versabatur, ibi actis illi gratiis iterum per suas naves Hang cheu (Hangzhou) usque deduxit. Interea patres socios omnes in destinatas stationes partiuntur, nimirum p. Philippum Couplet ad urbem Can cheu (Ganzhou) provinciae Kiangsi (Jiangxi) apud p. Jacobum Fabrum (Le Faure). Nankini vero reliqueram pp. Franciscum Ferrari (De Ferrariis), Albertum Dorville (D'Orville) et Ferdinandum Verbiest ad provinciam Xensi (Shaanxi) ituros, ad eius metropolim Sigan (Xi'an) stationem p. Provincialis Ignatii da Costa. P. vero Andream Ferram (Ferrao) Hoaiganum (Huai'an) misi ad p. Felicianum Pachecum; p. vero Dominicum Gabbiani ad p. Franciscum Brancati Xang hai (Shanghai), et mecum Hang cheu perduxi pp. Franciscum Rougemont et Prosperum Intorcettam et F. Coadiutorem Antonium Fernandez." Letter of Martini to the Jesuit general, July 11, 1659. ARSI, *Jap.-Sin.* 162, ff. 51-52.

13 Sebes, "M. Martini's Role", *op. cit.*, pp. 477-478.

14 Bolognani, *op. cit.*, pp. 86, 90.

for the last Ming emperor in Beijing in the early 1640s, and Verbiest also proved to have been a good pupil of Schall in this respect, when making over four hundred cannons for the Kangxi emperor more than thirty years later[15].

II.
Martini and Verbiest as Geographers

The scientific work of Verbiest consisted essentially in publicizing and applying in China many important facets of European science and technology of the times. The scientific work of Martini consisted essentially in digesting and publicizing in Europe some important aspects of China's history and geography. Thus the approach of the two authors was basically different, as required by the audience they were addressing, and inspired by the contents of their publications. Moreover, Martini's main achievements are in the field of China's geography and history, while Verbiest's major contribution to the China-Europe cultural interchange is found in the fields of mathematics and astronomy, and his work on geography is only secondary. However, one can detect in their books on geography some details worth comparing.

Martini's original contribution to the China-Europe cultural interchange in his *Atlas* was twofold: on the one hand, he collected and selected from Chinese sources significant data about China's geography, both physical and administrative, as well as economic and anthropological; on the other hand, he drew seventeen detailed maps, a general one of the Chinese empire, one of every Chinese province, one of Korea and Japan, they being a combination of European cartography techniques and of Chinese cartography and numerical data regarding the location and the respective distances of towns and of geophysical features. Here we refer mainly to the written text of Martini's *Atlas* running through 171 pages in the earlier Latin edition.

About Verbiest as a geographer, I shall here consider some points from his *Kunyu tushuo*, and I shall add a couple of remarks from his travel account about the configuration of the areas of "Tartary" he visited while accompanying the Kangxi emperor. In Part I of the *Kunyu tushuo*, Verbiest explains to the Chinese readers some elements from European general geography of that time, dealing with topics as the

---

15 Jonathan Spence, *The China Helpers* (London, 1965), p. 15.

perfect correspondence between the main notations of the celestial sphere and the earth globe, the sphericity of the earth, various theories on earthquakes, on the origin of mountains and rivers, on waves and tides, on the atmosphere, winds and rain, and on the four elements. He closes with some short notes on worldwide anthropological geography[16]. Part II of the *Kunyu tushuo* is a booklet of Giulio Aleni[17], reedited by Verbiest, in which the main features of various countries in the five continents are introduced to the Chinese[18].

As preliminary sectorial research on the proposed subject, I suggest a brief review of a few points as dealt with both by Martini and Verbiest in their books, as specified above. Let us start with some geographical coordinates in China. Martini's maps in the *Atlas* obviously propose graphically the geographical coordinates of China's localities. The *Atlas* also contains a detailed numerical Table of Latitudes and Longitudes

---

16 In the copy I examined from the National Library in Rome, Part One starts with five Chinese-style double pages of index (*mulu*), and runs through thirty-one double pages, under fifteen numbered headings and several unnumbered subheadings. On p. 31, one reads: *zhili lifa jixi Nan Huairen zuanzhu*, i.e., "compiled by Nan Huairen Jixi (Verbiest), in charge of the Calendar." On p. 1, Verbiest refers to four western colleagues (*tongxue xishi*) who before him published Chinese works on the subject, namely the Jesuits Ricci, Aleni, Vagnoni, and De Ursis.

17 In the copy I examined from the National Library in Rome, Part Two (*xia juan*) starts with a cover page, followed by ten pages of index and two pages of foreword, ending with the words "*Yuanxi Yesu huishi Ai Rulüe zhi*," i.e., "Signed by Ai Rulüe Yuanxi (G. Aleni) of the Society of Jesus," and runs through a text of fifty-six pages distributed under several unnumbered headings and subheadings, followed by a kind of appendix, not recorded in the index, of nineteen Chinese-style double pages (with numbering starting again from page 1), where pp. 1 to 11r have twenty-three drawings of rare animals all over the world with captions about their features and whereabouts, p. 11v has a drawing of an idealized large western cargo ship, and pp. 12-19 have drawings and accounts of the "seven wonders" of the world, plus the Coliseum in Rome.

18 The five continents include Asia, Europe, Libya (i.e. Africa), South and North America, and Magellanica (*Mowalanijia yi ming Marelanijia*); there follows a general paragraph on the "four seas," their features, inhabitants, products, and navigation. Ricci wrote about the "Megaglianica" in the first edition of his World Map: "It is entirely in the South; it emerges only at the South Pole, while the North Pole is always hidden from it. One does not dare to fix its boundaries, as they were not explored yet, except for its northern section, delimited by Java Major and Minor and by the Magellan Strait." Verbiest (*Kunyu tushuo* I, p. 3) repeats these words of Ricci. In the *Kunyu tushuo* II, p. 1 of the Appendix (see note 17), one reads that "The continent of *Mowalanijia* is a land around the South Pole."

covering all prefecture (*fu*) and district (*zhou* and *xian*) level cities and towns. A limited number of them were observed directly by Martini, who utilized them as reference points for his calculations on the rest, with the help of data on the distances between the various places, both numerical and cartographic which were usually offered in the Chinese sources[19].

In the chapter about the province of Beijing, Martini takes the meridian of Beijing as the prime meridian (*meridianus primus*), and in his table he notes the degrees of longitude, as east or west of Beijing. Moreover, in the relevant maps of the *Atlas*, he marks both the eastern and western values, taking the meridian of Beijing as zero, and he notes the longitudes as well, according to a prime meridian located at the Canary Islands. In this way he marks the longitude of Beijing as 145°, following his own astronomical calculation of longitudes in China[20].

Besides applying his personal findings in drawing his map, Verbiest gives a few numerical instances of latitudes and longitudes in the *Kunyu tushuo*. He keeps the prime meridian at Beijing, and notes various degrees of longitude according to a full circle of 360°, proceeding eastward from Beijing, so that the Chinese localities west of the capital are given values close to 360°. Thus the longitude of Taiyuan in Shanxi province is given as 355° and the longitude of Kaifeng as 357°. Corresponding notations in Martini's tables are 4°35' long. W. for Taiyuan, and 2°54' long. W. for Kaifeng. In both cases, Martini's numerical notations are slightly closer to modern calculations. Verbiest, however, might give approximate figures, as he is rather concerned with

---

19 The Jesuit geographers of the eighteenth century in China were able to make more direct observations and calculations; nevertheless, many of their data, according to Richthofen, were only collected from Chinese documentation. Cfr. Henri Bernard, "Les étapes de la cartographie scientifique pour la Chine et les pays voisins", *Monumenta Serica* I (1936), pp. 460-461.

20 One reads in Martini's *Atlas*: "Longitudo illius (Beijing) apud me nulla, nam hujus ego urbis meridianum, primum pono (nec enim apud Europaeos quicquam reperio certi & pudendum sane, in re tanta Cosmographos non convenire). Ex Eclypsibus autem Lansbergiana hypothesi à me calculatis & observatis, tum in hac urbe, tum in aliis aliquibus Sinarum, à Goesa Zelandiae distare comperi horis omnino 7.56' quae dant gradus aequatoris 119.0, unde cum Goesa (Goes) supponatur a Fortunatis (Canaries) insulis distare gradus 27, fiet huius urbis meridianus 146 grad. vel 149.20' si ponatur Goesa distare gradibus 30.20' juxta variam primi meridiani asumptionem." According to modern calculations, Hierro Island, the westernmost of the Canaries, is placed at 18° long.W. and Goes at 3°54' long.E., while to Beijing are assigned 116°30' long.E. Thus Beijing's distance from Goes is put at about 112°.

comparing some features in and outside China[21].

So, Verbiest writes the local time of Taiyuan, long. 355°, is two hours (one *chen*) ahead of that of Ceylon, long. 325°, and Kaifeng, long. 357°, has a twelve hours time difference with the Rio de la Plata area, long. 177°. Both values are not very far from modern calculations[22]. Verbiest's comparisons extend to some values of latitude, as we shall see below.

In the introductory section of the *Atlas*, Martini assumes that the Chinese mainland "extends for about 30°, from the promontory near Ningbo to Amasei (possibly the Kunlun) Mountains." Actually, in the relevant maps of the *Atlas*, the Kunlun Mountains are given a north-south direction beyond the western borders of Sichuan province, and the western borders of the province of Yunnan run close to the 126th (or 19° W.) meridian, about 25° from the promontory of Ningbo, which is placed at the 151th (or 6° E.) meridian. Verbiest's longitude values are close to Martini's.

Both Martini and Verbiest give the linear length of a degree of longitude as 250 Chinese *li*, as earlier calculated by Schall. The graphic rendering of a Chinese foot (*chi*) printed for reference in the *Atlas* measures 29.6 cm.; 250 *li* would thus be equivalent to 159.84 km., - too much at variance from any approximate equivalent of the accepted linear length of a degree of longitude. Verbiest, along with others, realized the different values given in China to the foot, which were at the origin of such variances of calculations. But in China, the problem as applied to the degree of terrestial longitude was faced only towards the turn of the seventeenth century, and solved in 1702 by measurement on the ground and by an imperial decree fixing the earth degree equivalent at the equator to 200 *li*, while at the same time standardizing the value of the Palace foot[23].

---

21 *Kunyu tushuo*, I, pp. 4-5. Modern calculations are 112°30' long.E. for Taiyuan, and 114°21' long.E. for Kaifeng. Corresponding values, according to Martini's notations, would be Verbiest's 355°25' for Taiyuan and 357°6' for Kaifeng.

22 In modern calculations, the island of Ceylon is located between 80° long.E., and Buenos Aires on Rio de la Plata is at 58°27' long.W., which in Verbiest's way of reckoning would correspond to about 324° to 326° long. and 184° long. respectively.

23 Pasquale d'Elia (ed.), *Fonti Ricciane*, II (Roma, 1942-1946), pp. 22-23; Bernard, "Les étapes", pp. 455-458; Osvaldo Baldacci, "The Cartographic Validity and Success of Martino Martini's Atlas Sinensis", in Melis, *op. cit.*, pp. 73-88. The modern linear value of the earth degree of longitude at

As for latitude values of some selected Chinese cities, recorded both by Martini and by Verbiest, the former's *Atlas* gives as 39°59' N. the latitude of Beijing city; about the same value (40° N.) is given in the latter's *Kunyu tushuo*, where it is compared roughly - to explain the correspondence of climatic zones - with the Cape of Good Hope's, given as 35° S. Verbiest also gives the Kaifeng latitude as 34° N., while Martini notes 35°50' for this city[24]. Moreover, Verbiest was able to calculate by himself, in his travels to "East Tartary," the longitude of Shenyang as 41°6'. Years later, however, the young Kangxi emperor still thought Shenyang's latitude was the same as Beijing's[25], and that of northwest Ula as c. 43°50'. Martini could not possibly give any such values for "East Tartary," as they did not appear in the Ming sources he was using.

Ferdinand Richthofen, writing in 1877 from his observations and studies, concluded that, in general, the geographical coordinates in China recorded by Martini "approximately correspond to the modern astronomic positions, provided that in the average one moves west the longitude by 47', and moves north the latitude by 28'"[26]. Verbiest upheld against critics the scientific value of Martini's work[27], and blamed occasional geographic lapses in the *Atlas* on the Chinese sources. He was also able to correct, thanks to information received from the Russian envoys, Martini's mistake about Samarcand, which in the *Atlas* was located, according to information from Chinese geo-

---

the equator is calculated approximately as of 111.32 km.

24 Modern values are 39°55' N. for Beijing, 34°51' N. for Kaifeng, 33°55' S. for Cape Town, and 34°52' S. for Cape Agulhas.

25 Jean-Baptiste du Halde, *Description de la Chine*, Tome I (Paris, 1735); Bernard, "Les étapes", *op. cit.*, p. 458.

26 Ferdinand P.W. Richthofen, *China, Ergebnisse eigener Reisen und darauf gegründeter Studien* (Berlin, 1877), 3 vols. with maps, as quoted by Bernard ("Les étapes", *op. cit.*, pp. 447-448), wrote: "Among the missionaries (of the seventeenth century) there were men - like Schall and Verbiest - of a mathematical and technical culture which, however, did not give them the "feeling" of the country and of its character. Actually the history of the whole mission during this (seventeenth) century, among hundreds (sic!) of fathers, numbers only one geographer, namely Martino Martini; and even during the seventeenth century, not only was he not overtaken, but he was barely equaled." Richthofen notes that the average adjustment of coordinates he proposes does not include the cases of Gansu, Guiyangfu and Fengxiangfu (*ibid.*, p. 448).

27 Josson and Willaert (eds.), *op. cit.*, pp. 314-315, quoted by Bernard, "Les Sources mongoles", *op. cit.*, p. 143.

graphic knowledge, quite close to the northwestern border of China, just across a narrow strip of desert[28]. For the rest, Verbiest did not improve much on Martini's data, nor did he mean to do so. As I already pointed out, he was rather providing his Chinese audience with general notions on geography and with some news about the world outside China, together with some comparative hints about Chinese and foreign geographical data. Such an approach only occasionally caused him to cross paths with Martini's geographical work on China.

In one instance, Verbiest's position at court seems to influence and condition his writings. "According to a map of the Yuan dynasty, the Yellow River runs at the start in a northeast direction across the Xifan, up to Lanzhou, for over 4,500 *li*. Then it enters China, and flows again in a northeast direction into the barbarian lands for a distance of over 2,500 *li*. It then turns eastward, and next it runs southward as far as Puzhou, over 1,800 *li* in a winding course." Verbiest's quotation and account on the Huanghe stops there, at the southern corner of the Shaanxi-Shanxi border, leaving out altogether the long middle and lower courses of the famous river. And among the rivers mentioned by him, the Yellow River is the only one of which he does not note the total length in *li*, for the entire course[29]. Verbiest's sketchy and incomplete description of the Yellow River's course, possibly for reasons of political opportunity, is given as documentary support for a Chinese source dated three centuries before.

On the contrary, Martini describes in the Preface to his *Atlas* the course of the Huanghe, quite in detail, from its sources in the "Kunlun Mountains" in the Sea of Constellations ("*Sing cieu mare*"), across four provinces (Shaanxi, Shanxi, Shandong, and Jiangnan), down to Huai'an and the East China Sea. This river is the second Chinese river, the first being to Martini the "Jiang" or "Yangzi jiang", also described at length in the Preface to the *Atlas* in an account, based on contemporary sources[30].

The different approaches of Martini and Verbiest in their works again show very clearly when they deal with the ultimate components of the material world. In his book on general geography[31], Verbiest writes

28 Bernard, "Les étapes", *op. cit.*, p. 453.

29 *Kunyu tushuo* I, p. 21. Among the main Asian rivers, Verbiest does not mention the Changjiang or Yangzi jiang.

30 Martini, *Atlas*, *op. cit.*, Praefatio, pp. 13-14.

31 *Kunyu tushuo* I, pp. 28-29.

about the Aristotelian theory of the four elements, taking it as an obvious truth. He explains how each natural element has a "natural" locus, according to a definite placement and a fixed order, depending on weight or lightness, and on mutual sympathy or repulsion, and how the elements have a strong tendency to move back to their natural place when dislocated. Moreover, the four elements are round in shape, because the whole universe is spherical and because, in the sublunar world, they are intermingled, and consequently any shape different from round would allow for empty holes - a situation which cannot be admitted, since "nature abhors vacuum."

Martini introduces to his European readers the Chinese theory of the five elements in his "Summary" of China's history, i.e., in his *Sinicae Historiae Decas Prima*[32].

> Suius (Suiren) was an excellent astronomer. From the contemplation of the stars and of our globe, he derived for the first time among Chinese philosophers the existence of five elements, namely metal, wood, water, earth, and fire; this last one he found in the air. ... We would not find the Chinese ridiculous for including wood and metal among the elements if we added briefly the kind of philosophy of the five elements found in the book *Xuking* (*Shujing*). The elements are five, we read: water, fire, wood, metal, and earth. Water dampens and goes down, fire dries and goes up, wood is erect and distorted, metal can be shaped and made to shine, earth receives seed and produces fruits. That which dampens and goes down gives salty (taste), that which dries and goes up gives bitter, that which is erect and distorted makes acid, that which can be shaped and made to shine makes biting (hot) or sour, that which receives seeds and produces fruits generates sweet. Such are the characteristic flavors of the five elements. These explanations I got entirely from the Chinese, and I added nothing of my own; I leave them to the reader to evaluate.

Writing for a European reading public, Martini thus introduces the five elements of the Chinese tradition. On the contrary, the five elements of the Chinese are not even mentioned in Verbiest's work on general geography. Instead one finds there a long paragraph on the "four elements" of Aristotelian cosmology. This topic is surely more of a philosophical nature, but is likewise conducive to the purpose of comparing the two men.

---

32 Martini, *Decas*, *op. cit.*, p. 19.

THE "EXCOMMUNICATION" OF FERDINAND VERBIEST

FRANCIS A. ROULEAU, S.J. (1900-1984)
EDWARD J. MALATESTA, S.J.[1]

Francis A. Rouleau, S.J., went to China in 1929. After his ordination to the priesthood, he was assigned to teach Church history to diocesan and Jesuit seminarians at Zikawei, Shanghai. For eight years he researched, especially in the archives of Rome, the history of the early Jesuit mission to China. He died on February 21, 1984 after publishing several scholarly articles and leaving behind a vast documentation and drafts for other projects, especially for a major study of the embassy to China of papal legate Carlo Tommaso Maillard de Tournon who in 1707 condemned the Chinese Rites.

Edward J. Malatesta, S.J., holds a doctorate in Sacred Scripture and taught biblical spirituality at the Gregorian University from 1966 to 1979. Since 1979 he has been engaged in the Chinese apostolate. He worked with Father Rouleau during the last seventeen months of the latter's life, and in June 1984 co-founded with Theodore N. Foss the Institute for Chinese-Western Cultural History at the University of San Francisco. Through the Institute he is attempting to carry forward the research begun by Father Rouleau while cooperating with the founding and development of an Institute for Religious Studies at Peking University.

The term "excommunication" is used by Henri Bosmans, S.J. when describing the opposition suffered by the great Flemish Jesuit in his position as head of the imperial Tribunal of Mathematics at Beijing. Bosmans's reference to the matter is brief and rather ambiguous: "Des envieux l'accusèrent (Ferdinand Verbiest) de brigue, d'ambition, allèrent même jusqu'à faire fulminer contre lui une sentence d'excommunication"[2]. The author does not develop the point, either at this place

---

1 The text of this essay is, in part, the same as the unpublished study with this title found among the papers of Francis Rouleau, S.J., with whom I worked from September 1982 until his death in February 1984. I have added some new material to the original text, made minor stylistic changes, and provided the references and footnotes on the basis of Rouleau's research and my own (EJM).

2 Henri Bosmans, S.J., "Ferdinand Verbiest Directeur de l'Observatoire de

or on p. 406, where it is again mentioned. We are not told, for instance, by whom and on what authority the "sentence of excommunication" was imposed. As a consequence, the expression might suggest to some that Ferdinand Verbiest was the subject of a canonical sanction imposed by the Church. This would be a grave misapprehension. Neither at Rome nor in China was there ever any ecclesiastical action against Verbiest. The issue was a purely domestic one within the Society of Jesus. It involved conflicting interpretations of the simple vow pronounced by the professed fathers of the Order on not aspiring to high dignities outside of the Order, as applied to Verbiest when raised by the emperor to the dignity of mandarin[3]. The following outline covers the principal

---

Pékin (1623-1688)", *Revue des questions scientifiques* 71 (1912), 234. In the reprinted edition of Louvain, 1912, pp. 42 and 112 respectively. Bosmans, in another article, tries unsuccessfully to explain his unfortunate expression. To the question "Was Verbiest excommunicated?", he gives a clear, correct, negative answer and then goes on to say: "L'excommunication à laquelle je fais allusion fut celle du visiteur de la Chine et du Japon, Louis da Gama". Da Gama might have thought that Verbiest had incurred an excommunication, but he had no ecclesiastical authority to pronounce one. "La notice nécrologique de Ferdinand Verbiest", *Annales de la société d'émulation* 64 (1914), 110-111. Roger A. Blondeau, *Mandarijn en Astronoom* (Brugge & Utrecht, 1970), pp. 400-405, treats the question of excommunication, which was discussed with much heat in Flanders, in 1913-1914, especially because of a play written by Rev. Cyriel Verschaeve, which caused ripples as late as 1935.

3 See Part X, nos. 817-818 of *The Constitutions of the Society of Jesus*. Translated, with an introduction and a commentary by George E. Ganss, S.J. (St. Louis, 1970), pp. 334-335. As in every Roman Catholic religious Order, the members of the Society of Jesus pronounce vows. According to the Jesuit Constitutions, vows are pronounced on two occasions. After the successful completion of the first probation, or novitiate, which lasts two years, one is allowed to pronounce the three customary vows of poverty, chastity and obedience. On this occasion, the person is understood to be offering himself to the Order to live his life therein. The second occasion is at least ten years later. At that time the Order puts its seal upon the person after so many years of testing. Verbiest pronounced his final vows in Macao on February 16, 1659, eight months after his arrival in East Asia, and eighteen years after his entrance into the Order. The final vows of Jesuits, according to their particular legislation, are of two kinds. Ignatius of Loyola, the founder of the Order, envisaged a core of priest-members called "*Societas Professa*" (the professed Society); the other members, whether brothers or priests, are called "*Coadiutores*" (helpers). The professed priests, in addition to the three vows of poverty, chastity and obedience, and a special vow of obedience to the pope, pronounce five other vows. They promise: 1) when legislating on the practice of poverty, only to make it stricter; 2) not to aspire to positions of authority or honors within the Order; 3) not to aspire to them outside the Order; 4) to report to superiors those whom they know to aspire to such positions; 5) if promoted outside the Order, to listen to the

phases of this conflict.

1. Dispute over the legitimacy of accepting the office of mandarin by a professed Jesuit first arose in the case of Johann Adam Schall von Bell, who was sharply criticized for his assumption of the honor. His most outspoken antagonists were two Beijing associates, Ludovico Buglio (1606-1682) and Gabriel de Magalhaes (1610-1677). They held, besides reasons of a practical nature, that the mandarinate, being a secular dignity, violated the spirit and express law of the Jesuit Institute regarding the professed. The quarrel agitated the China mission for years. It was only settled, provisionally, when Pope Alexander VII, consulted by the vicar-general, Gianpaolo Oliva, gave a *viva voce* indult (April 3, 1664) permitting Jesuits in China, although they are professed, to accept the *officium et dignitatem mandarini et mathematici regii*, since the title involved no exercise of jurisdiction[4]. In January 1660, Simón da Cunha, the visitor, had visited the Beijing fathers. He decided in favor of Schall, and sent a letter in that sense to all the missionaries of the Society in China. Less than a year after his arrival in Beijing, Verbiest had written a long treatise to Superior General Goswin Nickel defending Schall's holding of the presidency of the Tribunal of Mathematics[5].

---

advice of the General in important matters. The precise point at issue in regard to Verbiest, was whether or not he failed in his vow of not aspiring to positions of authority outside the Order.

4 See Louis Delplace, S.J., *Synopsis Actorum S. Sedis in causa Societatis Iesu 1605-1773* (Louvain, 1895), p. 381.

5 Verbiest to Nickel, Beijing, the beginning of 1661. ARSI, *Jap.-Sin. 143*, n° 8, *ff.* 161-179v, in H. Josson, S.J. and L. Willaert, S.J., *Correspondance de Ferdinand Verbiest* (Brussels, 1938), pp. 41-103 (hereafter *Correspondance*). In an unpublished communication to Nickel a little more than a year later, Verbiest writes another defence of Schall, affirming, probably with tongue in cheek, that if one professed father of the Society could wish a dignity for another, he would wish that Schall be made a bishop, for in that way, his honor would be restored. For after all, he observes, the natural God-given right to reputation takes precedence over the positive legislation of the Jesuits. Beijing, May 12, 1662, ARSI, *FG 730 I, ff.* 48v-49r; another copy dated May 20, 1662, *ff.* 50r.
See also Jacques Le Faure, Vice-visitor 1672-1674, who in 1664 wrote a *Tractatus de Tribunali Mathematico Pekinensi*, in defence of Schall, ARSI, *FG 722, fasc. 23.* The same Le Faure reports in 1688 that when Feliciano Pacheco, vice-provincial, and his consultors were in favor of the Beijing fathers receiving astronomers to help Verbiest, Luis da Gama, the visitor at the time, opposed. He even ordered Verbiest not to accept any such office. Le Faure to Oliva, Guangzhou, October 26, 1688, ARSI, *Jap.-Sin. 162, ff.* 223-225r.

In a letter of 1664, Oliva mentions information sent from China by da Cunha, the visitor, who had died in 1660, and by Schall. The General not only sees no incompatibility in holding the office, but he encourages Schall to prepare a Jesuit to succeed him. He imposes silence on those who assert the contrary opinion and threatens stronger measures, if necessary, for the sake of peace and harmony[6].

2. After Schall's disgrace and death (August 15, 1666), followed by several years of anti-Christian persecution, Verbiest gradually attracted the notice of the young Kangxi sovereign, as a result of a series of brilliant astronomical experiments. He was assisted by his two able collaborators, Buglio and Magalhaes. As a result, the direction of the Tribunal of Mathematics was restored to the European scientists and the emperor appointed Verbiest to the office. His prestige rising daily, Verbiest could foresee that sooner or later the Manchu lord would insist on his being raised to the grade of mandarin, as had been done to Schall in the preceding reign. It was probably at Verbiest's suggestion that Prospero Intorcetta, while in Rome in 1672 as China mission procurator, again brought the mandarinate issue to the Superior General's attention. "To remove every scruple," Oliva again approached the Holy See and petitioned for the faculty that the "Prefect of Mathematics in China ... be enabled to accept the mandarinate associated with the aforesaid Tribunal, with all the jurisdiction attached to that office." This faculty Pope Clement X "generously granted"[7].

3. Verbiest's prevision proved exact. He was created mandarin of the second class in 1674. This was soon followed by two successive promotions that elevated him to the highest official position below that of Imperial Secretary[8]. To an acceptance of these honors, Verbiest's senior assistants, the veterans Buglio and Magalhaes, remained as stubbornly opposed as they had been under Schall. Their opposition, we may benignly judge, was inspired mainly by two motives of no little weight, one practical, the other religious. There was, first of all, the fear that possession of superior rank on the part of a foreigner, in the

---

6 Oliva, Vicar General, to Grueber, Rome, 1664, ARSI, *FG no. 3, int. 5, ff.* 1r-v.

7 Verbiest to Gabiani, Beijing, 1681, "*Responsa ad varia dubia P. Navarrete*", *Correspondance*, pp. 318-319.

8 For a detailed discussion of Verbiest's positions and honors during his entire Beijing career, see the study of Prof. Dr. Willy Vande Walle, "Ferdinand Verbiest and the Chinese Bureaucracy", in the present volume.

treacherous and xenophobic capital of China, would arouse intense jealousy in official circles and eventually bring about Verbiest's downfall, with disastrous consequences for the entire China Church - as had recently happened under Schall. This apprehension was shared by a number of fathers and superiors at Macao, influenced no doubt by those mendicants, Franciscans and Dominicans, who had long strongly attacked - as, at least, a dangerous risk - Jesuit involvement in these functions at court.

As a second motive, the opponents could not sincerely reconcile the assumption of such dignities with the religious conscience of the professed members of the Society, despite oral assurances to the contrary at Rome[9]. Verbiest, however, could ignore this second objection, authorized as he was by papal indult and the approval of superiors in Rome. Moreover, he had the backing of the near unanimity of the China missionaries of his Order, and of many missionaries from other Orders[10]. His influence with the rulers of the empire, he and they reasoned, was the indispensable factor making for the toleration and progress of Christianity in the Middle Kingdom.

Account should be taken of a third probable motive: a spirit of competition, envy and embarrassment on the part of Verbiest's Jesuit colleagues at court - sentiments which can exist alongside very positive qualities in generous missionaries. Verbiest himself suggests this problem when, in a frank letter written in Flemish to his friend Philippe Couplet, he says that while he, Magalhaes, and Buglio work together and send gifts together, the work done by the three of them is most often attributed to him alone. In documents pertaining to public affairs, his name is written first, while the two others are placed as "an appendix." When they send gifts together, the mandarins reply only to him, etc.[11] Tomasz Ignacy Dunin-Szpot goes so far as to use the word *aemulatio* to describe the attitude of some of Verbiest's confreres[12].

---

9 Namely, the *viva voce* authorizations given by Pope Alexander VII and Clement X, as mentioned above.

10 See the eighteen extracts from letters sent to Verbiest by missionaries of other Orders, expressing their appreciation for the protection of their ministry obtained by reason of his service to the emperor. Verbiest to Rome (as promised to the General the preceding January 25), Beijing, April 28, 1684, ARSI, *Jap.-Sin. 145, ff.* 136r-137v; copy authenticated by Verbiest, *ibid., ff.* 71-72v; *Correspondance*, pp. 462-468.

11 Verbiest to Couplet, Beijing, January 23, 1670; *Correspondance*, pp. 161-162.

12 Dunin-Szpot, *Collectanea pro Historia Sinensi*, *Tomus* 2, ARSI, *Jap.-Sin.*

4. On the point of the mandarinate as such, therefore, no impeachment of Verbiest was possible. But at this juncture, there enters a new element, the only one that concerns our initial question of "excommunication." With Rome's blessing, Verbiest was on safe ground in accepting the successive promotions thrust on him by moral pressure from the throne; but the Jesuit vow explicitly forbade, under penalty of grave censure, the "ambition" for and "seeking" of such dignities. Verbiest's Beijing opponents claimed that he had, in fact, done this: namely, that he had procured his promotions through the distribution of *donaria* (bribes), thus clearly manifesting the ambition for secular dignities. Consequently, he was guilty of an overt violation of his vow, incurring, thereby, the censure of a *casus reservatus* (a reserved case) in the Society, a fault which normally could be absolved only by superiors[13].

5. Verbiest's ascent to the honors of being mandarin began during the superiorship of Andrea Lubelli, vice-provincial of China, who resided in Guangzhou. Lubelli was a "Macao" man, that is, a member of the province of Japan, not of the China vice-province; and, like some others of his group, he looked on Verbiest's civil titles with considerable suspicion and misgiving. Apparently informed from Beijing of the doubts entertained by the aged Buglio and Magalhaes, the vice-provincial made a trip to the capital to investigate the matter. When questioned, Verbiest explained the juridical basis that justified his assumption of the mandarinate through its various stages. But on the critical point of ambition, through dealings that suggested bribery (as charged), the vice-provincial gave full credence to the two complainers, whose veracity, he asserts, "it would be wrong to doubt." It does not appear, however, that he took any action at this time[14].

105, I, ff. 10rv. Magalhaes, who was for a time superior of the Beijing residence, was so opposed to Verbiest that the visitor and vice-provincial exempted Verbiest from his authority in what pertained to the dignities of his position. Pereira and Grimaldi were so hostile that, at one point, they suggested a separate residence.

13 See Augustine Oswald, S.J., *Commentarius in Decem Partes Constitutionum Societatis Jesu*, 1895, 2nd edition, printed for private circulation, pp. 340-347, 149.

14 Lubelli, visitor, to the General, Macao, December 15, 1683, ARSI, *Jap.-Sin. 163, ff.* 229-230r. Cited by Bosmans, "Ferdinand Verbiest Directeur", as indicated above in note 2, p. 406, note 3. Rouleau lists another letter of Lubelli dated January 17, 1682, in ARSI, *FG 730*.

6. Within two years after this event, there was a change of positions. Lubelli took over the government of the province of Japan and took up residence in Macao (July 26, 1677). Verbiest succeeded Lubelli as vice-provincial of China (May 19, 1677), remaining, of course, in Beijing. The General dispatched from India a new visitor for the Japan and China areas, Sebastiao de Almeida, who arrived in Macao on July 3, 1677. As an outsider, with no first-hand experience of Chinese conditions, Visitor de Almeida was not the best equipped to weigh and judge the Beijing situation. Probably briefed by Lubelli at Macao on the mandarinate case, he sent a report to the General criticizing Verbiest and apparently making his own the charge of ambition as Lubelli had accepted it during his exploratory visit in Beijing[15].

7. On the basis of Visitor de Almeida's report, received in Rome in 1680, Superior General Oliva judged that Verbiest "was guilty of ambition": he ordered the Beijing scientist "effectively to renounce the other titles conferred on him by the emperor," and to retain only that of "Prefect of Mathematics." He instructed the visitor to absolve the delinquent from the "reserved case" incurred by his infringement of the vow not to aspire to dignities. This decision was sent to the visitor, whose duty it was to execute it[16].

8. Oliva's drastic measure reached Macao in 1682. The visitor was no longer de Almeida, whose information had brought about the action, but the same Lubelli who had originally accepted the criticism at Beijing and who had succeeded de Almeida as visitor (August 9, 1680). Lubelli transmitted the General's orders to Verbiest. In his later report to Rome he stated that, regarding the absolution from the censure of ambition, "although the good Father had sinned in good faith, he most obediently did what he had been ordered"[17].

9. As Lubelli testifies, Verbiest was perfectly submissive to the humiliating sentence rendered against him. At the same time, he felt bound in conscience to warn his immediate China superior, vice-provincial Giovanni Domenico Gabiani (successor to Verbiest in this office, May 1, 1680), that renunciation of the dignities of mandarin, as demanded by Rome, would certainly anger the emperor, and thus result

---

[15] De Almeida, visitor, to the General, sometime after July 3, 1677, as reported by Lubelli, *op. cit.*

[16] Oliva, General, to the visitor, as reported by Lubelli, *ibid.*

[17] Lubelli, *ibid.*

in serious prejudice, not only to the position of the Beijing fathers (by their distinction and services the bulwark of the whole Mission), but to the China Church at large. Always a staunch supporter of Verbiest's scientific activity, Gabiani, acting on Verbiest's suggestion, immediately sounded out the China Jesuits "some in person, some by letter," and asked for their opinion in the matter. With only two or three exceptions, they were emphatic in advising the superior that, in the present precarious circumstances, the implementing of Oliva's order would strike an irreparable blow to the mission. Thus, in agreement with the near totality of the personnel in China, the vice-provincial felt justified in telling Verbiest to hold up the renunciation until the General could be fully informed of the critical situation[18].

10. Eight months before Gabiani's letter to Rome, Verbiest himself wrote to Charles de Noyelle, informing him of what the dignities are and what they mean, and sending him excerpts from letters written by missionaries of other Orders attesting to the importance for the mission of his position at court. But at the same time, he affirms that if, in accepting these dignities for the greater service of God, he had failed in any way against the vow taken by the professed, he should be deprived of active and passive voice both within the Society and outside of it. By accepting this punishment, he could thus more securely in his premature old age prepare himself for death[19].

---

18 Gabiani to the General, Suzhou, May 6, 1683. *Jap.-Sin. 163, ff.* 220r. From Oliva's letter of censure as reported in the two documents cited just above, it seems that the General did not order absolutely the renunciation of dignities on the part of Verbiest, but in a more moderate and conditional form said: "*se cupere ... si fieri possit*" (he desired it ... if it could be done). That is why Gabiani had no scruple, after consulting the missionaries, in telling Verbiest to hang on as before, and he simply states the fact to the General, as though he were within his limits in doing this. He interpreted the General's declaration not as an order, but as a wish. At the same time, Oliva undoubtedly considered that Verbiest had incurred a *casus reservatus* by aspiring to dignities. Hence, there were two distinct parts in Oliva's censure: the declaration of a *casus reservatus* necessitating absolution, for which Lubelli was empowered, and a directive for the future, recommending as prudence would dictate, a renunciation of dignities, "if it could be done."

19 Dunin-Szpot, *op.cit.*, *Jap.-Sin. 105, I*, 11v. The compiler adds: "We do not know what reply Ferdinand received to his letters. It can, however, be affirmed that the judgment made in Rome was in favor of Ferdinand, since afterwards none of those for whom those honorary titles caused a serious problem, dared to make an issue of those honors the emperor is accustomed to grant him. Nor do the records of the following years make any mention of Ferdinand's actions being blamed."

11. As it turned out, the anxious moves made in China were unnecessary. Oliva died on November 26, 1681. De Noyelle, assistant for Germany, assumed the government of the Society as vicar-general until his election to the generalate by the Twelfth General Congregation (July 5, 1682). Almost his first act as vicar-general was a fresh adjudication of Verbiest's case. Two events brought about this action. On December 3, 1681 Pope Innocent XI issued a laudatory apostolic letter to Verbiest, in which he praised the Flemish Jesuit's signal services to the Christian religion in China, and particularly his "use of the profane sciences for the salvation of the Chinese people and for the increase of the Christian faith"[20]. At the same time, Vicar-General de Noyelle received from Verbiest a copy of the Chinese petition which he, Verbiest, had composed, printed and circulated on the occasion of his royal promotion to the second dignity of mandarin. In this petition to the throne, he argued strongly against the honor which was being proposed, and begged to be excused from it. It was clear evidence disproving the charge of ambition which had been reported to Rome, and which had led Oliva to take his disciplinary decision against Verbiest.

On receipt of this evidence, the vicar-general instantly reversed Oliva's decision and declared Verbiest innocent of the charge and of the censure of ambition for which he had been punished. He was to continue in his official dignities as before, and without further molestation[21]. This ruling reached the Visitor, Lubelli, in the summer of 1683, just a year after Oliva's opposite one. Though still disagreeing personally with the favorable interpretation of Verbiest's case, Lubelli loyally fulfilled the orders that completely vindicated the Beijing Jesuit mandarin. The unfortunate misunderstanding was at an end and Lubelli, in informing de Noyelle of the outcome, closed his report with the statement: "Silence is imposed as regards the past." Just before this, he gave high praise to Verbiest's prudence and vigilance in governing the vice-province (1677-1680), mentioning how "he nurtures, protects and defends Christianity," and sincerely thanking him in the name of the vicar-general[22].

Perseverance in the exacting and fatiguing service of the emperor, because of the honors it entailed, caused opposition to and criticism of

---

20 Innocent XI to Verbiest, Rome, December 3, 1681, *Correspondance*, pp. 368-370.

21 De Noyelle, vicar-general, to Lubelli, visitor, as reported by Lubelli, *op. cit.*

22 *Ibid.*

Verbiest, not only from persons in Europe and in China (including, even, a few Jesuit superiors), but even from those who were his closest Jesuit collaborators at court, and with whom he lived in community: Magalhaes, Buglio, Claudio Filippo Grimaldi, and Tomas Pereira. For most of the twenty-eight years that Verbiest lived and worked in the capital, he had to endure, amidst so many other difficulties, this misunderstanding regarding positions and dignities, first in the case of his mentor Schall, whom he defended and supported, and then in regard to himself. Yet, in all of the sources we possess, Verbiest's pen does not produce even one word of bitterness, anger, or recrimination.

It seems reasonable to conclude that the motive which allowed Verbiest to support patiently some painful consequences of his service of the emperor, was the same motive that inspired that service itself. Shortly before his death, Verbiest wrote to the emperor and, in a few words, stated unequivocally the goal he had pursued:

> Sire, I die content, because I have employed almost all the moments of my life in the service of Your Majesty. But I very humbly pray you to remember, after my death, that in all that I have done, I had no other goal in mind than to procure, in the person of the greatest King in the Orient, a protector for the most holy religion of the universe[23].

---

23 *Correspondance*, p. 165.

FERDINAND VERBIEST AND THE CHINESE BUREAUCRACY

WILLY VANDE WALLE

Willy Vande Walle studied Japanese at the Osaka University of Foreign Studies and the history of Buddhism in China and Japan at the National University of Kyoto. In 1978 he became lector of Japanese at the Institute of Modern Languages of the Catholic University of Leuven and was appointed assistant professor, with tenure, at the Department of Oriental Studies of the same university. In 1989 he was appointed professor in charge of the chair of Japanese studies, at the same department. He has published books and articles on the history of Chinese and Japanese Buddhism, the history of the Jesuit mission in China, the history of Japanese art, Japanese classical literature, Japanese linguistics, as well as on the history of Sino-Japanese relations.

The sometimes laudatory descriptions of Ferdinand Verbiest's contributions to the spread of western science in China or enthusiastic analyses of his pivotal role in the history of the old missions in China must not make us lose sight of the fact that in his day-to-day routine he was constantly confronted with the metropolitan bureaucracy. Indeed he himself was - and had to be if he was to have any leverage on the decision-making process and the course of events - a bureaucrat. As a cog in the huge cogwheel of Chinese bureaucracy, he had to conform to the rules and regulations that governed it and master its trappings.

Among the numerous studies that have been devoted to one or other aspect of Verbiest's life and work, few make more than a perfunctory mention of his offices, ranks and promotions. From a western point of view, this may be one of the tedious topics concomitant with the study of Verbiest, but from a contemporary Chinese point of view, it was of overriding importance. Western science was introduced into China within the framework of an institutional environment, where the political implications of science, on the one hand, and political accommodation of scientific activity, on the other, were at issue. The institutional adaptation of science put heavy constraints on the development of a scientific movement in early modern China[1]. Similar constraints were at work in the field of religion and missionary activity. Thus the right understanding of Verbiest's position and of many of his actions in the Directorate of Astronomy (*qintianjian*) and at court is tied in with the global problem of the Qing administration during the initial decades of Manchu rule over China. The Shunzhi (1644-1661) and Kangxi (1662-1722) periods mark the gradual transition from Ming bureaucratic structure and procedure to one which was better adapted to the needs and constraints of Manchu-Chinese dyarchy, on the one hand, and to changing exigencies and priorities in the political and administrative configuration on the other[2]. During the interregnum of the regents (1661-1669) Manchu practices and procedures reasserted themselves temporarily, and the regents' administrative creations provided an important contribution to the format the Kangxi emperor would eventually adopt.

After the death of the Shunzhi emperor, anti-Christian literati, headed by the scholar and would-be astronomer Yang Guangxian[3], accused Johann Adam Schall von Bell, then acting head of the Directorate of Astronomy, of spying and instigation to revolt (KH 3:7:26, September 15, 1664)[4]. As a result, Catholicism was officially denounced, and Schall and his Chinese colleagues sentenced to death. Eventually the lives of Schall and two of his colleagues were spared[5],

---

1 See Jonathan Porter, "Bureaucracy and Science in Early Modern China: The Imperial Astronomical Bureau in the Ch'ing Period", *Journal of Oriental Studies*, 18 (1980), 61-76.

2 See Piero Corradini, "A propos de l'institution du Nei-ko sous la Dynastie des Ts'ing", *T'oung Pao*, 48 (1960), 416-424; and his, "Civil Administration at the Beginning of the Manchu Dynasty", *Oriens Extremus*, 9 (1962), 133-138.

3 Yang Guangxian was no astronomer in the strict sense and certainly could not manipulate astronomical instruments, Porter, *op. cit.*, p. 68. For astronomical calculations he enlisted the help of Wu Mingxuan. Even in astrometeorological forecasts he had no confidence in his own expertise, and was frantically looking for an expert. See *Xichao ding'an*, ARSI, *Jap.-Sin.* 67. Photostatic reproduction in Wu Xiangxiang (ed.), *Tianzhujiao dongchuan wenxian*, 1st series (Taibei, 1965), pp. 47a, 49a-b. We will come back to this point.

4 For the translation of this memorial, see Fu Lo-shu, *A Documentary Chronicle of Sino-Western Relations (1644-1820)* (Tucson, 1966), pp. 35-36. Dates for the Kangxi era are abbreviated as KH.

5 The State Council's sentence denouncing Catholicism and the imperial edict

but Catholicism was proscribed, the churches closed, and the missionaries banished to Canton, except for Schall, Verbiest, Gabriel de Magelhaes and Ludovico Buglio, who were put under house arrest in Beijing[6]. Yang Guangxian was appointed head of the Directorate of Astronomy[7].

While under house arrest, Verbiest devoted his enforced leisure to the study of mathematics and astronomy, until late in 1668 opportunity once again knocked at the door. Yang Guangxian's knowledge of astronomy and calendar computation was minimal. What little he knew, he had learned from the Muslim astronomer Wu Mingxuan, whom he had brought into the Directorate as his assistant. Both men, having misgivings about their own ability, sent up several petitions asking for additional experts. But these, apparently, could not be found. On KH 7:8:30 (October 5, 1668) Wu Mingxuan once again asked the Ministry of Rites to seek expert opinion on the calendars he and other astronomers had computed, because there were discrepancies to be found among them. The Ministry of Rites judged that Wu's calendar was the most accurate and should be issued as the official calendar for the ninth year of the Kangxi reign (1670). (The calendar for the eighth year, which was only a few months away, had already been issued.) By now, the emperor himself had grown suspicious of the so-called expertise of Yang Guangxian and Wu Mingxuan, and decided to carry out an investigation into the matter.

The young emperor went about it in a thorough manner. He issued an order to Yang Guangxian, Hu Zhenyue, Li Guangxian, Wu Mingxuan, Magelhaes, Buglio and Verbiest, that it was time to put the old enmity to rest and, in an impartial manner, to find out the most accurate method of calendar computation (KH 7:11:23, December 26, 1668). At the same time, he ordered them to repair the observatory to conduct an experimental verification of their respective methods.

The following day, the Grand Academician (*da xueshi*), Li Wei, and five Academicians (*xueshi*), together with the Minister of Rites (*libu shangshu*), accompanied Yang, Wu, Verbiest and the others to the observatory, in order to have them predict the length of the sun's

---

sparing Schall are translated in Fu Lo-shu, *op. cit.*, pp. 37-38.

6 See e.g. Jonathan Spence, *To Change China. Western Advisers in China 1620-1960* (Boston and Toronto, 1969), pp. 21-23.

7 The designation "Directorate of Astronomy" and all other Chinese bureaucratic terminology in this essay are based on Charles O. Hucker, *A Dictionary of Official Titles in Imperial China* (Stanford, 1985).

shadow at noon, and thus decide which of both methods - the traditional or the western - was the more accurate. Yang and Wu objected to the method proposed, because they were unable to do the calculation beforehand, but their protest was ignored. Verbiest, on the other hand, put up a gnomon with a length of eight feet, four and nine-tenths inches and calculated the length of the shadow it would cast at noon. To make this verifiable, he made a mark on the horizontal bar of the gnomon to indicate the furthest point of the shadow. His prediction proved to be correct, while his opponents appeared to have made a guess far off the mark. The emperor also ordered Verbiest to predict the length of the shadow a wooden gnomon of two feet and two inches long would cast on the next day. The experiment was carried out on street level in front of the southern gate of the imperial palace, and Verbiest's calculations once again proved to be correct. The following day (KH 7:11:26, December 29, 1668) Verbiest's prediction of the length of the shadow the gnomon atop the observatory would cast at noon was verified and found to be accurate.

On that same day the emperor had a copy of Wu Mingxuan's calendar forwarded to Verbiest for verification. The Jesuit studied it for one month, and found it to be riddled with errors. He submitted his findings some time during the twelfth month of KH 7 (January 1669). On KH 7:12:26 (January 27, 1669), a joint conference of the Princes of the Deliberative Council (*yizheng wang*), Beile, Grand Ministers (*dachen*), the Nine Chief Ministers (*jiuqing*), Supervising Secretaries and Censors (*kedao*), and seal-holding and non seal-holding officials, was called to weigh the charges Verbiest had leveled against Wu's calendar. The conference moved to set up a commission for verification. Kangxi named twenty men, headed by Tuhai, to the commission (KH 7:12:29, January 30, 1669). Assisted by Mahu, Manchu head of the Directorate of Astronomy, the commission checked the accuracy of Verbiest's prediction of two solar periods (*lichun* and *yushui*) and the phases of the moon, Mars and Jupiter. In all instances, Verbiest's predictions matched the actual occurrences, whereas Wu's calculations were all wrong. The commission submitted its report on KH 8:1:24 (February 24, 1669), recommending that the calendar for the next year (1670) be entrusted to Verbiest, and suggesting that the Ministry of Rites file a proposal about the official appointment of Verbiest.

The emperor was not satisfied with the report of the commission. He reprimanded them for their lack of critical judgment. Four years earlier they had put an end to Schall's method as being inaccurate, and now they were declaring it accurate. The commission submitted a new report on KH 8:2:5 (March 6, 1669). This time it spelled out in some technical detail the scientific reasons why Verbiest's and Schall's method was correct. This time the arguments seemed to satisfy the emperor. Thereupon, the Ministry of Rites proposed to cashier Yang Guangxian and turn him over to the Ministry of Justice (*xingbu*) in order that the latter could deliberate on a severe punishment (KH 8:2:5, March 6, 1669). To this, the emperor replied in a decree: "In accordance with the deliberation (i.e., of the Ministry of Rites) Yang Guangxian actually ought to be turned over to the Ministry of Justice and severely punished. But, since his former accusation of Adam Schall was true (*sic*), we order that he be cashiered in accordance with your deliberation. However, for the time being, we want to be lenient and spare him from extradition to the Ministry of Justice" (KH 8:2:7, March 8, 1669)[8].

Subsequently, the Ministry of Rites proposed to the emperor to appoint Verbiest head of the Directorate of Astronomy, since the position had become vacant after Yang Guangxian's dismissal (KH 8:2:10, March 11, 1669). The emperor, however, found this too high an appointment (KH 8:2:12, March 13, 1669). The Ministry of Rites then proposed him for the rank and class (*pinji*) of vice-director, "in charge of the affairs of the *yamen*." He could not be appointed to the position (or office) of vice-director, since there was no vacancy on that level (KH 8:2:16, March 17, 1669). The proposal was endorsed by the emperor (KH 8:2:22, March 23, 1669). As a result, the Ministry of Personnel (*libu*) proposed to appoint him to the "brevet rank"[9] of vice-

8 The preceding account is based on *Xichao ding'an*, ARSI, *Jap.-Sin. 67*, pp. 1-12b, and BNP *2907 I*, pp. 1-12b. The collection of memorials and edicts put together under the title *Xichao ding'an*, in the *Bibliothèque Nationale de Paris*, is listed under three consecutive numbers: BNP *2907*, BNP *2908* and BNP *2909*. However, these numbers do not correspond to any meaningful division in terms of contents or chronology. Within each number one can identify several "bundles" which I have called BNP *2907 I*, etc. A detailed analysis of the internal structure and kinship of the different bundles is to be found in Willy Vande Walle, "Problems in Dating the Writings of Ferdinand Verbiest. The *Astronomia Europaea* and the *Xichao Dingan*", *International Conference in Honor of Ferdinand Verbiest. Commemoration of the 300th Anniversary of his Death (1688-1988)* (Taibei, 1987), pp. 237-252. In the emperor's leniency towards Yang we may possibly detect the lingering influence of the regent Oboi, who was still in power at this moment of time. See Lawrence D. Kessler, *K'ang-hsi and the Consolidation of Ch'ing Rule 1661-1684* (Chicago, 1976), p. 63.

9 The original text (*Xichao ding'an*, BNP *2907 I*, p. 14a) has the term *zhi-*

director in charge of the affairs of the Directorate (KH 8:2:29, March 30, 1669), which was endorsed by the emperor (KH 8:3:1, April 1, 1669)[10]. The Ministry of Personnel's memorial added that Verbiest could be appointed to the formal position of vice-director as soon as there would be a vacancy. Such a vacancy, however, is never mentioned in any of the subsequent memorials. As a result, Verbiest was never even formally appointed to the position of vice-director. After Yang Guangxian had been cashiered, the Ministry of Rites had proposed to fill the vacancy by promoting the then vice-director of the Left Hu Zhenyue as director, and vice-director of the Right Li Guangxian as his deputy (KH 8:3:7, April 17, 1669), presumably to create the vacancy to which Verbiest could be formally appointed. More than a year later (KH 9:6:9, July 25, 1670), the emperor made his decision: he would leave the position of director open[11]. This does not mean there was no director at the Directorate, for there was a Manchu, by the name of Mahu, but the position of the Chinese director was formally left vacant. Verbiest, however, was put in charge of the management of the Directorate, thus continuing a period of European dominance of the Directorate of Astronomy, which had been started by Schall and which, in all, would cover 185 years. Yet, neither Verbiest nor his direct successors formally held the position of director. The first European to do so was in point of fact Ignaz Koegler, who was appointed in 1725[12].

After the imperial endorsement of his appointment, Verbiest lost no time in addressing a personal memorial (*zou*) to Kangxi imploring the emperor to revoke the appointment, while at the same time stressing his

---

xian, which one would be tempted to translate as "real and brevet rank," but in view of the context, it simply has to be translated as "brevet rank." *Xian* is defined as follows: "sometimes used specifically to indicate that an official's rank (on the basis of which he was paid) was not appropriate to and, normally, was lower than the office he actually occupied (*guan*); sometimes used, e.g., when an official had been promoted to a higher post, but his promotion had not yet been confirmed by the appropriate authorities ..." (Hucker, *op. cit.*, pp. 240-241, n. 2493). At any rate, both meanings point in the same direction of a discrepancy between the functional office actually held and the nominal rank. This indicates that, though Verbiest was not appointed director *de iure*, he was considered such *de facto*.

10 See *Xichao ding'an*, ARSI, *Jap.-Sin. 67*, p. 13 and BNP *2907 I*, p. 13. Fu Lo-shu, *op. cit.*, p. 452, n. 64 mistakenly gives KH 8:3:17 (April 17, 1669) as the date of Verbiest's appointment.

11 *Xichao ding'an*, BNP *2908 I*, p. 1.

12 See Porter, *op. cit.*, p. 68; Kessler, *op. cit.*, p. 189 n. 52 says that institutionally, the Jesuits replaced the Chinese director, while there always remained a Manchu director in addition.

willingness to go on fulfilling the tasks of managing the Directorate without the formal rank (KH 8:3:15, April 15, 1669). In this memorial, Verbiest calls himself "in charge of the method of calendar calculation," without the attributive of his newly acquired rank of vice-director of the Directorate of Astronomy. The emperor first denied Verbiest's request (KH 8:3:17, April 7, 1669), but after a second plea (KH 8:5:?, June 1669), he finally granted it (KH 8:6:15, July 12, 1669). At the instruction of the emperor, the Ministry of Rites proposed to pay him the salary of a vice-director (KH 8:6:26, July 23, 1669)[13]. If we take into account that, in 1667, under Yang Guangxian's directorship, the rank of director had been raised from the fourth to the third grade, the rank of vice-director must have been of the fourth grade. This did not last long however, for in 1670, shortly after Verbiest had been put in charge of the Directorate, the ranks of director and of vice-director were reduced two grades[14]. It is not clear whether the upgrading - and subsequent downgrading - had any influence on Verbiest's emolument. At any rate, the emperor granted him a yearly salary of a hundred taels of silver and twenty-five bushels of rice, to be paid from the treasury of the Ministry of Revenue (*hubu*) (KH 8:6:29, July 26, 1669).

In a letter in Dutch addressed to Philippe Couplet, dating from 1670[15], Verbiest relates that it took him three months to convince the emperor to revoke his appointment, which corresponds with the Chinese sources above. The Dutch letter also explains that it took him so long because his enemies were still influential at court, and they would not accede to Verbiest's request of revocation. "They (i.e., his enemies) wanted to insult me with the title of vice-director, although the Tribunal of Courtesy (i.e., the Ministry of Rites) designated me director, i.e. head of the astronomers"[16], he complains to his friend. At the end of the sixth chapter of the *Astronomia Europaea*[17], Verbiest again says that

---

13 *Xichao ding'an*, ARSI, *Jap.-Sin. 67*, p. 15 and BNP *2907 I*, p. 15.

14 Porter, *op. cit.*, pp. 68-69. According to Zhao Erxun *et al.* (eds.), *Qingshi gao* (Beijing, 1977), p. 3323, the director was initially of the fourth grade and the vice-director of the fifth. In 1667 they were raised to third and fourth grade, but in 1670 lowered to fifth and sixth grade.

15 H. Josson, S.J. and L. Willaert, S.J., *Correspondance de Ferdinand Verbiest de la Compagnie de Jésus (1623-1688), directeur de l'Observatoire de Pékin* (Brussels, 1938), pp. 160-163.

16 "majoraal van de schole der sterrekieckers."

17 Ferdinand Verbiest, S.J., *Astronomia Europaea sub Imperatore tartaro-sinico Cam-hy appellato ex umbra in lucem revocata* (Dillingen, 1687), p. 18. Dutch translation in Noël Golvers and Ulrich Libbrecht, *Astronoom van*

he had been entrusted with the (computation) of the calendar of the whole empire and the restoration of astronomy, in his capacity of highest prefect of the Tribunal of Mathematics (i.e., the Directorate of Astronomy). In the beginning of the fifth chapter of the *Astronomia Europaea*[18], he mentions in passing that the Directorate of Astronomy used to be subordinate to the highest Ministry of Rites at the time of the dramatic showdown between Yang Guangxian's clique and himself, but that this is no longer so. This, possibly, implies that it was made independent from the Ministry of Rites when he was put "in charge of the calendar" or "in charge of the affairs of the *yamen*," to use the wording of the memorials quoted above. In the fifteenth year of the Shunzhi era, it had been detached from the Ministry of Rites but reverted to its jurisdiction in the second year of the Kangxi era. Verbiest was holder of the seal (*zhangyin*) of the Directorate, just as Schall had been before him[19]. The term seal-holder normally preceded the title of an official who was the senior among equals in an office, or who could not regularly be the head of the office in question[20]. We will have occasion to point out other analogies between Schall and Verbiest further on.

In the series of memorials concerning the restoration of the meteorological instruments on top of the observatory[21], there are two memorials in which Verbiest is explicitly referred to as vice-director. One is from the head of the Directorate, Mahu (KH 8:5:12, June 10, 1669) and the second one from the Ministry of Rites (KH 8:6:5, July 2, 1669). Both are, therefore, prior to the revocation of his appointment. From KH 8:6:15 (July 12, 1669) on, we find no further reference of vice-director, but consistently the attributive "in charge of the calendar." Thus, it would appear that he was in charge of one of the three sections of the Directorate, namely the Calendar Section (*shixian ke*). This section comprised the Five Offices (*wu guan*), a cluster of five groups of calendrical specialists, each headed by a director (the "Five Regulators")[22], with rank 6b, i.e., equal in rank to the vice-directors.

---

de Keizer. Ferdinand Verbiest en zijn Europese Sterrenkunde (Louvain, 1988), p. 144.

18 Verbiest, *Astronomia Europaea*, p. 13. Translation in Golvers and Libbrecht, *op. cit.*, p. 137.

19 Fu Lo-shu. *op. cit.*, p. 37.

20 Hucker, *op. cit.*, s.v. *zhangyin*, p. 113, n. 217.

21 *Xichao ding'an*, ARSI, *Jap.-Sin. 67*, p. 28a.

22 Porter, *op. cit.*, p. 64.

From the viewpoint of hierarchical precedence, Verbiest stood on a par with the Five Regulators. He presumably started his work in the calendrical section, and gradually expanded his range of activity and control over the other sections. This seems to be confirmed in his description of the three sections in the *Astronomia Europaea*. In the beginning of chapter eight he writes: "After restoring the calendar of that year (i.e., KH 8) (and bringing it into accord) with itself and heaven, I turned my mind to the remaining matters that pertained to the Mathematical Tribunal"[23]. He then proceeds with a description of three tribunals involved in astronomy: the first situated in the eastern part of the city, where the observatory stood; the second located in the western part of the city, adjacent to the Jesuits' residence, where the method of computation and the theory of astronomy were taught; the third in the center of the city, not far from the imperial palace, where the major public affairs related to mathematics were performed. The latter "public affairs" refer, no doubt, to the official presentation of the calendar to the emperor. Next, Verbiest distinguishes three "mathematical classes" which, from the subsequent description, appear to be the three sections of the Directorate of Astronomy. "The first section is the one responsible for the yearly edition of the ephemerides and the solar and the lunar eclipses, as well as all other things related to computation." In chapter ten of the *Astronomia Europaea*, Verbiest alludes to the astro-meteorological reports he has to file eight times a year. Many of these are preserved in the Diyi lishi dang'an guan in Beijing. They are invariably signed by Mahu, director, and Verbiest, "in charge of the calendar." These predictions fall within the province of the second class, or section: the astronomical section (*tianwen ke*), which performed astronomical observations, recorded omens and portents, compiled meteorological records and trained astronomy students in the use of the observational instruments[24], a strange mixture of experimental science and astrological superstition. The third class, corresponding to the water-clock section (*louke ke*), prescribes the right place and time for public works, and the construction of houses and sepulchres. Wherever they go, the officials of this class carry with them a magnetic needle and a clockwork of some sort. They are also in charge of the water-clock in the observatory and of proclaiming the watches in the imperial palace

---

23 Verbiest, *Astronomia Europaea*, p. 21. Golvers and Libbrecht, *op. cit.*, p. 150.

24 Porter, *op. cit.*, p. 63.

and the city[25]. After taking control of the calendar section, Verbiest also brought the astronomical section within the range of western science. He succeeded in convincing the Ministry of Rites and, subsequently, the emperor, that astro-meteorology (*houqi*) had nothing to do with the computation of the calendar, and that this pseudo-science, which had a strong advocate in Yang Guangxian, might as well be abolished[26]. *Houqi*, which can be translated as "watching for the ethers," was a time-honored cosmic magic dating back as far as the Han dynasty. According to the descriptions we find in sources, twelve pitch-pipes made of bamboo, jade, or bronze, and graduated in length according to a mathematical progression, were vertically thrust into the soil, with their open upper ends even with the surface or projecting slightly above it. The pipes had to be arranged in a circle in accordance with the twelve lunar months and their corresponding compass directions. The open ends had then to be filled with fine ashes. The theory has it that, at each month, the *yin* or *yang* ether of that month would rise from under the ground to the surface. As it did, it entered the lower end of the particular pitch-pipe of that month (and of none other), thus blowing the ashes forth from the upper end. According to Derk Bodde, however, ever since the sixth century, hardly anyone had ever tried to prove or disprove the theory through actual experiment. It was simply passed on from generation to generation as part of the Chinese cosmological heritage[27]. If that is indeed the case, one wonders why Yang Guangxian was looking for a specialist "watcher of the ethers," and why Verbiest proposed to abolish the practice of watching for the ethers at the offices of the Shuntian prefecture. Xing Yunlu's assertion, quoted by Bodde from his *Gujin lüli kao*, was perhaps less fanciful than Bodde would lead us to believe. According to the passage quoted by Bodde, officials of the Directorate of Astronomy would go to the offices of Shuntian prefecture (located in the northeast section of Beijing, east of the Drum Tower) and, by using some contraptions, "stage" a rigged watching for the ethers with ashes. Whether this was really the case or not, Verbiest can be linked to the official abolition of "watching for the ethers," if not

---

25 Verbiest, *Astronomia Europaea*, p. 33. Golvers and Libbrecht, *op. cit.*, p. 162.

26 Memorial dated KH 8:6:29, 26 July 1669; imperial endorsement of KH 8:7:1, July 28, 1669: in *Xichao ding'an*, ARSI, *Jap.-Sin. 67*, p. 49b.

27 See Derk Bodde, "The Chinese Cosmic Magic Known as Watching for the Ethers", in Else Glahn (ed.), *Studia Serica Bernhard Karlgren Dedicata* (Copenhagen, 1959), pp. 14-35.

in theory, at least as a practice within the framework of the activities of the Directorate of Astronomy.

By 1670 Verbiest had a firm hold on the two main sections of the Directorate. This is borne out by a decree of KH 9:6:9 (July 25, 1670) to which we already have alluded[28]: "Since the calendar and astronomy are already being managed by Verbiest, there is no need to fill the vacant position of head of the Directorate of Astronomy"[29]. Thus, Verbiest was carrying out a gradual but radical reform of the Directorate - a silent revolution. The political implications were potentially far-reaching. By proving Chinese calculations and predictions to be erroneous, western astronomy challenged the legitimacy of Chinese traditional cosmological models. This put the Chinese in a dilemma: correct prediction and mastery of the natural order, or adherence to traditional cultural assumptions and autonomy[30]. Yang Guangxian had clearly opted for the latter alternative, even at the expense of accuracy. His strong opposition against Schall's methods reflected the reaction of all those who had vested interests in the status quo. Since western science was a challenge to the traditional cultural order, it was also felt to be a threat to the existing social order, all the more so because the Jesuits intended their methods of superior accuracy to be a spearhead for Christianity. This was something that could not be tolerated. Therefore, the impact of western science had to be contained as much as possible. This could be achieved in two ways: first, by treating western science as a mere method, a technique, and secondly, by leaving the bureaucratic structure undisturbed. This may have been the rationale for the emperor's not appointing Verbiest to the position of director. Appointing him would have been too patent a deviation from bureaucratic usage, which could hardly tolerate westerners. Moreover, it is noteworthy that the job of calendar-making seems to have fallen in low esteem in the beginning of the Qing, at least if we are to believe Verbiest: "The calculation of the calendar and the clarification of time constituted the prime office under Yao and Shun, and ever since has always been classified among the pure offices. It was not until the end of the Ming that it was looked down on and its officials held in low esteem"[31]. As already mentioned, it was not until 1725 that a westerner

---

28 Cf. *supra*.

29 *Xichao ding'an*, BNP *2908 I*, p. 1b.

30 Porter, *op. cit.*, p. 70.

31 *Xichao ding'an*, BNP *2907 I*, p. 75a.

was officially installed as director. Presumably by this time, western astronomy had been sufficiently assimilated and its methods routinized to be considered harmless.

In his capacity of *de facto* manager of the Directorate, Verbiest had a considerable staff working with or under him. The calendar section comprised five groups, headed by one Manchu and one Chinese director (or regulator) each, besides about twenty erudites (*boshi*). The astronomical section was staffed principally by eight directors of the Imperial Observatory (*lingtai lang*) and six erudites. Verbiest relates that every day twenty or more "mandarin" observers were sent to the observatory from the major "tribunal" located in the vicinity of the imperial palace[32]. The water-clock section had a staff of four supervisors of the water-clocks and six erudites. Verbiest had numerous disciples. In the already quoted Dutch letter to Couplet, dated January 23, 1670, he says his disciples number one hundred or more[33]. In the *Astronomia Europaea*, he says that at the time he proposed to delete the intercalary month from the calendar of the current year 1669, his disciples numbered 160. Elsewhere in the *Astronomia*[34] he says that their number varies between 160 and 200. In his letter to Couplet he also stresses that his disciples are "esteemed like mandarins." Indeed, they received the same salary as an official of the ninth rank[35] or were at least allowed to wear the cap of an official of that rank. We know that in 1670 the emperor placed students of the banner schools in the calendar section of the Directorate of Astronomy to study alongside ninety-four Chinese students. Six Manchus and four Chinese were picked from each of the eight banner schools and sent to college[36]. This amounts to a total of 174 students, which roughly corresponds with the figure mentioned by Verbiest. These students were lodged in the tribunal that was situated in the vicinity of the imperial palace. On fixed days they came to the tribunal adjacent to the Jesuits' residence to listen to Verbiest's lectures[37]. Among these pupils, more than a hundred are

---

32 Verbiest, *Astronomia Europaea*, p. 49. Golvers and Libbrecht, *op. cit.*, pp. 181-182.

33 Josson and Willaert, *Correspondance*, p. 162.

34 Verbiest, *Astronomia Europaea*. Golvers and Libbrecht, *op. cit.*, pp. 146, 181.

35 Zhao Erxun, *Qingshi gao*, p. 3323.

36 Kessler, *op. cit.*, p. 153.

37 Josson and Willaert, *Correspondance*, p. 224. Golvers and Libbrecht, *op. cit.*, pp. 146, 181.

said to have held an official rank[38]. We find Verbiest particularly caring for his subordinates. In the fall of 1670 (KH 9:7:23, September 6, 1670) he personally presented a long memorial to the emperor, asking for a raise in the rank and the salaries of the Manchu and the Chinese officials in the Directorate, while, for the students of astronomy "who attend lectures according to a rotating roster and take turns doing observations," he asked that they be allowed to wear a rank button (*dingdai*) - to be worn as an insignia of rank atop their headgear - when the Directorate presented the new calendar to the emperor[39]. What he requested amounted to exceptions to the statutory provisions to which, therefore, the Ministries of Rites and of Personnel reacted negatively. The Kangxi emperor, however, granted the request (KH 9:9:30, November 12, 1670).

The work at the Directorate kept growing, and Verbiest had two of his confreres who were well-versed in the calendar, Fathers Christian Herdtrich and Claudio Filippo Grimaldi, summoned to Beijing (KH 11:4:4, April 30, 1672)[40]. A few months later, the former received the imperial permission (secured by Verbiest) to return for a few months to his two former churches in Taiyuanfu and Jiangzhou and get the books on astronomy and calendrical science, as well as the instruments he had left there when he had been exiled to Canton in 1666 (KH 11: intercalary 7:16, September 7, 1672)[41]. According to Jesuit sources, he availed himself of this opportunity to do missionary work in his former mission stations[42].

During the restoration of the old calendar and his tenure as head of the Directorate of Astronomy (1665-1669), Yang Guangxian initiated a purge which made many victims, not in the least among the students who had been trained in the western method of calendar computation. As late as KH 10:6:27 (August 1, 1671), Verbiest pleaded for the rehabilitation of the erudite Bao Yingqi, who, like many others, had been sacked by Yang Guangxian and banished on charges of bribery.

---

38 Arnold H. Rowbotham, *Missionary and Mandarin. The Jesuits at the Court of China* (Berkeley and Los Angeles, 1942), p. 95 (unspecified source but possibly based on the letter to Couplet, *Correspondance*, p. 162).

39 *Xichao ding'an*, BNP *2907 I*, pp. 76-77.

40 *Ibid.*, p. 74a-b.

41 *Ibid.*, p. 101a-b.

42 Louis Pfister, S.J., *Notices biographiques et bibliographiques sur les Jésuites de l'ancienne mission de Chine 1552-1773* (Shanghai, 1932-34), p. 365.

Bao was alleged to have received twenty taels of silver from a paper-maker, who wanted to promote his product for use in the printing of the calendar. An official involved in a bribery of more than ten taels was punished with forty beatings and sent into exile with his family. Verbiest explicitly states that he asked the emperor for pardon "from the viewpoint of the need of manpower in the calendar and astronomical (sections)"[43]. In this case Verbiest made himself the spokesman of the two younger brothers of the hapless Bao Yingqi, who tried to have their brother rehabilitated. He clearly had gained a good measure of confidence, for he argued his case in much detail, thus exceeding the number of characters allowed in a petition for personal matters (*zou*), and presented his petition in person. Verbiest succeeded in having Bao called back from exile, and his punishment commuted to a fine (KH 10:11:23, December 23, 1671). Incidentally, in the memorial settling this matter, the Ministry of Rites mentions no less than thirty persons who had been affected by Yang Guangxian's purges, but most of them had, in the meantime, been rehabilitated[44].

The last pocket of resistance towards Verbiest and the calendrical science he stood for, was wiped out as late as 1672. In quite a vindictive tone Verbiest denounced a certain Yang Zhu'nan, a man living in the area of Wu, south of the Yangzi, for privately issuing a calendar. This calendar, entitled *Zhenli yan* (An Explanation of the True Calendar) was allegedly a compilation of terms culled from books in the vein of Yang Guangxian's. In a vehement tone - which is rarely seen elsewhere - and invoking *raison d'état*, Verbiest requested a severe punishment for the imposter (KH 11:7:22, August 14, 1672). Yang Zhu'nan had had the nerve to request the Office of Transmission (*tongzheng shisi*) to present a memorial on his behalf, challenging Verbiest's calendar. The matter was evidently referred to the Ministry of Rites, which had him condemned to a hundred whippings, forty beatings, three years deportation and had the woodblocks of his book burned[45].

When the construction of the six new astronomical instruments was completed in 1674 Verbiest offered "books and drawings" (i.e., *Yixiang zhi* and *Yixiang tu*)[46] to the emperor (KH 13:1:29, March 6, 1674). Riding the crest of his initial successes, Verbiest had managed to secure

43 *Xichao ding'an*, BNP *2907 I*, p. 81a.

44 *Ibid.*, p. 87 b.

45 *Ibid.*, pp. 91a-100b.

46 *Xichao ding'an*, BNP *2907 II*, p. 46b.

imperial permission for the building of these new instruments - and the platform to build them on - as early as 1669 (KH 8:8:16, September 10, 1669)[47]. The *Xinzhi lingtai yixiang zhi*, including fourteen chapters of explanations and calculations (i.e., *Yixiang zhi*) and two chapters of drawings (i.e., *Yixiang tu*), was compiled by thirty-one persons[48]. It served as the specifications for casting the instruments and constructing their bases. The emperor was very pleased with the result, and ordered that Verbiest be rewarded with a high promotion. However, true to form, the Jesuit was not interested in any rewards for himself, but once again he showed himself a good superior. From a long memorial by one of the Grand Secretaries (*da xueshi*), dated KH 13:3:28 (May 3, 1674), we learn that Verbiest had asked his superior ministry, the Ministry of Rites, to reward all the officials of the Directorate of Astronomy who had collaborated on the construction of the six new astronomical instruments. The Ministry of Rites was at a loss as to what promotion to propose for endorsement to the emperor. Neither in the archives (*dang'an*) nor in the statutes (*huidian*) could they find precedents of promotions proposed as a reward for constructing astronomical instruments. However, in the archives they did find the precedent of Schall. In the third year of the Shunzhi era the Ministry of Rites had proposed to award him the title of vice-minister of the Court of Sacrificial Worship (*taichang si shaoqing*), in addition to his position of acting head of the Directorate of Astronomy, as a reward for his contribution in the establishment of the new calendrical method, *xinfa* (formerly called *xifa*, designation forbidden since the prosecution by the Four Regents during the infancy of the Kangxi emperor). This proposition was, in its turn, based on a precedent of the Ming dynasty. The Ministry of Rites had found an inscription on a funerary stele, mentioning a head of the Directorate of Astronomy who had been awarded the title of vice-minister of the Court of Sacrificial Worship. Referring to this precedent, the Ministry of Personnel had proposed to the Shunzhi emperor to award Schall the same title, since Schall was then acting head of the Directorate of Astronomy. The emperor had endorsed this proposition. Now (KH 13), the Ministry of Rites proposed to repeat the same procedure. However, there were several problems. The premises were different. Verbiest had obviously not asked to be

---

47 *Xichao ding'an*, ARSI, *Jap.-Sin. 67*, p. 34b; also Golvers and Libbrecht, *op. cit.*, p. 177.

48 Yi Shitong and J. Kistemaker, *Chinese Stars and Constellations in 22 Maps* (Amsterdam, s.d.), p. 6.

promoted, but to award his collaborators. Besides, he had requested that his drawings of the six instruments and their supports, i.e., the drawings that make up the work known as the *Xinzhi lingtai yixiang zhi* (sixteen scrolls) be printed under imperial auspices (i.e., to be financed by the Ministry of Rites). The latter request was granted. For the promotion of the collaborators, the matter also had to be referred to the Ministry of Personnel. Collaborators were numerous: two officials from the Directorate who had supervised all personnel involved in the work; two officials from the Ministry of Works, who had supervised the construction of the instruments; seven lower-ranking officials of the Directorate of Astronomy, measuring and checking the instruments under construction with the drawings; fourteen still lower-ranking officials of the Directorate (including students), in charge of placing the constellations (on the celestial globe, presumably), and adjusting the instruments, and making out copies of the drawings. Besides, Verbiest had left many lower-ranking officials of the Ministry of Works unmentioned. He therefore suggested to the Ministry of Rites, that it ask a list of all officials who had been involved. The Ministry of Works is a big ministry, he apologizes, so that he could not list all of them himself. But he added thirteen more Manchu officials of the Directorate who also deserved a reward for their collaboration. He did not specify, however, what their contribution had been.

The whole list was transmitted to the Ministry of Personnel for deliberation. The Ministry of Personnel first addressed the problem of the promotion of Verbiest. It stated that, since Verbiest did not formally hold an office - referring to KH 8:6, when Verbiest's plea to revoke his appointment of vice-director was granted - he should be appointed head of the Directorate and entitled with the brevet rank (*zhixian*) of minister of the Court of Sacrifical Worship. So, the Ministry of Personnel followed the proposal of the Ministry of Rites to refer to the precedent of Schall's promotion. However, the Ministry of Personnel was at a loss as to the request to reward all collaborators. No such precedent was to be found. In the case of Schall, only he had been promoted. So, the Ministry concluded, it proposed to do the same: only promote Verbiest. Two days later the emperor's reply came: Verbiest was to receive the title of minister of the Court of Sacrificial Worship, but no mention is made of the Directorate. He was to remain in charge of the calendar, the decree says, but this was not to constitute an official position. The emperor also ordered the Ministry of Works to deliberate on the promotion of all the other collaborators as well, thereby ignoring precedent. Verbiest declined the appointment, but was forced to accept it. Subsequently, most of the collaborators were promoted one additional degree[49].

An unequivocal proof of Verbiest's self-confidence during the first half of the 1670s, is the fact that he almost consistently used private petitions which he personally and orally presented to the emperor. When, in 1673, he asked the emperor that Giovanni Domenico Gabiani be allowed to transport the body of the deceased Giovanni Francisco de Ferrariis to Xi'an, and stay there to attend to the grave, he was particularly impatient: "Unable to wait until your command reached us, we have already petitioned the Ministry of Rites to submit a memorial to His Majesty on our behalf. But, since (the Ministry) has not yet done so, we present a petition to you in our own person" (KH 12:8:2, September 12, 1673)[50]. Verbiest is seen here taking considerable liberties with administrative procedure. By doing so, he could avoid the editing of his petition by the Grand Secretariat (*neige*) or the Offices of Scrutiny (*liuke*). Editing implied, among other things, that the number of characters was limited - as when using a yellow-label petition - but Verbiest liked to state his case in detail. This gives a personal touch to his petitions, though - as with all Chinese official documents - they still hide more than they reveal. It should be noted that Schall, too, was permitted to hand over his official memorials in person, and was even excused from the customary prostrations and allowed to sit in the imperial presence[51].

In chapter eleven of his *Astronomia Europaea*, Verbiest relates how the emperor had asked him to calculate the tables of eclipses and planets for a time-span of two thousand years, the so-called "Perpetual Calendar of Kangxi." These were extensive computations, and Verbiest divided the work among the "mandarins" of the Directorate. To each table he added a method to extrapolate the computations into the infinite. This is what he calls in Chinese "the table of the year zero" (*lingnian biao*). This required some knowledge of mathematics on top of computational techniques. Such experts were hard to find. Verbiest blamed this dearth to the lack of career prospects for astronomers. Within the Directorate of Astronomy, opportunities for promotion were very scarce because Manchu staff were regularly transferred from other ministries or

---

49 *Xichao ding'an*, BNP *2907 II*, p. 46a.

50 *Ibid.*, p. 62a-b; also Pfister, *op. cit.*, p. 317.

51 Spence, *op. cit.*, p. 19.

agencies to fill the better positions within the Directorate, such as the position of director of the Five Offices, or higher, although these Manchu officials did not have the least astronomical background. Verbiest asked to discontinue this practice so as to open the path of promotion for the astronomy students. For those with the rank of *lingtai lang* or lower, he asks that they receive additional grades on an annuity basis. He further asks that the officials and students of the Directorate be eligible for detachment on commission to other offices, which was a regular practice in other ministries and agencies. Detachment on commission could constitute a kind of promotion. During the early Kangxi reign, student (*shengyuan*) quotas were drastically reduced. Under normal conditions, this would only have increased the chances of appointment or promotion in the bureaucracy for the students, but in view of the alien nature of the dynasty, Manchu and Chinese bannermen had a considerable advantage over Han Chinese when seeking official appointment. Moreover, to offset military expenditures that ran high during the 1670s - when campaigns to put down the rebellion of the Three Feudatories had to be financed - the government adopted a policy of selling titles and offices. Notably on the lower echelons, holders of purchased degrees were crowding out graduates of prefectural, provincial and metropolitan examinations[52]. As a rule, officials of the Directorate could not be promoted outside their own office. This greatly reduced opportunities for promotion. Services were only rewarded by raising one's brevet rank[53]. Verbiest also regretted that the printing of the calendar in the provinces was in the hands of the registry (*jingli*) of the provincial administration commission (*buzheng si*), who published it without verification by the Directorate of Astronomy, which often resulted in faked calendars. He asked that when vacancies were reported in the provincial administration commissions, officials of the Directorate would be appointed, for this would serve a double purpose: better control of the calendar and career opportunities for officials of the Directorate (KH 14:12:9, March 1, 1676)[54]. We do not know whether his proposal received imperial endorsement.

On August 27, 1678 (KH 17:7:11), Verbiest presented the "Perpetual Calendar of Kangxi" to the throne[55]. He had requested the paper on

[52] Kessler, *op. cit.*, pp. 156-157.

[53] Zhao Erxun, *Qingshi gao*, p. 3325.

[54] *Xichao ding'an*, BNP *2907 IIb*, p. 5-6.

[55] *Xichao ding'an*, BNP *2907 IIe*.

which to print the voluminous work, on KH 15:1:17 (January 23, 1676)[56]. It was a book in thirty-two volumes, all bound in silk. The emperor was naturally delighted, and wished to reward Verbiest with a promotion and, equally naturally, the Ministry of Rites, which had to make a proposal, referred to the precedent of Schall. The brevet rank finally proposed by the Ministry of Personnel and endorsed by the emperor was that of commissioner of the office of transmission (*tongzheng shisi tongzheng shi*, rank 3a) with one additional degree[57]. As could be expected, Verbiest declined the promotion, but Kangxi ignored the Jesuit's protests. A copy of the calendar was sent and stored in the Hanlin Academy. The emperor also conferred by imperial patent upon Verbiest the honorary title of *tongfeng dafu*, a civilian title of the rank 2b[58]. Since, in the Chinese scheme of things, the recipient of this kind of honor could not properly perform the rites of ancestor worship unless the parents were ennobled in a similar way, the grandparents and parents of Verbiest received similar titles[59]. There is some uncertainty about the date of these diplomas. Since they are not mentioned in the petitions we have to rely on the *Astronomia Europaea*. There they are mentioned in the context of the year 1678, but the diploma given to Verbiest expressly mentions the installation of the crown prince, which took place in 1676 (24 February, 1676)[60]. It will be remembered that Schall received a similar title, that of *tongyi dafu*, on the occasion of the Shunzhi emperor's personal assumption of power (February 1, 1651)[61].

The crowning achievement in the career of Verbiest as a Chinese official was his promotion to the brevet rank of vice-minister of the right in the Ministry of Works (*gongbu youshilang*), which was a 3a rank in the early Qing (KH 21:02:10, March 18, 1682). He received this rank as a reward for casting 130 guns and presenting the *Shenwei tushuo*, his treatise on the casting of guns, to the emperor[62].

The ranks Verbiest held were considerably high within the echelons

---

56 *Xichao ding'an*, BNP *2907 IIb*, pp. 5-6.

57 *Xichao ding'an*, BNP *2907 IIe*.

58 See Pierre Hoang, *Mélanges sur l'administration* (Shanghai, 1902), p. 94.

59 Verbiest, *Astronomia Europaea*, p. 35. Golvers and Libbrecht, *op. cit.*, p. 164.

60 Zhongguo diyi lishi dang'an guan (ed.), *Kangxi qiju zhu*, 3 vols. (Beijing, 1981), 1:244.

61 Hoang, *op. cit.*, p. 98.

62 *Xichao ding'an*, BNP *2908 II*, pp. 183b-184.

of the imperial bureaucracy. Yet we must not forget that they were brevet ranks, and did not constitute substantive offices. Arguably, the Jesuit father drew as much power and influence from his routine access to the inner court and his personal contacts with the emperor, as from his position as an official. Be that as it may, no Jesuit father before or after him climbed the "ladder of success" as high as he did, although we must recognize that Schall's precedent did help him considerably. In a state where precedent was of such importance, Schall leveled the path for his success. Schall could rise as high as he did while his activities coincided with a period of ferment and change, when the Qing dynasty was still groping for an appropriate formulation of bureaucratic structure.

Table 4

Organizational Diagram of the Board of Astronomy in Verbiest's Time

Director (*jianzheng*) 1 M (1 C)

Senior vice-director (left) [*zuo jianfu*] 1 M, 1 C

Junior vice-director (right) [*zhubu ting*] 1 M, 1 C

Recorder's Office
(*zhubu ting*)
1 M, 1 C

Calendar Section (*shixian ke*)	**Astronomical Section** (*tianwen ke*)	**Water-clock Section** (*louke ke*)
Directors of the 5 offices (*wuguan zheng*) 2 M, 2 Mg, 1 Cm	Keepers of the Imperial Observatory (*wuguan lingtai lang*) 3 M, 1 Cm, 4 C	Supervisors of Water-clocks (*qiehu zheng*) 2 M, 2 C
Director of Spring (*chuguan zheng*)1 C Director of Summer (*xiaguan zheng*)1 C Director of Autumn (*qiuguan zheng*)1 C Director of Winter (*dongguan zheng*)1 C	Directors of Calendrical Calculation (*baozhang zheng*) 2 C (till 1675)	Timekeepers (*sichen*) 1 Cm,7 C (from 1675)
Compiler (*sishu*)1 C	Astronomical Observer (*wuguan jianhou*) 1 C	Clerks (*bitieshi*) 11 M, 4 Mg, 2 Cm
Erudites (*boshi*) 6 M, 2 Cm, 22 C	Erudites (*boshi*) 4 M, 2 C	

Office of Instruction
Preceptor (*zhujiao*) 1
Instructors (*jiaoxi*) 2

Students
Astronomy students (*tianwen sheng*) 16 M, 16 Mg, 8 Cm, 24 C
Stipendiary Astronomy students (*shiliang tianwen sheng*) 56 C
Stipendiary *yinyang* Students (*shiliang yinyang sheng*) 10 C

Abbreviations: C = Chinese; Cm = Chinese military; M = Manchu; Mg = Mongol

Verbiest

Impact in China and in Europe

A GLIMPSE OF THE POST-VERBIEST PERIOD: JEAN-FRANÇOIS FOUCQUET'S *LIFA WENDA* (DIALOGUES ON CALENDRICAL TECHNIQUES) AND THE MODERNIZATION OF CHINESE ASTRONOMY OR URANIA'S FEET UNBOUND

JEAN-CLAUDE MARTZLOFF

Jean-Claude Martzloff, a sinologist, is a researcher of the "Centre National de la Recherche Scientifique" (C.N.R.S.). He is the author of a history of Chinese mathematics (Histoire des mathématiques chinoises, Paris, 1987).

Le troisième Prince[1] *proposa à Sa Majesté*[2] *de faire dans l'Astronomie des PP. Adam (Schall) et Verbiest un changement et de corriger une erreur qu'il se vantoit d'avoir découvert par ses observations quoiqu'il nous conste à tous qu'il n'a pu scavoir cela que par l'ouvrage du P. Foucquet.*[3]

(Archivum Romanum Societatis Iesu, *Jap.-Sin.* 182, f. 398r.)

I.
Introductory Remarks

The history of seventeenth- and eighteenth-century Chinese astronomy is often described as the result of the intrusion of obscurantist religious forces into the affairs of an open-minded and receptive China. Chinese astronomers were ready to accept modern astronomy, but the missionaries purposely concealed the truth and taught them an antiquated pre-Copernican astronomy because it fitted better with the dogmas of their religion. Once taught, the revealed truths of any kind were unchangeable[4]:

> Down to the very end of the mission, the Jesuits were prisoners of their limited motive ... Any acceptance of Copernicanism would have

1 I.e. the third son of the Kangxi emperor, Yinzhi.

2 I.e. the Kangxi emperor.

3 This refers to the *Lifa wenda*, ARSI, *Jap.-Sin.* 142, f. 189r.

4 J. Needham, *Science and Civilisation in China* (Cambridge, 1959), 3:450.

> raised doubts about all of Ricci's teachings. In fact, the penalty of enlisting *live science*[5] in the service of fixed doctrine was to inhibit its development - Urania's feet were bound.

Such a judgment reveals much about Joseph Needham and his disciples' conception of history, and seems to be the result of a reaction to "Jesuit historiography" (as exemplified in Henri Bernard-Maitre and Pasquale M. d'Elia's works[6]) rather than the result of historical investigation. In fact, as John Witek has noted in his biography of Jean-François Foucquet, "(The) story (is) more complex"[7] and the reasons for such a situation are numerous.

During the seventeenth and eighteenth centuries, the fate of Chinese astronomy was neither exclusively linked to, nor limited to, the Manichaean problem of the introduction, or non-introduction, of heliocentrism. Even a casual glance at the *Xinfa suanshu* (Johann Adam Schall von Bell's astronomical encyclopedia) reveals that the mid-seventeenth-century Chinese astronomy was at the same time Copernican, Galilean and Ptolemaic, Keplerian and Tychonic. Moreover, at the end of the seventeenth century, even in Europe, the best astronomers - Philippe de la Hire, Jean Picard, Giovanni Domenico Cassini - were rather not inclined to accept uncritically the revolutionary new theory. They often spoke of the "Copernican hypothesis" on a level of equality with the "Tychonian hypothesis" and, more often than not, were disposed to attach importance to experimental evidence, not to theory[8]. A priori, they rejected nothing. The process of transition between medieval and modern science took place in the Braudelian *longue durée* and Needham's implicitly alleged Bachelardian *rupture épistémologique* simply did not happen.

Among the Jesuits themselves no consensus really existed. Some of

---

5 Italics added.

6 See Henri Bernard, "L'encyclopédie astronomique du P. Schall", *Monumenta Serica*, 3 (1938), 35-77, 441-527; Pasquale M. d'Elia, *Galileo in Cina* (Rome, 1947).

7 John W. Witek, S.J., *Controversial Ideas in China and in Europe. A Biography of Jean-François Foucquet, S.J. (1665-1741)* (Rome, 1982), p. 181.

8 See René Taton, "La Hire, Philippe de," *Dictionary of Scientific Biography*, 14 vols. (New York, 1970-1976), 7:576-579 and Pierre Costabel, "La réception de la cosmologie nouvelle à la fin du XVIIe siècle", *Avant, avec, après Copernic. La représentation de l'univers et ses conséquences épistémologiques* (Paris, 1975), p. 261-266.

them, such as Kilian Stumpf[9], the president of the Board of Astronomy, wanted to maintain the *status quo*. According to them, not an iota was to be added to the works of Schall and Ferdinand Verbiest. But others, such as Foucquet, sharply criticized the views of their predecessors and, in the light of the new science, convincingly pointed out what was wrong with the theories advocated by the honorable fathers. Moreover, they were a priori taken seriously by the Chinese because, in their eyes, the source of authority was not limited to the sole president of the Board of Astronomy.

Besides maintaining themselves in China by means of their astronomical *savoir-faire*, the French Jesuits also had in mind another scientific purpose: the cartography of China. And, for obvious reasons, the completion of this geographical program was to be realized according to the highest scientific standards then available. These standards were of course not independent of the development of astronomy. In this respect, the situation was somewhat different from what it had been a century before. For some scientifically-trained scholars, it was merely intolerable to keep unaltered the old astronomical methods, so obviously outdated by the development of instrumentation, physical theory and experimentation.

From a Chinese point of view on the whole, what was at issue was not a theoretical and cosmological problem but, rather, practical needs: to predict eclipses successfully, to fix beforehand the length of the days of the year, eclipses and equinoxes, latitudes and longitudes, and other such goals.

Last but not least, it seems difficult not to believe that the key to our problem does not lie on the Chinese side. Had the missionaries or other Europeans overtly introduced heliocentrism in China - and in fact, Foucquet had, even if it was a fictive form of heliocentrism, as I will show later in this essay - would they have accepted it unreluctantly and accomplished their own scientific revolution? This is dubious. After all, even in Beijing, Chinese astronomers had almost everything at hand to understand what was going on in Europe. But were they willing to study European languages and science in their original garments? Would the Chinese version of *Rangaku* have appealed to them? Why not ? - or, on the contrary, were their goals of a different nature: political,

---

9 The German monograph by Sebald Reil, *Kilian Stumpf, 1655-1720* (Münster, 1978), does not emphasize the scientific aspects of his life. See Joseph Dehergne, S.J., *Répertoire des Jésuites de Chine de 1552 à 1800* (Rome and Paris, 1973), p. 261.

pragmatical, technical? With the writings of the Kangxi emperor[10] and other Chinese authorities, one is inclined to think that the last supposition is more correct than incorrect.

II.
Foucquet's *Lifa wenda*

1. Foucquet's Education and Intellectual Background

Foucquet studied at the Collège Louis-le-Grand in Paris[11], one of the leading French government-supported institutions which were responsible for the scientific successes registered in Old Regime France under the reign of the Sun King[12]. In September 1681, Foucquet became a novice in the Jesuit order and later began three years of philosophical studies - the third year of which was devoted to mathematics (i.e., arithmetic, geometry, music and astronomy)[13], under the guidance of Jean de Fontaney[14], a now almost-forgotten personage who was, nevertheless, considered an important scientist both during his lifetime and in the following century[15]. Later on, before leaving for China,

---

10 In China, the position of the emperor was not exactly comparable to the position of a European monarch. The Kangxi emperor, for example, was "the great educator," a learned man in all domains of knowledge, and believed to be able, in such matters, to show the right path to his subjects - even in a domain such as astronomy.

11 Witek, *op. cit.*, p. 77.

12 The other main institutions were the Academy of Sciences, the Paris Observatory, the School of Military Engineering at Mezières and the Jardin du Roi. See R. Hahn, *The Anatomy of a Scientific Institution. The Paris Academy of Sciences 1666-1803* (Berkeley, 1971) and also Gustave Dupont-Ferrier, *La vie quotidienne d'un Collège parisien pendant plus de trois cent cinquante ans - du Collège de Clermont au Lycée Louis-le-Grand*, 3 vol. (Paris, 1921). In contrast to what happened in other European countries, the university in seventeenth-century France entered a period of decay and played no prominent role in the development of the sciences. See Stephen d'Irsay, *Histoire des universités françaises et étrangères du XVIe siècle à nos jours*, 2 vols. (Paris, 1935), 2:61.

13 C. de la Rochemonteix, *Le Collège de la Flèche*, 4 vols. (Paris, 1889), 1:36.

14 On Fontaney see Henri Bernard, "Le voyage du Père de Fontaney au Siam et à la Chine 1685-1687 d'après des lettres inédites", *Bulletin de l'Université de l'Aurore*, 3rd ser., 3 (1942), 227-280.

15 See Lagrange's eulogy of Fontaney quoted in Dupont-Ferrier, *op. cit.*, 1:185, n. 18.

Foucquet taught at La Flèche, another illustrious Jesuit college. The astronomy which was taught there was[16] "un composé plus ou moins heureux des systèmes d'Aristote, de Ptolémée et de Tycho Brahe. L'hypothèse de Copernic qui fait du soleil le centre des mouvements célestes (y était) regardée comme fort ingénieuse, mais pas conforme à la vérité."

In 1711, in China, the failure of a solstice-prediction caused the emperor to order both his Chinese and European astronomers to undertake new scientific investigations[17]. Foucquet decided to place his knowledge at the disposal of the Kangxi emperor. Although he was not a professional astronomer, he worked very hard in order to become instrumental in the handling of the astronomical live science of his time. His numerous French and Latin manuscripts still preserved in the Biblioteca Apostolica Vaticana show beyond any doubt that he studied and understood well not only antiquated Ptolemaic science, but also the most recent findings of the late seventeenth century published in the French annual *Comptes-rendus de l'Académie des Sciences*, de la Hire's tables (published in 1703), Cassini's reports and Picard's notes, among other things. He then gradually persuaded himself of the ineluctability *à terme* of a reformation of seventeenth-century Chinese astronomy, and he explained in Chinese what kind of changes were needed and why they were needed. His labors resulted in the production of a two volume potential delayed-action bomb: the *Lifa wenda*. A bomb, because it was almost a contestation *sans appel* of previous Jesuit science. Potential and delayed because, for complex reasons, it did not explode as could have been expected. But, eventually, in 1716, the Chinese were confronted with two distinct and contradictory systems of European astronomy!

2. The *Lifa wenda*

Two manuscript copies of the *juan* 1 and 2 of the *Lifa wenda* are preserved in the Biblioteca Apostolica Vaticana, but other copies of it may still be extant in Beijing (presumably in the Imperial Palace, Gugong) or elsewhere. Foucquet wrote them in the form of a dialogue because, he explained, according to him, such a form was more adapted to Chinese taste and culture than arid and unlively theoretical con-

---

16 Rochemonteix, *op. cit.*, 4:39.

17 Witek, *op. cit.*, p. 182.

siderations[18]. He put in comparison, qualitatively and quantitatively, four astronomical systems: the Ptolemaic system; the Tychonic system; the Keplerian system (*cf. infra*, note 18); the new system (*xinfa*)[19].

But his main line of attack was directed against Tycho Brahe[20], the very heart of Schall and Verbiest's astronomy (even if they also took heterogeneous elements from Galileo Galilei and Johann Kepler)[21].

In light of the latest European discoveries, Foucquet methodically expounded what was wrong with Brahe: his inaccurate determinations of time (he then quoted Christiaan Huygens's newly invented pendulum[22]); his false conception of the refraction wrongly considered to be negligible beyond 45° (he showed - not quite correctly according to

---

18 See J. F. Foucquet, "Relation de ce qui s'est passé à Pékin par raport (= rapport) à l'astronomie européane (= européenne) depuis le mois de juin 1711 jusqu'au commencement de novembre 1716", Archivum Romanum Societatis Iesu, *Jap.-Sin. II, 154, ff.* 12 where he states: "il commença cet ouvrage formé de dialogues sur l'idée de l'Epitome (a minor work by Kepler) ou abrégé de l'astronomie Copernicienne composé par Kepler, idée qu'il avait proposé à l'empereur et que l'empereur avait approuvé. Le P. Foucquet l'avait préféré à d'autres manières parce qu'il savoit que quelques livres de Tchin uin cong (I have not been able to identify this proper name) philosophe célèbre ... sont fort estimés des Chinois et parce qu'il espérait dans les demandes et les réponses du dialogue pouvoir exposer plus facilement les difficultés et les subtilités de l'astronomie."

19 I.e. the "system" of the French Académie des Sciences.

20 Foucquet, Bibliotheca Apostolica Vaticana, *Borgia lat. 566, f.* 177v : "Autrefois dans l'astronomie on ne suivait que Tycho. Toute l'Europe le regardoit comme le maistre, mais on s'aperçut avec le temps que ses regles ne s'accordoient pas parfaitement avec le ciel, car dans l'astronomie il y a particulièrement cinq points dont les conséquences sont grandes. Le premier est l'obliquité de l'écliptique; le deuxième, les réfractions; le troisième, les diamètres des luminaires; le quatrième, leurs parallaxes; le cinquième enfin, leur éloignement de la terre. Il n'y a pas aujourd'hui un astronome habile qui ne schache que par rapport à ces cinq points la doctrine de Tycho n'est pas tout à fait exacte." See also N. Sivin, "Copernicus in China", *Studia Copernicana*, 6 (1973), 63-122.

21 See works on the subject by Henri Bernard and Pasquale M. d'Elia, cited in note 6 above. See also Yabuuchi Kiyoshi, *Chûgoku no tenmon rekihô* (Chinese Calendrical Astronomy) (Tokyo, 1969); N. Sivin, "Wang Hsi-shan," *Dictionary of Scientific Biography*, 14:159-168; Hashimoto Keizo, "Galileo's Invention and Kepler's Optical Astronomy at the Compilation of the Ch'ung-Chen Li-shu Astronomical Encyclopedia: 1629-1635", *Kansai daigaku Shakaigakubu kiyô*, 18:1 (November, 1986), 255-281, and his *Hsü Kuang-ch'i and Astronomical Reform. The Process of the Chinese Acceptance of Western Astronomy, 1629-1635* (Osaka, 1988).

22 *Zhuizibiao*.

modern standards - that the refraction was independent of the season, of day and night)[23]; and his false determinations of the parallaxes of the sun, of the moon and of the planets. Taken together, these inaccuracies resulted in certain determinations that were too rough: that of the obliquity of the ecliptic, that of the apparent diameters and distances of the sun and the moon, and that of the latitudes and the longitudes, affecting eclipse-prediction, as well as the correctness of geographical maps and the like.

As a consequence, he proposed to replace the old tables by new ones. His choice fell on La Hire's tables[24]. Moreover, he devised two eclipse-prediction methods, the first one based on computations, the second, on an instrument[25].

Moreover, incredible as it may seem, Foucquet was not reluctant to introduce heliocentrism, even though he pretended to believe that it was nothing more than a false hypothesis[26]:

> 天文家有二說。一則以地不動。日與月旋轉繞地而行。此固一定之理也。一則設日不動。月與地旋繞日而行。此乃新天文家所設想者。非真有此理也。二說雖皆可推求交食之始中終。并各容像。因新學喜由日不動之一說。爲其奇特。亦爲便于推未。今亦從之。
>
> The astronomers (in Europe) have two (astronomical) patterns (*li*). According to the first pattern, the earth is immobile, while the sun and the moon both revolve around the earth. This pattern is really certain. According to the other pattern, the sun is immobile, while the earth and the moon revolve around the sun: this is an hypothesis[27] (*shexiang*) devised by the new astronomers, one which cannot be considered as a principle really existing[28]. Nevertheless, both patterns can be used in order to predict eclipses ... But the new pattern with its fixed sun is marvelous and is particularly well-adapted to

---

[23] But he did not mention the law of sines. In fact, he maintained the mathematical apparatus to a minimum.

[24] La Hire, *Tabulae Astronomicae*, Paris, 1702.

[25] La Hire's instrument.

[26] *Lifa wenda*, *juan* 2, *f.* 39v-40r.

[27] Hypothesis: this is the very term still used by Jacques Cassini (1667-1756) to qualify the system of Copernicus as late as 1740. See his *Éléments d'astronomie* (Paris, Imprimerie Royale, 1740), pp. 9-10.

[28] Presumably Foucquet was cautious because he feared attacks coming from Stumpf's quarters.

> these predictions (*ri bu dong zhi yi shuo. wei qi qite yi. wei bian yu tuiqiu*). Accordingly, I shall follow it here (*jin yi cong zhi*).

Foucquet went even further and suggested the existence of Keplerian ellipses[29]. But he did not pursue the question, not so much for theological reasons, but rather because he was hampered by the mathematical side of the question. Given the Chinese mathematical background of the time, the amount of mathematics to be translated into Chinese would have been truly formidable!

3. The Influence of Foucquet's Works

Foucquet's "revolutionary" theories met with active opposition on the Jesuit side, and the fact that he was a fervent upholder of Figurism played against him: his *Lifa wenda* was never printed and was therefore doomed to oblivion in Europe. Nevertheless, as Pierre Jartoux tells us, Foucquet's works were conscientiously studied by the Kangxi emperor's third son, Yinzhi, and his team of astronomers[30]:

> (Les manuscrits de Foucquet) sont aujourd'hui entre les mains du troisième Prince qui pour se passer des Jésuites fait calculer actuellement par des Chinois toutes nos tables du Ciel, des sinus, des logarithmes, etc. (Foucquet) a donné aux Chinois malgré nous toutes les nouvelles tables de la Hire réduites à ce méridien aux époques chinoises.

Consequently, we are bound to conclude that the reasons for the non-acceptance - or rather, of the Chinese indifference to heliocentrism - are not only, and not even essentially, to be looked for in Jesuit attitudes toward the new theory, but rather in the Chinese social and intellectual background itself.

---

29 He called them *danxingxian* (lit.: egg-shaped lines); cf. *Lifa wenda*, *juan* 1, *f*. 68.

30 ARSI, *Jap.-Sin.*, 182, *f*. 398.

APPENDIX I

Lifa wenda[31]

Preface

(ff. 1) In calendrical astronomy nothing is more important than lunar and solar eclipses ... In the past, methods of eclipse (prediction) have already been devised. Special books are devoted to the subject[32]. Why should we repeat them? ... Since Tuolemi (Ptolemy) (he lived at the end of the Han dynasty), generations of famous scholars appeared in succession. A hundred years before, appeared Digu (Tycho Brahe), whose knowledge was more extended than anybody else before him ... But instruments devised to measure time with a great degree of precision were still needed. Brahe could not obtain such instruments. ... Recently, new instruments from far away countries have been brought to China[33] ... Let us discuss them in what follows ...

***juan* 1**

Concerned with principles related to eclipses. Explaining the size of the sun and the moon. The precise method of eclipse-computation will be dealt with in *juan* 2.

Contents

1. During the past sixty years new methods superseding the old ones have appeared.
2. Detailed discussion of the incompleteness of the techniques of eclipse prediction according to the school of Brahe.
3. Why the Tychonic determination of the distances between the two luminaries and the center of the earth is inaccurate.
4. Why the Tychonic determinations of the apparent diameters of the two luminaries are inaccurate.
5. Detailed discussion of refraction[34] and of the parallaxes which are considerably important with respect to calendrical astronomy.

---

[31] Bibliotheca Apostolica Vaticana, *Borg. cin.* 319 (3) entitles in Italian, "Metodo di calcolare l'Eclipse".

[32] Allusion to *Chongzhen lishu: Jiaoshi lizhi*, *juan* 64-70; *Jiaoshi biao*, *juan*, 72-80).

[33] Allusion to the instruments offered to the Kangxi emperor by Louis XIV.

[34] *Mengqi cha*. In the present context, *mengqi* refers to the terrestial atmosphere as opposed to the ether, *qingqi*.

6. Discussion of the true principle of refraction.
7. The lunar parallax.
8. Evaluation of the distance between the moon and the center of the earth.
9. Determination of the true value of the apparent diameter of the moon.
10. The true parallax of the sun.
11. Determination of the distance between the sun and the center of the earth.
12. Determination of the true apparent diameter of the sun.

juan **2**

1. In the case of an eclipse, what are the characteristics of the shadows of the moon and of the earth? a: according to the ancient theories; b: according to the new theories.
2. Determining whether an eclipse of the sun or of the moon will happen or not. Two methods: a: using definite numbers; b: using an instrument.
3. What principles must one know in order to predict eclipses?
4. Determination of an eclipse of the moon according to the new science (*xin xue*).
5. Explaining in detail the principle of an earth-eclipse rather than of a sun-eclipse.
6. The same according to the new science: a: using calculations; b: using graphical methods.
7. Explanation of the observation of eclipses according to the new science.
8. Using eclipses in order to correct geographical maps: a: using lunar eclipses; b: using solar eclipses.

Appendix: New tables devised by Layi'er (La Hire) in order to predict eclipses.

APPENDIX II

Proper Names

I have been able to identify (for the first time for some of them, to the best of my knowledge) the following authorities repeatedly quoted by Foucquet[35]:

Afuxi	Alphonse (Borelli ?)
Asu	(Adrien) Auzout[36]
Bijia'er	(Jean) Picard[37]
Ding Shi *	(Christopher) Clavius
Gaxini *	(Giovanni Domenico) Cassini
Gebaini *	(Nicolaus) Copernicus
Gudi	(Paul) Guldin
Hale	(Edmund) Halley[38]
Jiasengdi	(Pierre) Gassendi[39]
Kebai'er *	(Johannes) Kepler
Layi'er *	(Philippe de) la Hire[40]
Lishi'er	(Jean) Richer[41]
Lizhuoli	(Giambattista) Riccioli[42]
Luomo'er	(Ole) Römer[43]
Mouduo	(Gabriel) Mouton (?)
Shalexi	(Charles François Millet de) Chales[44]
Xurini	(Christiaan) Huygens

---

35 (?) indicates that the identification remains problematic. The (*) shows proper names identified in the past.

36 1622-1691; see *Dictionary of Scientific Biography*, 1:341-342. His career centered on astronomical instruments.

37 1620-1682; see *Dictionary of Scientific Biography*, 10:595-597. He perfected the movable micrometer and had access to Tycho Brahe's registers of observation.

38 Observations of southern hemisphere stars.

39 See "Gassendi, Pierre, 1592-1655", *Dictionary of Scientific Biography*, 5:284-290.

40 Philippe de La Hire (1640-1718), the author of the *Tabulae Astronomicae*, 1st ed. (Paris, 1702). See *Dictionary of Scientific Biography*, 7:576-579.

41 He is the French astronomer who went to Cayenne in 1672-1673 and participated in the earth measurement program. See *Dictionary of Scientific Biography*, 11:423-425 and John W. Olmsted "The Scientific Expedition of Jean Richer to Cayenne (1672-1673)", *Isis*, 34 (1942-1943) 117-128.

42 "Riccioli, Giambattista", *Dictionary of Scientific Biography*, 11:411-412.

43 *Ibid.*

44 A professor of mathematics in the famous Parisian Collège de Clermont.

THE FRENCH MISSION AND FERDINAND VERBIEST'S SCIENTIFIC LEGACY

CATHERINE JAMI

Catherine Jami studied at the École Normale Supérieure, Paris. At present, she is working at the Centre National de la Recherche Scientifique (CNRS), Paris. Her current research is on mathematics in late imperial China, and on scientific contacts with Europe. She has published "Les Méthodes Rapides pour la Trigonométrie et le Rapport Précis du Cercle (1774). Tradition chinoise et apport occidental en mathématiques" (Paris, 1990).

I.
Introduction

The aims of the Jesuits in China were primarily missionary, and it is well known that from the beginning they used science as a means of arousing Chinese scholars' interest in Christianity. However, it can be said that their influence in China was effective mostly in the field of science. This situation must be taken into account when studying the impact of the parts of western scientific knowledge that they introduced in China. In that respect, the reign of the Kangxi (1662-1722) emperor is a specially interesting period. Due to the emperor's policy (in particular, to his interest in western science and his tolerance towards Christianity), it was a very prosperous time for the mission. I will summarize briefly the origins of this policy. In 1629 the calendar reform according to western methods had been set out by Johann Adam Schall von Bell and other Jesuits under the supervision of Xu Guangqi[1]. It had remained in force during the reign of the first Manchu emperor

---

[1] Cf. Henri Bernard-Maitre, "L'Encyclopédie astronomique du Père Schall", *Monumenta Serica*, 3 (1938), 35-77, 441-527; K. Hashimoto, *Hsü Kuang-ch'i and Astronomical Reform. The Process of the Chinese Acceptance of Western Astronomy, 1629-1635* (Osaka, 1988).

(1644-1662). The Jesuits had been dismissed from the Imperial Board of Astronomy (*qintianjian*) during the regency (in 1664), following Yang Guangxian's accusations; these accusations were concerned both with scientific matter and with alleged seditious implications of Jesuit activities. In 1668, the emperor, finding the calendar unsatisfactory, sought Ferdinand Verbiest's opinion on it. Verbiest pointed to a number of errors. The comparison between the Chinese and western methods then revealed that the latter were far more accurate for prediction, and in February 1669 calendar-making was entrusted to the Jesuits again. The emperor's admiration for western science rose from this affair. From the Chinese point of view, this episode was part of the emperor's efforts towards ensuring his own political authority against the regent Oboi[2]. Following this, the Kangxi emperor began to study western science. Verbiest taught him mathematics and astronomy[3]. During the rest of his life he supervised the scientific activities of the Board of Astronomy and he was the emperor's teacher in western science at the same time. So it was Verbiest who played a key role in restoring the Jesuits to the emperor's favor, as well as to their institutional position concerning scientific activities.

The first Jesuits of the French mission arrived in Beijing on February 7, 1688, a few days after Verbiest's death[4]. Along with Antoine Thomas, two of them - Joachim Bouvet and Jean-François Gerbillon - were to succeed Verbiest as the emperor's teachers in western science. The French mission had a particular status, in several respects. It arose from several factors, both political and scientific, and was to result in important dissensions among the Jesuits in China, which in turn had effects on the transmission of science.

---

2 L. Kessler, *K'ang-hsi and the Consolidation of Ch'ing Rule 1661-1684* (Chicago, 1976), pp. 58-64. The political and symbolic importance of the calendar in China must be kept in mind according to Joseph Needham, *Science and Civilisation in China* (Cambridge, 1954), 3:186-194.

3 Kessler, *op. cit.*, p. 137. This will to be learned in "Western knowledge" was part of the emperor's conception of what an emperor should be: "an exemplar of the Confucian monarch, the grand patron of learning," also accomplished in classical Chinese studies.

4 They were de Fontaney, Le Comte, Visdelou, Bouvet and Gerbillon. Cf. Louis Pfister, *Notices biographiques et bibliographiques sur les Jésuites de l'ancienne Mission de Chine 1552-1773* (Shanghai, 1932-1934).

II.
Origins of the French Mission

The creation of the French Jesuit Mission of China was the result of a conjunction of circumstances which can briefly be summarized. Many times during the two centuries that the mission lasted, the Jesuits of China felt the need to reinforce the mission with capable mathematicians, able to take up the task of astronomers. In 1678 Verbiest had written a letter to all the Jesuits of Europe on that matter. When Philippe Couplet visited France, he brought a letter from him to Father Jean de Fontaney, and was introduced to Louis XIV by his confessor, Father François de La Chaise. The latter was in charge of organizing the French mission[5]. Sending French Jesuits to China was also a way for Louis XIV to send an "embassy" to the Kangxi emperor, and thus to develop French influence in the Far East. This gave rise to a conflict, as all the missionaries of China were supported by the king of Portugal up to that time. It seems that giving the Jesuits the title of "the King's Mathematicians" was partly a means to evade this diplomatic difficulty[6]. Nevertheless, this title given to the French Jesuits before their departure for China had a more significant meaning. They were also sent there to make observations for the French Académie Royale des Sciences, founded by Jean-Baptiste Colbert in 1666, and supervised by him until his death in 1683. According to de Fontaney's account, this scientific expedition was Colbert's idea:

> Science, Father, does not deserve that you should bother to cross the seas and be reduced to live in another world, far away from your fatherland and friends. But as the wish to convert infidels and to win souls over to Jesus Christ often leads your Fathers to embark upon such travels, I should wish that they take the opportunity, at the time when they are not so busy preaching the Gospel, to gather on the spot a number of observations that we lack for the perfection of arts and sciences[7].

This idea, which apparently arose when Father de La Chaise showed Verbiest's letter of 1678 to Colbert, fit into an important geographical

---

5 *Ibid.*, pp. 308, 420.

6 V. Pinot, *La Chine et la formation de l'esprit philosophique en France* (Paris, 1932), pp. 40-47, 422-423; Yves de Thomaz de Boissierre, *Un Belge mandarin à la cour de Chine aux XVIIe et XVIIIe siècles. Antoine Thomas 1644-1709. Ngan To P'ing Che* (Paris, 1977), pp. 38-39.

7 Pfister, *op. cit.*, p. 421: quotation of a letter from Father de Fontaney.

project which was being carried out at the time, one involving several surveys and measurements in different parts of the world. And it seemed that the easiest way to have these made in China was to have recourse to the Jesuits[8].

This meant that the "King's Mathematicians" had to be capable of carrying out experiments and surveys according to the criteria and methods of the scholars of their time, as they were to participate in "European science." Moreover, they were to keep in contact with some of the most eminent scientists for that purpose. This is something new in the Jesuits' scientific activities in China. It is this aspect which was emphasized in the decree which gave the missionaries the title of "the King's Mathematicians."

> Being very glad to contribute from our part to everything that can increasingly establish safety in navigation and improve science and arts, we have thought that, to be more sure to match these needs, it was necessary to send to the Indies and to China some learned persons capable of making observations from Europe; and judging that we could not make a better choice for this purpose than Father N ..., knowing his extraordinary abilities, because of that and other reasons, ... we hereby appoint him our mathematician[9].

III.
Arrival in China

The arrival in China of the five "Mathematicians" was marked by several difficulties. They did not pass through Macao, as they had not taken the oath of allegiance to the king of Portugal. When they arrived at Ningbo (Zhejiang) they were not allowed to go on to Beijing by the local authorities, and it was only thanks to Verbiest's intercession with the emperor that they were finally allowed to proceed to the capital. In this episode two things are noteworthy. In the first place, Verbiest interceded against the will of the Portuguese fathers, including his own superior, "so that no king should be offended, and for the greater advantage of the mission"[10]. Secondly, in order to convince the emperor, he told him about the scientific skills of the French. It must have been a decisive argument, as the counter-order sent by the emperor to prevent the French from being expelled shows. "It is not improbable

---

8 *Ibid.*, p. 420.

9 *Ibid.*, p. 423.

10 Thomaz de Boissierre, *op. cit.*, p. 40.

that among the foreigners, Hong Ruo (de Fontaney) and others, are those who may know the method of calendar-making. If so, we order them to repair to Beijing and wait for Our employment ...!"[11].

After having received them, the emperor kept Bouvet and Gerbillon, and allowed de Fontaney, Louis Le Comte and Claude Visdelou to go to the provinces. The French had brought scientific instruments, some of which were given to the emperor as presents from the king of France.

An interesting account of these events is given by Father Marcel Le Blanc[12]. According to it, the Portuguese fathers tried to confiscate the instruments brought from France, and to forbid the French from making any observations when they traveled. Moreover, the French were ordered to conceal their scientific skills from the emperor. This was, of course, incompatible with the French plans[13]. They were supported by Louis XIV precisely in order to make observations, and as scientific envoys to the emperor. This was only part of an attempt to prevent the French from setting up an autonomous mission, and also from obtaining the emperor's special favor for their scientific abilities. It seems that the French Jesuits' scientific competence appeared dangerous to the others insofar as it granted them imperial favor and independence. From this, the Jesuits of China appear to have been aware that science was the main motivation of the emperor's interest and tolerance towards them.

Bouvet and Gerbillon wrote several letters in which they described both their scientific practice and their teaching to Kangxi. They also sent reports to the French Academy. For instance, in a letter to Father Le Comte[14], Bouvet mentions a few things that he is sending him for the French Academy:

> All I could send you is 1) an anatomical description of a tiger with the figure, 2) a translation of the Chinese description of this animal, from their natural history, 3) several remarks about the elephants of Siam ... 4) descriptions of some singular fishes that we saw in the Gulf of Siam ... I am sending you the ephemerides of the daily

---

[11] Fu Lo-shu, *A Documentary Chronicle of Sino-Western Relations (1644-1820)* (Tucson, 1966), p. 93.

[12] It is the manuscript *Jap.-Sin. 127*, ff. 127-170 of the ARSI in Rome; the author says that his account is based on letters which have been lost.

[13] For instance, de Fontaney intended to set up an observatory in Nanjing, from where he would have kept in touch with the observatories in Beijing and Paris (*Jap.-Sin. 127*, f. 57).

[14] Written from Beijing on October 20, 1691; Le Comte was then in Fuzhou. ARSI, *Jap.-Sin. 165*, ff. 100-102.

> changes that occurred in the air since November 12, 1690 until the end of October 1691, which I have observed in Beijing this year. These observations are of four kinds: serenity of the weather, wind, heat and heaviness of the air. I have made them everyday at three different times, i.e. around four in the morning, around eight in the evening and at noon, the latter being equally distant from the two former times. I missed a good part of the noon observations, that is to say a good part of the days, when I had to go to the palace. The observations of the heaviness of the air have been made on a double barometer filled partly with quicksilver and partly with *acqua fortis*. The observations of the heat of the air have been made on a thermometer 18 inches long, which gives the degrees of heat and cold by rarefaction and condensation of the liquor or spirit that it contains. These two instruments are made by Mr Lubin. For some time I have also marked the degree of humidity or dryness of the air by means of several hairs from a horse tail which I had tied together about five or six fathoms long, and hung in my room with a lead at the bottom which, going up or down depending on the degrees of humidity or dryness, allowed me to note the differences everyday ...

He goes on to explain that he gave up these last observations after realizing that the measuring-device he had worked out was not reliable.

In the same letter, Bouvet mentioned that they were teaching philosophy to the emperor:

> ... After having completed this introduction to philosophy in Tartar, which we made as briefly and clearly as we could, - suppressing all the involved terms and pure chicanery, according to the style of the Moderns - we started with physics, and knowing that this prince has a very high opinion of European medicine, and that he most wishes to know the structure of the human body, we start this part of philosophy by the science of the human body, in which, in addition to a short anatomy, with all the figures and their explanations, and all the beautiful discoveries of the authors, both ancient and modern, we will put, with the help of God, all the curious observations that we have here in the first part of the memoirs of the gentlemen of the Academy of Sciences on the animals, and all those that we will get from these gentlemen on the same subject ...

Bouvet's intention was to praise the works of the academicians, a way of glorifying the king of France. He wants to promote the relations between Louis XIV and the Kangxi emperor.

> If these two great monarchs knew each other, the mutual esteem they would have for each other's royal virtues could not but prompt them

> to tie a close friendship, and demonstrate it to each other, if only by an intercourse in matters of science and literature, by a kind of exchange between the two crowns of everything that has been invented until now in the way of arts and sciences in the two most flourishing empires of the Universe. If Heaven graced us with the achievement of this goal, we would feel that we did a lot for the good of Religion which, under the auspices and protection of two such powerful princes, could not fail to progress considerably in this empire[15].

It is clear that in Bouvet's mind, his triple activity as missionary, correspondent of the French Academy, and teacher of the emperor, correspond to a coherent project in the context of "cultural exchanges" between France and China; these exchanges would lead "naturally" to the evangelization of the latter.

IV.
Scientific Practice: Disagreements Among the Jesuits

The conflicts between the French mission and the other Jesuits in China lasted throughout the Kangxi reign; they were mainly concerned with problems of authority. One of these conflicts, on the other hand, is clearly focused on scientific issues. It can be stated in one question: should the Jesuits take the progress of European science into account in their scientific practice and teaching in China? I will give a brief account of this conflict, according to one of its main protagonists, Father Jean-François Foucquet[16].

Louis Pfister mentions this incident very briefly in Foucquet's biography (wrongly dating it in 1710):

> Father Foucquet, together with other French Fathers, had suggested a few modifications in the astronomical computations used in Beijing by their predecessors. They mostly wanted to introduce La Hire's tables[17]. Without pronouncing on the value of these tables, the Father-visitor and other Fathers did not allow it, for fear of giving the impression of blaming what their predecessors had had so much difficulty in establishing, and of giving a handle to new accusations against religion[18].

---

15 *Ibid*.

16 It is the manuscript *Jap.-Sin. II 154* of the ARSI. As far as I know, it is the only detailed account of the affair available.

17 Philippe de La Hire, *Tabulae astronomiae* (Paris, 1702).

18 Pfister, *op. cit.*, p. 551; John Witek, S.J., *Controversial Ideas in China and*

Joseph Needham considers this as an illustration of the fact that "the Jesuits were the prisoners of their limited motive"[19].

The account I will now give relies mostly on Foucquet's report. In 1711 the emperor measured the summer solstice himself, and found out that it had been miscalculated in the tables supplied by the Jesuits of the Board of Astronomy (none of the French Jesuits who were the emperor's teachers worked there at the time); these tables were based on the Tychonic theory. The Board of Astronomy was then directed by Kilian Stumpf. Foucquet reports a dialogue between this missionary and a Chinese, concerning a miscalculation which the Europeans of the Board of Astronomy refused to acknowledge: "The Palace mandarin named Vang Taohoa told Father Stumpf ... that the emperor might be offended, and give this Board away to some others." "Well let him give it away if he pleases," Father Stumpf answered, and added "I did not come to China for that. What do I care?" "I know," the mandarin said, "that even if you were dead, it would not matter. But that the Board should be withdrawn from Europeans is a matter of consequence to others. Your religion itself depends on this. Father Verbiest would never answer as you do"[20].

Following that, the emperor commanded the Jesuits to explain astronomy to him, that is to say "a doctrine by means of which a man who knows geometry and arithmetic could make astronomical tables"[21].

Foucquet was chosen for that purpose, together with Father Franz Thilisch[22], who was then Stumpf's assistant. They completely disagreed on the doctrine that was to be explained to the emperor:

> Concerning the astronomical doctrine, (Father Thilisch) was less of a help than an obstacle and an impediment, since being attached to Father Stumpf, and being of the same nationality - and, moreover, subjected to the Portuguese vice-province - he was determined to uphold the Tychonic theory which is followed at the Board, and to which Father Foucquet saw great objections[23].

This is the crucial problem: Foucquet thought that both Tycho Brahe's

---

Europe: A Biography of Jean-François Foucquet, S.J. (1665-1741) (Rome, 1982), pp. 181-191. Witek gives an account of the whole "affair."

19 Needham, *op. cit.*, p. 450.

20 *Jap.-Sin. II 154*, f. 2.

21 *Ibid.*, pp. 4-5.

22 Franz Thilisch's biographical notice is in Pfister, *op. cit.*, pp. 624-625.

23 *Jap.-Sin. II 154*, f. 12.

system, and the tables worked out by Verbiest on that basis about sixty years earlier, were outdated; and once the Chinese themselves found this out, this system could not be defended. This is why he suggested to use Philippe de La Hire's tables.

During almost five years, Foucquet wrote several treatises on astronomy[24]. According to him, the emperor seemed to approve them, while he was surprised by the differences between Verbiest's astronomy and what Foucquet was explaining:

> After having read a booklet on the theory of the sun, the emperor, more pleased than usual, asked why Father Verbiest had not explained this doctrine to him. The answer was that, since the time of Father Verbiest, many interesting things had been discovered, which Father Foucquet meant to present to His Majesty.

Among the novelties he had included in his treatises were the adoption of elliptic orbits and a corrected value for the ecliptic angle.

In a letter to Father General Michelangelo Tamburini, dated October 9, 1716[25], Stumpf reports that Foucquet was trying to promote a reform of astronomy; and, as he puts it, this is the opportunity to glorify French science that the French have been waiting for during the last thirty years. So he interpreted this as one of the attempts made by the French mission to gain the emperor's special favor, and as a "chauvinistic" behavior. And, indeed, when in 1716 the emperor demanded that the Jesuits come to an agreement about astronomical theories, all the French supported Foucquet at first. Stumpf, who was then "President of the Tribunal of Mathematics"[26] and visitor, considering that he had in this Orient all the powers belonging the general of the Society of Jesus, chose to use all the authority of his office to support a doctrine that the strength of truth could not support[27].

The argument used against "modernization" was that it could endanger the mission by throwing doubt on the illustrious predecessors, and especially on Verbiest. But Foucquet's argument too was tactical. If the Chinese found out by themselves the shortcomings of a theory, the Jesuits, by supporting it, would throw discredit on the mission. On the

---

[24] The copies of some of these treatises are extant in the Vatican Library in Rome: *Borgia Cinese* 319 (1), (2) & (3).

[25] ARSI, *Jap.-Sin. 177*, ff. 114-123.

[26] This is the name given by the Jesuits to the Imperial Board of Astronomy.

[27] *Jap.-Sin. II 154*, ff. 37-38.

other hand, if they taught the Chinese new and more difficult things, they could "reduce them to the condition of pupils of the Europeans. With the six first books of Euclid, and a few propositions from Archimedes on solids that have been added since, it was not to be feared that they should be able to penetrate by themselves the secrets of conic sections, or a higher geometry; the modern theories of the sun, the moon, and other planets would thus be a hermetic language to them; and following that way, the Europeans could not fail to be employed in a reformation of astronomy if the emperor undertook it ..."[28]. Of course Foucquet's aim was not to promote scientific progress, but to maintain the Chinese in a state of dependence, and thus to make sure that the Jesuits would not be expelled.

So the argument was not really about scientific matters, but about the best way of using science for missionary purposes. However, it reveals quite different conceptions of science on both sides. The "conservative" attitude seems to imply a conception of truth in science which was very much inspired by religion. It must appear as eternal and immovable, and by contradicting predecessors, one can throw discredit on them. This suggests a somewhat low opinion of Chinese scientific abilities and of science's intrinsic value. Foucquet's point of view on science, on the other hand, could be considered as more "modern," although his motivations appear tactical and non-scientific. Both attitudes were subordinate to the interest of the mission.

The controversy on astronomy among the Jesuits ended at Foucquet's disadvantage. Because he would not give up his figurist ideas, the French stopped supporting him on astronomy grounds, although they did not object to his scientific ideas.

V.
Conclusion

Verbiest had done his best to ensure that there would be competent successors to take up his scientific tasks in Beijing, being well-aware of the vital importance of those for the mission. The French mission only took up part of them, whereas the Board of Astronomy was run by Jesuits who had already been there for some time. Calendar-making required

---

28 The study of conic sections - which dates back to Greek mathematics in the western tradition (T. Heath, *A History of Greek Mathematics*, Oxford, 1921) - was not introduced into China before the nineteenth century, although they are mentioned in some of the Jesuits' works. Their study is, of course, a prerequisite to an understanding of Kepler's astronomical theory.

some continuity, and the Jesuits of the Board of Astronomy had no reason to modernize their methods, which would have required another reform. As has been said about the Chinese imperial astronomers[29], they had no interest in such a change. Those who taught the emperor, on the other hand, had to arouse his curiosity, and introducing novelties was a good way of doing so. They relied on their own scientific background. The fact that the French used the works of the Academy should not be seen merely as chauvinistic behavior aimed at glorifying French science. They were drawing their information from the sources they knew best, according to the scientific style with which they were familiar. Their scientific activity in China consisted mostly in gathering observations and information. This was in accordance with the type of work undertaken by the Academy, which was oriented towards experimentation rather than theoretical speculation[30]. Bouvet had the hope of setting up an equivalent of the Academy under imperial auspices, and works such as the making of the map of China are reminiscent of some of those undertaken by the French Academy[31].

On the other hand, the more up-to-date scientific background of the French Jesuits seems to have had little influence on the science practised in China. They proposed only few and still marginal changes according to scientific progress in Europe, which were not always adopted; and there was no agreement on the subject among the Jesuits. In astronomy, the condemnation of Galileo Galilei and of heliocentrism did not, of course, encourage innovation - but anyway, the Jesuits' aim was not to promote modern science. And even though the French claimed they wanted to do so, there was never any attempt made for the systematic updating of scientific knowledge and practice by the Jesuits of China, some of whom worked as full time scientists for the emperor. As can be seen from Foucquet's attempt to innovate in the field of astronomy, scientific matters were implied in much broader issues among the Jesuits than they were among the Chinese[32]; on both sides it was not

---

29 This argument has often been used to explain the "stagnation" of Chinese science. It is already found in a letter by Father Dominique Parrenin dating from 1730. See I. & J.L. Vissière (eds.), *Lettres édifiantes et curieuses de Chine par les missionnaires jésuites (1702-1776)* (Paris, 1979), pp. 359-366.

30 R. Hahn, *The Anatomy of a Scientific Institution. The Paris Academy of Sciences, 1666-1803* (Berkeley, 1971), p. 15.

31 *Ibid.*, p. 21.

32 The arguments used by Yang Guangxian in the calendar controversy during

scientific criteria that prevailed, but "political" ones. The emperor's attitude appears as remarkably modern in that respect.

The absence of cohesion among the Jesuits, and their disagreements, corresponded to nationalities and to different functions as scientists: official astronomers or imperial teachers. Verbiest was serving as both. He had, at the same time, an undisputed scientific authority among the Jesuit missionaries, and the emperor had much consideration for him (which was essential, considering the emperor's personal involvement in scientific matters). In that respect, Verbiest had no successor. After his death, due to the increasing discrepancies between European science and that practised by the Jesuits, as well as to nationality dissensions, there was no consistency in the scientific knowledge made known to the Chinese.

the 1660s are a good illustration of this on the Chinese side. See Kessler, *op. cit.*, pp. 58-60.

THEOPHILUS SIEGFRIED BAYER (1694-1738) ON FERDINAND VERBIEST

KNUD LUNDBAEK

Previously professor of medicine at the Aarhus University, Knud Lundbaek is at present research fellow at the Institute of East Asian Studies, Aarhus University. His latest publication deals with "Joseph de Prémare (1666-1736), S.J. - Chinese Philology and Figurism" (Aarhus, 1991).

Theophilus (Gottlieb) Siegfried Bayer (1694-1738) was a German classical scholar who spent most of his short life studying Chinese. He was born into a poor but pious Protestant family. At school he showed a great aptitude for languages, and was eventually sent to the university to read Latin, Greek and Hebrew. While on a short study tour to other German universities, the then twenty year-old Bayer found the Chinese-Latin vocabulary of the Dominican Francesco Diaz, as well as a number of manuscripts dealing with the Chinese language. Back in Königsberg, he published his first book in 1717, the *De eclipsi sinica*. This little work, with his *praeceptiones* about the Chinese language, contains nine pages. In 1726 he was called to the chair of classical languages and literature at the newly opened Imperial Academy in St. Petersburg. He was to stay there for the rest of his short life, occupied most of the time with studies of the Chinese language, and especially with the Chinese script.

In 1730 he published his *Museum Sinicum* in two volumes, a textbook of Chinese - primitive, to be sure, but nevertheless the first to appear in Europe. A few years later he entered into correspondence with some of the learned Jesuits at the imperial court in Beijing: Dominique Parrenin, Antoine Gaubil, Ignaz Koegler, André Pereira, and Karl Slavicek. Through their letters, these men - Parrenin in particular - helped him in his endeavors to understand the mysteries of the Chinese language. This help was invaluable to Bayer, working, as he was, at a time when there were no Chinese in Europe, and no printed dictionaries or grammars of the Chinese language. During these years he published a book called *De horis sinicis*, dealing with a Chinese sundial, and revealing his notions about the system of the *Book of Changes*, as well as a number of long sinological articles, much more mature than was his *Museum Sinicum*. One of them contains the first description of a Chinese dictionary. Most of them were printed in the annual of the Academy, the *Commentarii academiae scientiarum imperialis metropolitanae*; others appeared in the *Miscellanea Berolinensia*, published by the *Societas Regia Scientiarum* in Berlin, of which he was a member. During all these years, he also worked on an immense Chinese-Latin dictionary which, however, was never completed. After Bayer's death, his many manuscripts, drafts, letters received and written, and the manuscripts and Chinese books he received from his Beijing friends came to the University Library in Glasgow[1].

Bayer's contribution to the Verbiest literature is an article printed in the *Miscellanea Berolinensia* in 1740, two years after his death. Several years before, in a letter to Parrenin dated May 12, 1736, he had told him that he was working on something about Philippe Couplet's catalogue of China Jesuits and the Chinese work on which it was based. He mentioned, at the same time, Verbiest's world map, which Parrenin had sent him a few years earlier. The article - twelve pages in length with three additional pages of Chinese characters - is entitled *De Ferdinandi Verbiestii S.J. scriptis, praecipue vero de eius globo terrestri sinico*[2]. It begins:

> Three years ago Father Dominique Parrenin in Beijing sent me a Chinese world map. This gift made me very happy and I began at once to translate the geographical names on it. I wanted very much to know the Chinese names for the oriental regions. I needed them for a certain project of mine[3].

Parrenin had only said in the letter enclosed with the map that it had been made for the use of the Chinese by earlier missionaries. Bayer writes: "In one of the margins of the map I found the name of its author, written with large characters: Nan Huairen. This was the Chinese name of Ferdinand Verbiest, famous in China for his erudition and the high posts he held, as well as in Europe." He does not need,

---

1 K. Lundbaek, *T.S. Bayer (1694-1738) - Pioneer Sinologist* (London, 1986).

2 *Miscellanea Berolinensia*, VI, 1740, 180-192.

3 He probably meant his *Historia ecclesiarum Orientis*. As a young man he had planned to write a book with this title, and in a letter to Antoine Gaubil in 1735, he lists the titles of its many chapters. In the Glasgow University Library there is a fifty-page draft of one of them.

therefore, to write here about the life and works of this man. However, he will relate things from a Chinese book called *Shengjiao xinzheng* (Sincere Testimony of the Sacred Doctrine), a copy of which he had received from Beijing. He already possessed Couplet's *Catalogus Patrum S.J. qui post obitum Sti. Francisci Xaverii... in imperio Sinarum Jesu Christi fidem propagarunt...* (Paris, 1686), which was based on the *Shengjiao xinzheng*.

Bayer presents the short life of Verbiest, Chinese style, with the titles of some of his works from the Chinese book, together with Couplet's rendering in his catalogue. The Chinese texts are given in transliterations, with a word-for-word translation (the characters, 300 in all, are found in three tables at the end of the volume)[4].

The twelve works listed include four religious tracts, four astronomical treatises, a work on the thermometer, the explanation of the world map and the world map itself, and two star maps. For each of the works, Bayer referred to the Latin translation in Verbiest's *Astronomia Europaea* (Dillingen, 1687). About the second work on this list - called by Couplet *Ordo proponendi mysteria fidei* - Bayer writes that he owns a copy of it. "It is printed with the same incredible elegance as are all the other Chinese works published by the Jesuits." It may be the book which Koegler, Pereira and Slavicek sent him on September 12, 1732. In their letter they do not mention the title, but refer to it as "a small Chinese work (written in colloquial style) by Ferdinand Verbiest, whose Chinese name is Nan Huairen," adding that "those among us who begin to learn the Chinese style usually first train themselves by translating such works."

Number ten on the list is the world map, *Kunyu quantu*, which Couplet called the *Mappa totius mundi terrestris*. It consists of two hemispheres. The diameter of each of them is said to be 4½ Rhenish feet. They are so arranged that China is situated just about in the middle, as the Chinese want it to be, so that it conforms with their name for China: *Zhongguo*, the Middle Kingdom (similar, Bayer says, to the Greek notion of the world's navel being in Delphi). He worked hard with the map, and tried to translate some of the large and small inserts on it. Among the Bayeriana in the Glasgow University Library, there is a thirteen folio-page manuscript draft showing his attempts. As was to be expected, he did not get very far in his translations, and there are a number of mistakes. In the published article he translates the two

4 The transliterations and translations of twelve characters dealing with the honors bestowed on him by the Kangxi emperor are missing on p. 182.

subtitle-like phrases in the margins of the map, but one of them went awry, presumably because at an initial stage he copied the character *yue*, moon, as *ri*, sun. The same phrase contains two characters he says he has been unable to translate. No wonder: it is an astronomical word, the name of the twelfth Jupiter station, used here, perhaps, to designate the twelfth month of the year[5]. He discusses the meridian system, beginning at Beijing, the notes about the length of the day in the polar areas, the real and fabulous animals drawn in the southern parts, etc... He notes that the eastern part of the Great Wall is drawn above Liaodong province, reaching into Korea. However, he says, this is not Verbiest's fault, but the fault of ignorant Chinese persons who misled him.

There is a lengthy section about the names of cities and countries in the West - *Luoma*, Rome, *Lixibo*, Lisbon. They are given approximate Chinese sounds, often on the basis of their Latin forms. He deplores that the names of the regions west of China, in which he was particularly interested, are made unreadable by the decay of the paper in these spots. Some of the names are written with characters that give a meaning - such as *Luxiya*, Russia - which can be read as the "Road to the West." He had looked hard at the little insert in the region of Palestine. It says: "Judea. God (*Tianzhu*) was incarnated here; therefore it is called the Holy Land." But, as luck would have it, he misread one character for another very similar one, thereby partly spoiling the meaning of the sentence. He had sent this little specimen of his Chinese learning to Parrenin some years before, and Parrenin had corrected this "small error" in his usual kind way. Bayer, however, had already sent the manuscript of the Verbiest article to Berlin, and so did not have time to change it.

In all his writings, Bayer was in the habit of putting in some interesting *obiter dicta*. There are two examples of this stylistic device in the Verbiest article. One is a gibe at Andreas Müller (c. 1630-1694), the fabulous Berlin clergyman who published a number of small works on Chinese matters, and who had astonished the learned world by declaring that he had invented a *clavis sinica* by which one could learn Chinese in the course of a few days. In his youth Bayer had admired him, but later came to regard him as an erudite charlatan. In the Verbiest article, there is half a page about a world map printed in Müller's *Geographia Mosaica* (Berlin, 1689). Müller had heard something about Verbiest's map and wished to show it. His map also consists

5 See J. Needham, *Science and Civilisation in China* (Cambridge, 1959), 3:403-406.

of two hemispheres, but China is placed in the middle of the "eastern" one, and the meridians begin in the traditional way from the Insulae Fortunatis. "I cannot believe," Bayer says, "that Müller took this rude map from that of a China Jesuit; I think he was making it up." The other aside is about two maps he had in his private collection. One of them was a copy of a map brought to Italy centuries ago by the Venetian Niccolò de' Conti, and kept in the monastery of S. Michele in Murano. Müller had obtained a copy of it which was then in the Public Library of Stettin, and Bayer had received a copy of the copy[6]. The other was a Japanese world map belonging to the famous collector Hans Sloane in London. Bayer had obtained a copy of this map also. He takes the opportunity to tell about two shipwrecked Japanese sailors, by name Zose and Gansu, who had been brought to visit him in St. Petersburg, and who had told him that east of Japan there was a country inhabitated by *obitogun*, or gigantic naked cannibals[7]. Bayer does come back to his main subject, and he ends his article with some notes on earlier China Jesuits' maps, beginning with that of Matteo Ricci.

Bayer's Verbiest article is of interest to us as showing how the work of a great Jesuit scholar in China was presented half a century after his death by a European Protestant scholar who had had personal contacts with the Beijing Jesuits[8]. To Bayer himself, writing this article and sending it to the *Miscellanea Berolinensia* may have meant something personal. In 1737, after ten years in the Academy of Science in St. Petersburg, he tendered his resignation and was honorably discharged.

---

6 S. Michele monastery of Murano: famous Benedictine monastery on what is now the cemetery island of S. Michele, north of Venice. Demolished 1799. Niccolò de' Conti, Venetian explorer, traveled in India and Indonesia between 1419 and 1441. On his return, he told his story to Poggio Bracciolini, the papal secretary, who included it in Book IV of his *Historia de Varietate Fortunae*, written in 1447 or 1448. Printed editions and many translations of this work appeared in the last years of the fifteenth century. De' Conti was never in China, but he obtained information about it on his way. The great cartographer Fra Mauro, who lived in the S. Michele monastery, used some of it for his *Mappa mundi* (c. 1459). The map Bayer had may have been a copy of Fra Mauro's planisphere. (I am grateful to amanuensis Guiseppe Torresin, Institute of Ancient and Medieval History, Aarhus University, for this information, correcting my suggestion in note 55 on p. 187 of the book mentioned in note 1).

7 *Obito*, Japanese, a tall man.

8 In a new edition of Moreri's famous *Grand dictionnaire historique* ..., published in Paris in 1759, there is an entry about Ferdinand Verbiest. Referring to the article in the *Miscellanea Berolinensia*, it gives a summary of Bayer's presentation of Verbiest's works, 10:531-532.

He had sent his valuable books and manuscripts ahead by sea, and was ready to return to his homeland at the beginning of the next year. He was a highly respected scholar, a classicist, a historian, and an orientalist. He must have expected to be summoned to more than one German university, and may already have begun negotiating with one of them. Composing a paper like the Verbiest article, and sending it to the foremost academic journal in Germany, may have been his bid for a chair after his heart, but fate had decided otherwise. On February 11, 1738 he caught an acute disease, and died ten days later. It is not for a historian to ask the question: What would the history of sinology have been if Bayer had lived on for another twenty or thirty years? - but it is hard not to do so.

SELECT BIBLIOGRAPHY

Actes du III° Colloque international de sinologie. Appréciation par l'Europe de la tradition chinoise à partir du XVIIe siècle. Paris, 1983.

Aiton, E.J. "Galileo and the Theory of the Tides", *Isis*, 56 (1965), 56-61.

Aizier, Emmanuel. "Une faillite en Extrême-Orient (*Failure in the Far East*). Du nouveau sur la 'Querelle des Rites'", *Chine-Madagascar*, 65 (June 1957), 111-114.

Aleni, Jules. "Matthieu Ricci" (traduit par L. Desbuquois), *Revue d'Histoire des Missions*, 1 (1924), 52-70.

Alessandri, Bruno. "L'Archivio Generale Agostiniano di Roma e le missioni", *Euntes Docete*, 21 (1968), 177-189.

Allan, Charles W. *Jesuits at the Court of Peking*. Shanghai, 1935.

Andrews, Michael C. "The Study and Classification of Medieval Mappae Mundi", *Archaeologia*, 75 (1925), 61-67.

Arickx, Valère. *Geschiedenis van Pittem*. Pittem, 1951.

Arnauld, Antoine. *Morale pratique des Jésuites*. 8 vols. Cologne, 1669-1695.

"Astronomie. Réparation d'un ancien observatoire", *Beijing Information* (November 1979), 29-30.

Atkinson, Geoffroy. *Les relations de voyages du XVIIe siècle et l'évolution des idées*. Paris, 1924.

"Une automobile à Pékin du XVIIe siècle", *Bulletin catholique de Pékin*, 5 (1918), 531-532.

Avril, Philippe. *Voyages en divers états d'Europe et d'Asie entrepris pour découvrir un nouveau chemin à la Chine*. Paris, 1692.

Ayuzawa, Shintarô 鮎澤信太郎 . *Mateo Ritchi no sekaizu ni kansuru shiteki kenkyû* マテオ・リツチの世界圖に關する史的研究 (A Historical Study of Matteo Ricci's World Map). Yokohama, 1953.

Ayuzawa, Shintarô. "The Types of World Maps Made in Japan's Age of National Isolation", *Imago Mundi*, 10 (1953), 123-128.

Ayuzawa, Shintarô. "Geography and Japanese Knowledge of World Geography", *Monumenta Nipponica*, 19 (1964), 274-294.

Baddeley, John F. "Father Matteo Ricci's Chinese World Maps", *Geographical Journal*, 50 (1917), 254-270.

Baddeley, John F. *Russia, Mongolia, China*. 2 vols. London, 1919.

Bagrow, Leo. *History of Cartography*. London, 1964.

Bailly, Jean-Sylvain. *Lettres sur l'origine des sciences et sur celle des peuples de l'Asie, adressées à M. de Voltaire par M. Bailly et précédées de quelques lettres de M. de Voltaire à l'auteur*. London, 1777.

Baker, John N.L. *A History of Geographical Discovery and Exploration*. London, 1931.

Baker, John N.L. "Some Dutch Charts of the XVIIth Century", *Imago Mundi*, 2 (1937), 36.

Baker, John N.L. "Some Original Maps of the East India Company", *Tijdschrift van het Koninklijk Nederlandsch Aardrijkskundig Genootschap Amsterdam*, 53 (September 1936), 665-667.

Bangert, William W. *A History of the Society of Jesus*. St. Louis, 1972.

Barten, C. "Hollandse kooplieden op bezoek bij concilievaders", *Archief voor de Geschiedenis van de Katholieke Kerk in Nederland*, 1 (April 1970), 75-120.

Barthold, Vasilii V. *La découverte de l'Asie. Histoire de l'orientalisme en Europe et en Russie*. Paris, 1947.

Baruzi, Jean. *Leibniz et l'organisation religieuse de la terre*. Paris, 1907.

Baudier, Michel. *Histoire de la cour du roy de la Chine*. Paris, 1984.

Baudiment, Louis. *François Pallu, principal fondateur de la Société des Missions Étrangères*. Paris, 1934.

Baudry de Saunier, Louis. *Histoire de la locomotion terrestre*. Paris, 1936.

Bauer, Wolfgang. *China und die Fremden. 3000 Jahre Auseinandersetzung in Krieg und Frieden*. München, 1980.

Bayer, Theophilus S. "De Ferdinandi Verbiesti scriptis, praecipue vero de ejus Globo Terrestri Sinico", *Miscellanea Berolinensia* (1740), 180-192.

Bayer, Theophilus S. *Museum Sinicum in quo Sinicae linguae et litteraturae ratio explicatur*. 2 vols. St. Petersburg, 1730.

Beckmann, Johannes. "Genfer Drucke in der alten Jesuitenbibliothek von Peking", *Zeitschrift für schweizerische Kirchengeschichte*, 61 (1967), 301-340.

Beckmann, Johannes. "Handschriften zur Missionsgeschichte Chinas in Britischen Museum", *Neue Zeitschrift für Missionswissenschaft*, 1 (1945), 108-112.

Beckmann, Johannes. "Die lateinische Bildung des chinesischen Klerus im 17. und 18. Jahrhundert", in *Der Einheimische Klerus in Geschichte und Gegenwart. Festschrift P. Dr. Laurenz Kilger, OSB*. Schöneck-Beckenried, 1950, pp. 163-187.

Beckmann, Johannes. "Die Peitang Bibliothek in Peking. Ihre missionsgeschichtliche Bedeutung", *Neue Zeitschrift für Missionswissenschaft*, 4 (1948), 275-279.

Bernard, Henri. "L'église catholique des XVIIe et XVIIIe siècles et sa place dans l'évolution de la civilisation chinoise", *Monumenta Serica*, 1 (1935), 155-167.

Bernard, Henri. "L'encyclopédie astronomique du Père Schall", *Monumenta Serica*, 3 (1937), 35-77; 441-527.

Bernard, Henri. "Ferdinand Verbiest, continuateur de l'oeuvre scientifique d'Adam Schall", *Monumenta Serica*, 5 (1940), 103-140.

Bernard, Henri. "Galilée et les jésuites des missions d'Orient", *Revue des questions scientifiques*, 4th series, 38 (1935), 356-382.

Bernard, Henri. *Matteo Ricci's Scientific Contribution to China*. Peking, 1935.

Bernard, Henri. "Ricciana III. Un jésuite chinois, Li Ma-teou, le Père Matthieu Ricci", *Bulletin catholique de Pékin*, 22 (1935), 147-151; 197-202.

Bernard, Henri. "Ricciana V. Comment s'est posé le problème du vocabulaire chrétien en Chine", *Bulletin catholique de Pékin*, 22 (1935), 469-481.

Bernard, Henri. "Notes sur l'histoire du christianisme en Extrême-Orient", *Monumenta Serica*, 1 (1935), 478-486.

Bernard-Maitre, Henri. "Chine", *Dictionnaire d'histoire et de géographie ecclésiastiques*, 12, 693-730.

Bernard-Maitre, Henri. "Chinois (Rites)", *Dictionnaire d'histoire et de géographie ecclésiastiques*, 12, 731-741.

Bernard-Maitre, Henri. "Le Père Ricci et les missions de Chine (1578-1644)", *Histoire universelle des missions catholiques. Les missions modernes*. Edited by S. Delacroix. 4 vols. Paris, 1957, 2:19-37.

Bernard-Maitre, Henri. "Les réactions européennes à l'oeuvre missionnaire", *Histoire universelle des missions catholiques. Les missions modernes* Edited by S. Delacroix. 4 vols. Paris, 1957, 2:353-362.

Bernard-Maitre, Henri. *La science européenne au tribunal astronomique de Pékin (XVIIe-XIXe siècles). Conférence faite au Palais de la Découverte, le 16 juin 1951*. Paris, 1951.

Bernard-Maitre, Henri and E. Jarry, "Les Missions de Chine après 1644", *Histoire universelle des missions catholiques. Les missions modernes*. Edited by S. Delacroix. 4 vols. Paris, 1957, 2:165-180.

Bertuccioli, Giuliano. "A Lion in Peking. Ludovico Buglio and the Embassy to China of Bento Pereira de Faria in 1678", *East and West*, 26 (March-June, 1976), 223-238.

Blondeau, Roger A. *Ferdinand Verbiest (1623-1688). Als Oost en West elkaar ontmoeten*. Tielt, 1983.

Blondeau, Roger A. *Mandarijn en Astronoom. Ferdinand Verbiest, S.J. (1623-*

1688) aan het hof van de Chinese keizer. Brugge, 1970.

Blondeau, Roger A. "Westerse wetenschap in China in de 17e eeuw. Was dit verouderde wetenschap?", *Mededelingen van de Koninklijke Academie voor Wetenschappen, Letteren en Schone Kunsten van België*, 47 (1985), 87-102.

Bontinck, François. *La lutte autour de la liturgie chinoise aux XVIIe et XVIIIe siècles*. Louvain, 1962.

Bornet, Paul. "Chounntcheu, Kanghi et le christianisme", *Bulletin catholique de Pékin*, 34 (1947), 310-322.

Bornet, Paul. "Au service de la Chine. Schall et Verbiest maîtres fondeurs", *Bulletin catholique de Pékin*, 33 (May-June 1946), 1-25.

Bortone, Fernando. *I Gesuiti alla corte di Pechino, 1601-1813*. Rome, 1969.

Bosmans, Henri. "Documents relatifs à Ferdinand Verbiest. Les lettres annuelles de la vice-province de la Compagnie de Jésus en Chine, année 1669 par Adrien Grélon", *Annales de la Société d'Émulation de Bruges*, 62 (1912), 15-61.

Bosmans, Henri. "Les écrits chinois de Verbiest", *Revue des questions scientifiques*, 74 (1913), 272-298.

Bosmans, Henri. "Ferdinand Verbiest, directeur de l'observatoire de Pékin (1623-1688)", *Revue des questions scientifiques*, 71 (1912), 195-273; 375-464.

Bosmans, Henri. "Sur les lettres manuscrites des PP. Verbiest et Thomas analysées dans le catalogue n° 455 de la librairie Maggs Bros. de Londres", *Annales de la Société scientifique de Bruxelles* (Serie A.: Mathématiques), 47 (1927), 14-19.

Bosmans, Henri. *A la mémoire de l'astronome Verbiest, S.J., directeur de l'observatoire de Pékin*. Brussels, 1913.

Bosmans, Henri. "Notes et documents. A propos de l'état politique de la Chine au temps du Père Verbiest", *Annales de la Société d'Émulation de Bruges*, 67 (1924), 181-195.

Bosmans, Henri. "La notice nécrologique de Ferdinand Verbiest, par son secrétaire, Antoine Thomas de Namur", *Annales de la Société d'Émulation de Bruges*, 64 (1914), 102-133.

Bosmans, Henri. "Sur une particularité de l'astronomie chinoise au XVIIe siècle", *Annales de la Société scientifique de Bruxelles*, 27 (1902-1903), 122-125.

Bosmans, Henri. "Le problème des relations de Verbiest avec la Cour de Russie", *Annales de la Société d'Émulation de Bruges*, 63 (1913), 193-323; 64 (1914), 98-101.

Bouvet, Joachim. *Portrait historique de l'empereur de Chine*. Paris, 1697.

Bouvet, Joachim (Bo Jin) 白晉. *Qing Kang Qian liang di yu Tianzhujiao chuanjiao shi* 清康乾兩帝與天主教傳教士 (The Kangxi and Qianlong emperors and the Catholic missionaries). Translated from Japanese by Feng Zuomin 馮作民. Taibei, 1966.

Boxer, Charles R. "Macao as a Religious and Commercial Entrepot in the 16th and 17th Centuries", *Acta Asiatica*, 26 (1974), 64-90.

Brahe, Tycho. *Astronomiae instauratae mechanica*. Wandesburg, 1598.

Bretschneider, Emil V. *Medieval Researches from Eastern Asiatic Sources*. 2 vols. London, 1910; reprint, New York, 1967.

Brodrick, James. *Saint Francis Xavier (1506-1552)*. London, 1952.

Brucker, Joseph. *La Compagnie de Jésus, esquisse de son institut et de son histoire, 1521-1773*. Paris, 1919.

"Bureau de la société belge d'astronomie: en l'honneur du P. Verbiest, S.J.", *Ciel et Terre*, 34 (1913), 255-261.

Burke, Thomas J., ed. *Beyond All Horizons. Jesuits and the Missions*. New York, 1957.

Carton, Charles-Louis. "Biographie du R.P. Ferdinand Verbiest, missionnaire en Chine", *Album biographique des Belges célèbres*. Brussels, 1845.

Carton, Charles-Louis. *Notice biographique sur le père Ferdinand Verbiest, missionnaire à la Chine*. Bruges, 1839.

Chapman, Allan. "Tycho Brahe in China: The Jesuit Mission to Peking and the Iconography of European Instrument-making Processes", *Annals of Science*, 41 (1984), 417-443.

Ch'en, Kenneth. "Matteo Ricci's Contribution to and Influence on Geographical Knowledge in China", *Journal of the American Oriental Society*, 59 (1939), 325-359; 509.

Cibot, Pierre-Martial. "Observations de physique et d'histoire naturelle de l'empereur Kang Hi", *Mémoires concernant l'histoire, les sciences, les arts, les usages, etc. des Chinois*. 15 vols. Paris, 1777-1791, 4:452-483.

Collani, Claudia von. *P. Joachim Bouvet, S.J. Sein Leben und sein Werk*. Monumenta Serica Monograph Series XVII. Nettetal, 1985.

Combridge, John H. *Ferdinand Verbiest, S.J. (1623-1688). Compendium Latinum*. London, 1985.

Cordier, Henri. *L'imprimerie sino-européenne en Chine. Bibliographie des ouvrages publiés en Chine par les Européens au XVIIe et XVIIIe siècles*. Paris, 1901; reprint, San Francisco, 1979.

Cox, G.A. "De Chinese wereldkaart van Pater Ferdinand Verbiest, S.J., 1674", *Jaarverslag 1962, 1963, 1964 van de Vereeniging Nederlandsch Historisch Scheepvaart Museum*. Amsterdam, 1965, pp. 52-54.

Cummins, James S. *The Travels and Controversies of Friar Domingo Navarrete (1618-1686)*. 2 vols. Cambridge, 1962.

Dainville, F. de. *L'éducation des jésuites (XVIe, XVIIe et XVIIIe siècles)*. Paris, 1978.

Damry, A. "L'astronome Verbiest, S.J. et l'astronomie sino-européenne", *Ciel et Terre*, 34 (1913), 13-37.

Damry, A. "Le père Verbiest et l'astronomie sino-européenne", *Ciel et Terre*, 34 (1913), 215-239.

Danckaert, L. *Belgica in Orbe. Belgen in de wereld*. Brussels, 1977.

Da Qing shengzu ren (Kangxi) huangdi shilu 大清聖祖仁(康熙)皇帝實録 (Veritable Records of the Kangxi emperor). 6 vols. Taibei, 1964.

De Borchgrave, Emil, J.Y.M. "Flandre et Chine au XVIIe siècle, Ferdinand Verbiest", *Revue générale*, 94 (1911), 3-19; 849-856.

Debergh, Minako. "Une carte oubliée du P. Ferdinand Verbiest (1674) dans la collection Sturler de la Bibliothèque Nationale de Paris", *Journal Asiatique*, 277 (1989), 159-220.

Dehergne, Joseph. "Les archives des Jésuites de Paris et l'histoire des missions aux XVIIe et XVIIIe siècles", *Euntes Docete*, 21 (1968), 191-213.

Dehergne, Joseph. "Les chrétientés de Chine de la période Ming (1581-1650)", *Monumenta Serica*, 16 (1957), 1-136.

Dehergne, Joseph. "Les lettres annuelles des missions jésuites de Chine au temps des Ming (1581-1644)", *Archivum Historicum Societatis Jesu*, 49 (1980), 379-392.

Dehergne, Joseph. "Lettres annuelles et sources complémentaires des missions jésuites de Chine (suite)", *Archivum Historicum Societatis Jesu*, 51 (1982), 247-284.

Dehergne, Joseph. "La mission de Pékin à la veille de la condamnation des Rites", *Neue Zeitschrift für Missionswissenschaft*, 9 (1953), 91-108.

Dehergne, Joseph. *Répertoire des Jésuites de Chine de 1552 à 1800*. Rome and Paris, 1973.

De Jaegher, Karel. "Le Père Verbiest, auteur de la première grammaire mandchoue", *T'oung Pao*, 22 (1923), 189-192.

Descamps, Edouard E.F. *Histoire générale comparée des missions*. Paris, 1932.

De Schepper, R. "Verheerlijking van Pater Verbiest in 1912", *Annales de la Société d'Émulation de Bruges*, 61 (1911), 325-327.

Du Halde, Jean-Baptiste. *Description géographique, historique, chronologique, politique et physique de l'empire de la Chine et de la Tartarie chinoise*. 4 vols. Paris, 1735.

Dunne, George H. *Generation of Giants. The Story of the Jesuits in China in the Last Decades of the Ming Dynasty*. Notre Dame, 1962.

d'Elia, Pasquale. *Il mappamondo cinese del P. Matteo Ricci, S.I.* Vatican City, 1938.

d'Elia, Pasquale. "La reprise des missions catholiques en Chine à la fin des Ming (1579-1644)", *Cahiers d'Histoire mondiale*, 5 (1959-1960), 679-699.

d'Elia, Pasquale. "Recent Discoveries and New Studies (1938-1960) on the World Map in Chinese of Father Matteo Ricci, S.J.", *Monumenta Serica*, 20 (1961), 82-164.

D'Orleans, Pierre Joseph. *History of the Two Tartar Conquerors of China*. London, 1854.

Enoki, Kazuo 榎一雄. "Shokuhô gaiki no kampon ni tsuite" 職方外紀の刊本について (On the original edition of the *Zhifang waiji*), *Iwai hakushi koki kinen tenseki ronshû* 岩井博士古稀記念典籍論集 (Essays on classical literature in honor of the seventieth birthday of Dr. Iwai Hirosato), Tokyo, 1963, pp. 136-147.

Enoki, Kazuo. "Shokuhô gaiki no chûô Ajia chiri" 職方外紀の中央アジア地理 (Descriptions of Central Asia in the *Zhifang waiji*), *Wada hakushi koki kinen Tôyôshi ronsô* 和天博士古稀念東洋史論叢 (Collected essays in Oriental history to commemorate the seventieth birthday of Dr. Wada Sei), Tokyo, 1960, pp. 211-222.

Etiemble, R. *Les Jésuites en Chine. La Querelle des Rites (1552-1773)*. Paris, 1966.

Fang Hao 方豪. *Fang Hao liushi zi ding gao* 方豪六十自定稿 (The collected works of Maurus Fang Hao revised and edited by the author). 3 vols. Taibei, 1969.

Fang Hao. *Zhongguo tianzhujiaoshi renwu zhuan* 中國天主教史人物傳 (Biographies of Catholic historical figures in China). 3 vols. Hongkong, 1967-1973.

Fang Hao. *Zhongxi jiaotong shi* 中西交通史 (History of Sino-Western Relations). 5 vols. Taibei, 1954; reprint, 1983.

"La fermeture du vieil observatoire de Pékin", *Bulletin catholique de Pékin*, 21 (1934), 71-72.

Fonvielle, Wilfrid de. "L'astronomie en Chine: l'observatoire de Pékin", *La Nature*, 16 (1888), 406-410.

"Die Freigebung der christlichen Mission unter Kaiser Kanghi. Eine Episode aus der älteren chinesischen Missionsgeschichte", *Die Katholischen Missionen*, 33 (1904-1905), 4-7; 56-63.

Fu, Lo-shu. *A Documentary Chronicle of Sino-Western Relations (1644-1820)*. Tucson, 1966.

Fuchs, Walter. "Der russische-chinesische Vertrag von Nertchinsk vom Jahre 1689. Eine textkritische Betrachtung", *Monumenta Serica*, 4 (1939-1940), 546-593.

Gallagher, Louis. *The China That Was. China as Discovered by the Jesuits at the Close of the Sixteenth Century*. Milwaukee, 1942.

Gernet, Jacques. *Chine et Christianisme. Action et Réaction*. Paris, 1982.

Gernet, Jacques. "La politique de conversion de Matthieu Ricci en Chine", *Archives de Sciences sociales de religions*, 36 (1973), 71-89.

Gezelle, Guido and William H.J. Weale. "P. Verbiest van Pithem", *Rond den Heerd*. 2 vols. Bruges, 1866-1867, pp. 346-349.

"Glorification du Père Verbiest, S.J. à Pitthem", *Jesuitica Biographica*, 1 (Pitthem: Warnez, n.d.).

Golvers, Noël, and Ulrich Libbrecht. *Astronoom van de Keizer. Ferdinand Verbiest en zijn Europese Sterrenkunde*. Leuven, 1988.

Gueluy, Albert. "Un mandarin belge. Ferdinand Verbiest, missionnaire et astronome (Pitthem 1623- Peking 1688)", *Bulletin de la Société belge d'études coloniales*, 18 (1911), 824-841.

Hao, Zhenhua 郝鎮華. *Yi wu wu ling nian qian de Zhongguo jidujiao shi* 一五五零年前的中國基督教史 (A history of Christianity in China before 1550). Beijing, 1984.

Hao, Zhenhua. "Analysis of Spathary's 'Report of the Diplomatic Mission to the Qing Empire'", *Ferdinand Verbiest Foundation. China and Europe Yearbook 86*. Leuven, 1986, pp. 110-139.

Hardeman, Renatus. "Ferdinand Verbiest (1623-1688)", *De Bode van het Heilig Hart*, 45 (July 1913), 250-253.

Hucker, Charles O. *A Dictionary of Official Titles in Imperial China*. Stanford, 1985.

Humbertclaude, Pierre. "A propos de la mappemonde du P. Ricci", *Monumenta Nipponica*, 3 (1940), 643-647.

Hummel, Arthur, ed. *Eminent Chinese of the Ch'ing Period*. Washington, 1943-1944; reprint, Taipei, 1967.

Ibelings, Karel. "Adam Schall", *Jezuïetenmissies*, 30 (February-March 1939), 407-450.

Ickx, Jacques. *Ainsi naquit l'automobile*. Lausanne, 1961.

International Conference in Honor of Ferdinand Verbiest. Commemoration of the 300th Anniversary of His Death (1688-1988). Taipei, 1987.

Ivanivskij, A. "Het gezantschap van Spafari", *Zapiski vostoonogo kago obscestva*, 2 (1887), 81-124; 195-220.

Janssen, J. "Ferdinand Verbiest, S.J., missionaris in China (1659-1688), *Het Missiewerk*, 18 (1936-1937), 75-90.

Jennes, Joseph. "A propos de la liturgie chinoise", *Neue Zeitschrift für Missionswissenschaft*, 2 (1946), 241-254.

Jiang Weitang 姜緯堂. "Nan Huairen duzao guanxiang yi niandai kao" 南懷仁督造觀象儀年代考(A study on the supervision of astronomical observation by Ferdinand Verbiest), *Shiyuan* 史苑, 1 (1982), 62-67.

Josson, Henri and Leopold Willaert. *Correspondance de Ferdinand Verbiest de la Compagnie de Jésus (1623-1688)*. Brussels, 1938.

Kessler, Lawrence D. *K'ang-hsi and the Consolidation of Ch'ing Rule, 1661-1684*. Chicago, 1976.

Keunig, Johannes. "Blaeu's Atlas", *Imago Mundi*, 14 (1959), 74-89.

Keunig, Johannes. "Nicolaas Witsen as a Cartographer", *Imago Mundi*, 11 (1954), 95-110.

Koeman, Edward. *Joan Blaeu and His Grand Atlas*. London, 1970.

Kosibowicz, Edward. "Un missionnaire polonais oublié. Le Père Jean Nicolas Smogulecki, S.J. Missionnaire en Chine au 17e siècle", *Revue d'Histoire des Missions*, 6 (1929), 335-360.

Lai, Yi'en 賴詒恩 (Thomas Ryan, S.J.). *Yesu huishi zai Zhongguo* 耶穌會士在中國 (The Jesuits in China). Hongkong, 1965.

La Servière, Joseph de. "Le Père Adam Schall d'après un ouvrage nouveau", *Revue d'Histoire des Missions*, 11 (1934), 494-529.

Latourette, Kenneth S. *A History of Christian Missions in China*. London, 1929; reprint, Taipei, 1970.

Lawlor, Richard. *The Basic Strategy of Matthew Ricci, S.J. in the Introduction of Christianity in China*. Rome, 1951.

Lefevre, B. "Thomas Antoine", *Biographie nationale belge*, 25, 41-86.

"Lettre du P. Ferdinand Verbiest", *Précis historiques. Mélanges littéraires-scientifiques*, 7 (1858), 409-422; 433-444.

Lettres édifiantes et curieuses de Chine par des missionnaires jésuites, 1702-1776. Edited by Isabelle and Jean-Louis Vissière. Paris, 1979.

Li Madou 利瑪竇 and Xu Guangqi 徐光啓. "Jihe yuanben" 幾何原本 (Elements of Geometry), in Li Zhizao, ed. *Tianxue chuhan*. 6 vols. *Zhongguo shixue congshu* 中國史學叢書, no. 23, Taibei, 1965, 4:1921-2522.

Li Zhizao 李之藻. *Tianxue chuhan* 天學初函. 6 vols. *Zhongguo shixue congshu*. Edited by Wu Xiangxiang 吴相湘. Taibei, 1965.

Liang, Si-ing. *La rencontre et le conflit entre les idées des missionnaires chrétiens et les idées des Chinois en Chine depuis la fin de la dynastie des Ming*. Paris, 1940.

Libbrecht, Ulrich J. *Transmission of Western Science to China*. Presented to the

XVIIth International Congress of the History of Science. Leuven, August 6, 1985.

Lin Jinshui. "Matteo Ricci's Activities and Influence in China", *Ferdinand Verbiest Foundation. China and Europe Yearbook 86*. Leuven, 1986, pp. 140-156.

Lin, Tong-yang. *Études sur l'introduction de méthodes et de connaissances géographiques européennes en Chine (1583-1718)*. 2 vols. Paris, 1982.

Lo Kuang, Stanislaus, ed. *International Symposium on Chinese-Western Cultural Interchange in Commemoration of the 400th Anniversary of the Arrival of Matteo Ricci, S.J. in China*. Taipei, 1983.

Luk, Hung-kay Bernard. "Thus the Twain Did Meet? The Two Worlds of Giulio Aleni", Ph. D. dissertation. Indiana University, 1977.

Luk, Hung-kay Bernard. "A Study of Giulio Aleni's *Chih-fang wai-chi*", *Bulletin of the School of Oriental and African Studies*, 40 (1977), 58-84.

Mancall, Mark. *Russia and China. Their Diplomatic Relations to 1728*. Cambridge, 1971.

Martin, Gillian. "The Verbiest Map", *The Globe*, 7 (1977).

Martini, Martinus. *Novus Atlas Sinensis*. Amsterdam, 1665; reprint, Trent, 1981.

Mayers, William F. *The Chinese Government. A Manual of Chinese Titles Categorically Arranged and Explained*. 3rd ed. Shanghai, 1897; reprint, Taipei, 1970.

Minamiki, George. *The Chinese Rites Controversy from Its Beginning to Modern Times*. Chicago, 1985.

Mu, Qimeng 穆啓蒙 (Joseph Motte). *Zhongguo tianzhujiao shi* 中國天主教史 (A History of the Catholic Church in China). Taibei, 1981.

Mulder, W.Z. "De Chinese wereldkaart van pater Ferdinand Verbiest, S.J., 1674", *Jaarverslag van 1962, 1963, 1964 van de Vereeniging Nederlandsch Historisch Scheepvaart Museum*. Amsterdam, 1965, pp. 55-58.

Mungello, David E. *Curious Land. Jesuit Accommodation and the Origins of Sinology*. Stuttgart, 1985.

Nan, Huairen 南懷仁 (Ferdinand Verbiest). "Dadan lüxing ji" 韃靼旅行紀 (Notes on a voyage to Tartary), *Qingdai xiren jianwen lu* 清代西人見聞録. Beijing, 1983, pp. 69-84.

Nan, Huairen (Ferdinand Verbiest). *Xichao dingan* 熙朝定案 (Memorials to the Kangxi emperor). 2 vols. Vol. I in *Tianzhujiao dongchuan wenxian* 天主教東傳文獻 (*Zhongguo shixue congshu*, n. 24); vol. II in *Tianzhujiao dongchuan wenxian xubian* 天主教東傳文獻續編 (*Zhongguo shixue congshu*, no. 40, part 3). Edited by Wu Xiangxiang 吴相湘. Taibei, 1965-1966.

Navarette, Domingo F. "An Account of the Empire of China: Historical, Political, Moral and Religious", in Awnsham and John Churchill, *A Collection of Voyages and Travels*, 3rd ed. 6 vols. London, 1744-1746, 1:1-311; 6:751-824.

Needham, Joseph. *Within the Four Seas. The Dialogue of East and West*. London, 1969.

Olbrechts, Frans M. *Vlaanderen zendt zijn zonen uit*. Leuven, 1942.

Pan, Jixing 潘吉星. "*Kangxi di yu xiyang kexue*" 康熙帝與西洋科學 (The Kangxi emperor and western science), *Ziran kexue shi yanjiu* 自然科學史研究 (Studies in the history of natural sciences), 3 (1984), 177-188.

Panaitescu, Petre P. "Nicolas Spathar Milescu (1636-1708)", *Mélanges de l'école roumaine en France*, 1 (1925), 33-180.

Pelliot, Paul "La *Brevis Relatio*", *T'oung Pao*, 23 (1924), 355-372.

Peng, Rita Hsiao-fu. "The Kang-hsi Emperor's Absorption in Western Mathematics and Astronomy and his Extensive Application of Scientific Knowledge", *Lishi xuebao* 歷史學報 (Bulletin of historical research), 3 (February 1975), 349-422.

Pfister, Louis. *Notices biographiques et bibliographiques sur les Jésuites de l'ancienne Mission de Chine*. 2 vols. Shanghai, 1932-1934; reprint, Nendeln, 1971; San Francisco, 1976.

Piaget, Edouard. *Essai sur l'organisation de la Compagnie de Jésus*. Leiden, 1893.

Planchet, Jean-Marie. "La Mission de Pékin", *Bulletin catholique de Pékin*, 1 (1914), 58-60; 89-92; 129-133; 173-174; 211-213; 249-251; 291-294; 331-334; 492-495.

2 (1915), 36-40; 77-80; 118-120; 158-160; 197-200; 230-233; 270-276; 280-286; 312-315; 353-356; 392-394; 429-433; 471-474.

3 (1916), 34-37; 73-75; 121-126.

Pledge, Humphrey T. *Science since 1500. A Short History of Mathematics, Physics, Chemistry, Biology*. London, 1966.

Porter, Jonathan. "Bureaucracy and Science in Early Modern China: The Imperial Astronomical Bureau in the Ch'ing Period", *Journal of Oriental Studies*, 18 (1980), 61-76.

Qinming chuanjiao yue shu 欽命傳教約述 (A short explanation of the imperial decree on preaching Christianity), MSS., Xujiahui Library.

Rabbaey, Emil. *Zendeling en sterrekundige of E.P. Ferdinand Verbiest, S.J.*

van Pitthem, gestorven te Peking in 1688. Bruges, 1903; 2nd ed., Pitthem, 1911.

Rabbaey, Emil. "Ferdinand Verbiest", *Jesuitica Biographica*, 1.

Reil, Sebald. *Kilian Stumpf, 1655-1720. Ein Würzburger Jesuit am Kaiserhof zu Peking*. Münster, 1978.

Ricci, Matteo. *Fonti Ricciane*. Edited by Pasquale M. d'Elia. 3 vols. Rome, 1942-1949.

Ricci, Matteo. *The True Meaning of the Lord of Heaven* (*T'ien-chu shi-i*). St. Louis, 1985.

Rowbotham, Arnold. *Missionary and Mandarin. The Jesuits at the Court of China*. Berkeley, 1942.

Scheel, J.D.. "Precursor from Peking. Meet Ferdinand Verbiest, seventeenth-century missionary, diplomat, astronomer ... and automotive engineer", *Automobile Quarterly*, 23:3 (1985), 268-275.

Schrameier, Wilhelm L. "On Martin Martini", *Journal of the Peking Oriental Society*, 2 (1888), 99-119.

Sebes, Joseph. *The Jesuits and the Sino-Russian Treaty of Nerchinsk (1689). The Diary of Thomas Pereira, S.J.* Rome, 1961.

Sebes, Joseph. "The Summary Review of Matteo Ricci's *T'ien-chu shih-yi* in the *Ssu-k'u ch'üan-shu tsung-mu t'i-yao*", *Archivum Historicum Societatis Jesu*, 53 (January-February 1984), 371-391.

Sebes, Joseph. "Ferdinand Verbiest: Missionary or Spy?" *Proceedings of the Tenth International Symposium on Asian Studies*. Hongkong, 1988, vol. 1 (China), pp. 367-376.

Sommervogel, Carlos. *Bibliothèque de la Compagnie de Jésus*. 12 vols. Brussels, 1890-1932.

Sourie, Louis. *Ferdinand Verbiest, een groot zendeling*. Bruges, 1951.

Spence, Jonathan. *To Change China. Western Advisers in China, 1620-1960*. Boston, 1969; revised edition, Harmondsworth, 1980.

Spence, Jonathan and John E. Wills, eds. *From Ming to Ch'ing. Conquest, Region and Continuity in Seventeenth-Century China*. New Haven, 1979.

Standaert, Nicolas. *Yang Tingyun, Confucian and Christian in Late Ming China. His Life and Thought*. Leiden, 1988.

Swings, P. "Verbiest, Ferdinand", *Biographie nationale belge*, 26, 589-591.

Tsao, Kai-fu. "K'ang-hsi and the *San Fan* 三藩 War", *Monumenta Serica*, 31 (1974-1975), 108-130.

Übelhör, Monika. "Hsü Kuang-ch'i (1562-1633) und seine Einstellung zum

Christentum", *Oriens Extremus*, 15 (1968), 191-257; 16 (1969), 41-74.

Van De Maele, P. *De blanke mandarijn. De jezuïet Ferdinand Verbiest, 1623-1688*. Brussels, 1955.

Van Hee, Louis. *Ferdinand Verbiest, écrivain chinois*. Bruges, 1913.

Van Hee, Louis. "Le P. Jean-Adam Schall, S.J.", *Bulletin de l'union du clergé en faveur des missions*, 10 (1930), 16-24.

Van Hee, Louis. "Liste des lettres du P. Verbiest retrouvées jusqu'ici", *Revue missionnaire*, 5 (1931), 275-276.

Van Hee, Louis. "Missionnaires fondeurs des canons", *Bulletin de l'union du clergé en faveur des missions*, 10 (1930), 57-63.

Van Hee, Louis. "La stèle funèbre de Verbiest", *Annales de la Société d'Émulation de Bruges*, 66 (1923), 35-38.

Van Nieuwenborgh, Marcel. *Ferdinand Verbiest, missionaris of spion?* Bruges, 1984.

Väth, Alfons. *Un jésuite en Chine. Adam Schall, astronome et conseiller impérial (1592-1666)*. Brussels and Paris, 1936.

Verbiest, Ferdinand. "Voyages de l'empereur de la Chine dans la Tartarie, 1682-1683", *Recueil de voyage au Nord*. Amsterdam, 1715, 3:301-340.

Verbiest, Ferdinand. *Astronomia Europaea sub Imperatore tartaro-sinico Cam-hy appellato*. Dillingen, 1687.

Verhaeren, Hubert. *Catalogue de la Bibliothèque du Pet'ang*. Peking, 1949; reprint, Paris, 1969.

Wei Tsing-sing Louis. "Louis XIV et K'ang-hi. L'épopée des missionnaires français du Grand Siècle en Chine", *Neue Zeitschrift für Missionswissenschaft*, 19 (1963), 93-109; 182-194.

Wills, John E. "Some Dutch Sources on the Jesuit China Mission, 1662-1687", *Archivum Historicum Societatis Jesu*, 54 (1985), 267-294.

Wills, John E. *Embassies and Illusions. Dutch and Portuguese Envoys to K'ang-hsi, 1666-1687*. Cambridge, 1984.

Witek, John W. *Controversial Ideas in China and in Europe. A Biography of Jean-François Foucquet, 1665-1741*. Rome, 1982.

Witsen, Nicolaas. *Noord en Oost Tartary*. Amsterdam, 1705.

Witsen, Nicolaas. *Noord en Oost Tartary: behelzende eene beschrijving van verscheidene tarteische en nabuurige gewesten...* 2 vols. Amsterdam, 1785.

Wong, George H. C. "China's Opposition to Western Religion and Science during the Late Ming and Early Ch'ing", Ph D. dissertation. University of Washington, 1958.

Wong, George H. C. "China's Opposition to Western Science during the Late Ming and Early Ch'ing", *Isis*, 54 (1963), 29-49.

Xu Zongze 徐宗澤. *Mingmo Qingchu guanshu xixue zhi weiren* 明末清初灌輸西學之偉人 (The imparting of Western learning during the late Ming and early Qing). *Jindai Zhongguo shiliao congkan* 近代中國史料叢刊, no. 555, Taibei, 1970.

Xu Zongze. *Ming Qing jian Yesu huishi yizhu tiyao* 明清間耶穌會士譯著提要 (An annotated bibliography of works written or translated by Jesuits during the Ming and Qing dynasties). Taibei, 1958; reprint, Beijing, 1989.

Yazawa, Toshihiko 矢澤利彦. "Seiyô rekihô no fukkatsu" 西洋暦法の復活 (The rebirth of the Western calendar) in *Suzuki Shun kyôju kanreki kinen Tôyôshi ronsô* 鈴木俊教授還暦記念東洋史論叢 (Essays on Oriental History in honor of Professor Suzuki Shun). Tokyo, 1964, pp. 685-698.

Yazawa, Toshihiko, review of H. Josson and L. Willaert, *Correspondance de Ferdinand Verbiest*. *Toyo gakuho*, 39 (March 1957), 102-114.

Young, John D. *Confucianism and Christianity. The First Encounter*. Hong Kong, 1983.

Young, John D. "Science and Religion. The K'ang-hsi Emperor and Christianity", *Proceedings of the International Ch'ing Archives Symposium*. Taipei, 1978, pp. 411-413.

Zhang, Dexin 張德鑫. *Nan Huairen. Zhong-Bi jiaowang shi shang de yi wei xianquzhe* 南懷仁中比交往史上的一位先軀者 (Ferdinand Verbiest. Forerunner in the history of Chinese-Belgian Relations). *Dili zhishi* 地理知識 (Geographical knowledge), 5 (1984), 27-28.

Index with Glossary

Compiled and edited by
JOHN W. WITEK, S.J., ROMAN MALEK, S.V.D., and BARBARA HOSTER

MONUMENTA SERICA MONOGRAPH SERIES

(ISSN 0179-261X)
Edited by ROMAN MALEK, S.V.D. ▪ Institut Monumenta Serica
Arnold-Janssen-Str. 20, D-53754 Sankt Augustin, Germany

I. ANTOINE MOSTAERT, C.I.C.M., *Textes ordos recueillis et publiés avec introduction, notes morphologiques, commentaires et glossaire*, Peiping 1937, H. Vetch. [Out of print/vergriffen]

II. MARC VAN DER VALK, *An Outline of Modern Chinese Family Law*, Peking 1939, H. Vetch. [Out of print/vergriffen]

III. WOLFRAM EBERHARD, *Untersuchungen über den Aufbau der chinesischen Kultur. II. Lokalkulturen im Alten China.* Teil 2: *Die Lokalkulturen des Südens und des Ostens*, Peking 1942, Fu Jen Catholic University Press. [Out of print/vergriffen]

IV. WALTER FUCHS, *Der Jesuitenatlas der Kanghsi-Zeit. Seine Entstehungsgeschichte nebst Namenindices für die Karten der Mandjurei, Mongolei, Ostturkestan und Tibet mit Wiedergabe der Jesuiten-Karten in Originalgröße*, Peking 1943, Fu Jen Catholic University Press. [Out of print/vergriffen]

V. ANTOINE MOSTAERT, C.I.C.M., *Dictionnaire Ordos*, T. I-III, Peking 1941-1944, Fu Jen Catholic University Press. [Out of print/vergriffen]

VI. A. DE SMEDT, C.I.C.M. - A. MOSTAERT, C.I.C.M., *Le Dialecte Monguor parlé par les Mongols du Kansou occidental.* IIe Partie. *Grammaire*, Peking 1945, Fu Jen Catholic University Press. [Out of print/vergriffen]

VII. EUGEN FEIFEL, *Geschichte der chinesischen Literatur und ihrer gedanklichen Grundlage.* Nach NAGASAWA KIKUYA *Shina Gakujutsu* übersetzt von EUGEN FEIFEL, Peking 1945, Fu Jen Catholic University Press. [Out of print/vergriffen]

VIII. WALTER FUCHS, *The "Mongol Atlas" of China by Chu Ssu-pen and the Kuang-yü-t'u.* With 48 facsimile maps dating from about 1555, Peking 1946, Fu Jen Catholic University Press. [Out of print/vergriffen]

IX. KARL BÜNGER, *Quellen zur Rechtsgeschichte der T'ang-Zeit*, Peiping 1946, Fu Jen Catholic University Press. Erweiterte Neuauflage: St. Augustin - Nettetal [In preparation/in Vorbereitung].

X. WALTHER HEISSIG, *Bolur Erike "Eine Kette aus Bergkristallen". Eine mongolische Chronik der Kienlung-Zeit von Rasipungsug (1774-75)*, Peiping 1946, Fu Jen Catholic University Press. [Out of print/vergriffen]

XI. ANTOINE MOSTAERT, C.I.C.M., *Folklore Ordos. Traduction des "Textes oraux Ordos"*, Peiping 1947, Fu Jen Catholic University Press. [Out of print/vergriffen]

XII. JOSEPH JOHN SPAE, *Itô Jinsai. A Philosopher, Educator and Sinologist of the Tokugawa Period*, Peiping 1947, Fu Jen Catholic University Press. [Out of print/vergriffen]

XIII. W. LIEBENTHAL, *The Book of Chao. A Translation from the Original Chinese with Introduction, Notes and Appendices*, Peking 1948, Fu Jen Catholic University Press. [Out of print/vergriffen]

XIV. NOEL BARNARD, *Bronze Casting and Bronze Alloys in Ancient China.* Published Jointly by The Australian National University and Monumenta Serica, Nagoya 1961. [Out of print/vergriffen]

XV. CH'EN YÜAN, *Western and Central Asians in China under the Mongols. - Their Transformation into Chinese.* Translated and annotated by CH'IEN HSING-HAI and L. CARRINGTON GOODRICH, Los Angeles 1966, 328 pp. Reprint: St. Augustin - Nettetal 1989 (paperback). ISBN 3-8050-0243-2

XVI. YEN YÜAN, *Preservation of Learning. With an Introduction on His Life and Thought.* Translated by MANSFIELD FREEMAN, Los Angeles 1972, 215 pp.

XVII. CLAUDIA VON COLLANI, *P. Joachim Bouvet S.J. - Sein Leben und sein Werk*, St. Augustin - Nettetal 1985, 269 S., Abb. ISBN 3-87787-197-6

XVIII. W. SOUTH COBLIN, *A Sinologist's Handlist of Sino-Tibetan Lexical Comparisons*, St. Augustin - Nettetal 1986, 186 pp. ISBN 3-87787-208-5

XIX. GILBERT L. MATTOS, *The Stone Drums of Ch'in*, St. Augustin - Nettetal 1988, 497 pp., Illustr. ISBN 3-8050-0194-0

MONUMENTA SERICA MONOGRAPH SERIES

XX. LIVIA KÖHN, *Seven Steps to the Tao: Sima Chengzhen's "Zuowanglun"*, St. Augustin - Nettetal 1987, 205 pp. ISBN 3-8050-0195-9

XXI. KARL-HEINZ POHL, *Cheng Panch'iao. Poet, Painter and Calligrapher*, St. Augustin - Nettetal 1990, 269 pp., Illustr. ISBN 3-8050-0261-0

XXII. JEROME HEYNDRICKX (ed.), *Philippe Couplet, S.J. (1623-1693). The Man Who Brought China to Europe*. Jointly published by Institut Monumenta Serica and Ferdinand Verbiest Foundation, Leuven, St. Augustin - Nettetal 1990, 260 pp., Illustr. ISBN 3-8050-0266-1

XXIII. ANNE S. GOODRICH, *Peking Paper Gods. A Look at Home Worship*, St. Augustin - Nettetal 1991, 501 pp., Illustr. ISBN 3-8050-0284-X

XXIV. MICHAEL NYLAN, *The Shifting Center: The Original "Great Plan" and Later Readings*, St. Augustin - Nettetal 1992, 211 pp. ISBN 3-8050-0293-9.

XXV. ALFONS VÄTH S.J., *Johann Adam Schall von Bell S.J. Missionar in China, kaiserlicher Astronom und Ratgeber am Hofe von Peking 1592-1666. Ein Lebens- und Zeitbild.* Neue Auflage mit einem Nachtrag und Index. Eine gemeinsame Veröffentlichung des China-Zentrums und des Instituts Monumenta Serica, St. Augustin - Nettetal 1991, 421 S., Abb. ISBN 3-8050-0287-4

XXVI. JULIA CHING - WILLARD G. OXTOBY, *Moral Enlightenment. Leibniz and Wolff on China*, St. Augustin - Nettetal 1992, 288 pp. ISBN 3-8050-0294-7

XXVII. MARIA DOROTHEA REIS-HABITO, *Die Dharani des Großen Erbarmens des Bodhisattva Avalokitesvara mit tausend Händen und Augen. Übersetzung und Untersuchung ihrer textlichen Grundlage sowie Erforschung ihres Kultes in China*. St. Augustin - Nettetal 1993, 487 S., Abb. ISBN 3-8050-0296-3

XXVIII. NOEL GOLVERS, *The "Astronomia Europaea" of Ferdinand Verbiest, S.J. (Dillingen, 1687). Text, Translation, Notes*. Jointly published by Institut Monumenta Serica, Sankt Augustin and Ferdinand Verbiest Foundation, Leuven, St. Augustin - Nettetal 1993, 547 S. ISBN 3-8050-0327-7

XXIX. GERD WÄDOW, *T'ien-fei hsien-sheng lu. "Die Aufzeichnungen von der manifestierten Heiligkeit der Himmelsprinzessin". Einleitung, Übersetzung, Kommentar*, St. Augustin - Nettetal 1992, 374 S., Abb. ISBN 3-8050-0310-2

XXX. JOHN W. WITEK, S.J. (ed.), *Ferdinand Verbiest (1623-1688): Jesuit Missionary, Scientist, Engineer and Diplomat.* Jointly published by Institut Monumenta Serica, Sankt Augustin and Ferdinand Verbiest Foundation, Leuven, St. Augustin - Nettetal 1994, 602 S., Illustr. ISBN 3-8050-0328-5

XXXI. DONALD MACINNIS, *Religion im heutigen China. Politik und Praxis*. Deutsche Übersetzung herausgegeben im China-Zentrum von ROMAN MALEK. Eine gemeinsame Veröffentlichung des China-Zentrums und des Instituts Monumenta Serica, Sankt Augustin - Nettetal 1993, 619 S. ISBN 3-8050-0330-7

XXXII. PETER WIEDEHAGE, *Das "Meihua xishen pu" des Song Boren aus dem 13. Jahrhundert. Ein Handbuch zur Aprikosenblüte in Bildern und Gedichten*, St. Augustin - Nettetal [In preparation/in Vorbereitung].

XXXIII. D. E. MUNGELLO (ed.), *The Chinese Rites Controversy: Its History and Meaning*. Jointly published by Institut Monumenta Serica, Sankt Augustin and The Ricci Institute for Chinese-Western Cultural History, San Francisco, St. Augustin - Nettetal 1994, 356 S. ISBN 3-8050-0348-X

In preparation:

- JEONGHEE LEE-KALISCH, *Das Licht der Edlen. Der Mond in der chinesischen Landschaftsmalerei.*
- SECONDINO GATTA, *Il natural lume de Cinesi. Il ms. Iap-Sin 131 dell'ARSI del P. Philippe Couplet S.I. (1623-1693) missionario in Cina. Edizione critica e studio dei metodi di inculturazione.*
- RICHARD C. RUDOLPH, *Comprehensive Bibliography of Manchu Studies*. Enlarged and edited by HARTMUT WALRAVENS.
- CH'EN YÜAN, *The "Shih-hui chü-li". On Avoiding the Tabooed Names in China.*
- KARL BÜNGER, *Das Gesetzbuch der T'ang.*
- *"Lettres et Mémoires" d'Adam Schall S.J. édités par le P.* HENRI BERNARD *S.J. - "Relation Historique". Texte latin avec traduction française du P.* PAUL BORNET *S.J.* (Tientsin 1942). Mehrsprachige Studienausgabe.

MONUMENTA SERICA MONOGRAPH SERIES
Vol. XVII

Claudia von Collani

P. Joachim Bouvet S.J.

Sein Leben und sein Werk

Steyler Verlag, Nettetal 1985
269 S., Illustr.
ISBN 3-87787-197-6 ■ ISSN 0179-261X

Die vorliegende Arbeit beschäftigt sich mit dem Figurismus in China, der zu den interessantesten Kapiteln der chinesischen Missionsgeschichte gehört.

Die Arbeit von C. von Collani ist die bisher ausführlichste Würdigung des Figurismus als Theorie in der Chinamission. Neben der Biographie des Initiators des Figurismus, P. Joachim Bouvet S.J. (1656-1730), werden die wichtigsten Aussagen und Ziele des chinesischen Figurismus zusammengestellt.

Die Verfasserin weist nach, daß viele Vorwürfe theologischer wie sinologischer Art von Zeitgenossen, aber auch von späteren Autoren (z.B. Virgile Pinot und René Etiemble) zu Unrecht vorgebracht wurden. Claudia von Collani stellt fest, daß die Figuristen mit ihrer Theorie in vieler Hinsicht ihrer Zeit weit voraus waren und sie deshalb zwangsläufig in einen für die damalige Zeit unlösbaren Konflikt mit der herrschenden Meinung gerieten.

*

"Ms. Collani does an excellent job of raising Bouvet from the limbo of scholarly eccentricity to the status of a profoundly creative pioneer in Sino-European cultural exchange." (D.E. MUNGELLO)

*

Aus dem Inhalt: Einführung (1-9); Der Figurismus im Leben des P. Joachim Bouvet S.J. (9-97); Das figuristische Gedankensystem Joachim Bouvets (97-203) - hier u.a.: China im europäischen Weltbild des 17. Jahrhunderts - Die Göttliche Offenbarung in China - Die Spuren der Uroffenbarung in China - Der Erlöser der Welt in der chinesischen Literatur u.a.; Würdigung (203-212): Bouvets Persönlichkeit - Bouvets Werk; Literaturverzeichnis, Manuskripte und Briefe; Zeittafel; Glossarium und Register (mit chinesischen Zeichen).

MONUMENTA SERICA MONOGRAPH SERIES
Vol. XX

Livia Köhn

Seven Steps to the Tao

Sima Chengzhen's *Zuowanglun*

Institut Monumenta Serica, Sankt Augustin
Steyler Verlag, Nettetal 1987
205 pp., Illustr.
ISBN 3-8050-0195-9 ▪ ISSN 0179-261X

The *Zuowanglun*, "Treatise on Sitting in Oblivion," by the Taoist patriarch Sima Chengzhen of the 8th century, is an excellent example for the mysticism aspired to and practiced in medieval China. The text itself outlines the development of the human mind and body toward complete oneness with the Tao in seven steps or stages. Its appendix, found independently also under the title *Dingguan jing*, "Canon on Concentration and Observation," provides practical meditation instructions.

Livia Köhn's book offers complete and thoroughly annotated translations of these most revealing documents on medieval Chinese mysticism. More than that, it places the materials in a historical context and discusses underlying religious and psychological concepts within the framework of traditional Chinese thought. It provides, moreover, an understanding of Taoist spiritual realization in terms of 20th-century psychology. As alien as the ancient tradition of Taoism may ever remain to our experience, in this book we can yet appreciate its understanding of human salvation.

*

"Drawing from the most recent scholarship from China, Japan, and Europe, this slim volume makes important contributions to the history of Chinese religions, T'ang studies, and our understanding of meditation. ... This reader found *Seven Steps to the Tao* an enormously valuable study, full of fresh insights on the great classical traditions." (SUZANNE CAHILL)

*

From the contents: Dating and Lineage of the Text; The Seven Steps of Sitting in Oblivion: 1. Respect and Faith, 2. Interception of Karma, 3. Taming the Mind, 4. Detachment from Affairs, 5. True Observation, 6. Intense Concentration, 7. Realizing the Tao; Concepts Underlying the Seven Steps; Steps and Stages in the Mystical System; Modern Psychological Understanding.

MONUMENTA SERICA MONOGRAPH SERIES
Vol. XXI

Karl-Heinz Pohl

Cheng Pan-ch'iao

Poet, Painter and Calligrapher

Institut Monumenta Serica, Sankt Augustin
Steyler Verlag, Nettetal 1990
269 pp., Illustr.
ISBN 3-8050-0261-0 ■ ISSN 0179-261X

Cheng Hsieh (1693-1765), popularly known under his pen-name Cheng Pan-ch'iao, is grouped, as a painter of bamboo and orchids, with the "Eight Eccentrics of Yang-chou" and ranks among the most renowned literati of late Imperial China. His eminence rests not solely on his literary and artistic achievements but also on his integrity and concern for the common people as a scholar-official. Due to his unconventional behaviour and peculiar combination of poetry, painting, and calligraphy, Cheng gained the reputation of an unrestrained bohemian and eccentric artist. A careful analysis of his writings on literature and art shows, however, that his aesthetic ideals are rooted in the tradition of the literati aesthetics beginning with Su Shih in the Sung period.

The present study attempts to comprehend Cheng's so-called "eccentricity" mainly in the context of the socio-political conditions and changes of the early Ch'ing dynasty (i.e., Ming loyalism, rise of the salt-merchants and their domination of the cultural milieu of Yang-chou) and interprets it as an expression of his political dissent and indignation with declining literati values.

This monograph on Cheng Pan-ch'iao is the first comprehensive study of his life and works in the three Chinese scholarly arts in western language. It contains a biography, an exposition and discussion of his views on literature, calligraphy, and painting, as well as translations (his "Inscriptions on Paintings" are entirely translated), interpretations, and illustrations of representative works.

MONUMENTA SERICA MONOGRAPH SERIES
Vol. XXII

Jerome Heyndrickx (ed.)

Philippe Couplet, S.J. (1623-1693)
The Man Who Brought China to Europe

Jointly published by Institut Monumenta Serica, Sankt Augustin
and Ferdinand Verbiest Foundation, Louvain

Steyler Verlag, Nettetal 1990
260 pp., Illustr.
ISBN 3-8050-0266-1 ■ ISSN 0179-261X

The book contains the proceedings of the 1986 International Conference on Philippe Couplet in Louvain, organized by the Ferdinand Verbiest Foundation, China-Europe Institute at the Catholic University Leuven. Through his *Confucius Sinarum Philosophus,* Philippe Couplet was the first in history to introduce Chinese Confucian thinking to Europeans in an European language.

"These articles on Couplet constitute an indispensable body of literature for the study of this great Belgian Jesuit missionary ..." (*Sino-Western Cultural Relations Journal*)

Contents:

MONUMENTA SERICA MONOGRAPH SERIES
Vol. XXIII

Anne S. Goodrich

Peking Paper Gods
A Look at Home Worship

Institut Monumenta Serica, Sankt Augustin
Steyler Verlag, Nettetal 1991
501 pp., Illustr.
ISSN 0179-261X ■ ISBN 3-8050-0284-X

Paper gods are thin pieces of paper on which the images of different deities of the Chinese pantheon have been printed, usually by carved wooden blocks. In North China commonly known as "paper horses" (*chih-ma*), these prints form an important part of Chinese popular religious cults. Widely spread among the population, from peasants to high officials, they were hung in homes, shops, and places of work as a protection against evil spirits. The prints were also burnt as a means of transporting the deity represented to heaven.

Mrs. Anne Goodrich has collected a large number of those "paper horses" during the years she spent in Peking in the 1930s. The book presents numerous specimens of her unique collection in black-and-white illustrations, arranged according to the different types of gods, among them Gods of Wealth (Ts'ai-shen), Gods of Medicine, Patron Deities, and Household Gods. The prints show deities worshiped in Chinese popular beliefs as well as Taoist and Buddhist deities. The author analyses each print, its symbolism and iconography, and gives a detailed explanation of the specific deity's significance and the rituals and festivals connected with this deity.

"There can be no question as to the necessity of this work to all who would understand an aspect of popular mentality as it existed until recent years in Chinese cities, perhaps still survives there, and certainly must persist in country places ... Not often does a work on Chinese art lend itself so irresistibly to browsing." (WILLIAM WATSON)

Contents: Popular Religion of China; Paper Gods; New Year Prints; Ts'ai-shen; Birth and Children; Gods of Medicine; Exorcists; Nature Deities; Sky Powers; Patron Deities; Household Deities; Taoists Gods; Buddhist Deities; After Death; Time; Kuan-ti; Miscellaneous Gods; Pantheons; Several Appendices; Bibliography; Index.

MONUMENTA SERICA MONOGRAPH SERIES
Vol. XXIV

Michael Nylan

The Shifting Center

The Original "Great Plan" and Later Readings

Institut Monumenta Serica, Sankt Augustin
Steyler Verlag, Nettetal 1992, 211 pp.
ISBN 3-8050-0293-9 ▪ ISSN 0179-261X

The "Great Plan" ("Hung fan") chapter of the *Book of Documents* (*Shu ching*) is a major source of Confucian political, cosmological, and social theory in early China.

In her study of the "Great Plan" and its major commentaries, Michael Nylan explores the exegetical tradition of this significant text, thus following the "shifting center" of Chinese orthodoxy.

The first chapter attempts to reconstruct the original message of the "Plan", its concern with the unification of the state and its focus on techniques for maximizing the ruler's power. The second chapter examines the interpretive changes which accompanied the elevation of the "Plan" from administrative manual to canonical status in the Western Han (206 B.C. - A.D. 8), when it became one of the cornerstones of the *Yin-yang* Five Agents theory. The third chapter demonstrates still another shift in the center of Chinese orthodoxy in eleventh century Sung China. The appendix reviews scholarly efforts to date the "Plan", concluding with the author's own tentative proposal of the date for the work.

Ms. Nylan's careful examination of the commentaries of the "Great Plan" convincingly illustrates the fact that the Chinese Classics are by no means absolutes but texts which exist as functions of different historical contexts. Her study is of great importance for anyone interested in Chinese thought.

The book includes the Chinese text of the "Plan" according to Karlgren and, in addition, a reproduction of the Chinese text and its translation by Legge, a bibliography, and an index.

MONUMENTA SERICA MONOGRAPH SERIES
Vol. XXV

Johann Adam Schall von Bell S.J.
Missionar in China, kaiserlicher Astronom
und Ratgeber am Hofe von Peking 1592-1666

Ein Lebens- und Zeitbild

von

ALFONS VÄTH S.J.

Unter Mitwirkung von LOUIS VAN HEE S.J.

Neue Auflage mit einem Nachtrag und Index

Eine gemeinsame Veröffentlichung
des China-Zentrums und des Instituts Monumenta Serica, Sankt Augustin
Steyler Verlag, Nettetal 1991, 419 S., Abb.
ISSN 0179-261X ■ ISBN 3-8050-0287-4

"What we have before us is an outstanding missionary biography: the life of a person through which one is introduced also to the time and society in which he lived, and to the movement of which he was a part. Schall has been dealt with briefly elsewhere; (...) But this is a full scale treatment by a first rate scholar on the basis of very full primary sources. (...) It is a work of love. It is not uncritical of the subject but it is partisan, and perhaps that is as a biography should be. Written sixty years ago it raises fewer questions about the weaknesses of the missionary movement than a book written today would. It is also more certain of its rightness." (ARNE SOVIK)

"What is remarkable is how this work remains to this day a grand tale of narrative history and a veritable mine of information that is unsurpassed in treating its subject. Those interested in its contents would extend beyond specialists to include a semi-popular audience. The scholarly value of the book has been enhanced in the new edition by the addition of a bibliographical addendum by Ms. C. von Collani, in which the bibliography of the first edition is enlarged and brought up to date. In addition, a new index was prepared by Fr. R. Malek, S.V.D. that includes Chinese characters. Finally, a new genealogical tree featuring Schall's relationship to his family was contributed by a contemporary descendant of Fr. Schall, the Graf Schall-Riaucour. This is a surprisingly inexpensive, well-produced hardbound work that would be a valuable addition to any personal or institutional library in the field of Sino-Western cultural relations." (D.E. MUNGELLO)

MONUMENTA SERICA MONOGRAPH SERIES
Vol. XXVI

Julia Ching - Willard G. Oxtoby

Moral Enlightenment

Leibniz and Wolff on China

Institut Monumenta Serica, Sankt Augustin
Steyler Verlag, Nettetal 1992, 288 pp., Illustrations, Facsimile
ISBN 3-8050-0294-7 ▪ ISSN 0179-261X

Eighteenth-century Europe, commonly referred to as the Age of Enlightenment, witnessed a growing interest in China on the part of many great thinkers, inspired by reports of the Jesuit missionaries.

The German philosophers Gottfried Wilhelm Leibniz (1646-1716) and Christian Wolff (1679-1754) were among the admirers of Chinese thought and civilization. Leibniz' contribution to the Western understanding of China was mainly metaphysical and religious. His younger contemporary and friend Wolff focused on Chinese ethics, concentrating on the practical morality and political ideals of Confucius.

Julia Ching and Willard G. Oxtoby present English translations of important texts related to China by Leibniz and Wolff, accompanied by two introductory essays on the philosophical and historical context. The epilogue sketches the reversal of the European opinion on China in the succeeding centuries, as reflected in the writings of Kant and Hegel.

The book not only illustrates European understanding and knowledge of China at an early stage of intercultural contact but also sheds light on the European society of the time. It will be of interest for students of comparative philosophy and for all who are interested in East-West intellectual exchange.

Texts included:

Leibniz on China: Letter to Father Grimaldi (1692); "The Secret of Creation": New Year's Letter to Duke Rudolph August (1697); On the Civil Cult of Confucius (1700); An Explanation of Binary Arithmetic (1703); Discourse on the Natural Theology of the Chinese: Letter to Nicolas Rémond (1716).
Wolff on China: Discourse on the Practical Philosophy of the Chinese (1721); On the Philosopher King and the Ruling Philosopher (1730).

Place order with:

MONUMENTA SERICA MONOGRAPH SERIES
Vol. XXVII

Maria Dorothea Reis-Habito

Die Dhāraṇī des Großen Erbarmens des Bodhisattva Avalokiteśvara mit tausend Händen und Augen
Übersetzung und Untersuchung ihrer textlichen Grundlage sowie Erforschung ihres Kultes in China

Institut Monumenta Serica, Sankt Augustin
Steyler Verlag, Nettetal 1993, 450 S., Illustr., Facsimile
ISBN 3-8050-0296-3 ■ ISSN 0179-261X

Im Māhāyana-Buddhismus hat der Bodhisattva Avalokiteśvara, chin. Kuan(-shih)-yin, seit jeher eine besondere Bedeutung als Sinnbild der buddhistischen Barmherzigkeit.

Ein häufig rezitierter Gebetstext, der sich an die esoterische Manifestationsform des Avalokiteśvara mit tausend Händen und tausend Augen richtet, ist das "Sūtra der großen, vollkommenen, ungehinderten Dhāraṇī des großen Erbarmensgeistes des Bodhisattva Avalokiteśvara mit tausend Händen und tausend Augen" (*Ch'ien-shou ch'ien-yen Kuan-shih-yin p'u-sa kuang-ta yüan-man wu-ai ta-pei-hsin t'o-lo-ni ching*), das um 650 n. Chr. von dem Inder Bhagavaddharma ins Chinesische übersetzt wurde. Diese Dhāraṇī des Großen Erbarmens, die aufgrund ihrer Popularität auch als das "buddhistische *Pater Noster*" bezeichnet wurde, ist als Gebetstext das Mittel, um die helfenden Kräfte des Bodhisattva wirksam zu machen. Ihre Rezitation ist eng mit Beerdigungs- und Totenriten sowie dem Bußritual und Ahnenkult verbunden.

Die Autorin verfolgt zunächst die Entwicklung der Vorstellungen über Avalokiteśvara anhand der wichtigsten Themen der Dhāraṇī. Der zweite Teil behandelt die Einführung des Textes nach China. Durch Textvergleiche weist die Autorin die Besonderheit dieses Textes innerhalb des "gemischten esoterischen Buddhismus" auf. Der dritte Teil enthält die erste vollständige deutsche Übersetzung des *Ch'ien-shou ching*, sowie das Vorwort des Yung-lo-Kaisers (reg. 1403-1425), den Amoghavajra (705-774) zugeschriebenen Kommentar und die scholastische Interpretation des T'ien-t'ai-Meisters Chih-li (960-1028). Der vierte Teil beleuchtet die kultische Praxis der Dhāraṇī-Rezitation in China anhand verschiedener Quellen: Mönchsbiographien der T'ang-Zeit, Ch'ing-zeitliche Sammlungen von Wundergeschichten und monastische Vorschriften der Ch'an-Schule. Die Übersetzung von Chih-lis Bußritual beschließt diesen Teil.

Neben einer Konklusion und einer Bibliographie beinhaltet die vorliegende Monographie einen ausführlichen Index mit chinesischen Zeichen sowie einen Anhang mit der Reproduktion relevanter Texte und Abbildungen.

MONUMENTA SERICA MONOGRAPH SERIES
Vol. XXVIII

Noel Golvers

The *Astronomia Europaea* of Ferdinand Verbiest, S.J. (Dillingen, 1687)

Text, Translation, Notes

Jointly published by Ferdinand Verbiest Foundation, Leuven,
and Institut Monumenta Serica, Sankt Augustin
Steyler Verlag, Nettetal 1993, 547 pp., 43 Illustr., Facsimile
ISBN 3-8050-0327-7 ■ ISSN 0179-261X

In his *Astronomia Europaea*, published 1687 in Dillingen, Ferdinand Verbiest (1623-1688), the then *coryphée* of the Jesuit Mission in China, describes concisely yet detailed, how European astronomy, namely the European way to calculate the Chinese calendar and to predict lunar and solar eclipses etc., succeeded in resuming its former position after the persecution under the Oboi regency (1664-1669), viz. the Yang Kuang-hsien period, thanks to some very convincing experiments made by himself.

In the second part, Verbiest reports on the achievements of the Jesuits in fourteen distinct mathematical and mechanical sciences (gnomonics, ballistics, hydragogics, mechanics, optics, catoptrics, perspective, statics, hydrostatics, hydraulics, pneumatics, music, horologic technology, meteorology) during the decisive decade, 1669-1679, when a strong base was laid for the revival and the flowering of the Jesuit Mission.

This edition of the *Astronomia Europaea* presents: a photostatic reprint of the rare original edition; an English translation, annotated from contemporary first-hand and mostly unpublished documents, situating this treatise in its Chinese context and explaining it against its European and particular Jesuit background. An extensive introduction describes its complex growth process and defines its relation to the previous astronomical treatises of the same author between 1669 and 1678, both in Chinese and in Latin, including the recently discovered *Compendium Historicum* and *Mechanica*.

This work which also contains 43 illustrations, a bibliography, and an index with Chinese characters, meets the old desideratum already formulated by Paul Pelliot: "Une étude détaillée sur l'*Astronomia Europaea*, bien qu'amorcée par les recherches des PP. Van Hée and Bosmans, reste à écrire" (*T'oung Pao* 23[1924], p. 357, n. 5).

MONUMENTA SERICA MONOGRAPH SERIES
Vol. XXIX

Gerd Wädow

T'ien-fei hsien-sheng lu
"Die Aufzeichnungen
von der manifestierten Heiligkeit der Himmelsprinzessin"
Einleitung, Übersetzung, Kommentar

Institut Monumenta Serica, Sankt Augustin
Steyler Verlag, Nettetal 1992
374 S., Abb.
ISBN 3-8050-0310-2 ■ ISSN 0179-261X

The goddess T'ien-fei, "Heavenly Princess," usually known in the folk religious tradition by her popular name Ma-tsu, "Grandmother," since Sung times developed into one of the most important Chinese deities both of the folk religious tradition and the official state cult.

The text *T'ien-fei hsien-sheng lu* (*TF*), "Records of the Manifested Saintliness of the Heavenly Princess," is an anonymous compilation "in praise of the deity" that was composed in several phases and in its present form postdates 1727.

Due to the officious perception and the intention of the compilers to bring to prominence the deity's significance to the state and its affairs, the text sheds light on an often inadequately considered facet of one of the most remarkable Chinese deities which in the West is often only known as the "popular patron deity of the fishermen of South China." It provides evidence for the shifts in the pertaining specific attitudes of the relevant dynasties towards the realm of the deities of the state cult pantheon.

Its report period spanning four dynasties, the text registers official decrees for granting honorific titles to the goddess as well as imperial proclamations on the occasions of investitures and sacrifices to her. In addition, a second part of the text contains 55 miracle episodes which exemplarily constitute the core legend of the Heavenly Princess.

The work intends to contribute to the recently increasing Ma-tsu/T'ien-fei-research by presenting the first complete annotated translation of the *TF* including all forewords and epilogues as well as the editorial report of the Chinese edition used. It also contains the reproduction of the Chinese text of the *TF* and is supplemented by a series of seven Ch'ing-dynasty paintings from the holdings of the Asiatic Department of the Rijksmuseum in Amsterdam, Netherlands, on which some of the miraculous episodes of the legend are illustrated.

MONUMENTA SERICA MONOGRAPH SERIES

Vol. XXXI

Donald MacInnis

Religion im heutigen China

Politik und Praxis

Deutsche Übersetzung herausgegeben
im China-Zentrum von ROMAN MALEK

Eine gemeinsame Veröffentlichung des China-Zentrums
und des Instituts Monumenta Serica, Sankt Augustin

Steyler Verlag, Nettetal 1993, 619 S.
ISSN 0179-261X ■ ISBN 3-8050-0330-7.

Seit dem Erscheinen der amerikanischen Originalausgabe (*Religion in China Today. Policy and Practice*) im Jahre 1989 gehört die hier in vollständiger deutscher Übersetzung erschienene Dokumentation für alle am Thema Religion und Politik in der Volksrepublik China Interessierten zu den Standardwerken.

Der Autor, der sich seit über 40 Jahren mit der Materie befaßt, ist bereits mit einer Reihe einschlägiger Publikationen zum Thema hervorgetreten. Im zeitlichen Anschluß an diese früheren Publikationen wendet er sich hier nun konkret dem seit dem Ende der Kulturrevolution (1966-1976) und dem Tode Mao Zedongs (1893-1976) vergangenen Zeitraum zu, der durch ein erstaunliches Wiedererstarken des religiösen Lebens in China gekennzeichnet ist.

Im ersten Teil, "Die Religionspolitik nach der Kulturrevolution," sind offizielle Schlüsseldokumente der Partei und des Staates, Erklärungen der Religionspolitiker sowie Artikel und Presseberichte über die staatliche Religionspolitik bis Ende der achtziger Jahre zusammengetragen worden, in denen u.a. die folgenden Themen zur Sprache kommen: Bestimmungen der Provinzen für die Überwachung religiöser Aktivitäten; Religionspolitik und nationale Minderheiten; Erklärungen religiöser Führungskräfte zur Religionspolitik; Marxismus, Religion und Mitgliedschaft in der Kommunistischen Partei; Politik der Vermeidung von Verschwendung und Übermaß in religiösen Praktiken; Religion und Religionspolitik in einem Lehrbuch für Jugendliche; Definitionen von Religion im heutigen China.

Der zweite Teil, "Die religiöse Praxis seit der Kulturrevolution," bringt Interviews, wissenschaftliche Studien, Reden und Berichte religiöser Amtsträger, chinesische Presseberichte und Zeitschriftenartikel zur aktuellen Situation der einzelnen Religionen in China seit 1979, darunter Buddhismus, Daoismus, Islam, katholische, protestantische und russisch-orthodoxe Kirche, Judentum, Volksreligiosität, Konfuzianismus, "Ersatzreligionen" und das Thema "Jugend und Religion."

Ein eigens für die deutsche Ausgabe erstellter Index macht das Buch darüber hinaus zum benutzerfreundlichen Nachschlagewerk.

MONUMENTA SERICA MONOGRAPH SERIES
Vol. XXXII

Peter Wiedehage

Das *Meihua xishen pu* des Song Boren aus dem 13. Jahrhundert

Ein Handbuch zur Aprikosenblüte in Bildern und Gedichten

Institut Monumenta Serica, Sankt Augustin
Steyler Verlag, Nettetal (in Vorbereitung), ca. 400 S., Abb.
ISSN 0179-261X

Das in einem Nachdruck von 1261 vollständig erhaltene *Meihua xishen pu* des Song Boren stellt kunstgeschichtlich und kulturhistorisch ein Zeitdokument höchsten Ranges dar. Es enthält die in Holzdrucken reproduzierten einhundert Tuschzeichnungen, die den Lebenszyklus der Aprikose (*meihua*) vom ersten Knospen über ihr Erblühen bis hin zum Verwelken in acht Kapiteln nachzeichnen. Mit einhundert Bildüberschriften und begleitenden Kurzgedichten ist es die früheste erhaltene illustrierte Abhandlung dieser Art in China. Seine Konzeption wurde für Malanleitungsbücher folgender Dynastien beispielgebend.

Diese erste Gesamtdarstellung des *Meihua xishen pu* beleuchtet das Werk zunächst vor seinem zeitgeschichtlichen Hintergrund. Im zweiten Teil werden die einhundert Blütentypen einer stilistischen Analyse unterzogen und mit Blütenformen songzeitlicher Tuschmalereien zur *meihua* verglichen. Das Blütenvokabular des Werkes kann der seinerzeit bereits orthodoxen Maltradition der Yang-Tradition zugeordnet werden. Dies bedeutet, daß das *Meihua xishen pu* bereits zur Song-Zeit einen wichtigen Beitrag zur Kodifizierung der *meihua*-Tradition leistete.

Die Untersuchung der poetischen Bildtitel und kommentierenden vierzeiligen Kurzgedichte (*wuyan jueju*) zeigt die vielfältigen Bedeutungsebenen der *meihua* auf, die sie in den Augen eines typischen Literatenbeamten der Südlichen Song-Zeit besaß. Sie verdeutlicht, wie Song Boren die Blüte als Spiegel seines assoziativen Denkens und Sinnbild seiner Ideale bewußt als Mittel und Werkzeug seiner Kritik an Gesellschaft und Tagespolitik einsetzte.

Der dritte Teil der Arbeit untersucht die Auswirkungen des *Meihua xishen pu* auf andere, auch in China teilweise schwer zugängliche Malanleitungsbücher der Yuan-, Ming- und Qing-Zeit. Der Autor belegt für eine Anzahl anderer Werke, daß diese das *Meihua xishen pu* entweder als Ganzes nachgeahmt, einzelne Blütenformen zitiert oder seine Konzeption aufgegriffen und weiterentwickelt haben.

Das Buch enthält den chinesischen Originaltext mit sämtlichen Illustrationen des Originals sowie weiteren Abbildungen, ein Literaturverzeichnis und einen Index mit Glossar.

Anne Swann Goodrich

The Peking Temple of the Eastern Peak

The Tung-yüeh Miao in Peking and Its Lore

Appendix

Description of the Tung-yüeh Miao of Peking in 1927

by JANET R. TEN BROECK

Monumenta Serica, Nagoya, Japan, 1964
331 pp., Illustr.

The Tung-yüeh Miao in Peking is a veritable pantheon housing hundreds of deities with scores of attendants, presided over by the Great God of the Eastern Peak. Mrs. Goodrich, in her delightful narrative, relates the stories of these deities: how they came to be deified, how they secured the faith and trust of thousands of devotees, and how they managed to accommodate themselves to each other without interfering with the rights and responsibilities of the neighboring deities.

Here is the story of how a faithful people, common folk, gradually built up a religious, syncretistic system, happily free from any shackles of logic and historical accuracy. Yet it worked so well that the Chinese flocked to this temple at regular intervals. If there is a popular demand, a common need, a private hope or anxiety, there must be a deity who answers to it. If not, people will create such a deity and show that its worship is not useless and therefore justified.

Mrs. Anne S. Goodrich lived in China for seven years, spending most of her time in Peking, but making trips that took her as far north as Jehol, and as far south as Changsha and Canton. In 1923 she married L. Carrington Goodrich. With her husband she traveled extensively in Asia.

Janet R. Ten Broeck made detailed studies of the Tung-yüeh Miao, part of which she published together with Dr. Yiu Tung in *T'oung Pao* XL (1950-1951). She lived in Peking with her husband who for seven years (1920-1927) held a professorship at the Peking Union Medical College.

Anne Swann Goodrich

Chinese Hells

The Peking Temple of Eighteen Hells
and
Chinese Conceptions of Hell

Institut Monumenta Serica, Sankt Augustin
Second edition 1989
167 pp., Illustr.

The present book of Mrs. Anne S. Goodrich may be considered a sequel to *The Peking Temple of the Eastern Peak.* The Peking Temple of the Eighteen Hells, she deals with in the first part, was a much smaller institution opposite the Temple of the Eastern Peak. The notes on which the descriptions of both sites are based were taken during frequent visits in the years 1931 and 1932, when Mrs. Goodrich accompanied her husband, the late Professor L. Carrington Goodrich, on a study tour.

The commentaries on both places, provided by Mrs. Goodrich's teacher and guide, Mr. Shih, a resident of the area, preserve much of the folk beliefs concerning the world of gods and ghosts. It is very fortunate that Mrs. Goodrich has been able to illustrate her description with numerous photographs generously supplied by Professor Robert des Rotours.

In the second part of the book Mrs. Goodrich traces the origins of the Chinese ideas of Hell and gives a survey of Chinese descriptions of Hell, particularly those contained in stories of visits to Hell, and of the influence which the belief in an afterlife and in otherworldly punishment had on the practical conduct of people.

"*Chinese Hells* is indeed a valuable if belated contribution to sinology and to preserving a record of the folkways of old Peking as they persisted into the 1930s; furthermore, it and its earlier companion volume, *The Peking Temple of the Eastern Peak*, were clearly labors of love." (JON W. HUEBNER, *Chinese Culture* 34 [1992], 2)

Place order with:
Steyler Verlag, Postfach 24 60, D-41311 Nettetal, Germany
Tel.: 02157/12 02 20 * Fax: 02157/12 02 22

Monumenta Serica
Journal of Oriental Studies
Index to Volumes I-XXXV
(1935-1983)

Compiled and edited
by
ROMAN MALEK, S.V.D.

Institut Monumenta Serica,
Sankt Augustin 1993, 471 pp.
ISSN 0254-9948 ■ ISBN 3-8050-0312-9

To provide an easy access to the variety of materials included in *Monumenta Serica · Journal of Oriental Studies*, this general index to volumes I-XXXV (1935-1983) has been prepared.

The index covers the volumes published in the course of the journal's history of more than five decades. It contains the following parts:

I. Tables of contents of the single volumes in chronological order (I-XXXV);
II. General index of the volumes I-XXXV in alphabetical order of items;
III. Index of the titles and tables of contents of the "Monumenta Serica Monograph Series" (vols. I-XX). The general index also refers to the authors and main subjects of each of the monographs.

The general index of subjects includes all the titles of the articles, the names of authors of articles and reviews, and the key-words concerning the subjects. Every key-word is indicated in the language in which it was written. To each of the non-English key-words references to English terms are added.

The index is an indispensable tool for an efficient use of *Monumenta Serica* and provides access to a wealth of information gathered in fifty years of sinological studies.